Mycotoxins in Food, Feed and Bioweapons

Mahendra Rai · Ajit Varma
Editors

Mycotoxins in Food, Feed and Bioweapons

Springer

Editors
Mahendra Rai, Ph.D.
Professor and Head
Biotechnology Department
SGB Amravati University
Amravati-444 602
Maharashtra
India
mkrai123@rediffmail.com

Professor Dr. Ajit Varma
Director General
Amity Institute of Microbial Technology
Amity University Uttar Pradesh
& Vice Chairman
Amity Science, Technology & Innovation Foundation
Block A, Amity Campus, Sector 125
Noida, UP 201303
India
ajitvarma@aihmr.amity.edu

ISBN 978-3-642-00724-8 e-ISBN 978-3-642-00725-5
DOI: 10.1007/978-3-642-00725-5
Springer Heidelberg Dordrecht London New York

Library of Congress Control Number: 2009929167

Cover design: WMXDesign GmbH, Heidelberg, Germany

Printed on acid-free paper

Springer is part of Springer Science+Business Media (www.springer.com)

Foreword

Mycotoxins are a large and varied group of mold-secondary metabolites. Their common features are that they are all produced by fungi, and that they all have toxic effects against vertebrates and other organisms. Nevertheless, it is hard to generalize about mycotoxins. They are made by different biosynthetic pathways, and they have an extremely wide range of pharmacological effects. Human beings and domestic animals come in contact with them by different routes (diet, dermal contact, respiration); the fungi which produce them occupy different ecological niches, and they span a wide range of important fungal genera (*Aspergillus*, *Fusarium*, *Penicillium* and many others). Even the definition of "mycotoxin" is a matter of dispute. Not everyone includes zearalenone and the yeast killer toxins — both discussed in this monograph — within the mycotoxin rubric. Zearalenone, although biologically potent as an estrogen mimic, is not particularly toxic in the traditional sense of causing death. Yeast killer toxins are neither mold metabolites nor viewed as particularly poisonous environmental agents. They are best known for their use in biotyping and as immunological derivatives. Nevertheless, while experts differ about which compounds should be classified appropriately as mycotoxins, everyone agrees that mycotoxins impose an enormous economic burden. They frequently render agricultural produce unusable, they inflict untold suffering and death on human and animal populations, and their regulation has become a major factor in international trade.

Mycotoxins have been reviewed many times, but this monograph touches on fresh territory, including (1) new analytical methods for detection, (2) the adoption of an ancient Mexican process for detoxification of aflatoxins, (3) mycotoxin management in Ireland, Lithuania and South America, (4) mycotoxin reduction through plant breeding and integrated management practices, and (5) an overview of natural aflatoxin inhibitors from medicinal plants. The forthcoming chapters highlight recent research findings, with emphasis on aflatoxins, ochratoxins, selected trichothecenes, and zearalenone. Aflatoxin-like gene clusters are discussed, as are aspects of sclerotial development in *Aspergillus flavus* and *A. parasiticus*. Of particular interest are the chapters on the weaponisation of mycotoxins and the potential for use of mycotoxins as agents of bioterrorism. Similarly, the discussion of conjugated ("masked") mycotoxin derivatives is a topic rarely encountered in the mycotoxin literature.

This book will update readers on several cutting-edge aspects of mycotoxin research, and will stimulate new thinking on the need to develop therapeutic as well as preventative interventions for lowering their toxicological menace. With increasing globalization and world trade, the mycotoxin problem will grow only more significant in coming years.

Professor (Dr.) J.W. Bennett
Department of Plant Biology and Pathology
School of Biological and Environmental Sciences
Rutgers University
New Brunswick, New Jersey 08901, USA
profmycogirl@yahoo.com

Preface

Mycotoxins are produced by fungi as secondary metabolites in cereals and food. These chemical compounds are lipid-soluble, and easily absorbed by the intestines, airways and skin. In recent years, the risk due to mycotoxins is increasing particularly in food and feed, resulting in deterioration of human and animal health. The contamination of food by mycotoxins has become a matter of great concern, as these are responsible for many lethal diseases like cancer and other chronic diseases. According to an estimate, 25% of the world's crops are affected by toxigenic fungi. Mycotoxins have become part of the food chain. Reduction of these toxins requires a multifaceted approach, including action by farmers, government agencies, food processors and scientists. There are more than 400 compounds known as mycotoxins. Mycotoxins released by molds occurring in damaged buildings in an indoor environment are also hazardous to health. Among these, aflatoxins produced by *Aspergillus flavus* are well-known in the world. Other mycotoxins include citrinin, ochratoxin, patulin, trichothesene, and zearalenone.

Another area of mycotoxins which warrants attention is their use in bioweapons. The T-2 mycotoxins (trichothecene), usually produced by *Fusarium* species, are used as bioweapons during war, since mold-toxins are cheap, easy to access and can be applied for a small group of enemies. Other fungi which produce trichothecens are *Myrothecium Trichoderma,* and *Stachybotrys.* As a matter of fact, it is very important to understand the symptoms of mycotoxin infection for rapid identification, in order to inactivate its ill-effects on patients.

There is a pressing need for strict regulations in order to manage mycotoxins. A strategy should be developed collectively by farmers, scientists, clinicians and policy-makers to combat the mycotoxin menace for a sustainable future. In fact, the mycotoxin menace to human beings and animals has generated the idea to bring out this book, in which efforts have been made to gather diverse topics on mycotoxin with up-to-date information.

Broadly divided into three sections and 22 chapters, this book is aimed at the diverse backgrounds of students and researchers. Section I focuses on general topics, Section II concentrates on the role of mycotoxins in agriculture and food, and Section III deals with toxicity and the use of myotoxins as bioweapons.

The book would be essential reading for mycologists, agriculture scientists, clinicians, fungal technologists, microbiologists, both plant and human pathologists, pharmacologists and forensic scientists, who may be interested to unravel the mysteries of toxicity of mycotoxins in food, feed and bioweapons.

We thank Dr. Jutta Lindenborn, Springer, Heidelberg, Germany for her patience and suggestions. MKR is thankful to Professor Dr. György János Kövics, Head, Plant Protection Department, Debrecen Agriculture University, Debrecen, Hungary for providing necessary facilities for editing and finalizing the manuscript during my visit to his institute. Further, MKR wishes to thank Alka Karwa, Aniket Gade, Ravindra Ade, Avinash Ingle, Dnayeshwar Rathod, Vaibhav Tiwari, Alka Yadav and Jayendra Kesharwani for help in editing.

April 2009
Amravati, Maharashtra
and Noida, Uttar Pradesh

Mahendra Rai
Ajit Varma

Contents

Part I Biology and Biotechnology

1 **Artificial Systems for Molecular Recognition of Mycotoxins** 3
Claudio Baggiani, Laura Anfossi, and Cristina Giovannoli

2 **Molecular Mechanism of Detection of Aflatoxins and Other Mycotoxins** 21
Avishay-Abraham Stark

3 **The Destruction of Aflatoxins in Corn by "Nixtamalización"** 39
Doralinda Guzmán-de-Peña

4 **Mycotoxigenic Fungi on Baled Grass Silage in Ireland** 51
Martin O'Brien

5 **Aflatoxin-like Gene Clusters and How They Evolved** 65
Kenneth C. Ehrlich and Jiujiang Yu

6 **Aflatoxin Biosynthesis and Sclerotial Development in *Aspergillus flavus* and *Aspergillus parasiticus*** 77
Perng-Kuang Chang

7 **Interaction Between Aflatoxin B_1 and Other Risk Factors in Hepatocarcinogenesis** 93
Michael C. Kew

8 **Zearalenone and its Derivatives: Known Toxins in New Aspects** 113
Piotr Golinski, Agnieszka Waskiewicz, and Karolina Gromadzka

9 **Zearalenone: Undesirable Substance** 131
Maciej Gajęcki, Magdalena Gajęcka, Ewa Jakimiuk, Łukasz Zielonk, and Kazimierz Obremski

Part II Agriculture and Food

10 Mycotoxins in Lithuanian Cereals and Grain Products 147
Audrone Mankeviciene

11 Control of Mycotoxin Contamination in Cereals by Breeding 163
Ákos Mesterházy

12 *Fusarium* and Fumonisins in Maize in South America 179
Adriana M. Torres, Maria L. Ramirez, and Sofia N. Chulze

13 Prevention and Management of Mycotoxins in Food and Feed 201
Bulent Kabak

14 Mycotoxigenic Fungi and Mycotoxins in Animal Feed in South American Countries 229
Carina Elizabeth Magnoli, Lilia Renée Cavaglieri, Carlos Alberto da Rocha Rosa, and Ana María Dalcero

Part III Toxicology and Bioweapons

15 Nivalenol: The Mycology, Occurrence, Toxicology, Analysis and Regulation 253
Yoshiko Sugita-Konishi and Takashi Nakajima

16 Yeast Killer Toxins Technology Transfer 275
Walter Magliani, Stefania Conti, Laura Giovati, and Luciano Polonelli

17 Trichothecenes as Toxin and Bioweapons: Prevention and Control 291
Avinash Ingle, Ajit Varma, and Mahendra Rai

18 Risk Assessment of Ochratoxin A (OTA) 307
Diana Ringot and Abalo Chango

19 Natural Aflatoxin Inhibitors from Medicinal Plants 329
Mehdi Razzaghi-Abyaneh, Masoomeh Shams-Ghahfarokhi, Mohammad-Bagher Rezaee, and Shohei Sakuda

20 The Myco-Agents of Bioterrorism 353
Nathan P. Charlton and Christopher P. Holstege

21 The Weaponisation of Mycotoxins 367
Robert Russell Monteith Paterson and Nelson Lima

22 Masked Mycotoxins and Mycotoxin Derivatives in Food: The Hidden Menace 385
Chiara Dall'Asta, Gianni Galaverna, Arnaldo Dossena, Stefano Sforza, and Rosangela Marchelli

Index 399

Contributors

Laura Anfossi
Laboratory of Bioanalytical Chemistry, Department of Analytical Chemistry, University of Torino, Italy; via P. Giuria 5, 10125 Torino, Italy

Claudio Baggiani
Laboratory of Bioanalytical Chemistry, Department of Analytical Chemistry, University of Torino, Italy; via P. Giuria 5, 10125 Torino, Italy, claudio.baggiani@unito.it

Lilia Renée Cavaglieri
Departamento de Microbiología e Inmunología, Facultad de Ciencias, Exactas Físico-Químicas y Naturales, Universidad Nacional de Río Cuarto, Ruta 36, km 601, Río Cuarto, Córdoba, Argentina, lcavaglieri@exa.unrc.edu.ar, lcavaglieri@arnet.com

Perng-Kuang Chang
Southern Regional Research Center, Agricultural Research Service, U.S. Department of Agriculture, 1100 Robert E. Lee Boulevard, New Orleans, LA 70124, USA, perngkuang.chang@ars.usda.gov

Abalo Chango
Institut Polytechnique Lasalle, Beauvais, Rue Pierre Waguet, BP 30313, Beauvais, Cedex 60026, France, abalo.chango@lasalle-beauvais.fr

Nathan P. Charlton
Department of Emergency Medicine, Division of Medical Toxicology, University of Virginia School of Medicine, P.O. Box 800774, Charlottesville, VA 22908–0774, USA, npc8a@virginia.edu

Sofia N. Chulze
Mycology Section, Department of Microbiology and Immunology, National University of Río Cuarto. Ruta 36 km. 601 (5800) Río Cuarto, Córdoba, Argentina, schulze@exa.unrc.edu.ar

Stefania Conti
Dipartimento di Patologia e Medicina di Laboratorio, Sezione di Microbiologia, Università degli Studi Parma, Viale A. Gramsci, 14 – 43100 Parma, Italy

Ana María Dalcero
Departamento de Microbiología e Inmunología, Facultad de Ciencias, Exactas Físico-Químicas y Naturales, Universidad Nacional de Río Cuarto, Ruta 36, km 601, Río Cuarto, Córdoba, Argentina, adalcero@exa.unrc.edu.ar

Arnaldo Dossena
Dipartimento di Chimica Organica e Industriale, Viale G.P. Usberti 17A, Campus Universitario, Università di Parma, I-43100, Parma, Italy, arnaldo.dossena@unipr.it

Kenneth C. Ehrlich
Southern Regional Research Center, ARS, USDA, 1100 R.E. Lee Blvd. New Orleans, LA 70124, USA, ehrlich@srrc.ars.usda.gov

Magdalena Gajecka
Division of Veterinary Prevention and Feed Hygiene, Department of Veterinary Health Protection, Faculty of Veterinary Medicine, University of Warmia and Mazury in Olsztyn, St. M. Oczapowskiego 13, 10–719 Olsztyn, Poland

Maciej Gajecki
Division of Veterinary Prevention and Feed Hygiene, Department of Veterinary Health Protection, Faculty of Veterinary Medicine, University of Warmia and Mazury in Olsztyn, St. M. Oczapowskiego 13, 10–719 Olsztyn, Poland, gajecki@uwm.edu.pl

Gianni Galaverna
Dipartimento di Chimica Organica e Industriale, Viale G.P. Usberti 17A, Campus Universitario, Università di Parma, I-43100, Parma, Italy, gianni.galaverna@unipr.it

Cristina Giovannoli
Laboratory of Bioanalytical Chemistry, Department of Analytical Chemistry, University of Torino, Italy; via P. Giuria 5, 10125 Torino, Italy

Laura Giovati
Dipartimento di Patologia e Medicina di Laboratorio, Sezione di Microbiologia, Università degli Studi Parma, Viale A. Gramsci, 14 – 43100 Parma, Italy

Piotr Golinski
Department of Chemistry, University of Life Sciences, Wojska Polskiego 75, 60–625 Poznan, Poland, piotrg@up.poznan.pl

Karolina Gromadzka
Department of Chemistry, University of Life Sciences, Wojska Polskiego 75, 60–625 Poznan, Poland

Doralinda Guzmán-de-Peña
Departamento de Biotecnología y Bioquímica, Campus Guanajuato CINVESTAV-IPN, Km 9.6 Libramiento Norte, Irapuato, Guanajuato 36500, México, dguzman@ira.cinvestav.mx

Christopher P. Holstege
Department of Emergency Medicine, Division of Medical Toxicology, University of Virginia School of Medicine, P.O. Box 800774, Charlottesville, VA 22908–0774, USA, ch2xf@virginia.edu

Avinash Ingle
Department of Biotechnology, SGB Amravati University, Amravati 444 602, Maharashtra, India, avinash_ingle123@rediffmail.com

Ewa Jakimiuk
Division of Veterinary Prevention and Feed Hygiene, Department of Veterinary Health Protection, Faculty of Veterinary Medicine, University of Warmia and Mazury in Olsztyn, St. M. Oczapowskiego 13, 10–719 Olsztyn, Poland

Bulent Kabak
Department of Food Engineering, Faculty of Engineering, Hitit University, Corum, Turkey, bkabak@cu.edu.tr, bulentkabak@hitit.edu.tr

Michael C. Kew
Department of Medicine, University of Cape Town and Groote Schuur Hospital, Cape Town, South Africa and Department of Medicine, University of the Witwatersrand Medical School, Johannesburg, South Africa, Michael.C.Kew@uct.ac.za

Nelson Lima
Micoteca da Universidade do Minho, IBB/Centre for Biological Bioengineering, Campus de Gualtar, University of Minho, 4710–057 Braga, Portugal, nelson@iec.uminho.pt

Michał Luczynski
Department of Chemistry, Faculty of Environmental Management and Agriculture, University of Warmia and Mazury in Olsztyn, Łódzki Sq. 4, 10–957 Olsztyn, Poland

Walter Magliani
Dipartimento di Patologia e Medicina di Laboratorio, Sezione di Microbiologia, Università degli Studi Parma, Viale A. Gramsci, 14 – 43100 Parma, Italy, magliwal@unipr.it

Carina Magnoli
Departamento de Microbiología e Inmunología, Facultad de Ciencias, Exactas Físico-Químicas y Naturales, Universidad Nacional de Río Cuarto, Ruta 36, km 601, Río Cuarto, Córdoba, Argentina, cmagnoli@exa.unrc.edu.ar, carmagnoli@hotmail.com

Audrone Mankeviciene
Department of Plant Pathology and Protection, Lithuanian Institute of Agriculture, LT-58344 Akademija, Kėdainiai District, Lithuania, audre@lzi.lt

Rosangela Marchelli
Dipartimento di Chimica Organica e Industriale, Viale G.P. Usberti 17A, Campus Universitario, Università di Parma, I-43100, Parma, Italy, rosangela.marchelli@unipr.it

Ákos Mesterházy
Cereal Research non-profit Company, H-6701 Szeged, P.O.Box 391, Hungary, akos.mesterhazy@gabonakutato.hu

Takashi Nakajima
National Agricultural Research Center for Kyushu Okinawa Region, Suya 2421, Koshi, Kumamoto 861–1192 Japan, ntakashi@affrc.go.jp

Kazimierz Obremskio
Department of Veterinary Health Protection, Division of Veterinary Prevention and Feed Hygiene, Faculty of Veterinary Medicine, University of Warmia and Mazury in Olsztyn, St. M. Oczapowskiego 13, 10–719 Olsztyn, Poland

Martin O'Brien
Teagasc, Grange Beef Research Centre, Dunsany, Co. Meath, Ireland, martin.obrien@teagasc.ie

Robert Russell Monteith Paterson
Micoteca da Universidade do Minho, IBB/Centre for Biological Bioengineering, Campus de Gualtar, University of Minho 4710–057 Braga, Portugal, russell.paterson@deb.uminho.pt

Luciano Polonelli
Dipartimento di Patologia e Medicina di Laboratorio, Sezione di Microbiologia, Università degli Studi Parma, Viale A. Gramsci, 14 – 43100 Parma, Italy

Mahendra Rai
Department of Biotechnology, SGB Amravati University, Amravati 444 602, Maharashtra, India, mkrai123@rediffmail.com, pmkrai@hotmail.com

Maria L. Ramirez
Mycology Section, Department of Microbiology and Immunology, National University of Río Cuarto. Ruta 36 km. 601 (5800) Río Cuarto, Córdoba, Argentina, mramirez@exa.unrc.edu.ar

Mehdi Razzaghi-Abyaneh
Department of Mycology, Pasteur Institute of Iran, Tehran 13164, Iran, mrab442@pasteur.ac.ir, mrab442@yahoo.com

Mohammad-Bagher Rezaee
Department of Medicinal Plants, Institute for Research in Forests and Rangelands, Tehran, Iran

Diana Ringot
Institut Polytechnique Lasalle, Beauvais, Rue Pierre Waguet, BP 30313, Beauvais, Cedex 60026, France, diana.ringot@lasalle-beauvais.fr

Carlos Alberto da Rocha Rosa
Departamento de Microbiologia e Imunología Veterinária. Universidade Federal Rural do Rio de Janeiro. Instituto de Veterinária. Rio de Janeiro. 23890–000. Brazil, shalako1953@gmail.com, shalako@ufrrj.br

Shohei Sakuda
Department of Applied Biological Chemistry, Graduate School of Agricultural and Life Sciences, The University of Tokyo, 1-1-1 Yayoi, Bunkyo-ku, Tokyo 113–8657, Japan

Stefano Sforza
Dipartimento di Chimica Organica e Industriale, Viale G.P. Usberti 17A, Campus Universitario, Università di Parma, I-43100, Parma, Italy, stefano.sforza@unipr.it

Masoomeh Shams-Ghahfarokhi
Department of Mycology, Faculty of Medical Sciences, Tarbiat Modares University, Tehran 14115–111, Iran

Avishay-Abraham Stark
Department of Biochemistry, Tel Aviv University, Ramat Aviv, 60078 Israel, avis@tauex.tau.ac.il

Yoshiko Sugita-Konishi
Division of Microbiology, National Institute of Health Sciences, 1-18-1, Kamiyoga, Setagaya-ku, Tokyo 158–8501 Japan, ykonishi@nihs.go.jp

Adriana M. Torres
Mycology Section, Department of Microbiology and Immunology, National University of Río Cuarto, Ruta 36 km, 601 (5800) Río Cuarto, Córdoba, Argentina, atorres@exa.unrc.edu.ar

Ajit Varma
Amity Institute of Microbial Technology, Noida, U.P., India, ajitvarma@aihmr.amity.edu

Agnieszka Waskiewicz
Department of Chemistry, University of Life Sciences, Wojska Polskiego 75, 60–625 Poznan, Poland

Jiujiang Yu
Southern Regional Research Center, ARS, USDA, 1100 R.E. Lee Blvd.
New Orleans, LA 70124, USA

Łukasz Zielonk
Faculty of Veterinary Medicine, Department of Veterinary Health Protection, Division of Veterinary Prevention and Feed Hygiene, University of Warmia and Mazury in Olsztyn, St. M. Oczapowskiego 13, 10–719 Olsztyn, Poland

Part I
Biology and Biotechnology

Chapter 1
Artificial Systems for Molecular Recognition of Mycotoxins

Claudio Baggiani, Laura Anfossi, and Cristina Giovannoli

1.1 Introduction

Nowadays, public awareness about synthetic chemicals in food is high, and consumers continue to express concern about the health risks linked to the deliberate addition of chemicals to food. On the contrary, the perception of the health risks posed by food contamination due to mycotoxins is less marked. However, although effects are often difficult to link with a particular food, it is now largely accepted in academic circles and public health bodies that food and feed contamination from mycotoxins is a severe public health problem that can deeply affect health not only after a single massive exposure but, more often, after continuous exposure to low doses, and that such exposure can be related to several chronic diseases, including some types of cancer and serious hormonal dysfunctions (Lewis and Fenwick 1991; De Vries et al. 2002). Thus, good analytical protocols based on efficient analytical processes – sensitive, selective, fast, inexpensive and suitable for sample mass screenings – are required by legislation, health authorities and companies operating in the food market.

At present, commercially available rapid assays based on the use of immunoanalytical techniques are widely diffused, as these analytical techniques assure the feasibility of fast sample mass screenings in a more affordable fashion compared to the older thin layer chromatographic methods (van Egmond 2004). However, a sample which is positive to toxicant contamination should be validated by using more sophisticated analytical methods. These methods are usually based on instrumental separative techniques coupled with mass spectrometry detectors of varying complexity. They have the sensitivity required for contamination detection and quantification, but direct application of these techniques on food and feed samples

C. Baggiani (✉)
Laboratory of Bioanalytical Chemistry, Department of Analytical Chemistry, University of Torino, via P. Giuria 5, 10125 Torino, Italy
e-mail: claudio.baggiani@unito.it

M. Rai and A. Varma (eds.), *Mycotoxins in Food, Feed and Bioweapons*,
DOI: 10.1007/978-3-642-00725-5_1, © Springer-Verlag Berlin Heidelberg 2010

can be rarely performed. In fact, contaminants are usually present in food at low concentration (ng–μg/kg) levels, dispersed in highly complex (thousand of different components) and morphologically structured matrices, with an elevated degree of point-to-point and sample-to-sample variability. Thus, such a type of matrix introduces severe disturbances in the analytical separation step. Moreover, very "dirty" samples show the noxious property to influence strongly the background ion current in a mass spectrometry detector, reducing its sensitivity (Hajslova and Zrostlikova 2003; Niessen 2003). In consequence, quantitative analysis can be performed only after extensive clean-up and preconcentration steps (Buldini et al. 2002; Careri et al. 2002; Pichon et al. 2002; Hennion and Pichon 2003). Thus, economical, rapid and selective clean-up methods based on "intelligent" materials are needed.

Unfortunately, there are very few examples of supramolecular organic structures able to behave as receptors for mycotoxins. Cyclodextrins and chlorphyllin are well known to strongly bind aflatoxins, but without any selectivity (Vazquez et al. 1992; Breinholt et al. 1995). On the contrary, tailor-made, artificial systems, such as molecularly imprinted polymers and binding peptides obtained by combinatorial synthesis, are good candidates for circumventing the drawbacks typical of more traditional solid phase extraction techniques. In fact, these materials seem to be particularly suitable for extractive applications where analyte selectivity in the presence of very complex samples represents the main problem.

1.2 Molecularly Imprinted Polymers

Molecularly imprinted polymers (MIPs) are synthetic materials provided with artificial binding sites able to selectively recognise a target molecule (Sellergren 2001; Yan and Ramström 2004; Piletsky and Turner 2006). In brief, as illustrated in Fig. 1.1, these materials are obtained by polymerization around a template molecule of functional and cross-linking monomers chosen for their ability to interact with the functional groups of the template through non-covalent interactions

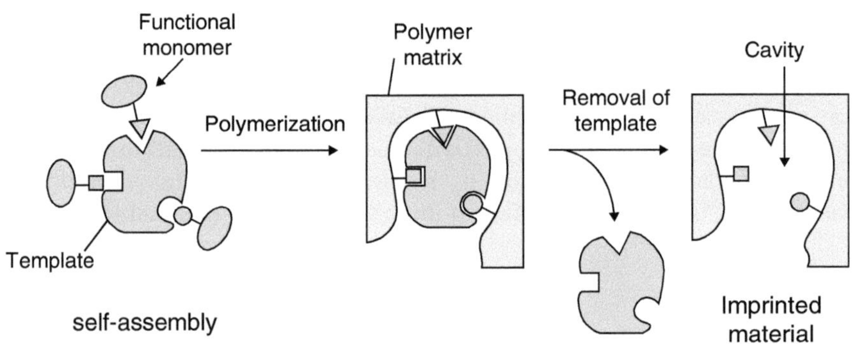

Fig. 1.1 Key steps in the molecular imprinting process

(non-covalent imprinting), reversible covalent bonds (covalent imprinting) or mixed combinations of the two methods (semi-covalent imprinting). Once polymerization has taken place, a highly cross-linked three-dimensional network polymer is formed and binding sites with shape, size and functionalities complementary to the template are established in the bulk of the polymer. These artificial binding sites have the same features as the antibody binding sites, showing binding reversibility, enhanced selectivity, high affinity constant and a significant polyclonality (non-covalent approach) or monoclonality (covalent approach). Nevertheless, imprinted polymers show large differences compared to antibodies. In fact they are macroscopic objects, stiff, and insoluble in any solvent, whereas their biological counterparts are nanoscopic objects, flexible, and soluble in water.

1.2.1 Molecular Imprinted Solid-Phase Extraction

In the last few years, a growing number of papers have been dedicated to the clean-up and preconcentration of analytes of clinical, pharmaceutical or food chemistry interest from several types of matrices. In fact, considering the number of papers published worldwide on peer-reviewed journals, molecular imprinted solid phase extraction (MISPE) is one of the fastest growing applications, with more than 280 papers published since 1994 (ISIS Web of Knowledge 2008; Society for Molecular Imprinting Database 2008).

The MISPE technique is very similar to the traditional solid phase extraction made on non-specific stationary phases. A small amount of imprinted polymer (typically 25–500 mg) is packed in an open column (for off-line applications) or in a short HPLC column (for on-line applications). Then, the usual steps of column conditioning, sample loading, column washing and analyte elution are carried out. Usually, the extraction protocol has been previously tested on artificial samples to consider the feasibility of the method, Less frequently, the same optimised protocol has been validated on real samples against a more commonly used method or published in literature, considering issues such as robustness, accuracy, precision, limits of quantification and determination (Olsen et al. 1999; Gallego-Gallegos et al. 2006; Urraca et al. 2006a; Zhang et al. 2006).

The main critical point associated with the development of a MISPE protocol is related to the residual template not being completely removed from the polymeric matrix and slowly leaking during loading, washing and elution operations. Such a template loss (polymer "bleeding") is usually detected at trace levels during the elution step, and it represents a significant source of interferences and systematic errors in trace analysis. Moreover, concern for the possible contamination of the analytical samples by the residual template released during the analyte elution is one of the main obstacles to a wider diffusion of the MISPE method in current sample treatment methods. Several methods have been proposed to overcome this drawback by efficiently removing the residual template, including thermal annealing of the imprinted polymer (Zander et al. 1998) and severe washing conditions

(Ellwanger et al. 2001), but, despite all efforts, it seems that to remove all the template molecules from the imprinted polymer will be extremely difficult if not impossible using the current technology.

Thus, the most successful strategy has been revealed to be the use of a mimic of the analyte as a template molecule. The so-called “template mimic” technique consists of the use of a structural analogue of the molecule of analytical interest as a template. The choice of this putative template requires a certain degree of creativity from the chemist (and a certain cleverness in organic chemistry), as it should be made in such a way as to obtain imprinted binding sites provided with good selectivity towards the analyte molecules. At the same time, this structural analogue should be different from the analyte in such a way that the analytical separation performed after the extraction step discriminates clearly between the analyte and the residual template molecules released by the imprinted material. Differences in molecular structure between the analyte and the putative template should be minimal and localised far from relevant structural motifs and substituents directly involved in non-covalent interactions with the binding sites. Thus, any modification of the target involving structures critical for molecular recognition should be discarded.

The choice of a mimic template is usually made through empirical considerations, but in some cases the selection has been made using computational methods, taking into account not only shape similarity, but also electronic and hydrophobic factors (Baggiani et al. 2002). Several different approaches have been described to conceive an efficient template mimic. When the target consists of a class of molecules, typically a major analyte and its metabolites, with minimal differences in the molecular structures, addition/subtraction of one or more carbon atoms to the molecule skeleton, especially if an aliphatic chain is present, seems to be particularly convenient (Turiel et al. 2001; Cacho et al. 2003). The same approach is valid not only for carbon atoms in aliphatic chains, but also for target analytes provided with substituents not directly involved in non-covalent interactions with the imprinted binding sites (Theodoridis et al. 2003). Alternatively, as the selection of a suitable mimic template is mainly driven by the similarity to the analyte, analogs with isosteric substituents not directly involved in the non-covalent interaction with the imprinted binding sites can offer an interesting possibility for preparing putative templates (Blomgren et al. 2002).

One of the main drawbacks of the mimic template technique is related to the difficulties of practically attaining some optimal templates. In fact, as they may be difficult to synthesise, expensive, or simply not available as commercial products, it could be necessary to use commercially available substances as mimic templates that are less strictly related to the target analyte, paying the price of a more limited molecular recognition effect. As a consequence, in many cases, structural differences between the analyte and the mimic template are significant, and similarity between molecules remains confined to the overall molecular shape and the preservation of substituents able to form non-covalent interactions with the binding sites (Urraca et al. 2006a). This strategy to use mimic template poorly related to the target analyte is brought to its extreme consequence in the so-called “fragmental imprinting" (Kubo et al. 2004), "epitope imprinting" (Rachkov and Minoura 2001) or "substructure imprinting" (Hall et al. 2006) approaches, where the mimic template is represented by a molecule largely different from the target analyte as a whole, but similar to one of the sub-structures in which the target molecule could be divided.

1.2.2 Imprinted Materials for Mycotoxin Analysis

As briefly seen in the introduction, mycotoxins are probably the most important food contaminants in terms of toxicity and width of diffusion. Nevertheless, people working in the molecular imprinting field have dedicated their attention only to these analytes in the recent years. As a consequence, compared to the very large number of research papers dedicated to the design, characterization and practical use of imprinted polymers for analytes such as drugs and pesticides, a relatively small number of papers dedicated to mycotoxins have been published up to the present day.

There are some reasons for this apparent lack of interest towards molecular imprinting of mycotoxins. First of all, immunoextraction represents an extremely valid competitor compared to any clean-up technique based on molecular imprinting (Haginaka 2005); and, in fact, immunoextraction is applied widely in mycotoxin analysis to perform extracts clean-up (Gilbert and Anklam 2002). These immunoaffinity-based materials have sufficient capacity to clean up heavily contaminated samples, and they usually efficiently remove compounds that could interfere in the determination of mycotoxins because the antibodies specifically recognise the toxin of interest. As a consequence, immunoextraction has been a consolidated technique in mycotoxin analysis for many years, and extraction cartridges are commercially available for many of these substances. Thus, even if imprinted materials are potentially competitive with immunoaffinity-based materials, the acceptance of this technology remains low. Generally speaking, the acceptance of new technology meets resistance from users of old technology, and this fact has a practical consequence on the development of imprinted materials competitive with immunoextraction cartridges. Probably, it is not just chance that no imprinted cartridges are commercially available for the extraction of mycotoxins, whilst they are for several drugs and pesticides.

Beside the concurrence of the immunoextraction technique, it is a widespread common opinion that mycotoxins are very difficult templates compared to other food contaminants (Baggiani et al. 2007). It should be noted that practical difficulties in preparing imprinted polymers for mycotoxins do not arise from lack of functional groups suitable for setting up non-covalent interactions during the imprinting process. In fact, all the most significant mycotoxins show many polar groups suitable for hydrogen bond or ion-pair interactions. Again, solubility in organic solvents commonly used in molecular imprinting is not a problem, and usually there is full chemical compatibility with the organic reagents used during the polymerization process.

Contrarily, difficulties can arise from the elevated toxicity of this class of food contaminants. In fact, it could be quite dangerous to directly manipulate the amounts of toxin necessary to synthesise a sufficient quantity of imprinted polymer to set up an extraction protocol, as contamination from minute amounts could cause serious health problems, including long-term effects. Another problem is related to the market availability of these compounds. In fact, while there are a lot of companies selling analytical standards of mycotoxins, it is quite difficult to purchase 100 mg-level amounts of the same toxins at an affordable price. In the experience of the authors of this review, besides all other possible considerations on the difficulty of obtaining truly functional imprinted polymers, several interesting metabolites of mycotoxins (see aflatoxin M1 and zearalenols as significant examples) cannot be used as templates to prepare imprinted polymers simply because of their inaccessible prices, provided that no alternative sources become available.

Nevertheless, it has been shown that most of the pitfalls described so far can be avoided through careful design of the imprinting process and, when necessary and feasible, the use of the mimic template approach. Thus, molecular imprinted polymers have been successfully prepared for several natural toxins, as reported in the next sections.

1.2.2.1 Ochratoxin A

Ochratoxin A (**1**) (see Fig. 1.2) was the first mycotoxin for which a successful molecular imprinting has been reported in literature. In fact, in 2002, the same synthetic approach – with minimal differences – was independently reported by our

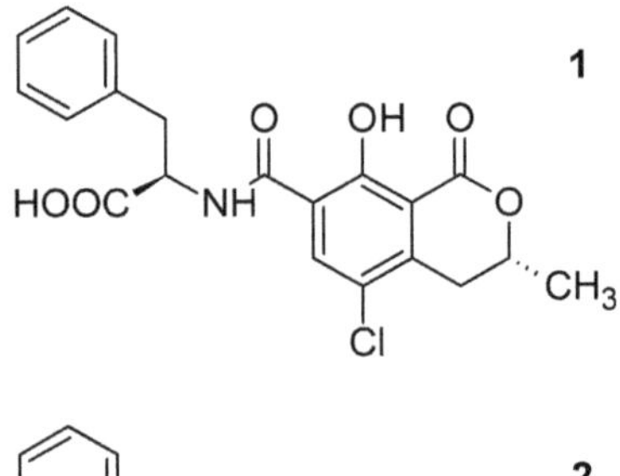

1

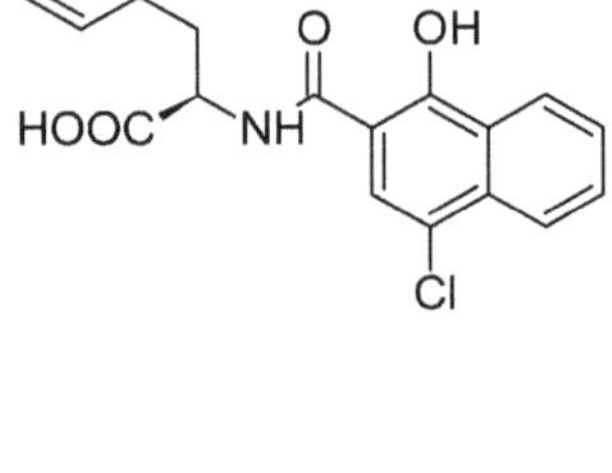

2

Fig. 1.2 Molecular structures of ochratoxin A (**1**) and its mimic template *N*-(4-chloro-1-hydroxy-2-naphthoylamido)-(L)-phenylalanine (**2**)

group (Baggiani et al. 2002) and Jodlbauer et al. (2002). A good mimic template was rationally designed to preserve the general structure of the target analyte, including the chirality of the amino acidic sub-structure and the planarity of the benzopiranic sub-structure. At the same time, the α-unsaturated lactone moiety characterizing many carcinogenic mycotoxins was eliminated, while the distinct points of potential interaction with functional monomers were maintained: the α-carboxyl of L-phenylalanine, the amido bridge, and the phenolic hydroxyl. The resulting mimic template, *N*-(4-chloro-1-hydroxy-2-naphthoylamido)-(L)-phenylalanine (**2**), showed almost complete overlapping of the two molecules, with a high degree of similarity not only as structures, but also as solvent accessible surfaces, electrostatic potential surfaces and lypophilic/hydrophilic surfaces. Different polymerization mixtures were considered to prepare the polymers. In fact, while we obtained an imprinted polymer using a traditional methacylic acid/ethylene dimethacrylate mixture, Maier and co-workers (2004) used a more exotic mixture with quinuclidine methacrylamide and *tert*-butylmethacrylamide as functional monomer and ethylene dimethacrylate as cross-linker. Anyway, both the approaches resulted valid, with the presence of specific molecular recognition effects due to hydrogen bond interactions and steric factors and good recognition of ochratoxin A compared to several analogs in polar (methanol, acetonitrile) and hydrophobic (chloroform) solvents.

The quinuclidine methacrylamide polymer reported by Maier et al. (2004) was used in a subsequent work to extract ochratoxin A from red wine before quantification by HPLC-fluorescence detection. The approach involved a two-stage sample clean-up protocol on coupled reversed-phase (C18-silica) and MISPE cartridges, where the use of the reversed-phase cartridge was crucial for the removal of the interfering acidic matrix compounds. The method provided recovery >90% and RSD <10%, with detection and quantification limits of 10 and 33 ng l^{-1} in spiked and commercial red wines. However, authors raised doubt on the effectiveness of the MISPE protocol, as similarly favourable performances were observed in control experiments in which the imprinted polymer was replaced by the corresponding non-imprinted material. These findings provided some evidence that under the employed experimental conditions analyte binding was mainly due to non-specific interactions with the polymeric matrix of the MISPE cartridge.

An imprinted polymer prepared by using ochratoxin A directly as a template molecule is described by Turner et al. (2004). In this case efficient functional monomers were identified by molecular mechanic simulations between the ochratoxin A molecule and 20 different functional monomers. From these simulations an equimolar mixture of methacrylic acid and acrylamide was selected as functional monomers, while ethylene dimethacrylate was used as cross-linker. The use of *N,N*-dimethylformamide as a porogenic solvent – uncommon in molecular imprinting – generated an imprinted polymer with excellent affinity and specificity for ochratoxin A in buffered aqueous solutions, while polar organic solvents such as acetonitrile suppressed the molecular recognition effects. Interestingly, as an effect of conformation change of the polymer matrix, low buffer concentration or basic pH caused a loss of the polymer recognition properties, while high buffer concentration or acidic pH enhanced the specific binding. Unfortunately, no applications in food clean-up for this polymer were reported by the authors.

An on-line MISPE with pulsed elution has been described for the rapid analysis of ochratoxin A in wheat extracts (Zhou et al. 2004). The imprinted polymer was synthesised in chloroform by using *N*-phenylacrylamide as functional monomer and trimethylolpropane trimethacrylate as cross-linker, crushed and packed in a HPLC micro-column for fluorescence detection. MISPE conditions were optimised for the loading of ochratoxin A extracts in methanol/acetic acid (99:1, v/v) and pulsed elution with methanol/triethylamine (99:1, v/v, 20 μl). Nearly quantitative binding could be achieved from one 20-μl injection of sample containing up to 30 ng of ochratoxin A, with a detection limit of 0.1 ng per injection (5.0 ng ml^{-1}) and a recovery of mycotoxin from wheat extracts was of 103 ± 3%. However, as in the case of the quinuclidine methacrylamide polymer reported by Maier et al. (2004), a non-imprinted polymer prepared with *N*-phenylacrylamide shows binding properties towards ochratoxin A and a significant level of selectivity compared to mycotoxins other than ochratoxin A, casting doubts on the real effectiveness of the imprinting process (Zhou and Lai 2004). In a completely different approach, on-line MISPE miniaturised devices based on thin layers of ochratoxin A-imprinted electropolymerised polypyrrole were described for detection and quantification of ochratoxin A in wine. The imprinted polypyrrole layers were supported on stainless-steel frits, directly fixed onto the porous steel surface (Yu et al. 2005) or adsorbed onto a layer of single-wall carbon nanotubes (Yu and Lai 2006, 2007). When relatively large amount of wine samples, up to 3 ml, were loaded onto the extraction devices, recovery up to 40% was obtained, with detection limits of 50 and 80 ng l^{-1} for a polypyrrole-imprinted layer supported by steel and single-wall carbon nanotubes respectively. This approach has been previously used by the same authors to set up a miniaturised surface plasmon resonance device for ochratoxin-A sensing in wine and wheat extracts with sensitivity down to 50 μg l^{-1} (Yu and Lai 2005).

1.2.2.2 Zearalenone

A first attempt to obtain imprinted polymers for mycotoxin zearalenone (**3**) (see Fig. 1.3) was down to Weiss et al. (2003). As the molecular structure of this mycotoxin brings a carbon–carbon double bond potentially interfering with the polymerization process, the authors used the flavonoid quercetin as putative template. Unfortunately, the polymer obtained showed some molecular recognition properties towards zearalenone, but a limited overall binding capacity, making it unsuitable for the development of a MISPE method.

In a different approach, Urraca et al. (2006a) used the cyclododecanoyl ester of resorcilic acid (**9**) as a mimic template to prepare imprinted polymers based on allylpiperazine as functional monomer, trimethylolpropan trimethacrylate as cross-linker and acetonitrile as porogenic solvent. As previously described, the choice of cyclododecanoyl resorcilate as mimic template was made on the basis of its

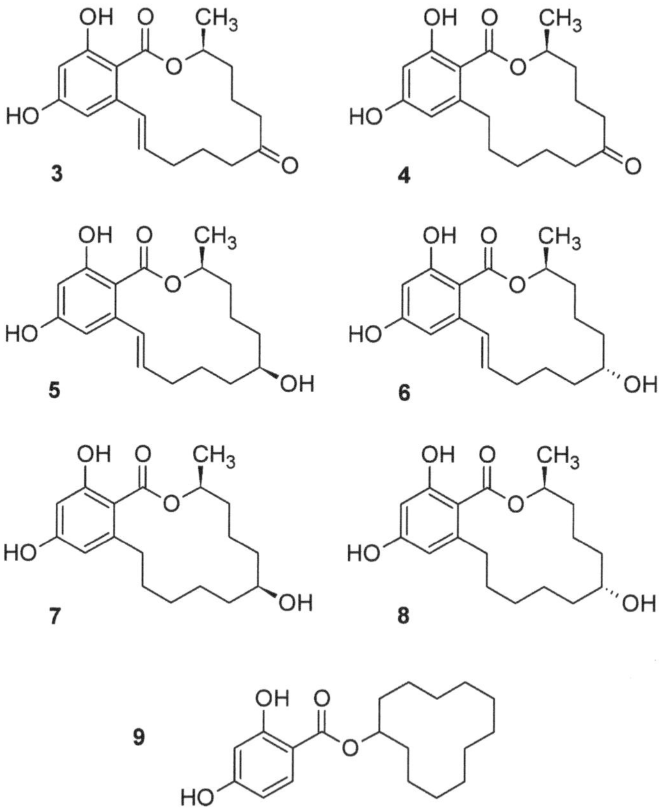

Fig. 1.3 Molecular structures of zearalenone (3), its metabolites zearalanone (4), α-zearalenol (5), β-zaralenol (6), α-zearalanol (7) and β-zearalanol (8), and its mimic template cyclododecanoyl resorcilate (9)

commercial availability and its resemblance to zearalenone in terms of size, shape and functionality, aiming to preserve the resorcilic sub-structure directly involved in the hydrogen bond interaction with functional monomers. The resulting imprinted polymer was evaluated by liquid chromatography, and good molecular recognition properties were observed not only towards the target molecule, but also towards its metabolites zearalanone (**4**), α-zearalenol (**5**), β-zearalenol (**6**), α-zearalanol (**7**) and β-zearalanol (**8**), while resorcylic acid and several steroidal estrogens were very poorly recognised. This imprinted polymer was successfully applied for the clean-up of zearalenone and its main metabolite α-zearalenol from cereal and swine feed sample extracts (Urraca et al. 2006b).

The analysis of these mycotoxins from the food samples was very efficiently accomplished using pressurised liquid extraction in organic solvent (methanol/acetonitrile 1 + 1, v/v). Clean-up was performed through the MISPE cartridge and quantitative determination by liquid chromatography with fluorescence detection. The method – validated using a corn reference material for zearalenone – gave recovery of 85–97% (RSD 2.1–6.7%) and 87–97% (RSD 2.3–5.6%) for α-zearalenol and zearalenone, respectively. The detection limits for the different matrices tested, based on a signal-to-noise ratio of 3:1, ranged between 1.7 and 2.4 ng g^{-1} for zearalenone and from 0.7 to 1.3 ng g^{-1} for α-zearalenol.

A MIP-based optode has been recently developed for zearalenone analysis (Navarro-Villoslada et al. 2007). The authors describe a semi-automated flow-through assay based on the displacement of tailor-made highly fluorescent tracers by the analyte from an imprinted polymer prepared with a mixture of cyclododecanoyl ester of resorcilic acid (template mimic), allylpiperazine (functional monomer) and trimethylolpropan trimethacrylate (crosslinker) in acetonitrile (porogen). A pyrene-containing fluorescent tracer, namely 2,4-dihydroxybenzoic acid 2-[(pyrene-1-carbonyl)amino] ethyl ester, was selected to optimise the zearalenone displacement fluorosensor, reaching a limit of detection (calculated as the analyte concentration for which 10% of fluorescent tracer was displaced from the polymer sensor) of 8 μg ml^{-1} in acetonitrile (RSD 2–6%), with a positive cross-reactivity for β-zearalenol, but not for related phenols such as resorcinol, resorcylic acid, 17-β estradiol, estrone or bisphenol-A. Unfortunately, no applications in food clean-up were reported by the authors as the sensitivity was not sufficient to reach the maximum analyte concentration levels in food samples that legislation allows.

1.2.2.3 Moniliformin

Imprinted polymers for the mycotoxin moniliformin (**10**) (see Fig. 1.4) have been described by Appell et al. (2007). A small library of polymers was prepared in dimethylformamide by varying template — functional monomer molar ratio, template structure, functional monomer and cross-linker. It was found that significant differences in moniliformin binding by the polymers were dependent on polymer composition, and these differences were highly dependent on the template used to imprint the polymer. The best binding polymer was obtained by using 3,4-diethoxy-3-cyclobuten-1,2-dione (moniliformin diethylether, **11**) as mimic template, *N,N*-dimethylaminoethyl methacrylate as functional monomer and trimethylolpropan trimethacrylate as cross-linker in molar ratio 1:8:40. This polymer was evaluated as a sorbent for MISPE of acetonitrile corn extracts. The method was not optimised, but clean chromatograms by liquid chromatography with UV detection at 229 nm and quantitative recoveries with corn samples spiked with 0.5 μg l^{-1} of moniliformin were obtained.

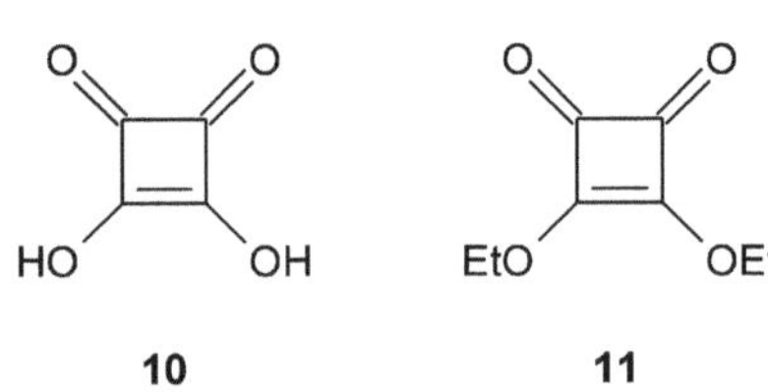

Fig. 1.4 Molecular structures of moniliformin (**10**) and its mimic template 3,4-diethoxy-3-cyclobuten-1, 2-dione (moniliformin diethylether, **11**)

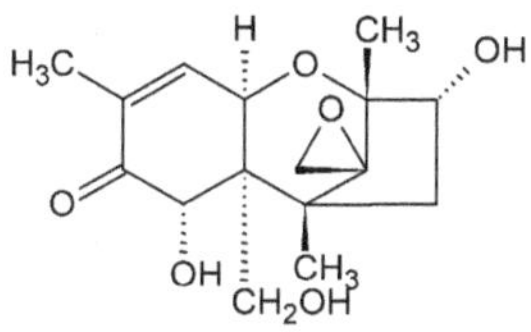

12

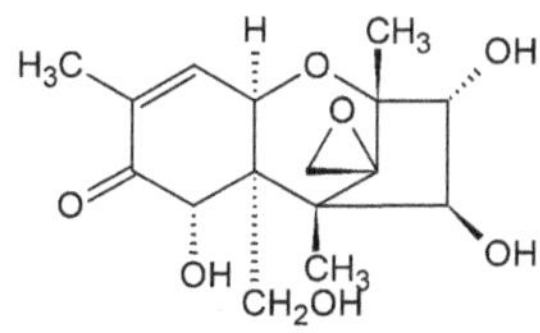

13

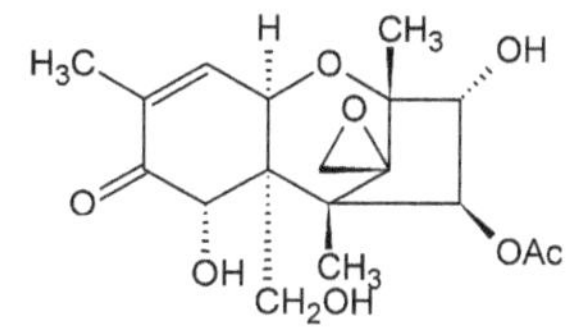

14

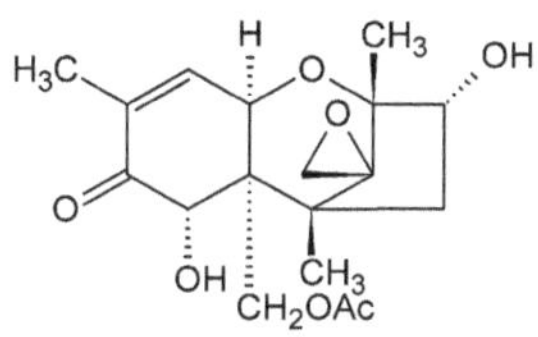

15

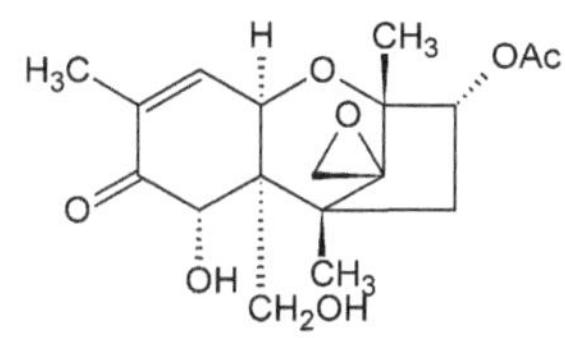

16

Fig. 1.5 Molecular structures of deoxynivalenol (**12**), nivalenol (**13**) and deoxynivalenol metabolites fusarenon-X (**14**), 15-acetyl deoxynivalenol (**15**) and 3-acetyldeoxynivalenol (**16**)

1.2.2.4 Deoxynivalenol

Weiss et al. (2003) described an imprinted polymer for mycotoxin deoxynivalenol (**12**) (see Fig. 1.5). The synthesis was performed in acetonitrile, by using the mycotoxin directly as a template, methacrylic acid as functional monomer and ethylene dimethacrylate as cross-linker. Interestingly, when selectivity was measured in acetonitrile, the polymer recognised nivalenol (**13**) slightly better than the template, while the analogs fusarenon-X (**14**), 15-acetyl deoxynivalenol (**15**) and 3-acetyldeoxynivalenol (**16**) were recognised only marginally. No analytical applications were given.

1.3 Combinatorial Peptides

As stated in the previous section, one of the main disadvantages of the molecular imprinting technology is the necessity of a relatively large quantity of template with a high degree of purity. This aspect, above all, limits the use of this technique for preparing imprinted polymers with several mycotoxins which are very toxic, commercially available with low purity or very expensive compounds.

An alternative approach for obtaining synthetic molecular recognition systems could be combinatorial chemistry applied to peptide synthesis, which allows us to prepare very large libraries of peptides with well-characterised binding properties. The roots of combinatorial organic synthesis stem from the development of solid-phase peptide synthesis by Merrifield (1963). Solid-phase peptide synthesis consists of sequentially coupling amino acid monomers onto a growing peptide chain, which is immobilised on small polymeric beads. The primary advantages of solid-phase synthesis are a large excess of reagent that can be used to drive peptide coupling reactions to completion, and purification protocols greatly facilitated by covalent attachment of the intermediate products to an insoluble solid-phase. Thus, nowadays solid-phase peptide synthesis provides a powerful means of preparing significant amounts of peptides with defined sequences. Well-designed peptide combinatorial libraries have great potential as capturing molecular recognition agents because of their varied chemical properties and functional groups, their different physical properties, and their well-known and sufficient number of synthetic protocols. Moreover, many amino acids could be easily purchased at low cost, even if some specific sequences could be obtained only by difficult syntheses that need expensive reagents, above all to prepare a sufficient amount of peptide.

1.3.1 Combinatorial Peptides for Mycotoxins

The present literature is principally related to the search for sequences with molecular recognition properties towards biomacromolecules with biotechnological or medical implications (Ding and Ho 2004; Aina et al. 2007), while peptide libraries with recognition towards small ligands are relatively rare and mainly directed towards the use of peptides for separative processes (Tozzi et al. 2003a, b; Li 2006). Thus, it is not surprising that it is very difficult to find literature on peptides with molecular recognition towards mycotoxins.

On this premise, our laboratory has recently developed a strategy for the preparation of synthetic peptide libraries for small-molecular-mass analytes such as estrogenic endocrine disruptors (Tozzi et al. 2002) and mycotoxins (Tozzi et al. 2003a, b; Giraudi et al. 2007). This strategy is based on the sequential development of a spatially addressable parallel library where – after each peptide elongation step – the binding properties towards the target molecule (and possible interfering substances) are measured, and only the peptide with the best binding behaviour and selectivity is retained and selected as a scaffold for the successive elongation step, while the other sequences are discarded (see Fig. 1.6). One of the main advantages of this approach is the possibility of selecting peptide sequences with good binding behaviour and high selectivity towards the target ligand starting from a relatively small library dimension. This possibility is due to the screening protocol for peptides with good molecular recognition during the library synthesis, which is made *during* the library synthesis, and not after the completion of it. Moreover, the use of chromatography grade poly(styrene-*co*-divinylbenzene) beads as solid phase synthesis support make it possible to use the peptides directly in an solid phase extraction method for the isolation of the target molecule.

1.3.1.1 Aflatoxins

Peptides with molecular recognition properties towards aflatoxins (see Fig. 1.7) were identified in a 8 $\times$ 8 library (Tozzi et al. 2003a, b). The chosen amino acids

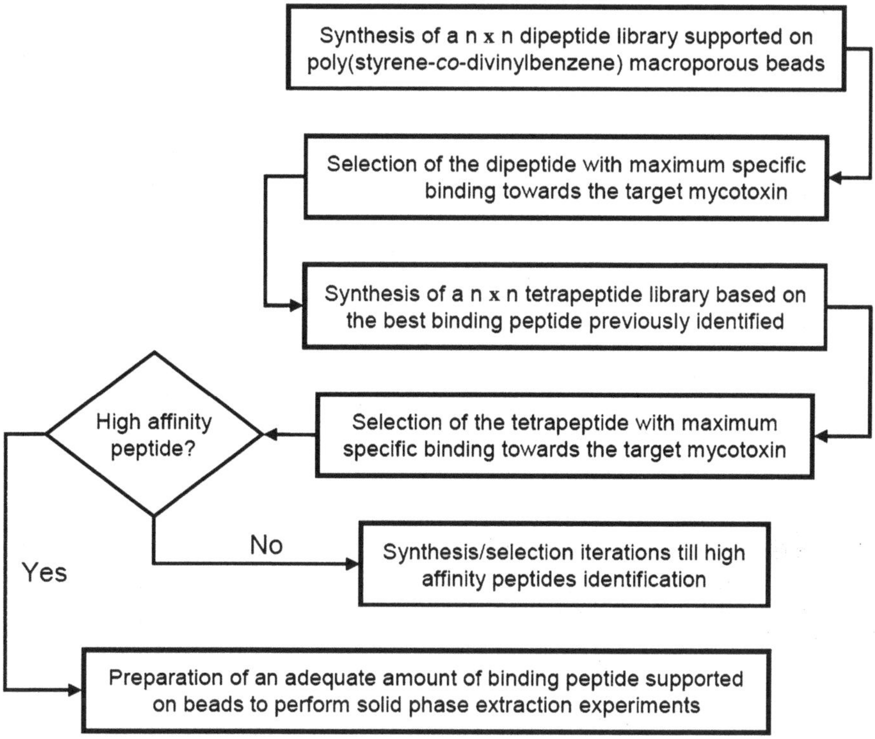

Fig. 1.6 Scheme of the sequential development of a spatially addressable parallel peptide library for mycotoxins

(Arg, Ser, Pro, Val, Leu, Gln, Gly, Ala) were selected because of their chemical properties (i.e. solubility, different functional groups), their easy availability and their low cost. In fact, these amino acids show functional groups in their lateral chains that cannot react under synthesis conditions, in this manner simplifying the preparation of the libraries.

The first amino acid library composed of 64 different dipeptides showed as best binding sequence the dipeptide Leu–Leu. An affinity constant measured on-beads for the target mycotoxins resulted in 7.9×10^3 and 8.8×10^3 M^{-1} for aflatoxin B1 (**17**) and B2 (**18**), and 0.9×10^3 and 0.7×10^3 M^{-1} for aflatoxin G1 (**19**) and G2 (**20**). Thus, a tetrapeptide library was prepared by using the sequence Leu–Leu as scaffold. From the second library the tetrapeptide with the sequence Leu–Leu–Ala–Arg showed the highest binding towards aflatoxins. Affinity-constant measured on-beads for the target mycotoxins resulted in 1.2×10^4 and 2.9×10^4 M^{-1} for aflatoxin B1 and B2, and 0.7×10^3 and 1.9×10^3 M^{-1} for aflatoxin G1 and G2, with a net increase of affinity for aflatoxins B1, B2 and G2. The binding constants towards aflatoxins B1 and B2 were rather high considering that the binding constants of the human serum albumin for compounds with a similar structure — such as coumarins — range between 10^4 and 10^6 M^{-1} (Shobini et al. 2001). It should be noted that in the

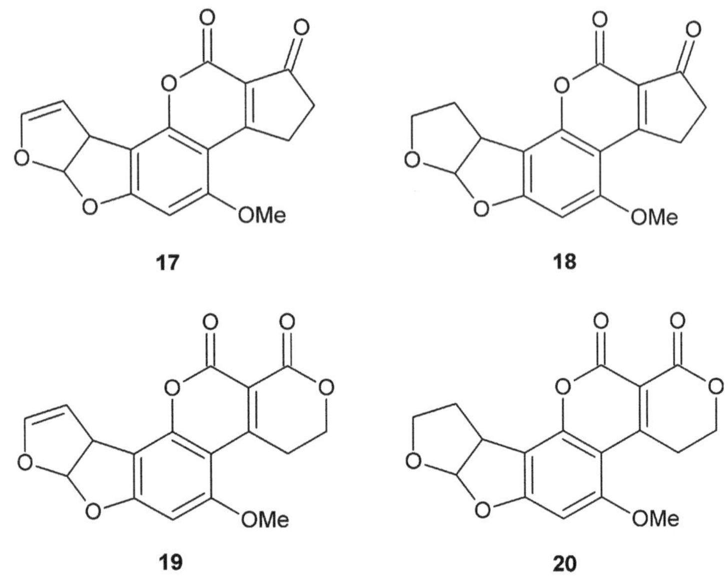

Fig. 1.7 Molecular structures of aflatoxin B1 (**17**), aflatoxin B2 (**18**), aflatoxin G1 (**19**) and aflatoxin G2 (**20**)

tetrapeptide library several sequences were found to be selective towards some of the mycotoxins considered. In fact, many peptides showed good affinity towards aflatoxin B1 and aflatoxin B2, with affinity constant greater than $1.0 \times 10^4\ M^{-1}$, but, beside the sequence Leu–Leu–Ala–Arg, only Leu–Leu–Ala–Pro showed good affinity for aflatoxin G1 (affinity constant $1.9 \times 10^3\ M^{-1}$) but not for aflatoxin G2, while Leu–Leu–Val–Pro and Leu–Leu–Gly–Ser showed good affinity for aflatoxin G2, with affinity constants of 2.0×10^3 and $2.3 \times 10^3\ M^{-1}$, but a limited affinity for aflatoxin G1.

Beads grafted with the sequence Leu–Leu–Ala–Arg were used to demonstrate the feasibility of a solid phase extraction method based on combinatorial peptides. Extraction of 1 ml of a buffered solution containing 5 μM of each aflatoxin was performed successfully, with recoveries of 76, 89 and 70% for aflatoxin B1, B2 and G2, respectively. As aflatoxin G1 was not efficiently recovered, an extraction cartridge loaded with beads grafted with the sequence Leu–Leu–Ala–Pro was used to extract this mycotoxin, with a recovery of 79%.

1.3.1.2 Ochratoxin A

Peptides with molecular recognition properties towards ochratoxin A were identified in a 12 × 12 library containing the amino acids Ala, Arg, Asn, Phe, Gly, His, Leu, Lys, Pro, Ser, Trp and Val (Giraudi et al. 2007).

The first amino acid library composed of 144 different dipeptides showed as best binding sequence the dipeptide Ser–Asn, with an affinity constant measured on-beads for the target mycotoxin of $7.9 \times 10^3\ M^{-1}$. A tetrapeptide library developed by using the sequence Ser–Asn as scaffold showed a general increase in the affinity for ochratoxin A, and the sequence Ser–Asn–Leu–His, with an affinity constant of $1.3 \times 10^4\ M^{-1}$, performed better than the other sequences in ochratoxin binding. The binding performances were developed further by preparing an hexapeptide library where the peptide Ser–Asn–Leu–His–Pro–Lys showed an affinity of $3.4 \times 10^4\ M^{-1}$.

Ochratoxin binding by the hexapeptide was found to be regulated by the solution pH. In fact, the maximum binding was observed for mildly acidic buffers (pH 3–4), while more acidic buffer (pH 2.2) reduced the binding and weak, neutral or alkaline buffers (pH 5–9) showed a marked binding inhibition. As ochratoxin can be considered a very weak acid ($pK = 7.1$) fully protonated below pH 6, this pH effect can be attributed to the protonation/deprotonation of the hexapeptide. Considering its sequence, it is clear that the only amino acids which ion charge mutates in the considered range of pH are histidine, whose imidazolic ring has a $pK \sim 6.0$, and lysine, whose α-carboxyl has a $pK \sim 2.2$, while both the α- ($pK \sim 8.9$) and ε-amine ($pK \sim 10.5$) of lysine are constantly protonated in the considered pH range. Thus, it can be concluded that in ochratoxin maximum binding is strongly related to the contemporary presence of (a) the mycotoxin in its neutral, fully protonated form, (b) the histidine side arm in its cationic protonated state and (c) the zwitterionic form of lysine.

It should be noted that OTA recognition by the selected hexapeptide in its fully protonated form involves a binding mechanism completely different from the one involved in the recognition by natural proteins. In fact, in this case, OTA is recognised in a neutral buffer as dianion by human albumin in a binding pocket formed by a tryptophan residue (Trp_{214}) and several other positively charged amino acids (Arg_{218}, Arg_{222}, His_{242} and Arg_{257}) (Il'ichev et al. 2002). It is in any case remarkable as, beside deep differences in the amino acid composition between the human albumin OTA binding pocket and the artificial hexapeptide, the presence of positively charged amino acid residues seems to be necessary for OTA binding.

Beads grafted with the sequence Ser–Asn–Leu–His–Pro–Lys were used to set up a solid phase extraction method for the ochratoxin A in wines at concentration levels down to 0.10 $\mu g\ l^{-1}$. Several different red, white and rosé wine samples fortified with the mycotoxin showed recovery of 95 and 98% at 2.0 and 4.0 $\mu g\ l^{-1}$, respectively, without any effect on the extraction efficiency of the matrix. The efficacy of this approach was successfully tested by comparison with an immunoaffinity extraction performed on commercial cartridges.

1.4 Conclusions

As shown in the previous sections, molecular imprinting can be successfully used to prepare intelligent materials for detection, clean-up and preconcentration of mycotoxins in complex samples. From the examples reported, imprinted polymers are potential competitors with traditional solid phase extraction materials for their selectivity, and with immunoaffinity extraction, for their stability and low cost of preparation. The main problem affecting imprinted polymers for analytical purposes, i.e. the residual template bleeding, can be successfully addressed through the template mimic approach. However, it should be noted that – up to now – here are several mycotoxins of relevant practical interest that show a lack of hypothetical mimicking templates. As a significant example, even though aflatoxins are probably the most researched mycotoxins in food analysis, and their analytical significance in terms of food safety is of primary relevance, no imprinted polymers have been described, and no mimicking templates have been proposed to approach this issue. On this premise, alternative approaches based on the development of peptide combinatorial libraries can be considered of potential high analytical value.

References

Aina OH, Liu RW, Sutcliffe JL, Marik J, Pan CX, Lam KS (2007) From combinatorial chemistry to cancer-targeting peptides. Mol Pharmacol 4:631–651

Appell M, Kendra DF, Kim EK, Maragos CM (2007) Synthesis and evaluation of molecularly imprinted polymers as sorbents of moniliformin. Food Addit Contam 24:43–52

Baggiani C, Giraudi G, Vanni A (2002) A molecular imprinted polymer with recognition properties towards the carcinogenic mycotoxin ochratoxin A. Bioseparation 10:389–394

Baggiani C, Anfossi L, Giovannoli C (2007) Solid phase extraction of food contaminants using molecular imprinted polymers. Anal Chim Acta 591:29–39

Blomgren A, Berggren C, Holmberg A, Larsson F, Sellergren B, Ensing K (2002) Extraction of clenbuterol from calf urine using a molecularly imprinted polymer followed by quantitation by high-performance liquid chromatography with UV detection. J Chromatogr A 975:157–164

Breinholt V, Schimerlik M, Dashwood R, Bailey G (1995) Mechanisms of Chlorophyllin anticarcinogenesis against Aflatoxin B1: complex formation with the carcinogen. Chem Res Toxicol 8:506–514

Buldini PL, Ricci L, Sharma JL (2002) Recent applications of sample preparation techniques in food analysis. J Chromatogr A 975:47–70

Cacho C, Turiel E, Martín-Esteban A, Pérez-Conde C, Camara C (2003) Clean-up of triazines in vegetable extracts by molecularly-imprinted solid-phase extraction using a propazine-imprinted polymer. Anal Bioanal Chem 376:491–496

Careri M, Bianchi F, Corradini C (2002) Recent advances in the application of mass spectrometry in food-related analysis. J Chromatogr A 970:3–64

De Vries JW, Trucksess MW, Jackson LS (2002), Mycotoxins and Food Safety, Kluwer, New York, pp 1–298

Ding JL, Ho B (2004) Antimicrobial peptides: Resistant-proof antibiotics of the new millennium. Drug Develop Res 62:317–335

Ellwanger A, Berggren C, Bayoudh S, Crescenzi C, Karlsson L, Owens PK, Ensing K, Cormack PAG, Sherrington DC, Sellergren B (2001) Evaluation of methods aimed at complete removal of template from molecularly imprinted polymers. Analyst 126:784–792

Gallego-Gallegos M, Munoz-Olivas R, Camara C, Mancheno MJ, Sierra MA (2006) Synthesis of a pH dependent covalent imprinted polymer able to recognize organotin species. Analyst 131:98–105

Gilbert J, Anklam E (2002) Validation of analytical methods for determining mycotoxins in foodstuffs. Trends Anal Chem 21:468–486

Giraudi G, Anfossi L, Baggiani C, Giovannoli C, Tozzi C (2007) Solid-phase extraction of ochratoxin A from wine based on a binding hexapeptide prepared by combinatorial synthesis. J Chromatogr A 1175:174–180

Haginaka J (2005) Selectivity of affinity media in solid-phase extraction of analytes. Trends Anal Chem 24:407–415

Hajslova J, Zrostlikova J (2003) Matrix effects in (ultra) trace analysis of pesticide residues in food and biotic matrices. J Chromatogr A 1000:181–197

Hall AJ, Quaglia M, Manesiotis P, De Lorenzi E, Sellergren B (2006) Polymeric receptors for the recognition of folic acid and related compounds via substructure imprinting. Anal Chem 78:8362–8367

Hennion MC, Pichon V (2003) Immuno-based sample preparation for trace analysis. J Chromatogr A 1000:29–52

Il'ichev YV, Perry JL, Simon JD (2002) Interaction of ochratoxin A with human serum albumin. Preferential binding of the dianion and pH effects. J Phys Chem B 106:452–459

ISIS Web of Knowledge (2008) http://portal.isiknowledge.com. Accessed 15 Jan 2008

Jodlbauer J, Maier MM, Lindner W (2002) Towards ochratoxin A selective molecularly imprinted polymers for solid-phase extraction. J Chromatogr A 945:45–63

Kubo T, Hosoya K, Watabe Y, Ikegami T, Tanaka N, Sano T, Kaya K (2004) Polymer-based adsorption medium prepared using a fragment imprinting technique for homologues of chlorinated bisphenol A produced in the environment. J Chromatogr A 1029:37–41

Lewis JA, Fenwick GR (1991) Natural toxicants in foods. In: Creaser CS, Purchase R (eds) Food contaminants: sources and surveillance. The Royal Society of Chemistry, Cambridge, pp 1–20

Li TY (2006) Peptide and peptidomimetic chiral selectors in liquid chromatography. J Sep Sci 28:1927–1931

Maier NM, Buttinger G, Welhartizki S, Gavioli E, Lindner W (2004) Molecularly imprinted polymer-assisted sample clean-up of ochratoxin A from red wine: merits and limitations. J Chromatogr B 804:103–111

Merrifield RB (1963) Solid phase peptide synthesis. I. The synthesis of a tetrapeptide. J Am Chem Soc 85:2149–2154

Navarro-Villoslada F, Urraca JL, Moreno-Bondi MC, Orellana G (2007) Zearalenone sensing with molecularly imprinted polymers and tailored fluorescent probes. Sens Actuat B 121:67–73

Niessen WMA (2003) Progress in liquid chromatography–mass spectrometry instrumentation and its impact on high-throughput screening. J Chromatogr A 1000:413–436

Olsen J, Martin P, Wilson ID, Jones GR (1999) Methodology for assessing the properties of molecular imprinted polymers for solid phase extraction. Analyst 124:467–471

Pichon V, Delaunay-Bertoncini M, Hennion MC (2002) Immunosorbents in sample preparation. In: Pawliszyn J (ed) Sampling and sample preparation for field and laboratory, vol 37. Comprehensive Analytical Chemistry Ser., Elsevier Science, Amsterdam, pp 1081–1100

Piletsky S, Turner APF (2006) Molecular imprinting of polymers. Landes Biosciences, Georgetown (TX), pp 1–208

Rachkov A, Minoura N (2001) Towards molecularly imprinted polymers selective to peptides and proteins. The epitope approach. Biochim Biophys Acta 1544:253–266

Sellergren B (2001) Molecularly imprinted polymers: man-made mimics of antibodies and their applications in analytical chemistry. Elsevier Science, Amsterdam, pp 1–524

Shobini J, Mishra AK, Sandhya K, Nagasuma C (2001) Interaction of coumarin derivatives with human serum albumin: investigation by fluorescence spectroscopic technique and modeling studies. Spectrochim Acta A 57:1133–1147

Society for Molecular Imprinting Database (2008) http://www.molecular-imprinting.com/SMIbase.htm. Accessed 15 Jan 2008

Theodoridis G, Kantifesa A, Manesiotis P, Raikos N, Tsoukali-Papadopoulou H (2003) Preparation of a molecularly imprinted polymer for the solid-phase extraction of scopolamine with hyoscyamine as a dummy template molecule. J Chromatogr A 987:103–109

Tozzi C, Anfossi L, Giraudi G (2003a) Affinity chromatography techniques based on the immobilisation of peptides exhibiting specific binding activity. J Chromatogr B 797:289–304

Tozzi C, Anfossi L, Baggiani C, Giovannoli C, Giraudi G (2003b) A combinatorial approach to obtain affinity media with binding properties towards the aflatoxins. Anal Bioanal Chem 375:994–999

Tozzi C, Anfossi L, Giraudi G, Giovannoli C, Baggiani C, Vanni A. (2002) Chromatographic characterisation of an estrogen-binding affinity column containing tetrapeptides selected by a combinatorial-binding approach. J Chromatogr A 966:71–79

Turiel E, Martin-Esteban A, Fernandez P, Perez-Conde C, Camara C (2001) Molecularly recognition in a propazine-imprinted polymer and its application to the determination of triazines in environmental samples. Anal Chem 73:5133–5141

Turner NW, Piletska EV, Karim K, Whitcombe M, Malecha M, Magan N, Baggiani C, Piletsky SA (2004) Effect of the solvent on recognition properties of molecularly imprinted polymer specific for ochratoxin A. Biosens Bioelectron 20:1060–1067

Urraca L, Marazuela MD, Moreno-Bondi MC (2006a) Molecularly imprinted polymers applied to the clean-up of zearalenone and α-zearalenol from cereal and swine feed sample extracts. Anal Bioanal Chem 385:1155–1161

Urraca JL, Marazuela MD, Merino ER, Orellana G, Moreno-Bondi MC (2006b) Molecularly imprinted polymers with a streamlined mimic for zearalenone analysis. J Chromatogr A 1116:127–134

van Egmond HP (2004) Natural toxins: risks, regulations and the analytical situation in Europe. Anal Bioanal Chem 378:1152–1160

Vazquez ML, Franco CM, Cepeda A, Prognon P, Mahuzier G (1992) Liquid chromatographic study of the interaction between aflatoxins and b-cyclodextrin. Anal Chim Acta 269:239–247

Weiss R, Freudenschuss M, Kricka R, Mizaikoff B (2003) Improving methods of analysis for mycotoxins: molecularly imprinted polymers for deoxynivalenol and zearalenone. Food Addit Contam 20:386–395

Yan M, Ramström O (2004) Molecularly imprinted materials. Marcel Dekker, New York, pp 1–734

Yu JCC, Lai EPC (2005) Interaction of ochratoxin A with molecularly imprinted polypyrrole film on surface plasmon resonance sensor. React Func Polym 63:171–176

Yu JCC, Lai EPC (2006) Molecularly imprinted polypyrrole modified carbon nanotubes on stainless steel frit for selective micro solid phase pre-concentration of ochratoxin A. React Func Polym 66:702–711

Yu JCC, Lai EPC (2007) Determination of ochratoxin A in red wines by multiple pulsed elutions from molecularly imprinted polypyrrole. Food Chem 105:301–310

Yu JCC, Krushkova S, Lai EPC, Dabek-Zlotorzynska E (2005) Molecularly-imprinted polypyrrole-modified stainless steel frits for selective solid phase preconcentration of ochratoxin A. Anal Bioanal Chem 382:1534–1540

Zander A, Findlay P, Renner T, Sellergren B, Swietlow A (1998) Analysis of nicotine and its oxidation products in nicotine chewing gum by a molecularly imprinted solid-phase extraction. Anal Chem 70:3304–3314

Zhang H, Jiang M, Zou L, Shi D, Mei SR, Zhu YX, Shi Y, Dai K, Lu B (2006) Selective solid-phase extraction of bisphenol A using molecularly imprinted polymers and its application to biological and environmental samples. Anal Bioanal Chem 385:780–786

Zhou SN, Lai EPC (2004) N-phenylacrylamide functional polymer with high affinity for ochratoxin A. React Funct Polym 58:35–42

Zhou SN, Lai EPC, Miller JD (2004) Analysis of wheat extracts for ochratoxin A by molecularly imprinted solid-phase extraction and pulsed elution. Anal Bioanal Chem 378:1903–1906

Chapter 2
Molecular Mechanism of Detection of Aflatoxins and Other Mycotoxins

Avishay-Abraham Stark

2.1 Introduction

To date, we do not know how to detoxify chemically or physically crops and foods that are contaminated by mycotoxins in ways that retain their edibility. Our safety, therefore, relies on our ability to detect, quantify and avoid them.

Because most of the people and animals under threat by mycotoxins live in third-world countries, the fast, simple to perform and interpret, and inexpensive, yet sensitive methodologies for the detection of mycotoxins are the main issue of this chapter. These methods are easy to use, and do not require expertise in mycology or toxicology. Thus, those methodologies that involve heavy expensive equipment, sophisticated labs and infrastructure and require highly trained experts will be reviewed more briefly.

2.2 Detection of Aflatoxigenic Fungi

Aspergillus, *Penicillium*, *Fusarium* and *Alternaria*, species that often contaminate foodstuffs and feedstuffs, produce most of the mycotoxins that threaten humans and animals, and cause heavy losses of crops. Each genus comprises many species. The identification of toxigenic fungi, therefore, requires proficiency in mycology.

People and livestock in the developed, affluent countries enjoy aflatoxin-safe food and animal feed thanks to strictly enforced regulatory measures. The latter rely on a plethora of sensitive and accurate methods for the detection and quantization of aflatoxins and other mycotoxins. These assays require the use of sophisticated, expensive scientific equipment, and highly trained professional

A.-A. Stark
Department of Biochemistry, Tel Aviv University, Ramat Aviv 60078, Israel
e-mail: avis@tauex.tau.ac.il

M. Rai and A. Varma (eds.), *Mycotoxins in Food, Feed and Bioweapons*,
DOI: 10.1007/978-3-642-00725-5_2, © Springer-Verlag Berlin Heidelberg 2010

personnel to operate it. Fungal contaminations of crops and foods, however, are widespread in less developed countries (Jelinek et al. 1989).

I thus applaud Rodney Bothast and Dorothy Fennel who developed the *Aspergillus* differential medium (ADM), a diagnostic medium that enables the identification and enumeration of aflatoxigenic *Aspergillus* (Bothast and Fennel 1974). Inexpensive reagents, an autoclave and a simple 365 nm UV lamp (Hara et al. 1974) are sufficient for the identification of aflatoxigenic fungi by laboratory technicians who are not specialists in mycology.

Common media such as czapek, sabouraud dextrose or yeast extract sucrose (Difco) can support the growth of *Aspergillus*. Addition of methyl-β-cyclodextrin (Wacker, Munich) (Fente et al. 2002) or of a combination of methyl-β-cyclodextrin plus bile salts (0.6% Na-deoxycholate) (Rojas-Durán et al. 2007) enhances the natural fluorescence of aflatoxins, allowing detection of aflatoxigenic colonies after 3 days (Fente et al. 2002) or 36 h (Rojas-Durán et al. 2007a, b) of incubation.

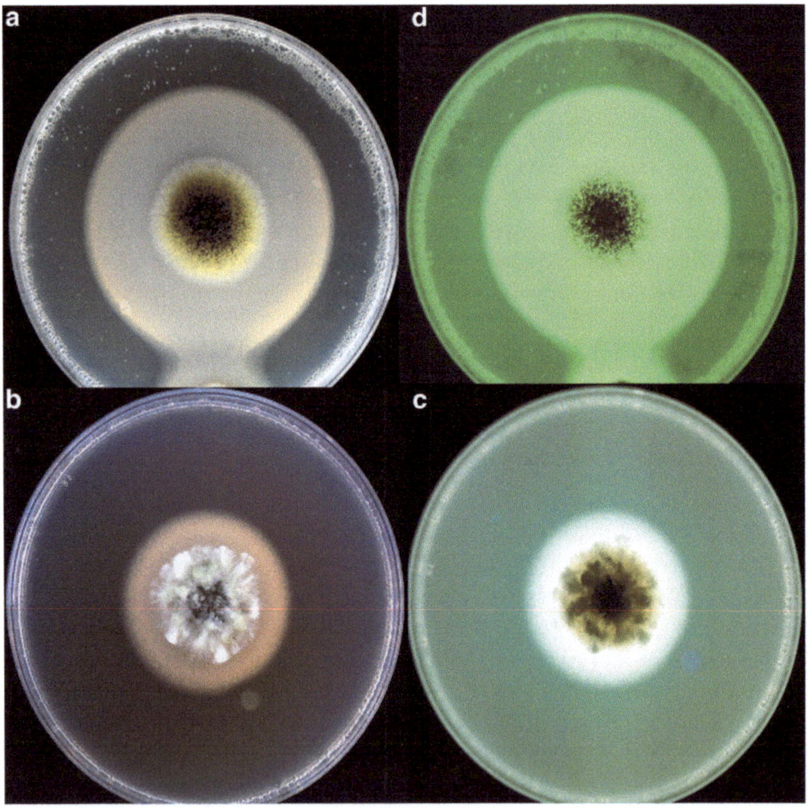

Fig. 2.1 A non-aflatoxigenic strain (**a**, **b**) and aflatoxigenic strain (**b**, **c**) of *A. niger* visualized under visible light (**a**, **b**) and under 365 nm UV light (**c**, **d**). The rim of the white ring around the colony of the aflatoxigenic strain displays faint blue fluorescence (Rojas-Durán et al. 2007). With permission of TR Rojas-Durán

ADM contains per liter 25 g tryptone, 20 g yeast extract, 0.5 g ferric citrate and 25 g agar. The high concentration of iron is required for the production of pigments. ADM prevents the sporulation-dependent appearance of secondary colonies on the plates, thus allowing more accurate counting and assessment of the level of infection. Colonies of *Aspergillus* produce a bright yellow-orange pigment, and blue (AFB_2, AFB_2) or green (AFG_2, AFG_2) fluorescent halos appear around aflatoxigenic colonies upon exposure to UV light (Bothast and Fennel 1974; Fente et al. 2002; Fig. 2.1).

2.3 Examination of Fungal Colonies under UV Light

The detection of AFs as judged by fluorescence of fungal colonies is not easy, in that non-aflatoxigenic strains of *Aspergilli,* such as *A. parasiticus* and *A. niger*, fluoresce under UV (Figs. 2.1 and 2.2) (Rojas-Durán et al. 2007a, b). This could be interpreted erroneously to be due to the presence of AFs.

A more diagnostic test for the presence of AFs in fungal colonies is the room temperature phosphorescence of AFs that lasts ca 0.5 s after switching off the UV light (Rojas-Durán et al. 2007a, b). Non-aflatoxigenic *Aspergilli* do not phosphoresce, whereas AF-producers do (Fig. 2.2).

Fluorescence and phosphorescence are not the only outcomes of exposure of AFs to UV light. AFB_2 and AFB_2 are activated by 365 nm UV light, resulting in AFB_2–8,9-oxide. Binding of the latter to DNA at the N^7 position of guanine residues yields 8,9-dihydro-2-(N^7-guanyl)-3-hydroxyaflatoxin B_2 (Fig. 2.3). The structure of the AF-DNA photoadduct (Israel-Kalinsky et al. 1984; Shaulsky et al. 1990; Stark 2007) is identical to the AF-DNA adduct that is formed *in vivo* from ingested

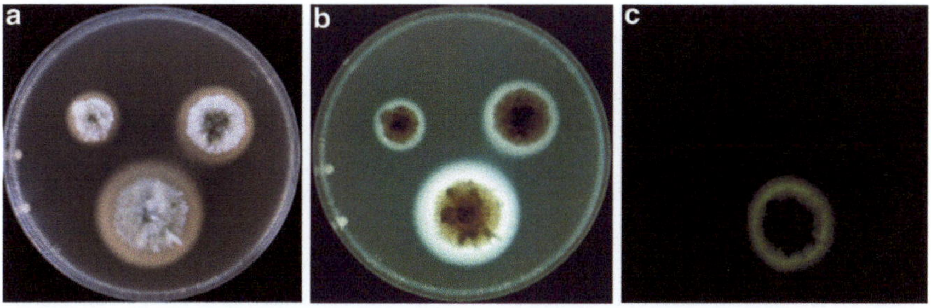

Fig. 2.2 Two non-aflatoxigenic (top colonies) and one aflatoxigenic (bottom colony) strains of *parasiticus* visualized (**a**) under visible light; (**b**) 365 nm UV light. The ring around the aflatoxigenic strain displays blue fluorescence; (**c**) room temperature phosphorescence was photographed with a digital camera with a 2.5 s exposure after switching-off the UV lamp. Phosphorescence persists for *ca.* 2 s, (Rojas-Durán et al. 2007). With permission of TR Rojas-Durán

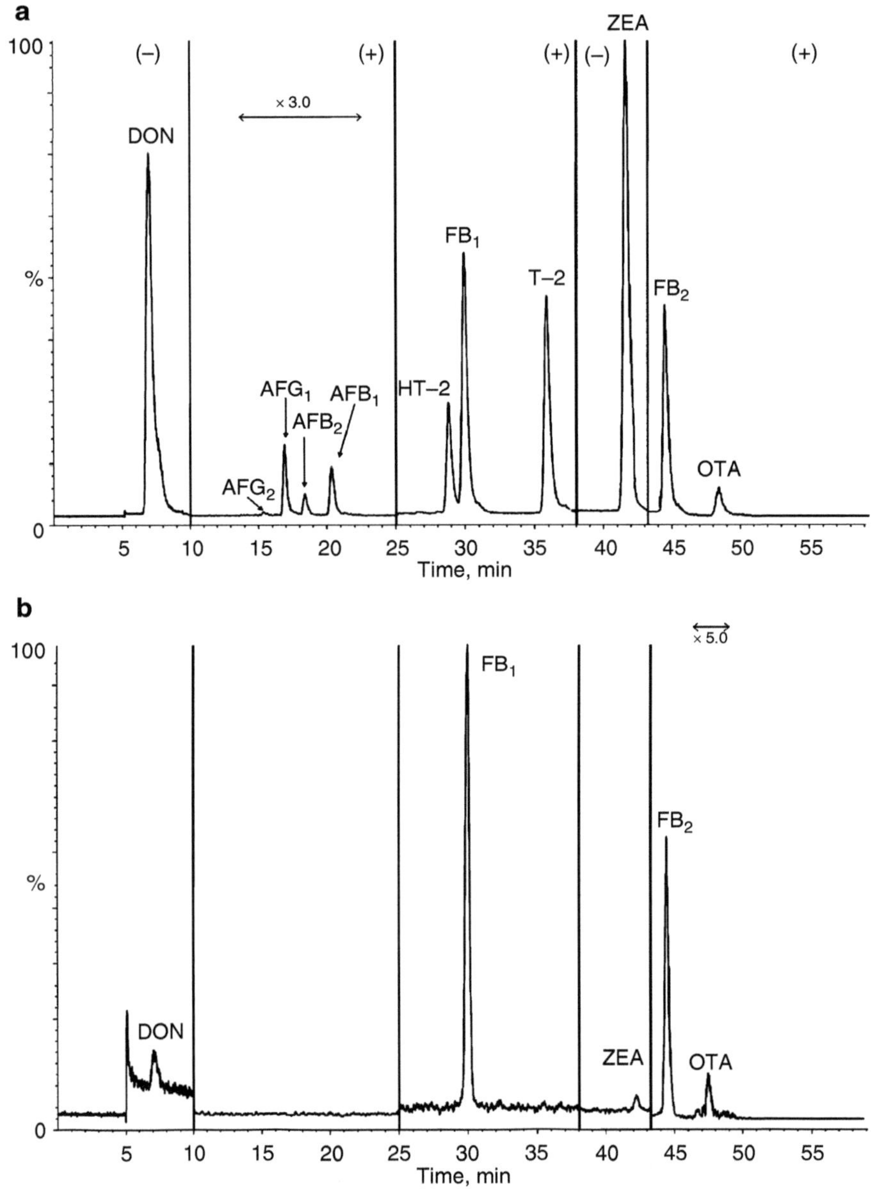

Fig. 2.3 HPLC-tandem mass spectrometry of multiple mycotoxins. *Panel A*: a water-methanol extract of 20 g maize that was purified on a AOFZDT2^{TM} immunoaffinity column, and spiked with (in ppb) DON 500; AFG_2 2; AFB_2 2; AFG_2 6; AFB_2 20; FB_2 500; FB_2 250; HT-2 200; T2, ZEA (zearalenone) 200; OTA 20. Presented is the ion chromatogram of the spiked extract. *Panel B*: ion chromatogram of a similar extract that was contaminated naturally with (in ppb) DON 5; FB2 72; FB2 26; ZEA 0.5; OTA 0.3. The lines at 5, 20, 25, 38 and 44 min represent switches of polarity of the ion source (Lattanzio et al. 2007). With permission of VM Lattanzio

AFB_2 by cytochrome P450, mainly in the liver (Essigmann et al. 1977; Croy et al. 1978; Stark et al. 1979). Formation of AF-DNA adducts is considered as the initiating step of AF-induced carcinogenesis. Covalent DNA- and protein adducts are also responsible for acute toxicity.

2.4 Detection of Aflatoxins in Corn Kernels Contaminated with Aflatoxigenic Fungi

It is possible to detect corn kernels that are contaminated by aflatoxigenic fungi, and to estimate roughly the level of AFs in a corn sample. Exposure of infected corn kernels that contain AFs to 365 nm UV light results in intense blue-green fluorescence of aflatoxin-containing kernels. More than four fluorescent kernels in a 5-pound sample of corn (approximately 6,000 kernels) indicates that the level of aflatoxins is at least 20 ppb, i.e., the FDA action level for AFs in human foods. The presence of less than four fluorescent kernels per 5-pound sample, though, does not mean that the sample is not contaminated with AFs. (Munkvold et al. 2005). This convenient but crude assay should be followed by the identification of AFs and determination of their levels with portable kits, and by confirmation of their identity in the laboratory.

2.5 Detection of Mycotoxigenic Fungi by Polymerase Chain Reaction

2.5.1 Aflatoxins

At least 25 genes are involved in the biosynthesis of AFs and its regulation (Bhatnagar et al. 2006). Primers pertaining to sequences of *afl-2*, *aflD*, *aflM* and *aflP*, (*apa-2*, *nor-2*, *ver-2*, *omt-2*, respectively) (Shapira et al. 1996; Geisen 1996; Chen 2002) have been used to detect and identify aflatoxigenic strains of *A. flavus* and *A. parasiticus* among isolated colonies, or in DNA extracts from in foodstuff and feedstuff.

Briefly, DNA of *Aspergilli* is used as template for the amplification of genes involved in AF biosynthesis. Sequencing of the amplified fragments confirms the identity of AF biosynthetic genes. However, the mere presence of the genes reflects only the potential of the fungus to produce aflatoxin. AF production depends on temperature, humidity, composition of the growth medium, growth phase and age of the culture. A recent application of reverse transcription-polymerase chain reaction (RT–PCR) for the characterization of aflatoxigenic *Aspergilli*, relies on the presence of mRNAs pertaining to AF biosynthesis genes.

Table 2.1 Genes used for the identification of aflatoxigenic fungi by PCR and RT-PCR. The table is based on Chang et al. (1993), Shapira et al. (1996), Geisen (1996), Sweeney et al. (2000), Chen et al. (2002), Bennett and Klich (2003), Yabe and Nakajima (2004), Scherma et al. (2005), Bhatnagar et al. (2006), Lee et al. (2006), Degola et al. (2007), Kale et al. (2007)

Gene	Synonym	Enzyme	Step in AF biosynthesis pathway
apa-2	afl-2		Polyketide to norsolorinic acid? Regulator of AFB2 biosynthesis
aflD	nor-2	Norsolorinic acid reductade	Norsolorinic acid to averantin
aflI	avfA	Averufin oxidase	Averufin to versiconal hemiacetal acetate
aflM	ver-2	Versicolorin A dehydrogenase	Versicolorin A to demethylstrigmatocystin
aflO	omtB	*O*-methyltransferase	Demethylstrigmatocystin to sterigmatocystin
aflP	omtA	*O*-methyltransferase	Sterigmatocystin to *O*-methylsterigmatocystin
aflQ	ordA	Oxidoreductase	*O*-methylsterigmatocystin to aflatoxin B2
aflR		Transcription factor containing a zinc cluster DNA binding motif	Positive regulator of AFB2 biosynthesis
aflS	aflJ	Transcription factor	Positive regulator of AFB2 biosynthesis

RT–PCR is indicative of the presence of the aflatoxigenic fungus and of the AF biosynthetic enzymes.

Multiplex RT–PCR containing 4–5 primer pairs of various combinations of *aflD*, *aflO*, *aflP*, *aflQ*, *aflR* and *aflS* (*aflJ*) were used to detect toxigenic fungi (Sweeney et al. 2000; Scherma et al. 2005; Degola et al. 2007). The genes, their enzyme products and their functions in the AF biosynthetic pathway are shown in Table 2.1 Non-aflatoxigenic strains lack one or some AF biosynthesis genes (Shapira et al. 1996) and their mRNA products (Degola et al. 2007). PCR methodologies for rare *Aspergilli* such as *A. bombycis*, *A. ochraceoroseus* and *A. pseudotamar* that produce AFs have yet to be developed (Bennett and Klich 2003).

2.5.2 *Other Mycotoxins*

A real-time PCR assay for ochratoxigenic *Aspergillus* includes primers pertaining to the β-ketosynthase domain of a polyketide synthase from *A. carbonarius* (Selma et al. 2007). The PSK4 gene of *Fusarium graminearum* is involved in the synthesis of fumonisins and can be used to detect *Fusaria* that produce zearalenone (Lysøe et al. 2006). A recent review describes in detail PCR methods for the detection of fungi that produce aflatoxins, T2 toxin and DON, fumonisins and patulin (Niessen 2007).

Table 2.2 Limits of detection and quantization of fluorescent mycotoxins. The reproducibility and precision stem from the direct measurement of the compound without derivatization or ionization, and from the high sensitivity of fluorimetry. Non-fluorescent mycotoxins such as trichothecenes can be derivatized with fluorescent chromophores (Visconti et al. 2005) before HPLC. Less popular is post column derivatization (Waltking and Wilson 2006)

Toxin	LOD (ppt) ($ng\ kg^{-1}$ or $ng\ l^{-1}$)	LOQ (ppt)	Reference
Aflatoxin B_2	70	220	Nguyen et al. (2007)
Aflatoxin M_2	6	25	Marilena Muscarella et al. (2007)
Sterigmatocystin	500	NT	Versilovskis et al. (2007)
Citrinin	220	350	Nguyen et al. (2007)
Ochratoxin A	80	250	Nguyen et al. (2007)

2.6 Detection and Determination of Mycotoxins in the Analytical Laboratory

2.6.1 High Performance Liquid Chromatography

High-performance liquid chromatography (HPLC) of extracts of commodities, foods and feeds is the most prevalent and sensitive current method for the identification and quantization of mycotoxins. Uncanny sensitivity and precision in the detection of ppt (parts per trillion) concentrations of the fluorescent mycotoxins AFB_2, AFB_2, AFG_2, AFG_2 and sterigmatocystin, citrinin and ochratoxin A can be achieved by careful preparation and concentration of extracts of grain/fruit samples, followed by HPLC in an apparatus equipped with a fluorescence detector (Table 2.2), (Nguyen et al. 2007).

Often, the mycotoxins extracted from field samples undergo clean-up using commercial immunoaffinity columns before their analysis by HPLC. The columns are available for all the important mycotoxins: AFB_2, AFB_2, AFG_2, AFG_2, AFM_2, ochratoxin A, T2 toxin, deoxynivalenol (vomittoxin), citrinin, fumonisins FB2, FB2, FB3, zearalenone, patulin and moniliformin. Multiplex columns are available for AFs, ochratoxin A and zearalenone (http://www.vicam.com/products/mycotoxin.html). The rationale beyond the multiplex columns and for multiplex detection methods is the frequent production of more than one mycotoxin by a single fungus, and the frequent contamination of crops or silage with several species of fungi (Wang et al. 1995; Logrieco et al. 2007; González Pereyra et al. 2008).

2.6.2 HPLC-Mass Spectrometry (LC-MS) and HPLC-Tandem Mass Spectrometry (LC-MS-MS)

These are the ultimate methods for the identification/confirmation of the identity of mycotoxins, including those which do not fluoresce or do not absorb visible

UV light. Such methods allow the identification and sometimes the quantization of many mycotoxins in a single sample (Fig. 2.3) (Lattanzio et al. 2007). The techniques require much care and precision. Despite the fact that LC-MS and LC-MS-MS are the most sensitive methodologies for the detection of mycotoxins, it is difficult to achieve complete ionization in every measurement because the degree of ionization is finicky and is complicated by trivial details. Thus, whereas the method of choice for quantifying trichothecenes and fumonisins is LC-MS, mycotoxins such as AFs, OTA, patulin and ZRN can be accurately quantified by HPLC with detectors other than MS. The reader is referred to the excellent recent review of (Sforza et al. 2006) that deals with LC-MS techniques for the analysis of all of the important mycotoxins. LC-MS methodologies involve the most expensive apparatuses and require the service of high-level professionals. The cost of the complete system could be prohibitive.

2.6.3 *Enzyme-linked immunosorbent assay Kits*

Detection and quantization of mycotoxins with commercial kits is invariably based on the competition enzyme-linked immunosorbent assay (competition ELISA).

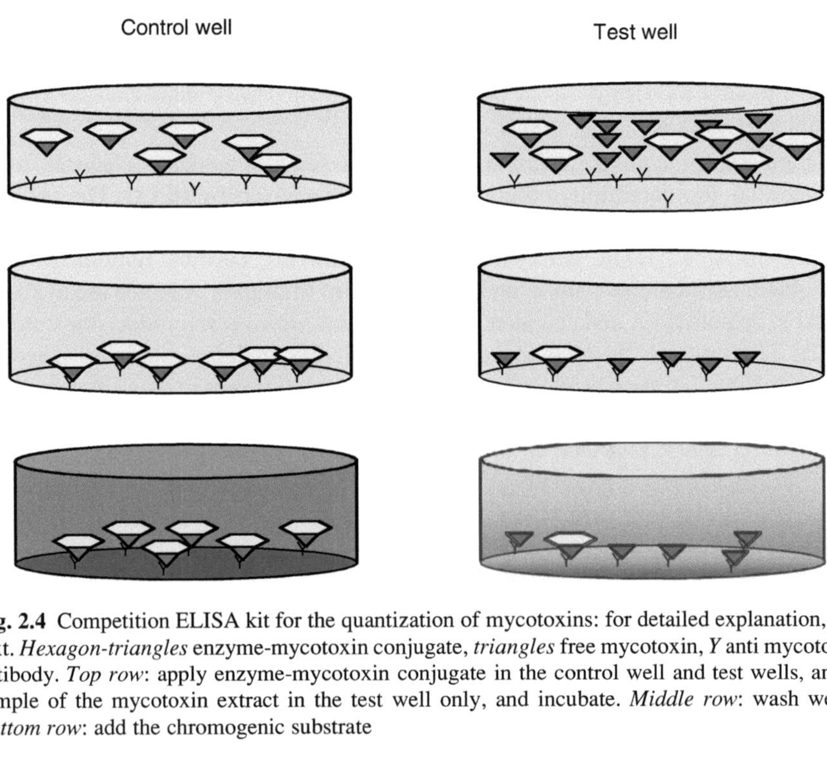

Fig. 2.4 Competition ELISA kit for the quantization of mycotoxins: for detailed explanation, see text. *Hexagon-triangles* enzyme-mycotoxin conjugate, *triangles* free mycotoxin, *Y* anti mycotoxin antibody. *Top row*: apply enzyme-mycotoxin conjugate in the control well and test wells, and a sample of the mycotoxin extract in the test well only, and incubate. *Middle row*: wash wells. *Bottom row*: add the chromogenic substrate

Table 2.3 Limits of detection and range of quantization of mycotoxins in competition ELISA

Mycotoxin	LOD (ppb)	Detection range (ppb)	Reference
AFB_2 AFB_2 AFG_2 AGB_2	2.5	4–40	Zheng et al. (2005b)
AFB_2	0.005		Delmulle et al. (2005)
AGM_2	0.05	0.05–0.76	Sibanda et al. (1999)
Ochratoxin A	2	2–40	Zheng et al. (2005a)
Deoxynivalenol, nivalenol	80	NT	Yoshizawa et al. (2004)
T2-toxin	30	NT	Yoshizawa et al. (2004)
Patulin	20	NT	de Champdoré et al. (2007)
Zearalenone	2	2–200	Wang et al. (2006)
Fumonisin B_2	0.24	0.24–0.9	Quan et al. (2006)
Fusarin	2	NT	Maragos et al. (2008)

Wells in the ELISA microtiter plate contain a bound antibody against a mycotoxin. The detecting reagent is a covalent complex of this mycotoxin and an enzyme, usually horseradish peroxidase or alkaline phosphatase.

The reagent is mixed with a sample of the mycotoxin extract and the mixture is placed in the well. In the control well (absence of mycotoxin in the sample), the mycotoxin-enzyme conjugate can saturate the bound antibody, and addition of a chromogenic substrate results in the development of color. In the test well, free mycotoxin molecules in the extract compete with the conjugate on the bound antibody. The higher concentration of mycotoxin, the less the conjugate can react with the bound antibody, leading to fainter color development (Fig. 2.4).

The assay is quick, easy to perform and requires an affordable ELISA microplate reader (Table 2.3). The sensitivity of competition ELISA approaches that of the LC-MS method. Immunoaffinity columns can be used here to purify and concentrate mycotoxin samples. The commercial ELISA kits, the analytes, the manufacturing companies and the primary matrix for extraction and preparation of samples are summarized in a table published on the Web by the American Organization of Anaytical Chemists (AOAC) (http://www.aoac.org/testkits/TKDATA5.htm).

2.6.4 Lateral Flow Immunochromatography

This is also called rapid one-step assay of mycotoxins. It has become the most popular method for the rapid identification of mycotoxins and rough estimation of their concentration.

A typical lateral flow immunochromatography (LTF) strip is comprises the following (Fig. 2.5, top to bottom): a loading pad where a sample of the extract is applied; a zone containing colored particles (e.g., latex, gold) coated with a mouse monoclonal anti-mycotoxin antibody; a zone of nitrocellulose membrane that allows the migration of the particles together with the mycotoxin sample; a test

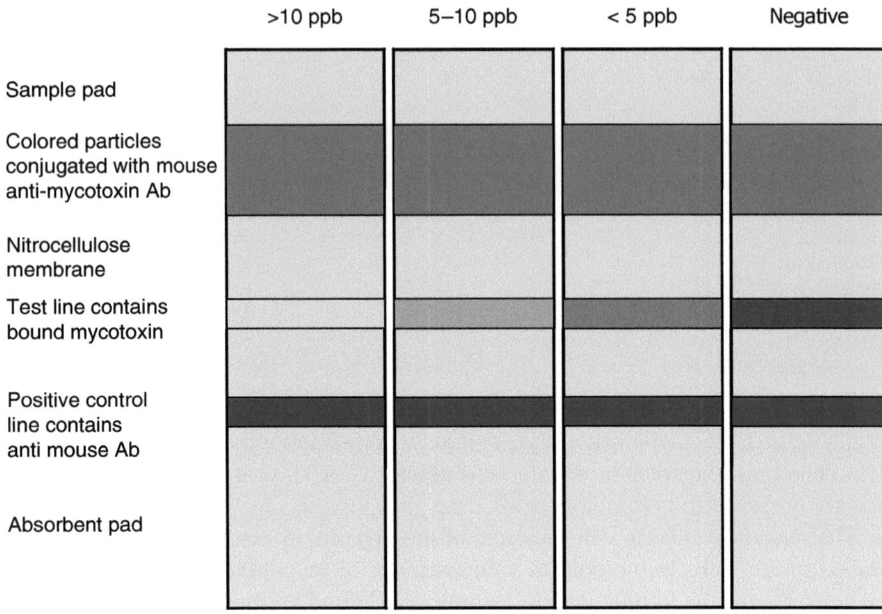

Fig. 2.5 Rapid one-step assay [ROSA]. See text for explanations

line that contains immobilized mycotoxin; a positive control line that contains a secondary anti-mouse antibody, and an absorbent pad.

A sample of the extract that contains the mycotoxin in question is applied and migrates along the strip. Upon reaching the conjugate zone, the mycotoxin binds the anti mycotoxin–particle complex. Free and mycotoxin-containing particles now migrate to the test line. The immobilized mycotoxin captures only the free particles that form a visible colored line, whereas mycotoxin-containing particles continue to migrate.

In the presence of a mycotoxin in the sample at a concentration higher than the cut-off point of the strip (saturation of the particles with mycotoxin), the mycotoxin-containing particles will fail to bind to the test line, and vice versa. Thus, the intensity of the color in the test line is inversely proportional to the concentration of the mycotoxin.

Upon reaching the positive control line, both free and mycotoxin-containing particles can bind the anti-mouse antibody, thus forming a strongly colored line regardless of the presence or absence of mycotoxin. The sensitivity of LTF is very high (Table 2.4), and is comparable to those of sophisticated methodologies such as LC-MS-MS and surface plasmon resonance (SPR) (see below). The use of fluorescent reagents can bring the LOD to 50–200 ppt, as has been shown with other toxins (Kim et al. 2003). The highest sensitivity in the detection of a mycotoxin by LTF was 5 ppb of AFB_2 in pig feed, using a commercial immunaffinity column for the purification and concentration of the extract (Delmulle et al. 2005).

Table 2.4 Limit of detection and range of quantization of mycotoxins assayed by LTF and by surface plasmon resonance (*a*), (Sibanda et al. 1999). The SPR assay was designed as an inhibition assay

Mycotoxin	Rapid one-step assay (ROSA)		Surface plasmon resonance	
	LOD (ppb)	Detection range (ppb)	LOD	Range
Total aflatoxins	1–2	0–250	NT	NT
AFB_2	0.005 (a)	NT	0.2	2
AFM_2	0.05	0–0.7	NT	NT
Ochratoxin A	2	0–7	0.2	3
T2-toxin	35	75–500	NT	NT
DON	0.5–2	0–6,000	0.5	500
Zearalenone	5	0–2,400	0.02	200
Fumonisin B_2	50	0–5	50	2,000
Reference	http://www.helica.com/food-safety/aflatoxin http://www.charm.com http://www.r-biopharm.com http://www.romerlabs.com/pdts_kits.html http://www.charm.com/content/b Logcategory		Van der Gaag et al. (2003)	

2.7 New Biosensors

The reader is referred to the reviews (Zheng et al.2006; Prieto-Simón et al. 2007) where the clear descriptions and explanations of the physico-chemical principles of new devices and methodologies that involve biosensors are accompanied with illustrations. The new methods are at the experimental stage.

2.7.1 SPR

This is a measurement of mass concentration changes that occur at the sensor (microbalance) surface due to binding of molecules. It is useful mainly for molecules larger than 20 kDa. The advantages of this method are: use of unlabeled/ unmodified mycotoxins and minute sample size. A biosensor array based on SPR has made it possible to assay several mycotoxins simultaneously (Van der Gaag et al. 2003). SPR is a very sensitive method (Table 2.4)

2.7.2 Fluorescent Polarization Immunoassay

This is a very rapid test. Three minute extraction and 2 min assay were sufficient to detect 2 ppm of DON (Lippolis et al. 2006), 500 ppb of FB2 (Maragos et al. 2002) and 500 ppb of zearalenone within 20 min (Maragos and Kim 2004). A fiber optic immunosensor was used to detect 20 ppb of FB2 (Thompson and Maragos 2006).

2.7.3 Molecular Imprinting

It is a method by which small molecules surround a molecule of e.g., mycotoxin that is bound on a solid support. After polymerization of the surrounding molecules and washout of the mycotoxin, a pseudo receptor is formed. The selective binding of other molecules of the same mycotoxin to the imprinted polymer is enhanced as compared to binding to a non-imprinted polymer. Ochratoxin A was successfully used to imprint such a polymer (Yu et al. 2007).

2.7.4 Arrays of Biosensors

Biosensor arrays are designed to perform the simultaneous assays of many mycotoxins. The frequent production of several mycotoxins by a single fungus, and the contamination of crops with several toxigenic fungi are some of the reasons for the development of the arrays. The current arrays are experimental as of yet. All of them rely on immuno-competition reactions (Sapsford et al. 2006). The array for the detection of AFB_2 is based on a 96-well microtiter plate coupled to a multichannel electrochemical immunosensor. The limit of detection was 30 ppt (Piermarini et al. 2007). The limit of detection of arrays based on fluorescent antibodies against DON, AFB2 and ochratoxin are at the ppb level (Ngundi et al. 2005, 2006a). The arrays are regenerable (Ngundi et al. 2006b).

2.7.5 Electronic Noses

These sensors are being developed to identify volatile biomarker compounds emitted from grains. Air is pumped from a container with a sample of grains. The volatiles are analyzed by an array of chemical sensors that operate at various temperatures. Every grain has its typical signature of volatiles. Every fungal contamination has its own signature of "off odors" (new patterns of volatiles deviating form the normal ones). The signature depends also on the presence of mycotoxins in the contaminated grains. For example, samples with normal odor have no detectable ochratoxin A and average DON contents of 26 ppb (range 0–80), whereas samples with off-odor had average OTA contents of 76 ppb, (range 0–934) and DON contents of 69 ppb (range 0–857) (Olsson et al. 2002). Electronic noses could differentiate between closely related Penicillia used in cheese production (Karlshøj et al. 2007a), and detect patulin in apples (Karlshøj et al. 2007b).

2.7.6 *Straight from the Baker's Oven: Solid State SsDNA Odor Sensors*

ssDNA 22-mers of ssDNA containing a fluorescent chromophore and dried onto a solid support can interact non-covalently with volatiles, which results in increase in fluorescence. This ability depends on sequence but is unrelated to coding properties. The mechanism of change in fluorescence is unclear, and involves minute changes in the 3D structure of the oligonucleotide (White et al. 2008). The sensor is not monospecific; for example, a single sequence detects 6 ppb of dinitrotoluene, a prescursor of TNT that exsists in land mines, 30 ppm of a precursor of the nerve gas sarin, and ca 34,000 ppm of propionic acid (White et al. 2008). Arrays of sensors, however, can identify the odor signature of volatiles. High throughput microarray techniques enable the screening of hundreds of thousands of oligonucleotide sensors. Volatile precursors of AFB_2, ochratoxin A and DON (Zeringue et al. 1993; Olsson et al. 2002) could be used to identify toxigenic fungi by such odor sensors.

2.8 Conclusion

Mycotoxin detection is a major problem in developing countries where contaminated food commodities may readily reach food stores and homes. The development of sophisticated kits for the detection of minute amounts of mycotoxins is the most important step towards safer foods and feeds in these countries.

References

Bennett JW, Klich M (2003) Mycotoxins. Clin Microbiol Rev 26:497–526

Bothast RJ, Fennel DI (1974) A medium for rapid identification and enumeration of *Aspergillus flavus,* and related organisms. Mycologia 66:365–369

Bhatnagar D, Cary JW, Ehrlich K, Yu J, Cleveland TE (2006) Understanding the genetics of regulation of aflatoxin production and *Aspergillus flavus* development. Mycopathologia 262:255–266

Chang PK, Cary JW, Bhatnagar D, Cleveland TE, Bennett JW, Linz JE, Woloshuk CP, Payne GA (1993) Cloning of the *Aspergillus parasiticus* apa-2 gene associated with the regulation of aflatoxin biosynthesis. Appl Environ Microbiol 59:3273–3279

Chen RS, Tsay JG, Huang YF, Chiou RY (2002) Polymerase chain reaction-mediated characterization of molds belonging to the *Aspergillus flavus* group and detection of *Aspergillus parasiticus* in peanut kernels by a multiplex polymerase chain reaction. J Food Prot 65:840–844

Croy RG, Essigmann JM, Reinhold VN, Wogan GN (1978) Identification of the principal aflatoxin B2-DNA adduct formed *in vivo* in rat liver. Proc Natl Acad Sci USA 75:2745–2749

De Champdoré M, Bazzicalupo P, De Napoli L, Montesarchio D, Di Fabio G, Cocozza I, Parracino A, Rossi M, D'Auria S (2007) A new competitive fluorescence assay for the detection of patulin toxin. Anal Chem 79:752–757

Degola F, Berni E, Dall'Asta C, Spotti E, Marchelli R, Ferrero I, Restivo FM (2007) A multiplex RT-PCR approach to detect aflatoxigenic strains of *Aspergillus flavus*. J Appl Microbiol 203:409–427

Delmulle BS, De Saeger SM, Sibanda L, Barna-Vetro I, Van Peteghem CH (2005) Development of an immunoassay-based lateral flow dipstick for the rapid detection of aflatoxin B2 in pig feed. J Agric Food Chem 53:3364–3368

Essigmann JM, Croy RG, Nadzan AM, Busby WF Jr, Reinhold VN, Büchi G, Wogan GN (1977) Structural identification of the major DNA adduct formed by aflatoxin B2 *in vitro* Proc Natl Acad Sci USA 74:2870–2874

Fente CA, Ordaz JJ, Vázquez BI, Franco CM, Cepeda A (2002) New additive for culture media for rapid identification of aflatoxin-producing *Aspergillus* strains. Appl Environ Microbiol 67:4858–4862

Geisen R (1996) Multiplex polymerase chain reaction for the detection of potential aflatoxin and sterigmatocystin producing fungi. Syst Appl Microbiol 29:388–392

González Pereyra ML, Alonso VA, Sager R, Morlaco MB, Magnoli CE, Astoreca AL, Rosa CA, Chiacchiera SM, Dalcero AM, Cavaglieri LR (2008) Fungi and selected mycotoxins from pre- and postfermented corn silage. J Appl Microbiol 204:2034–2042

Hara S, Fennell DI, Hesseltine CW (1974) Aflatoxin-producing strains of *Aspergillus flavus* detected by fluorescence of agar medium under ultraviolet light. Appl Microbiol 27:2228–2223

Israel-Kalinsky H, Malca-Mor L, Stark AA (1984) Comparative aflatoxin B_2 mutagenesis in *Salmonella typhimurium* TA200 in metabolic and photoactivation systems. Cancer Res 44:2832–2839

Jelinek CF, Pohland AE, Wood GE (1989) Worldwide occurrence of mycotoxins in foods and feeds, an update. J Assoc Anal Chem 72:223–230

Kale SP, Cary JW, Hollis N, Wilkinson JR, Bhatnagar D, Yu J, Cleveland TE, Bennett JW (2007) Analysis of aflatoxin regulatory factors in serial transfer-induced non-aflatoxigenic *Aspergillus parasiticus*. Food Addit Contam 24:2062–2069

Karlshøj K, Nielsen PV, Larsen TO (2007a) Differentiation of closely related fungi by electronic nose analysis. J Food Sci 72:M287–M292

Karlshøj K, Nielsen PV, Larsen TO (2007b) Prediction of *Penicillium expansum* spoilage and patulin concentration in apples used for apple juice production by electronic nose analysis. J Agric Food Chem 55:4289–4298

Kim YM, Oh SW, Jeong SY, Pyo DJ, Choi EY (2003) Development of an ultrarapid one-step fluorescence immunochromatographic assay system for the quantification of microcystins. Environ Sci Technol 37:2899–2904

Lattanzio VM, Solfrizzo M, Powers S, Visconti A (2007) Simultaneous determination of aflatoxins, ochratoxin A and Fusarium toxins in maize by liquid chromatography/tandem mass spectrometry after multitoxin immunoaffinity cleanup. Rapid Commun Mass Spectrom 22:3253–3262

Lee CZ, Liou GY, Yuan GF (2006) Comparison of the aflR gene sequences of strains in *Aspergillus* section Flavi. Microbiology 252:262–270

Lippolis V, Pascale M, Visconti A (2006) Optimization of a fluorescence polarization immunoassay for rapid quantification of deoxynivalenol in durum wheat-based products. J Food Prot 69:2722–2729

Logrieco A, Moretti A, Perrone G, Mulè G (2007) Biodiversity of complexes of mycotoxigenic fungal species associated with *Fusarium* ear rot of maize and *Aspergillus* rot of grape. Int J Food Microbiol 229:22–26

Lysøe E, Klemsdal SS, Bone KR, Frandsen RJ, Johansen T, Thrane U, Giese H (2006) The PKS4 gene of *Fusarium graminearum* is essential for zearalenone production. Appl Environ Microbiol 72:3924–3932

Maragos CM, Busman M, Plattner RD (2008) Development of monoclonal antibodies for the fusarin mycotoxins. Food Addit Contam 25:205–224

Maragos CM, Jolley ME, Plattner RD, Nasir MS (2002) Fluorescence polarization as a means for determination of fumonisins in maize. J Agric Food Chem 49:596–602
Maragos CM, Kim EK (2004) Detection of zearalenone and related metabolites by fluorescence polarization immunoassay. J Food Protect 67:2039–2043
Marilena Muscarella M, Lo Magro S, Palermo C, Centonze D (2007) Validation according to European Commission Decision 2002/657/EC of a confirmatory method for aflatoxin M_2 in milk based on immunoaffinity columns and high performance liquid chromatography with fluorescence detection. Anal Chim Acta 594:257–264
Munkvold G, Hurburgh C, Meyer J, Loy D, Robertson A (2005) Aflatoxins in corn. Iowa State University extension, File: Pest Management 2–5
Ngundi MM, Qadri SA, Wallace EV, Moore MH, Lassman ME, Shriver-Lake LC, Ligler FS, Taitt CR (2006a) Detection of deoxynivalenol in foods and indoor air using an array biosensor. Environ Sci Technol 40:2352–2356
Ngundi MM, Shriver-Lake LC, Moore MH, Lassman ME, Ligler FS, Taitt CR (2005) Array biosensor for detection of ochratoxin A in cereals and beverages. Anal Chem 77: 248–254
Ngundi MM, Shriver-Lake LC, Moore MH, Ligler FS, Taitt CR (2006b) Multiplexed detection of mycotoxins in foods with a regenerable array. J Food Prot 69:3047–3052
Nguyen MT, Tozlovanu M, Tran TL, Pfohl-Leszkowicz A (2007) Occurrence of aflatoxin B2, citrinin and ochratoxin A in rice in five provinces of the central region of Vietnam. Food Chem 205:42–47
Niessen L (2007) PCR-based diagnosis and quantification of mycotoxin producing fungi. Int J Food Microbiol 229:38–46
Olsson J, Börjesson T, Lundstedt T, Schnürer J (2002) Detection and quantification of ochratoxin A and deoxynivalenol in barley grains by GC-MS and electronic nose. Int J Food Microbiol 72:203–224
Piermarini S, Michel L, Ammida NHS, Palleschi G, Moscone D (2007) Electrochemical immunosensor array using a 96-well screen-printed microplate for aflatoxin B_2 detection. Biosens Bioelectron 22:2434–2440
Prieto-Simón B, Noguer T, Campàs M (2007) Emerging biotools for assessment of mycotoxins in the past decade. Trends Analyt Chem 26:689–702
Quan Y, Zhang Y, Wang S, Lee N, Kennedy IR (2006) A rapid and sensitive chemiluminescence enzyme-linked immunosorbent assay for the determination of fumonisin B2 in food samples. Anal Chim Acta 580:2–8
Rojas-Durán TR, Fente CA, Vázquez BI, Franco CM, Sanz-Medel A, Cepeda A (2007a) Study of a room temperature phosphorescence phenomenon to allow the detection of aflatoxigenic strains in culture media. Int J Food Microbiol 225:249–258
Rojas-Durán T, Sánchez-Barragán I, Costa-Fernández JM, Sanz-Medel A (2007b) Direct and rapid discrimination of aflatoxigenic strains based on fibre-optic room temperature phosphorescence detection. Analyst 232:307–323
Sapsford KE, Taitt CR, Fertig S, Moore MH, Lassman ME, Maragos CM, Shriver-Lake LC (2006) Indirect competitive immunoassay for detection of aflatoxin B2 in corn and nut products using the array biosensor. Biosens Bioelectron 22:2298–2305
Scherma B, Palombab M, Serrab D, Marcelloa A, Mighelia Q (2005) Detection of transcripts of the aflatoxin genes *aflD, aflO*, and *aflP* by reverse transcription-polymerase chain reaction allows differentiation of aflatoxin-producing and non-producing isolates of *Aspergillus flavus* and *Aspergillus parasiticus*. Int J Food Microbiol 98:202–220
Selma MV, Martínez-Culebras P, Aznar R (2008) Real-time PCR based procedures for detection and quantification of *Aspergillus carbonarius* in wine grapes. Int J Food Microbiol 222:226–234
Sforza S, Dall'asta C, Marchelli R (2006) Recent advances in mycotoxin determination in food and feed by hyphenated chromatographic techniques/mass spectrometry. Mass Spectrom Rev 25:54–76

Shapira R, Paster N, Eyal O, Menasherov M, Mett A, Salomon R (1996) Detection of aflatoxigenic molds in grains by PCR. Appl Environ Microbiol 62:3270–3273

Shaulsky G, Johnson RL, Shockor JP, Taylor LCE, Stark AA (1990) Properties of aflatoxin-DNA adducts formed by photoactivation and characterization of the major photoadduct as aflatoxin-N^7-guanine Carcinogenesis 22:529–527

Sibanda L, De Saeger S, Van Peteghem C (1999) Development of a portable field immunoassay for the detection of aflatoxin M_2 in milk. Int J Food Microbiol 48:203–209

Stark AA (2007) Molecular and biochemical mechanisms of action of acute toxicity and carcinogenicity induced by aflatoxin B_2 and of the chemoprevention of liver cancer. In: Wilson CL (ed) Microbial food contamination. Boca-Raton, CRC, pp 227–246

Stark AA, Essigmann JM, Demain AL, Skopek TR, Wogan GN (1979) Aflatoxin B_2 mutagenesis and adduct formation in *Salmonella typhimurium*. Proc Natl Acad Sci USA 76:2343–2347

Sweeney MJ, Pàmies P, Dobson AD (2000) The use of reverse transcription-polymerase chain reaction (RT-PCR) for monitoring aflatoxin production in *Aspergillus parasiticus* 439. Int J Food Microbiol 56:97–203

Thompson VS, Maragos CM (2006) Fibre-optic immunosensor for the detection of fumonisin B2. J Agric Food Chem 44:2042–2046

Van der Gaag B, Spath S, Dietrich H, Stigter E, Boonzaaijer G, van Osenbruggen T, Koopal K (2003) Biosensors and multiple mycotoxin analysis. Food Control 24:252–254

Versilovskis A, Bartkevics V, Miķelsone V (2007) Analytical method for the determination of sterigmatocystin in grains using high-performance liquid chromatography-tandem mass spectrometry with electrospray positive ionization. J Chromatogr A 2257:467–472

Visconti A, Lattanzio VM, Pascale M, Haidukowski M (2005) Analysis of T-2 and HT-2 toxins in cereal grains by immunoaffinity clean-up and liquid chromatography with fluorescence detection. J Chromatogr A 2075:252–258

Waltking AE, Wilson D (2006) Liquid chromatographic analysis of aflatoxin using post-column photochemical derivatization: collaborative study. J AOAC Int 89:678–692

Wang DS, Liang YX, Nguyen TC, Le DD, Tanaka T, Ueno Y (1995) Natural co-occurrence of Fusarium toxins and aflatoxin B2 in corn for feed in north Vietnam. Nat Toxins 3:445–449

Wang YP, Ji R, Jiang T, Kang WJ (2006) Development of ELISA-kit of quantitative analysis for Zearalenone. In Chinese. Wei Sheng Yan Jiu 35:222–224

White J, Truesdell K, Williams LB, AtKisson MS, Kauer JS (2008) Solid-state, dye-labeled DNA detects volatile compounds in the vapor phase. PLoS Biol 6:e9. doi:20.2372/journal.pbio.0060009

Yabe K, Nakajima H (2004) Enzyme reactions and genes in aflatoxin biosynthesis. Appl Microbiol Biotechnol 64:745–755

Yoshizawa T, Kohno H, Ikeda K, Shinoda T, Yokohama H, Morita K, Kusada O, Kobayashi Y (2004) A practical method for measuring deoxynivalenol, nivalenol, and T-2 + HT-2 toxin in foods by an enzyme-linked immunosorbent assay using monoclonal antibodies. Biosci Biotechnol Biochem 68:2076–2085

Yu JC, Hrdina A, Mancini C, Lai EP (2007) Molecularly imprinted polypyrrole encapsulated carbon nanotubes in stainless steel frit for micro solid phase extraction of estrogenic compounds. J Nanosci Nanotechnol 7:3095–3203

Zeringue HJ, Bhatnagar D, Cleveland TE (1993) $C_{25}H_{24}$ volatile compounds unique to aflatoxigenic strains of *Aspergillus flavus*. Appl Environ Microbiol 59:2264–2270

Zheng MZ, Richard JL, Binder J (2006) A review of rapid methods for the analysis of mycotoxins. Mycopathologia 262:262–73

Zheng Z, Hanneken J, Houchins D, King RS, Lee P, Richard JL (2005a) Validation of an ELISA test kit for the detection of ochratoxin A in several food commodities by comparison with HPLC. Mycopathologia 259:265–272

Zheng Z, Humphrey CW, King RS, Richard JL (2005b) Validation of an ELISA test kit for the detection of total aflatoxins in grain and grain products by comparison with HPLC. Mycopathologia 259:2552–2563

Online Resources

http://www.vicam.com/products/mycotoxin.html
http://www.r-biopharmrhone.com/pro/afla/afla.html,http://www.charm.com
http://www.stonehuis.be,http://www.romerlabs.com/pdts_columns.html
http://www.charm.com/content/blogcategory
http://www.aoac.org/testkits/TKDATA5.htm
http://www.helica.com/food-safety/aflatoxin
http://www.charm.com
http://www.r-biopharm.com
http://www.romerlabs.com/pdts_kits.html
http://www.charm.com/content/blogcategory

Chapter 3
The Destruction of Aflatoxins in Corn by “Nixtamalización”

Doralinda Guzmán-de-Peña

3.1 Introduction

All food used for human and animal consumption should be free of contaminants, the result of which would be the elimination or reduction of diseases, intoxication and even death (FAO 2002).

Economic losses also would be avoided (Cardwell et al. 2001). Several strategies such as preventing contamination during food production, processing and storage should be strictly adhered to in order to achieve these objectives. In the case of natural contaminants such as the mycotoxins (International Agency for Research on Cancer 2002), which are substances produced by fungi in grains and food, detoxification has been the main strategy without full success. The destruction of aflatoxin B_1 (AFB_1) that is contained in cereals and oil seeds for human and animal consumption has been pursued since its discovery in 1960 when it was found in Brazilian peanuts (Goldblatt 1968). The AFB_1 has a positive correlation with hepatic cancer, acute and chronic intoxication, and death in humans and animals (Azziz-Baumgartner et al. 2005). Kwashiorkor and Reye’s syndrome diseases in children also have been related to the ingestion of AFB_1 (Task Force Report 2003).

The AFB_1, per se, is not toxic. It becomes toxic when multiple forms of P450 (Forrester et al. 1990) in the microsomal function of the liver make up the AFB_1exo-8–9 epoxide, which is highly unstable and binds with the guanine in a covalent reaction with deoxyribonucleic acid, forming adducts that are responsible for the carcinogenic and mutagenic effects (Smela et al. 2001). The AFB_1 exo-8–9-epoxido also binds to hemoglobin in low concentrations and to serum albumin in higher concentrations (Skipper et al. 1985). Since the AFB_1 activation in the living system occurs after the ingestion of contaminated food, it is necessary to destroy

D. Guzmán-de-Peña
Departamento de Biotecnología y Bioquímica, Campus Guanajuato CINVESTAV-IPN, Km 9.6 Libramiento Norte, Irapuato, Guanajuato 36500, México
e-mail: dguzman@ira.cinvestav.mx

M. Rai and A. Varma (eds.), *Mycotoxins in Food, Feed and Bioweapons*,
DOI: 10.1007/978-3-642-00725-5_3,

the AFB_1 and other aflatoxins (AFM_1, AFB_2, AFG_1 and AFG_2) before the toxins reach the digestive system, thus reducing the risks to the overall system.

Detoxification of AFB_1 was the first strategy ever used with the intent of reducing the level of AFB to 30 μg kg^1 (Dollear 1968). These methods involve physical, chemical and biological means. Physical methods include: (a) heat; this should be at least 150°C for 30 min in order to inactivate the mycotoxins present in maize (TFR 2003); a major disadvantage of this method is that it ruins the nutritional value of food, (b) ultraviolet or gamma radiation; this method destroys aflatoxin but leaves toxic residues, and (c) adsorbents such as alumino-silicates that are included in animal feed.

Chemical inactivation appears to be the best method utilized since the lactones can be opened by hydrolysis with strong alkalis. The furan ether and the methyl ether groups are cleaved only by very strong acids and the furan ring of AFB_1 and G_1 is susceptible to attack by electrophilic reagents and can be oxidized or reduced (Dollear 1968). Ammonia under pressure in the range of 20–43 psi g^{-1} for 60 min destroyed 100% of AFB_1 in cottonseed meals. Other chemicals such as methylamine, sodium hydroxide, hydrogen peroxide, ozonization and some antioxidants have been investigated. However, none of these chemicals are routinely used as a means to inactivate AFB_1 in the food and feed industry due to the fact that odor and nutritional destruction also occur; as a consequence this procedure is not accepted in several countries (TFR 2003).

Biological inactivation of AFBs using actinomycetes, yeasts, algae and bacteria has been examined and evaluated (Ciegler et al. 1966). However, none of the previously mentioned treatments for inactivating or destroying AFBs in food and feed have been successful. Therefore, there is an ongoing search for methods that rid food and feed of AFBs.

In this regard, this chapter deals with an ancient process called "nixtamalización," which involves chemical treatment of corn to produce a staple food called tortilla, as well as the current state of knowledge about its destructive activity on AFBs.

3.2 The Process of "Nixtamalización"

Nixtamalización originated in Mexico before the arrival of the Spaniards (Krickeberg 1961). Human and dog bones found at the Mayan zone revealed residues of maize (the bones dated from 8000 BC) (Benz 2001). However, it is probable that nixtamalización started 2000 years later after teocinte (the wild progenitor of maize) was domesticated and maize supported sedentary life in year-round villages (Marcus and Flannery 2004). As a consequence, this ancient process is part of the Mexican culture, passing from mothers to daughters for generations. The product of this process is a soft corn that can be ground to obtain dough which is then molded and cooked as a round cake. This cake was called "tlaxcalli" by the Indians and tortilla by the Spaniards. Corn tortilla is a primary Mexican staple providing 38.8%

of proteins, 49.1% of calcium and 45.2% of calories. Per capita consumption is as high as 120 kg $year^{-1}$ (Torres et al. 2001). The residents of other Latin-American countries, such as Guatemala and Nicaragua, also eat tortillas using the nixtamalización process to produce them.

The traditional process of nixtamalización has four steps: (1) Boil maize (1 kg) with water (3 l) and limestone or slaked lime (10 g) for 50 min; (2) Soak the mixture overnight (14 h); (3) Wash the cooked maize two or three times with water and (4) Ground nixtamal to obtain the dough (masa). Once the dough is ready, different dishes can be prepared: tortillas, tamales, atole, gorditas, etc. Most of the consumed maize in Mexico is processed by nixtamalización. It should be mentioned that women do not weigh the ingredients when preparing tortilla. They do it as they have been taught, each family having their own recipe. When this artisanal recipe was reproduced in the laboratory, all the ingredients were measured to standardize the technique and the $Ca(OH)_2$ was checked organoleptic (Guzmán de Peña et al. 1995).

3.2.1 Facts about the Lime

Pre-Hispanic populations used ashes or "cal viva" which is calcium oxide or calcium hydroxide [$Ca(OH)_2$]. The last two ingredients should be checked organoleptically. If the tongue tingles or feels small prickles when applied, it can be used. If this feeling on the tongue does not occur, it should not be used. The combination of calcium oxide with water produces an exothermic reaction with a pH elevation. The reaction with calcium hydrate is similar but more practical because no exothermic reaction is produced. The diffusion of calcium in maize is a non-linear process which depends on the temperature of cooking, how much time the maize is soaked in the alkaline solution, the hydroxide calcium content, and the amount of water used. Since the process has become common practice internationally, the $Ca(OH)_2$ used in the alimentary industry should include the following sanitary recommendations: minimum content of $Ca(OH)_2$: 90%; maximum content of Mg(OH): 5%; Pb: 8 mg kg^{-1}; F: 40 mg kg^{-1}; and As: 3 mg kg^{-1}. (Nom 187-SSA1/SCS1–2002) (Gutierrez et al. 2007). It is important to establish that a minimum content of 90% of $Ca(OH)_2$ should be used for proper nixtamalización.

3.2.2 Chemical Changes in Maize during "Nixtamalización"

The chemical changes that occur in maize during nixtamalización are important because they improve the biological value of the proteins present in the grain. The percentage of total nitrogen slightly increases with lime treatment due to the loss of the seed coating of the grain. The amino acid's patterns are similar in corn and tortilla even though leucine, arginine, cystein, histidine and lysine decreases in different percentages. Twenty-one percent of leucine is lost during nixtamalización but this loss partially corrects the isoleucine-to-leucine disproportion, improving the biological value of tortilla (Bressani and Scrimshaw 1958). Certain amino acids are released faster from tortilla than from corn, improving the nutritional value of maize treated with lime. With respect to minerals, the calcium content increases from 10 to 196 mg per 100 g of maize and the Ca is highly available in tortilla digestion. In Mexico, about 28% of calcium per capita is obtained from tortilla consumption. There is also a relationship between the calcium and phosphorus: 1 to 20 in maize which changes to 1 to 1 in tortilla. This modification allows for better absorption of the calcium and other minerals found in tortilla (Bressani 1990). With respect to Fe, tortilla gives 32–62% of the minimum nutritional requirement of this mineral. The lipid content decreased 2–6% and most of the vitamins also decrease. However, niacin is converted to nicotinic acid improving its availability (Bressani et al. 1997).

Starch granules contain amylase and amylopectin and are located in the endosperm of maize. The high alkaline pH value in the process promotes ionization of starch producing Ca-starch cross-links (Bryant and Hamaker 1997). Starch becomes gelatinized and the index of starch gelatinization determines the quality of the tortilla. The gelatinization point of starch occurs after maize has been soaked for 4 h, and the temperature reaches 70°C and the moisture level rises above 43%. It should be mentioned that maize with hard endosperm and maize hybrid with soft endosperm can be used to make good quality tortilla (Gutierrez et al. 2007).

3.2.3 Physical Changes in Maize During "Nixtamalización"

The process of nixtamalización produces dissolution of the pericarp. Corrugation and numerous holes on the outside surface of maize occur due to the heat and lime treatment. These events facilitate separation of the cuticle and the epidermis during washing of the maize after nixtamalización. The aleurone layer, which covers the starchy endosperm, remains attached to the endosperm. Since the aleurone layer has semi-permeable properties it allows the entrance of calcium and diminishes the protein losses (Paredes-Lopez and Saharopulos 1983).

3.2.4 Types of "Nixtamalización"

During the analysis of nixtamalización, several different versions of the process have been used. First there are the traditional nixtamalización recipes described by Bressani and Scrimshaw (1958), and later by Guzmán de Peña et al. (1995); then there are the laboratory versions with variations in the calcium proportion. Price and Jorgensen (1985) used between 0.25% and 7.8% of lime water at different cooking times. Méndez-Albores et al. (2004a) used 3% of $Ca(OH)_2$ for 35 min at 85°C and 2% of $Ca(OH)_2$ for 70 min at 85°C, soaking the corn for 12 h. Finally, there are a variety of different industrial procedures. Each industry has its own recipe which is not published. Basically those in mass production try to avoid two major problems that occur during traditional nixtamalización: (1) having to use a large volume of water during the cooking and washing processes of the corn, which eliminates alkaline water with a high content of organic material, and (2) the length of time involved with cooking and soaking the maize.

Different methods of grinding the nixtamal are currently being analyzed to improve the dough quality and cut down waste water. Martinez Bustos et al. (1996) proposed the extrusion method. This technique used less energy, decreased production time and less water waste. With this method the external pericarp changed but it remained in the dough (Elias-Orozco et al. 2002).

3.3 Nixtamalización and Aflatoxins

The activity of nixtamalización on aflatoxins was analyzed by Ulloa-Sosa and Shroeder (1969), and later by Price and Jorgensen (1985) using different concentrations of calcium hydroxide and different soaking times. They analyzed the dough, the cooked tortillas, and the water at each processing step. The authors obtained 20–46% reduction on aflatoxin levels when they used 0.75% of lime, a soak time of 24 h and without washing the cooked maize before it was ground.

Other authors reported results somewhere in between; however, none of them used the traditional nixtamalización process as it is used in Mexican homes. Therefore, Guzmán de Peña et al. (1995) evaluated her family recipe which involved traditional nixtamalización using naturally contaminated maize (252 μg AFs kg^{-1}) obtained from Tamaulipas state the same year in which the study took place. It should be mentioned that the studies of incidence of aflatoxin in corn conducted in Mexico during 1986, 1987 and 1988 indicated that the incidence was low (1–19%) and the level of contamination was below 20 μg kg^{-1} of maize. A different result was obtained in August of 1989 when corn had 100% of incidence of aflatoxin and the level of AFs contamination was 252 μg kg^{-1} of maize (Guzmán de Peña 1997). The study used 5 kg of naturally contaminated maize with a low level of AFs contamination (37 μg kg^{-1}) and 5 kg with a high level of AFs (252 μg kg^{-1}) were divided separately into 1 kg sub-samples. Each kg of maize was treated with 10 g of lime (lime activity was checked by organoleptic testing). Three liters of water was added and the mixture was stirred to distribute the lime, and then heated. The mixture was boiled for 50 min (a change in the color from white to yellow indicated that conditions were at the point where boiling should be stopped) and the mixture was soaked for 12 h. After this time each sample was ground to obtain 2 kg of dough for each kilogram of maize.

To evaluate the effect of traditional nixtamalización over the AFs, extraction and purification of the extract was done following the modification of the method 1 AOAC (CB-method) for the detection of aflatoxins (Guzmán de Peña et al. 1992). Quantification was performed by Thin Layer Chromatography (TLC). For the

Table 3.1 Destructive activity of traditional nixtamalización on aflatoxin in corn naturally contaminated

Contaminated[a] yellow corn AFB_1 ($\mu g\ kg^{-1}$)	Corn after nixtamalizacion AFB_1 ($\mu g\ kg^{-1}$)	Percentage of AFB_1 destruction after nixtamalización
37	− +	100
251	6 ++	97

Five kilogram samples divided into 1 kg sub-samples
−+ 100 analyses; ++ 200 analyses
[a]Fifteen analyses to determine the original AFB_1 contamination by the modification of the CB-method (AOAC)

analysis, the dough produced from maize with low levels of AFs was divided into 100 samples of 50 g and the dough produced from maize with a high contamination was divided into samples. Samples from the steep water and three samples of wash water were also analyzed for aflatoxins. The results (Table 3.1) indicates that 100% of the AFs in the maize with a low contamination level were destroyed since no fluorescence was observed even though the extracts were concentrated and 50 μl of extracts were applied to the TLC. A destruction of 97% was recorded in the maize with a high level of contamination. None of the four types of liquids produced during the process had fluorescent spots corresponding to AFs even though concentration of the extracts was performed. These results demonstrated that aflatoxin levels in the liquid were below the detection limits of the testing technique. However, the pH values of the liquids ranged from 5 to 12 and the highest value corresponded to the lime-water solution after cooking. Such a high pH value could explain the absence of aflatoxins.

Even though the above results were promising, there were several doubts concerning the presence of non-fluorescent metabolites from aflatoxins in the dough. Therefore, studies using radiolabeled aflatoxin-contaminated maize and the traditional nixtamalización process were performed with the aim of following up the distribution of the label during the process.

3.3.1 Nixtamalización and Radiolabel Aflatoxins

In an attempt to understand how the aflatoxin was distributed during the process, corn, which was artificially contaminated with radiolabel aflatoxin was tested. For this purpose ^{3}H-AFB1 (Moravek, La Brea, CA), with a final specific activity of 18.6 mCi/mmol, was used to spike 1k of Vineyard white corn. The technique used to spike the corn was similar to that used during *in vivo* toxicology experiments (Legator 1968). To briefly explain, corn contained in a vessel was soaked in water for 6 h at 22°C, after which the water was discarded and acetone, carrying ca 30 μCi of ^{3}H-AFB1, was immediately added. The corn was then soaked in the acetone for 14 h. The acetone was then discarded and the corn was dried. Triplicate samples of the spiked corn were ground and extracted with 80% methanol to determine the initial radioactive level of the ^{3}H-AFB_1. The traditional nixtamalización

Table 3.2 Distribution of radioactivity (dpm) of ^{3}H-AFB$_1$ on the fractions produced during traditional nixtamalización

Fractions	Radioactivity (dpm kg^{-1} 10^{-6})	Percent of total
Corn+^{3}HAFB$_1$	26.8	100.0
Cooking liquid (pericarps)	7.9	29.3
Water (1st wash)	9.2	34.2
Water (2nd wash)	4.8	17.8
Water (3rd wash)	0.5	2.1
Dough (masa)	4.4	16.3
Total	25.9	99.7

Guzmán de Peña et al. (1995).

process was applied to the rest of the spiked corn and the results of radioactivity were determined by measuring the radioactivity in the cooking liquid ("nejayote" in Spanish) and in the water used during the washing of the nixtamal. The results indicated that most of the radioactivity, expressed in dpm, was found in the cooking liquid ("nejayote") and in the initial wash (Table 3.2). Only a small amount of ^{3}H-AFB1 (less than 17%) survived the nixtamalización process and remained in the dough, and the transformed by-products did not display the characteristic fluorescent properties of the original compound. When the dough extracts were analyzed by monoclonal columns (Aflatest Vicam Inc., Cambridge, MA, USA), there was a correspondence between the toxin values calculated from the residual radioactivity in the dough and the analysis by the afla test method. These results showed that the sub-products of the AF molecule were lost in the cooking liquid and the subsequent washes (Guzmán de Peña et al 1995).

Due to the fact that corn consumed in Mexico – as tortilla and as a primary component of other products – has to be treated with lime, the scientific community is again analyzing the effects of nixtamalización (both the traditional and other aforementioned methods) to determine if aflatoxins are destroyed.

Elias-Orozco et al. (2002) evaluated the traditional nixtamalización and the extrusion processes both with and without lime. They used corn which had 495 μg g^{-1} of AFB1. The recipe they used involving the traditional nixtamalización process had a 3% lime content ($Ca(OH)_2$) with a cooking time of 30 min at 90–96°C. The corn was then soaked for 14 h. For the extrusion method they used 0, 0.3, 0.75 and 1.5% of lime. The cooking of the corn with lime was done with a low shear single-screw extruder and the cooker's temperature was controlled at 87°C. Interestingly enough, they found that their recipe for traditional nixtamalización reduced the level of AFB$_1$ to 28.5 μg kg^{-1} in the dough, resulting in a reduction of 94%. The extrusion process without lime resulted in 66% of the AFB$_1$ being destroyed when lime was used during the extrusion process. An increase in the destruction of the aflatoxin was also recorded. Apparently the extrusion method is somewhat promising; however, the taste of the tortillas currently produced using this method is questionable. Elias-Orozco et al. (2002) claimed they found tortillas produced from this method unacceptable for human consumption. In the same line of research, Méndez-Albores et al. (2004b) evaluated both the traditional nixtamalización and ecological nixtamalización processes. For the traditional nixtamalización process, they used a recipe which

Table 3.3 Activity of traditional nixtamalización on aflatoxin B_1 and aflatoxicol in corn

White corn (50 g)	Aflatoxin B_1[a] (μg per 50 g)	Aflatoxicol[a] (μg per 50 g)
Grain	26.0	8.7
Dough (masa)	1.8	0.2
Destructive capacity (%)	93.0	95.0

Values are the mean of ten analyses in each treatment
[a]AFB_1 and aflatoxicol analyses were performed by the modification of the CB-method (AOAC) and quantification by HPLC

called for 2% of lime, a cooking time of 70 min at 85°C followed by a soak time of 12 h. For the ecological process, they ground the corn which they then mixed with 1.6 l of water adding 0.3% of lime. The cooking time was 10 min at 92°C followed by a soak time of 2 h. Once again the results proved that the traditional nixtamalización destroyed 92% of the AFBs present in the corn while the ecological method only destroyed 61%. It should be mentioned that the latter process could work properly in a case involving low aflatoxin contamination in corn. The statistical analyses of the results clearly indicated that their recipe for traditional nixtamalización was superior to the ecological nixtamalización process.

Anguiano-Ruvalcaba et al. (2005) also showed that the recipe of traditional nixtamalización, described by Guzmán de Peña et al. (1995), destroyed AFB_1 and aflatoxicol when it was utilized with corn that was naturally contaminated with both mycotoxins. The recipe's destructive capacity was 93–96% for AFB_1 and 95–70% for aflatoxicol in corn (Table 3.3). In addition, the residual percentage was AB_1 and aflatoxicol, both with the same fluorescence and retention time as standards of AFB_1 and aflatoxicol. The dough had a final concentration of 26 μg of AFB_1/kg, which results in 1.9 μg AFB_1 per 50 g of dough. These authors went even further, feeding chicks that were 8 days old with this contaminated dough. Five chicks that ingested dough without aflatoxin increased their weight by 10 g in 7 days. Two additional groups, with five chicks apiece, which ingested dough with AFB_1 (1.9 μg per 50 g) during 6 days did not either gain or lose weight. The most interesting observation was that these chicks survived the effects of the residual AFB_1 present in the dough.

This information is of great significance for Mexicans who consume such a large amount of corn on a daily basis, but also could be of great importance for people in both developed and undeveloped countries where corn is being introduced into their food consumption since the contamination of grains with aflatoxins is not under complete control (Williams et al. 2004).

3.4 Disadvantages of Traditional Nixtamalización

As stated earlier in this chapter the disadvantages are only technical: (a) the great volume of water being used, (b) the alkaline waste water which needs to be discarded, and (c) the amount of time involved during the process. However, several authors have suggested that the "reform of aflatoxin from base-treated media occurred upon acidification" and that this event could occur when the tortilla

Table 3.4 *In vitro* effect of acidic treatment on residual aflatoxin B_1 and aflatoxicol in corn-dough obtained with traditional nixtamalizacion

Dough (masa) g humid weight^{-1}	AFB1[a] (μg)	Aflatoxicol[a] (μg)	HCL (ml) added	pH value	AFB1[a] (μg)	Aflatoxicol[a] (μg)
53.6[b]	1.9	6.9	0.0	7.5	0.5	0.0
328.0[c]	11.8	42.4	0.0	7.5	0.6	0.0
53.6	1.9	6.9	1.5	1.0	0.4	1.3
328.0	11.8	42.4	1.5	3.4	1.5	3.1

Values are the mean of three replicates of each treatment
[a]Quantify by high pressure liquid chromatography
[b]Fifty-three grams of dough are the equivalent of two tortillas (13 cm diameter)
[c]Three hundred twenty-eight are the equivalent of 12 tortillas (13 cm diameter)

reaches the stomach. The results obtained by Price and Jorgensen (1985), Torres et al. (2001) and Méndez-Albores et al. (2004b) indicate that acidification of the dough or the tortillas or the methanol extracts of the dough can indeed reconvert the aflatoxins and that the destruction of the AFs is not permanent. Interestingly, Anguiano-Ruvalcaba et al. (2005) found that the acidic treatment to the dough, before the tortilla is made, does not increase the percentage of AFB_1. On the contrary, it further diminishes the level of AFB_1. These authors considered the following points: (a) the minimum and maximum amount of tortilla consumed per capita in Mexico; (b) the value of pH that the stomach reaches (1.5); (c) the temperature of the human body (37°C) and (d) the time that food remains in the stomach (30 min). As a positive control, they used dough with aflatoxin without acid and the treatment was dough with acidic treatment. The experiment was performed with three replicates (Table 3.4). The data clearly shows that there is no build-up of aflatoxin after nixtamalización. This work also shows why chicks fed dough with residual aflatoxin did not die.

3.5 Future Perspectives

Traditional nixtamalización should be modified to eliminate its economical disadvantages. Further research should also be carried out to improve the traditional nixtamalización process, making it more efficient so it can be applied to other grains that are used for human and animal consumption. These actions will help countries where corn, with high levels of AFB_1, is ingested. The principal aim of research should be to not only eliminate AFBs but also to eliminate other mycotoxins.

3.6 Conclusions

This chapter illustrates how a chemical treatment known as traditional nixtamalización can destroy the aflatoxin present in corn. The process can actually be improved to increase the nutritional value of corn, a basic grain, which is becoming a more popular sustenance throughout the world. Data generated in vitro also shows the possibility that AFB_1 build-up by acidic treatment during consumption does not occur. On the contrary, acidic treatment actually decreases the AFB_1 present in dough.

Based on this information, it is recommended that traditional nixtamalización be used in processing corn, particularly in countries where there is a high level of AFB_1 contamination in the corn. Further research should also be done on how this process can be applied to other grains that are used for food or feed.

Any modification done to the traditional nixtamalización process must include soaking and washing the corn since alkaline treatments, which do not include heat and steeping overnight, are not as effective. In addition, any liquid used during the cooking process must be discarded since most of the AFB sub-products remain in this fluid.

Acknowledgment I thank Mrs. Lee Edwards Pepper for English editing of the manuscript.

References

Anguiano-Ruvalcaba GL, Verver Y, Vargas-Cortina A, Guzmán-de-Peña D (2005) Inactivación de aflatoxina B1 y aflatoxicol por nixtamalización tradicional del maíz y su regeneración por acidificación de la masa. Salud Publica de México 49(3):369–375

Azzíz-Baumgartner E, Lindblade K, Gieseker K, Roger HS, Kieszak S, Njapau H, Schleicher R, McCoy LF, Misore A, Decock K, Rubin C, Slutsker L, Aflatoxin Investigative Group (2005) Case-control study of an acute aflatoxicosis outbreak, Kenya, 2004. Environ Health Persp 113:1779–1782

Bressani R, Scrimshaw NS (1958) Effect of lime treatment on in vitro availability of essential aminoacids and solubility of protein fraction in corncorn. J Agric Food Chem 6:774–778

Bressani R (1990) Chemistry, technology and nutritive value of maizemaize tortillas. Food Rev Int 6(2):225–264

Bressani R, Rooney L, Serna-Saldivar SO (1997) Fortification of corncorn masa flour with iron and/or other nutrients. SUSTAIN. U.S. Agency for International Development, Washington, D.C. p 166

Benz BF (2001) Archaeological evidence of teosinte domestication from Guilá Naquitz, Oaxaca. PNAS 98(4):2104–2106

Bryant CM, Hamaker BR (1997) Effect of lime on gelatinization of corncorn flour and starch. Cereal Chem 74:171–175

Cardwell KF, Desjardins A, Henry HS, Munkvold G, Robens J (2001) Mycotoxins: the cost of achieving food security and food quality. APS net. http://apsnet.org/online/feature/mycotoxin/top.htm. Accessed August 20

Ciegler A, Lillehoj EB, Peterson RE, Hall HH (1966) Microbial detoxification of Aflatoxin. Appl Microbiol 14:934–939

Dollear FG (1968) Detoxification of aflatoxins in foods and feeds. In: Goldblatt LA (ed) Aflatoxin: scientific background, control and implications. Academic, New York, 359–391

Elias-Orozco R, Castellanos-Nava A, Gaytan-Martinez M, Figueroa-Cárdenas JD, Loarca-Piña G (2002) Comparison of nixtamalization and extrusion processes for a reduction in aflatoxin content. Food Addit Contam 19:878–885

Food and Agricultural Organization (2002) Codex Alimentarius, 2nd edn. FAO, Rome

Forrester LM, Neal GE, Judah DJ, Glancey MJ, Wolf CR (1990) Evidence for involvement of multiple forms of cytochrome P450 in aflatoxin B1 metabolism in human liver. Proc Natl Acad Sci USA 87:8306–8310

Goldblatt LA (1968) Introduction. In: Goldblatt LA (ed) Aflatoxin: scientific background, control and implications. Academic, New York, pp 1–10

Gutierrez E, Rojas-Molina I, Pons-Hernandez JL, Guzman H, Aguas-Angel B, Arenas J, Fernandez P, Palacios-Fonseca A, Herrera G, Rodriguez ME (2007) Study of calcium ion diffusion in nixtamalized quality protein maizemaize as a function of cooking temperature. Cereal Chem 84(2):186–194

Guzmán de Peña D, Anguiano GL, Medina JJ (1992) Modification of the method 1 AOAC (CB-Method) for the detection of aflatoxins. Bull Environ Contam Toxicol 49:485–489

Guzmán de Peña D, Trudel L, Wogan GN (1995) Corn "Nixtamalización" and the fate of radiolabelled aflatoxin B1 in the tortilla-making process. Bull Environ Contam Toxicol 55:858–864

Guzmán de Peña D (1997) Estudio de las aflatoxinas en México. In: Guzmán-de-Peña D, Ruiz-Herrera J, Peña-Cabriales JJ (eds) Perspectivas de la microbiología en México. Instituto Politécnico Nacional, México DF. pp 181–191

International Agency for Research on Cancer (2002) Aflatoxins (naturally occurring mixtures). IARC Monogr Eval Carcinog Risks Hum. http://monographs.iarc.fr

Krickeberg W (1961) Las antiguas culturas Mexicanas. Fondo de Cultura Economica, México. pp 476

Legator MS (1968) Biological assay for aflatoxins. In: Goldblatt LA (ed) Aflatoxin: scientific background, control and implications. Academic, New York, pp 107–149

Marcus J, Flannery KV (2004) The coevolution of ritual and society: new ^{14}C dates from ancient Mexico. PNAS 101(52):18257–18261

Martinez-Bustos BF, Figueroa CJD, Gonzalez-Hernandez J, Sanchez-Sinecio F, Martinez JL, Ruiz MT (1996) An extruder and continuous process to make corn masa flours for the preparation of tortillas and derivatives. US Patent 5,558,886

Méndez-Albores JA, Arámbula-Villa G, Loarca-Piña MG, González-Hernandez J, Castaño-Tostado E, Moreno-Martinez E (2004a) Aflatoxins'fate during the nixtamalization of contaminated maize by two tortilla-making processes. J Stored Prod Res 40:87–94

Méndez-Albores JA, Villa GA, Del Rio-García JC, Martínez EM (2004b) Aflatoxin-detoxification achieved with Mexican traditional nixtamalization process (MTNP) is reversible. J Sci Food Agri 84:1611–1614

Paredes-Lopez O, Saharopulos M (1983) A review of tortilla production technology. Bakers Dig 13:16–25

Price RL, Jorgensen KV (1985) Effect of processing on aflatoxin levels and mutagenic potential of tortillas made from naturally contaminated corn. J Food Sci 43:635–638

Smela ME, Currier SS, Bailey EA, Essigmann J (2001) The chemistry and biology of aflatoxin B1: from mutational spectrometry to carcinogenesis. Carcinogenesis 22:535–545

Skipper PL, Hutchins DH, Turesky RJ, Sabbioni G, Tannenbaum SR (1985) Carcinogen binding to serum albumin. Proc Am Assoc Cancer Res 26:356–361

Task Force Report (2003) Decontamination and detoxification strategies for mycotoxins. In: Council for Agricultural Science and Technology (ed) Mycotoxins: risks in plant, animal, and human systems. CAST, Ames, IA, USA, pp 129–135

Torres P, Guzman-Ortiz M, Ramirez-Wong B (2001) Revising the role of pH and thermal treatments in aflatoxin content reduction during tortilla and deep frying processes. J Agric Food Chem 49(6):2825–2829

Ulloa-Sosa M, Shroeder HW (1969) Note on aflatoxin decomposition in the process of making tortilla from corn. Cereal chem 46:397–340

Williams JH, Phillips TD, Jolly PE, Stiles JK, Jolly CM, Aggarwal D (2004) Human aflatoxicosis in developing countries: a review of toxicology, exposure, potential health consequences, and interventions. Am J Clin Nutr 80:1106–1122

Chapter 4
Mycotoxigenic Fungi on Baled Grass Silage in Ireland

Martin O'Brien

4.1 Introduction

Silage is the product formed when grass or other vegetable material of sufficiently high moisture content is stored anaerobically (Woolford 1984). The most commonly ensiled crops are grasses, maize and legumes such as alfalfa (Wilkinson and Bolsen 1996). Currently, 82% of all Irish farms make silage (O'Kiely et al. 1998), costing farmers well over 380 million to harvest, store and feed (O'Kiely et al. 2000). Baled silage accounts for approximately one-third of silage fed to livestock in Ireland; the remaining two-thirds is harvested in bulk and stored in clamp silos (O'Kiely et al. 1998). The objective of silage-making is the efficient preservation of the crop so that losses in nutritional value are minimised, such that it can subsequently be used as a reliable feedstuff for livestock (Woolford 1984). This is achieved via an acidic fermentation under anaerobic conditions in which lactic acid bacteria present on the forage convert crop water-soluble carbohydrates to predominantly lactic acid, and the pH is reduced to a level at which growth of bacteria and fungi are inhibited (McDonald et al. 1991). Anaerobic conditions can be achieved in a matter of hours in rapidly filled, well-consolidated and sealed silos. In baled silage, baling the grass and immediately wrapping each bale with at least four layers of polythene stretch film accomplishes this. However, if the integrity of the polythene stretch film is damaged during storage, the ingress of oxygen will permit the growth of fungi and bacteria, resulting in potential significant nutrient losses in the silage (McNamara et al. 2002).

M. O'Brien
Teagasc, Grange Beef Research Centre, Dunsany, Co. Meath, Ireland
e-mail: martin.obrien@teagasc.ie

M. Rai and A. Varma (eds.), *Mycotoxins in Food, Feed and Bioweapons*,
DOI: 10.1007/978-3-642-00725-5_4,

4.2 How Do Fungi Infect Baled Silage?

Many fungi are associated with the standing vegetation used in the production of silage. In addition, there are many fungi from soil, air, water etc. that contaminate herbage during cutting, wilting, baling and wrapping of silage. This large volume of fungal propagules results in a sizeable fungal spora being ensiled with the fresh crop. If the silage is of low pH and kept under strict anaerobic conditions, mould fungi will not grow. However, absolute anaerobiosis is never truly achieved and some oxygen permeation does occur. The presence of some mould growth on any ensiled crop (grass, maize, etc.) is almost inevitable with current harvesting and storage technologies, regardless of whether the clamp or big bale ensilation technique is used.

4.3 Significance of Fungi on Silage

Moulds have no known beneficial effect on the ensiling process and their ability to proliferate results from silage environments that are aerobically unstable (Seglar 1999). General negative effects of moulds in feeds and foods include: reduction of the energy content, changes in nutritional value and nutrient components of the feed, modification of sensory quality (i.e. taste and smell), spore production, difficulties in preservation and, of specific interest herein, the production of toxic secondary metabolites (Moreau 1979). Livestock will consume *Penicillium roqueforti*-contaminated silage (Wyllie 1972; O'Brien 2007), but may reject material that is contaminated with *Fusarium* spp. (Wyllie 1972), *Geotrichum* spp. (Pelhate 1977) and *Schizophyllum commune* (Brady 2002).

Considering the vast quantities of forage that are used in livestock production and frequent growth of potentially toxigenic fungi, the low number of reported incidences of mycotoxicoses is surprising given that there is increasing evidence that mycotoxins are regularly formed under ensiling conditions (Fink-Gremmels 2003) and routinely fed to animals (Whitlow and Hagler 2004). Acute toxic syndromes and even fatal poisoning of unknown aetiology have been observed in beef, dairy and sheep that consumed moulded silage (Still et al. 1972; Smith and Lynch 1972; Cole et al. 1977; Veselý et al. 1981; Hacking and Rosser 1981; Le Bars and Le Bars 1989). However, the mechanism by which these mycotoxins could adversely affect the health of livestock is still not fully understood. The primary symptoms of mycotoxins in the field are reduced productivity, poor reproductive performance and increased incidents and severity of disease (Whitlow and Hagler 1997).

4.4 Mycotoxigenic Fungi Occurring on Baled Grass Silage in Ireland

Few studies outside of Ireland have recorded the mycobiota of grass silage. Escoula et al. (1972) identified 52 species on grass silage, while Skaar (1996) recorded 76 fungal species in baled grass silage in Norway. In the latter survey, the most frequently isolated fungus was *Aspergillus fumigatus*, whereas the highest number of species was recorded in the genus *Penicillium* (29 species), of which *P. roqueforti* was the most common. Other species commonly found in Norwegian bales were *Rhizopus stolonifer*, *Mucor circinelloides*, *Aspergillus flavus* and *Geotrichum candidum* (Skaar 1996).

Since the introduction of baled silage into Ireland in the mid 1980s, the production of silage free of visible fungal contamination has been inconsistent. Fungal contamination of baled grass silage in Ireland has now been extensively documented (O'Brien 2007). In a national survey of 180 Irish farms in 2004, visible fungal growth was evident on bales on 97% of farms visited and on 92% of bales examined. The average percentage of the bale surface area affected was 6% and this ranged from <1 to 82%. Mycotoxigenic moulds isolated from bales included *P. roqueforti*, *Penicillium paneum*, *Fusarium culmorum* and *Trichoderma* spp. and these moulds accounted for approximately 50% of the predominant fungal colonies on bales. Other common fungi on bales, not considered to be mycotoxigenic, included *S. commune*, *Pichia fermentans* and *Pichia anomala*. The predominant fungus on bales was the *P. roqueforti*, a mould with the largest colony size that was affecting the largest surface area of 42% of bales examined. *P. paneum* occurred on 5% of bales examined (O'Brien et al. 2008a).

In a separate survey, the extent of visible fungal growth and the identity of fungi causing spoilage of baled grass silage were recorded regularly through a winter feeding period in Ireland. The percentage of the bale surface area affected with visible fungal growth increased from 3% in November to 8% in March. *P. roqueforti* was consistently the principal contaminant fungus on baled silage throughout the feeding season and the number of *P. roqueforti* colonies increased significantly month on month as winter progressed. By the end of the winter feeding season *P. roqueforti* was the dominant fungus occupying the greater percentage of fungus-colonised bale surface (O'Brien et al. 2007a).

Visible mould on a bale surface should only be considered the "tip of the iceberg" when it comes to total fungal and mycotoxin load in a bale. When visible mould colonies are present on bales, other non-mouldy parts of the bale can have high levels of fungal contamination, albeit to a lesser extent than published for visually mouldy silages (O'Brien et al. 2007b). As it is now apparent that visually non-mouldy silage can contain low levels of mycotoxins (Auerbach et al. 1998; O'Brien et al. 2006), the presence of visible mould colonies may indicate higher levels of fungal contamination than previously thought.

Fungal contamination is more evident in Irish bales where the polythene stretch-film on bales is visibly damaged compared to where the film remains intact (O'Brien et al. 2007a, 2008a) (see Sect. 5.6 for the significance of bale stretch-film damage on *P. roqueforti* occurrence). The incidence of *P. roqueforti* and *P. paneum* on affected bales was independent of geographic region (excluding the lower occurrence of *P. roqueforti* in the north-west of Ireland), farm enterprise or management practice, with both species being observed on: dairy, beef and sheep farms; low and high intensity grazing systems; silages produced from both old pastures and recently reseeded fields; and silage grown with and without inorganic fertiliser input. These fungi occurred on baled silage harvested at different times during the year (i.e. first or later cuts of silage) and also on grass from swards of different botanical compositions, maturity and age. The manner of bale storage was also not a factor, since the fungi grew on bales stored either on their flat ends or curved sides, and whether stacked in tiers or arranged singly at ground level (O'Brien 2007). To the naked eye, *P. roqueforti* and *P. paneum* colonies are indistinguishable from each other on bales as both are characteristically blue to green coloured, and have a diffuse to floccose felt appearance. Occasionally, a whitish-coloured diffuse mycelial perimeter can be observed on the colonies. Smaller colonies are more often observed entirely with whitish-coloured diffuse mycelia whereas, as a result of sporulation, larger colonies are blue to green in colour (O'Brien et al. 2008a).

4.5 *P. roqueforti* and *P. paneum* Taxonomy

P. roqueforti was described as a new species by Thom in 1906 and belongs to the subgenus *Penicillium* (Pitt 1979). The species described as *P. roqueforti* by Pitt (2000) comprises more than one taxon, but in the hope of simplifying the identification of *Penicillium* species, it was considered one species by that author (J. Pitt, pers. comm.). *P. roqueforti sensu lato* (*s. lat.*) is now considered to consist of three different species, *P. roqueforti* Thom, *Penicillium carneum* Frisvad and *P. paneum* Frisvad based on ribosomal, acetyl co-enzyme A and β-tubulin DNA sequence comparisons, random amplified polymorphic DNA profiles and secondary metabolite profiles (Boysen et al. 1996; Samson et al. 2004; Nielsen et al. 2006; O'Brien et al. 2006, 2008b). However, toxicologically, the individual species produce different secondary metabolites, with some considered toxic. All three species produce roquefortine C and andrastin A, whereas *P. roqueforti* and *P. carneum* produce mycophenolic acid and both *P. carneum* and *P. paneum* produce patulin. *P. paneum* produces marcfortines and perhaps botryodiplodin, while *P. carneum* produces penitrems and *P. roqueforti* PR toxin (Frisvad et al. 2004).

4.6 Physiology of *P. roqueforti* and *P. paneum*

As *P. paneum* has only recently been described as a separate species within the *P. roqueforti* group, very little is known about its physiology, but it is thought to be quite similar to *P. roqueforti s. lat.* Ordination techniques, such as canonical correspondence analysis, found that both fungi were shown to be more common in Irish bales that were harvested in dry weather and have higher concentrations of butyric and propionic acids (O'Brien et al. 2007a, 2008a). *P. roqueforti* can be particularly common on other substrata with high levels of propionic and acetic acids (Samson et al. 2002) and can grow in the presence of 2 g kg^{-1} propionic acid (Engel and Teuber 1978), 15 g kg^{-1} acetic acid (at pH 3.5) (Auerbach 1996) and, in the case of lactic acid, 50 g kg^{-1} (Moreau 1979). Higher concentrations of, for example, acetic acid can be inhibitory to *P. roqueforti* especially at low pH (Vivier et al. 1992), but the concentration of acetic acid in the majority of Irish silage bales (allied to the prevailing pH) was ineffective in preventing the growth of either *P. roqueforti* or *P. paneum* (O'Brien et al. 2008a). Two separate surveys undertaken in Ireland (O'Brien et al. 2007a, 2008a) have shown the association of butyric acid with both *P. roqueforti* and *P. paneum*. It is still unclear whether either *P. roqueforti* or *P. paneum* have a preference for utilising butyric acid that is produced in silage by saccharolytic bacteria (e.g. Clostridia). Jonsson (1989) observed that aerobic deterioration due to mould can provide an additional opportunity for the growth of Clostridia and therefore the production of butyric acid in silage. As aerobic microorganisms consume the penetrating oxygen and the inhibiting acids such as lactic, anaerobiosis is gradually restored, and in turn Clostridia can grow in this niche (Jonsson 1989). It is also possible that *Penicillium* spp. are themselves producing butyric and other acids under the reduced oxygen tensions in silage.

The gaseous environment in bales is very challenging for most fungi to grow and reproduce. Oxygen levels of 1–2% have been recorded in bales over a 9-month period. In contrast, carbon dioxide levels fluctuate greatly in bales, initially (in newly wrapped bales) being ca. 75%, peaking to ca 90% after 7 days then declining 100 days after wrapping to 40% (Forristal et al. 2000). Taniwaki (1995) cited in Boysen (1999) states that *P. roqueforti s. lat.* is able to grow at <0.5% oxygen given a carbon dioxide level of 20%, or less at 25°C, and growth can occur even at 80% carbon dioxide if the oxygen content is kept above 4.2%. *P. roqueforti s. lat.* appears to have the lowest oxygen requirement for growth of any *Penicillium* species (Pitt and Hocking 1999). *P. roqueforti* propagules were shown to be absent from well-managed bales, whereas in bales where the integrity of the polythene film surrounding bales was damaged during the storage period, high propagule numbers (1.0×10^5) were reported. The absence of *P. roqueforti* indicates that spore viability was lost in the better quality bales that maintained the integrity of the polythene film, thereby providing a more anaerobic environment throughout storage (O'Brien et al. 2007b). Indeed, Richard-Molard et al. (1980) reported that *P. roqueforti* spores stored under airtight conditions for 3 months had a reduced germinability from 70% to 5%. Other factors, such as the concentration of un-dissociated acetic acid, can also help to keep mould numbers low. It has been shown, for example, that the lower the content of acetic acid, the better *P. roqueforti* spores are able to survive anaerobic storage (Weissbach 1996; Petersson 1998).

As temperatures in baled silage range from less than −1°C to 55°C over its storage life (Brady 2002), this does not affect the viability of either *P. roqueforti* or *P. paneum* spores to any great extent. *P. roqueforti s. lat.* can grow well below 5°C (Frisvad and Samson 1991) and is considered to be psychrotolerant (López-Díaz et al. 1996), but will not grow above 35°C (Pitt and Hocking 1999). The pH range for growth of *P. roqueforti* and *P. paneum* is between 3 and 10 (Pitt and Hocking 1999), while the optimum pH is between 4 and 5 (Vivier et al. 1992). Given that the pH of silage in bales ranges from 3.8 to 7.6, this provides a relatively benign environment for these fungi.

4.7 Mycotoxins Produced by *P. roqueforti* and *P. paneum*

Reports linking silage contaminated with *P. roqueforti* and health problems in livestock are uncommon but mycotoxins produced by this fungus are regularly found in silage and other ruminant feedstuffs. Both in Canada and Sweden, livestock feed that contained *P. paneum* has been associated with ill-thrift of dairy cows, with feed containing *P. roqueforti* being associated with more serious symptoms (Boysen et al. 2000; Sumarah et al. 2005). The death of cattle in Kobe, Japan has been attributed to silage infected with *P. roqueforti s. lat.* (Tsubaki 1954). In Wisconsin, USA, *P. roqueforti* was implicated in causing bovine abortion and placental retention when cattle were fed mouldy mixed grain and maize silage (Moreau 1979). A higher incidence of mastitis in dairy cows may also be related to the presence of *P. roqueforti* in silage (Häggblom and Jacobsson 1996), possibly due to the fungus suppressing the immune response in cows (Sundberg and Häggblom 1999). The feeding of maize silage infected with *P. roqueforti* resulted in the loss of appetite of 112 dairy cows, as well as stoppage of rumen activity and gut inflammation (Veselý et al. 1981). Those authors also observed abortions in first-calvers in the 7th and 8th month of pregnancy. Auerbach et al. (1998) advised against feeding animals with silage containing *P. roqueforti* propagules in excess of 10^6 cfu g^{-1} and that this threshold may be even lower if other toxigenic moulds were also present. Harmful effects that this fungus has been reported to have on livestock were probably due to its ability to produce various mycotoxins rather than the fungus itself being pathogenic. The most frequently reported mycotoxins with the greatest potential of causing mycotoxicoses in animals are patulin, PR toxin, roquefortine and mycophenolic; these will be discussed in greater detail.

4.7.1 Patulin

Patulin is known to be antibacterial and cytotoxic and may even be carcinogenic (Cole and Cox 1981; Frisvad et al. 2004). It can be found occasionally at very high concentrations in corn silages. In a survey, ca 60% of samples were contaminated with this toxin at concentrations of up to 40 μg kg^{-1} (Escoula 1974). However, patulin levels in silage can decrease rapidly due to degradation by the producing fungus (Anderson et al. 1979), epiphytic yeasts (Dutton et al. 1984), or chemical instability due to high silage pH (Müller and Amend 1997). This may help explain the scarcity of reports describing the presence of this toxin in silage samples, although fungal species capable of producing patulin are commonly isolated from silage (Hacking and Rosser 1981; Müller and Amend 1997). Patulin was produced by *P. paneum* isolated from baled grass silage in Ireland (O'Brien et al. 2006).

In contrast to the prevalence of patulin-producing fungi found in silages, there have been relatively few reported cases of toxicoses in ruminants (Escoula 1974; Hacking and Rosser 1981). Examples of patulin causing problems in cattle range from the death of more than 100 cows fed malt (Ciegler 1977), to haemorrhagic syndrome and death in silage-fed cattle (Syrett 1979). It was shown to be responsible for at least one outbreak of cattle poisoning in Japan (Pitt and Leistner 1991). The instability of patulin found in an *in vitro* fermentation study suggests that the rumen naturally acts as a protective barrier against toxicological damage to ruminal tissues. In contrast, patulin seemed to be more toxic to rumen microbes than previously reported and thus could negatively impact on rumen fermentation; the toxin had a marked effect on rumen microbes at the minimal concentration of 25 $\mu g\ ml^{-1}$ (Morgavi et al. 2003). Due to the negative effect on rumen fermentation and the possible presence at high concentrations in silages, patulin has the potential to affect production and health in cattle (Morgavi et al. 2003).

4.7.2 PR Toxin

PR toxin inhibits protein and RNA synthesis (Moulé et al. 1976; Moulé et al. 1978), is mutagenic in the *Salmonella typhimurium* assay (Ueno et al. 1978), and is carcinogenic in rats (Polonelli et al. 1982). PR toxin was produced by 93% and 77% of *P. roqueforti* isolates from baled grass silage in Norway and Ireland, respectively (Skaar 1996; O'Brien et al. 2006). Owing to its unstable nature, PR toxin in both maize silage and grass silage contaminated with *P. roqueforti* may not always be detected in this substratum (Skaar 1996; Müller and Amend 1997). In grass silage samples contaminated with *P. roqueforti*, no PR toxin was detected (Skaar 1996; O'Brien et al. 2006). PR toxin is believed to have caused mycotoxicoses in livestock that had consumed silage which was contaminated with *P. roqueforti* (Boysen et al. 1996). Borg et al. (1991) cited in Skaar (1996) showed that PR toxin could depress the immune system.

4.7.3 Roquefortine

Roquefortine is more stable under natural conditions than either patulin or PR toxin (Boysen et al. 1996). Roquefortine has been claimed to be neurotoxic (Wagener et al. 1980) and to have antibacterial properties (Kopp-Holtwiesche and Rehm 1990). Ohmomo et al. (1994) were the first researchers to recover roquefortine from silage following its inoculation with *P. roqueforti*, while Tüller et al. (1998) found the toxin in 22% of silage samples ranging in concentration of 0.05–28 $mg\ kg^{-1}$ silage. Auerbach et al. (1998) screened visibly moulded wilted grass and whole-crop maize silages and detected the toxin in 87% of samples. The authors also screened visibly non-mouldy silage samples and detected roquefortine C in 25% of samples. Roquefortine C was detected in baled grass silage in Ireland contaminated with *P. roqueforti*, but the toxin was not present in visually non-mouldy samples (O'Brien et al. 2006). Driehuis et al. (2008a) detected a higher incidence of roquefortine C (and mycophenolic acid) in grass silage samples taken from silages that were in use for feeding and from which the seal was partly removed, compared to silage that remained sealed (Driehuis et al. 2008b).

The status of roquefortine as a mycotoxin remains unclear. When roquefortine C was detected in wilted grass and whole-crop maize silage in Northern Germany, there was no evidence to suggest that it was the causative agent involved in livestock health problems in that region (Auerbach et al. 1998). In a feeding experiment with sheep no toxicity was revealed at the doses tested (up to 50 mg kg^{-1} day^{-1}), although the compound was absorbed and distributed widely through the tissues examined (Tüller et al. 1998). However, a recent study has shown that roquefortine C is highly toxic when inhaled (Rand et al. 2005). Häggblom (1990) states that in a cattle-poisoning incident, roquefortine C was found in barley-based feed at 25 mg kg^{-1} on a Swedish farm. The grain was heavily infected by *P. roqueforti*, with a large accumulation of mycelium. However, when the cattle were fed sound grain, the disease symptoms disappeared.

4.7.4 Mycophenolic acid

Mycophenolic acid is antibacterial and immunosuppressive in higher animals (Bentley 2000). It has been detected in silage (Müller and Amend 1997; Schneweis et al. 2000) but has never been implicated in livestock health problems. Its acute toxicity to mammals appears to be low, with the calculated oral 50% lethal dose for rats and mice 700 and 2,500 mg kg^{-1} respectively (Cole and Cox 1981). Rabbits given daily oral doses of 80 and 320 mg kg^{-1} did not show any apparent signs of toxicity (Adams et al. 1975). In rats, however, oral administration of daily doses of the order of 30 mg kg^{-1} brought about anaemia, with death following several weeks later. Monkeys receiving 150 mg kg^{-1} developed abdominal pains, diarrhoea with blood and anaemia after 2 weeks of feeding (Wilson 1971). The inhalation by mice of pure mycophenolic acid was shown to cause an inflammatory response (Rand et al. 2005).

Schneweis et al. (2000) found mycophenolic acid present in 32% of grass and maize silage samples in Bavaria at levels ranging from 0.02 to 35 mg kg^{-1} and O'Brien et al. (2006) detected the toxin in both visually mouldy (up to 20 mg kg^{-1}) and visually non-mouldy baled grass silage samples (<0.0–5 mg kg^{-1}). Dzidic et al. (2006) showed that when sheep were fed silage contaminated with mycophenolic acid, high concentrations were measured in the liver (2.6 μg g^{-1}) and kidneys (3.6 μg g^{-1}), and these authors concluded that high mycophenolic acid concentrations exert potential immunosuppressive properties. In contrast,

Mohr et al. (2007) dosed sheep daily for 44 days with 300 mg of mycophenolic acid (5.4 mg kg^{-1} body weight) without any obvious impact on the sheep's general state of health. Notwithstanding the current toxin status of this metabolite, consumption of immunosuppressive compounds increases the risk of infectious diseases in livestock.

In 2003, certain member states of the EU, as well as the European Parliament, drew the attention of the EU Commission to the need to assess newly discovered substances for inclusion as undesirable in animal feeds. The possibility of mycophenolic acid entering the human food chain via meat or milk could be a potential risk, resulting in the European Commission's interest in the metabolite. It is the Commission's view that mycophenolic acid may represent an emerging risk, but the ability to qualify and quantify this risk is currently unavailable (European Commission 2003).

4.8 Conclusions

It can be concluded that the majority of baled silage fed to livestock in Ireland is contaminated with mould fungi, primarily *P. roqueforti* and to a lesser extent *P. paneum*. On that basis, it is possible that "cocktails" of mycotoxins are produced in many bales, a conclusion also arrived at by Mansfield and Kuldau (2007) for maize silage. Large quantities of fungal-contaminated silage is consumed by livestock, with adverse health and productivity consequences that are not fully quantified or understood. Current practices of bale-making, transportation and storage need to be modified on many farms to reduce the incidence of fungal spoilage and the accompanying risk to both livestock and human health.

References

Adams E, Todd G, Gibson W (1975) Long-term toxicity study of mycophenolic acid in rabbits. Toxicol Appl Pharm 34:509–512

Anderson MS, Dutton MF, Harding K (1979) Production and degradation of patulin by *Paecilomyces* species, a common contaminant of silage. J Sci Food Agric 30:229–232

Auerbach H (1996) Bases of the procedure for lowering the risk of an infestation of silage by *Penicillium roqueforti* and contamination with a mycotoxin of this mould fungus. Landbauforschung Völkenrode 168 (special edition):1–167

Auerbach H, Oldenburg E, Weissbach F (1998) Incidence of *Penicillium roqueforti* and roquefortine C in silages. J Sci Food Agric 76:565–572

Bentley R (2000) Mycophenolic acid: a one hundred year odyssey from antibiotic to immunosuppressant. Chem Rev 100:3801–3825

Boysen ME (1999) Molecular identification and quantification of the *Penicillium roqueforti* group. PhD thesis. Swedish University of Agricultural Sciences, Uppsala, Sweden

Boysen ME, Jacobsson KG, Schnürer J (2000) Molecular identification of species from the *Penicillium roqueforti* group associated with spoiled animal feed. Appl Environ Microbiol 66:1523–1526

Boysen M, Skouboe P, Frisvad J, Rossen L (1996) Reclassification of the *Penicillium roqueforti* group into three species on the basis of molecular genetic and biochemical profiles. Microbiology 142:541–549

Brady KC (2002) The biology of *Schizophyllum commune* with special reference to its growth on big-bale silage in Ireland. PhD Dissertation. University College Dublin, Ireland

Ciegler A (1977) Patulin. In: Rodricks JV, Hessletine CW, Mehlman MA (eds) Mycotoxins and human health. Pathotox, Park Forest South, IL, pp 609–624

Cole RJ, Kirksey JW, Dorner JW, Wilson DM, Johnson JC Jr, Johnson AN, Bedell DM, Springer JP, Chexal KK, Clardy JC, Cox RH (1977) Mycotoxins produced by *Aspergillus fumigatus* species isolated from molded silage. J Agric Food Chem 25:826–830

Cole RJ, Cox RH (1981) Handbook of toxic fungal metabolites. Academic, New York

Driehuis F, Spanjer MC, Scholten JM, te Giffel MC (2008a) Occurrence of mycotoxins in feedstuffs of dairy cows and estimation of total dietary intakes. J Dairy Sci 91:4261–4271

Driehuis F, Spanjer MC, Scholten JM, te Giffel MC (2008b) Occurrence of mycotoxins in maize, grass and wheat silage for dairy cattle in the Netherlands. Food Addit Contam B 1:41–50

Dutton MF, Westlake K, Anderson MS (1984) The interaction between additives, yeasts and patulin production in grass silage. Mycopathologia 87:29–33

Dzidic A, Prgomet C, Mohr A, Meyer K, Bauer J, Meyer HHD, Pfaffl MW (2006) Effects of mycophenolic acid on inosine monophosphate dehydrogenase I and II mRNA expression in white blood cells and various tissues in sheep. J Vet Med A 53:163–169

Engel G, Teuber M (1978) Simple aid for the identification of *Penicillium roqueforti* Thom. Eur J Appl Microbiol 6:107–111

Escoula L, Le Bars J, Larrieu G (1972) Studies on the mycoflora of silage. 1. Mycoflora of opened silos of graminaceous fodder. Ann Rech Vétér 3:469–481

Escoula L (1974) Toxinogenic moulds in ensilaged forage. I. Presence of patulin in silage cutting fronts. Ann Rech Vétér 5:423–432

European Commission (2003) Mycotoxins. In: Opinion of the scientific committee on animal nutrition on undesirable substances in feed (adopted on 20 February 2003). European Commission, Health and Consumer Protection Directorate-General, Brussels, pp 1–24

Fink-Gremmels J (2003) Mould and mycotoxins in silage. In: The 2nd World Mycotoxin Forum, Noordwijk, The Netherlands, 17–18 February 2003, pp 49–51

Forristal PD, O'Kiely P, Lenehan JJ, Mescall MP (2000) The influence of the number of layers of film cover and stretch level on silage preservation, gas composition and mould growth in big bale silage. In: Proceedings of the Agricultural Research Forum, Teagasc, Dublin, Ireland, 14–15 March 2000

Frisvad JC, Samson RA (1991) Filamentous fungi in foods and feeds: ecology, spoilage and mycotoxin production. In: Arora DK, Mukerji KG, Marth EH (eds) Handbook of applied mycology: foods and feeds, vol 3. Marcel Dekker, New York, pp 31–68

Frisvad JC, Smedsgaard J, Larsen TO, Samson RA (2004) Mycotoxins, drugs and other extrolites produced by species in *Penicillium* subgenus *Penicillium*. Stud Mycol 49:201–241

Hacking A, Rosser WR (1981) Patulin production by *Paecilomyces* species isolated from silage in the United Kingdom. J Sci Food Agric 32:620–623

Häggblom P (1990) Isolation of roquefortine C from feed grain. Appl Environ Microbiol 56:2924–2926

Häggblom P, Jacobsson K (1996) The presence of *Penicillium roqueforti* in silage and its implication for the onset of bovine mastitis. In: Research and development for animal health. National Veterinary Institute, Uppsala, Sweden, p 48

Jonsson A (1989) The role of yeasts and clostridia in silage deterioration: identification and ecology. PhD Disertation, Swedish University of Agricultural Sciences, Uppsala, Sweden

Kopp-Holtwiesche B, Rehm HJ (1990) Antimicrobial action of roquefortine. J Environ Pathol Toxicol Oncol 10:41–44
Le Bars J, Le Bars P (1989) Fungal species in corn silage. Mycotoxin hazards. Rec Méd Vétér 165:433–439
López-Díaz TM, Santos J, Otero A, García ML, Moreno B (1996) Some technological properties of *Penicillium roqueforti* strains isolated from a home-made blue cheese. Lett Appl Microbiol 23:5–8
Mansfield MA, Kuldau GA (2007) Microbiological and molecular determination of mycobiota in fresh and ensiled maize silage. Mycologia 99:269–278
McDonald P, Henderson AR, Heron SJE (1991) The biochemistry of silage. Cambrian, Aberystwyth, UK
McNamara KM, O'Kiely P, Whelan J, Forristal PD, Lenehan JJ (2002) Simulated bird damage to the plastic stretch-film surrounding baled silage and its effects on conservation characteristics. Ir J Agr Food Res 41:29–41
Mohr AI, Lorenz I., Baum B, Hewicker-Trautwein M, Pfaffl M, Džidić A, Meyer HHD, Bauer J, Meyer K (2007) Influence of oral application of mycophenolic acid on the clinical health status of sheep. J Vet Med A 54:76–81
Moreau C (1979) Moulds, toxins and food. Wiley, Chichester, UK
Morgavi DP, Boudra H, Jouany JP, Graviou D (2003) Prevention of patulin toxicity on rumen microbial fermentation by SH-containing reducing agents. J Agric Food Chem 51:6906–6910
Moulé Y, Jemmali M, Rousseau N (1976) Mechanism of the inhibition of transcription by PR toxin, a mycotoxin from *Penicillium roqueforti*. Chem Biol Interact 14:207–216
Moulé Y, Jemmali M, Darracq N (1978) Inhibition of protein synthesis by PR toxin, a mycotoxin from *Penicillium roqueforti*. FEBS Lett 88:341–344
Müller HM, Amend R (1997) Formation and disappearance of mycophenolic acid, patulin, penicillic acid and PR toxin in maize silage inoculated with *Penicillium roqueforti*. Arch Anim Nutr 50:213–225
Nielsen KF, Sumarah MW, Frisvad JC, Miller JD (2006) Production of metabolites from the *Penicillium roqueforti* complex. J Agric Food Chem 54:3756–3763
O'Brien M, Nielsen KF, O'Kiely P, Forristal PD, Fuller HT, Frisvad JC (2006) Mycotoxins and other secondary metabolites produced *in vitro* by *Penicillium paneum* Frisvad and *Penicillium roqueforti* Thom isolated from baled grass silage in Ireland. J Agric Food Chem 54:9268–9276
O'Brien M (2007) The mycobiota of baled grass silage in Ireland. PhD Dissertation. University College Dublin, Dublin, Ireland
O'Brien M, O'Kiely P, Forristal PD, Fuller HT (2007a) Visible fungal growth on baled silage during the winter feeding season in Ireland and silage characteristics associated with the occurrence of fungi. Anim Feed Sci Technol 139:234–256
O'Brien M, O'Kiely P, Forristal PD, Fuller HT (2007b) Quantification and identification of fungal propagules in well-managed baled grass silage and in normal on-farm produced bales. Anim Feed Sci Technol 132:283–297
O'Brien M, O'Kiely P, Forristal PD, Fuller HT (2008a) Fungal contamination of big-bale grass silage on Irish farms: predominant mould and yeast species and features of bales and silage. Grass Forage Sci 63:121–137
O'Brien M, Egan D, O'Kiely P, Forristal PD, Doohan FM, Fuller HT (2008b) Morphological and molecular characterisation of *Penicillium roqueforti* and *P. paneum* isolated from baled grass silage. Mycol Res 121:921–932
Ohmomo S, Kitamoto HK, Nakajima T (1994) Detection of roquefortines in *Penicillium roqueforti* isolated from moulded maize silage. J Sci Food Agric 64:211–215
O'Kiely P, Forristal D, Lenehan JJ, Fuller H, Brady K, Barlow M (1998) Characteristics of baled silage on Irish farms. Farm Food 8:29–32
O'Kiely P, McNamara K, Forristal D, Lenehan JJ (2000) Grass silage in Ireland. Farm Food 10:33–38
Pelhate J (1977) Maize silage: incidence of moulds during conservation. Folia Vet Lat 7:1–16

Petersson S (1998) Yeast/mold interactions during airtight storage of high-moisture feed grain. PhD Dissertation, Swedish University of Agricultural Sciences, Uppsala, Sweden

Pitt JI (1979) The genus *Penicillium* and its telemorphic states *Eupenicillium* and Talaromyces. Academic, London, UK

Pitt JI, Leistner L (1991) Toxigenic *Penicillium* species. In: Smith JE, Henderson RS (eds) Mycotoxins and animal foods. CRC, Boca Raton, Florida, pp 81–99

Pitt JI, Hocking AD (1999) Fungi and food spoilage. Aspen, Gaithersburg, MD

Pitt JI (2000) A laboratory guide to common *Penicillium* species. Food Science Australia, North Ryde, NSW, Australia

Polonelli L, Lauriola L, Morace G (1982) Preliminary studies on the carcinogenic effects of *Penicillium roqueforti* (PR toxin) on rats. Mycopathologia 78:125–127

Rand TG, Giles S, Flemming J, Miller JD, Puniani E (2005) Inflammatory and cytotoxic responses in mouse lungs exposed to purified toxins from buildings isolated *Penicillium brevicompactum* Dierckx and *P. chrysogenum* Thom. Toxicol Sci 87:213–222

Richard-Molard D, Cahagnier B, Poisson J (1980) Wet grain storage under modified atmosphere. In: Shejbal J (ed) Controlled atmosphere storage of grains. Elsevier Scientific, Amsterdam, The Netherlands, pp 173–182

Samson RA, Hoekstra ES, Frisvad JC, Filtenborg O (2002) Introduction to food- and airborne fungi. Centraalbureau voor Schimmelcultures, Utrecht, The Netherlands

Samson RA, Seifert KA, Kuijpers AFA, Houbraken JAMP, Frisvad JC (2004) Phylogenetic analysis of *Penicillium* subgenus *Penicillium* using partial b-tubulin sequences. Stud Mycol 49:175–200

Schneweis I, Meyer K, Hörmansdorfer S, Bauer J (2000) Mycophenolic acid in silage. Appl Environ Microbiol 66:3639–3641

Seglar B (1999) Ruminant disorders associated with pathogens found within ensiled forages. In: Smith RA (ed) Proceedings of the 32nd annual conference for the American Association of Bovine Practitioners, American Association of Bovine Practitioners, Nashville, TN, pp 54–60

Skaar I (1996) Mycological survey and characterisation of the mycobiota of big bale grass silage in Norway. PhD Dissertation. Norwegian College of Veterinary Medicine, Norway

Smith DF, Lynch GP (1972) *Aspergillus fumigatus* in samples of mouldy silage. J Dairy Sci 56:828–829

Still PE, Wei R, Smalley EB, Strong FM (1972) A mycotoxin from *Penicillium roqueforti* isolated from toxic cattle feed. Fed Proc 31:733

Sumarah MW, Miller JD, Blackwell, BA (2005) Isolation and metabolite production by *Penicillium roqueforti*, *P. paneum* and *P. crustosum* isolated in Canada. Mycopathologia 159:571–577

Sundberg M, Häggblom P (1999) Microbiological status in tower silos. In: Proceedings of the 12th International Silage Conference, Uppsala, Sweden, pp 329–330

Syrett R (1979) Moulds and mycotoxins in animal foodstuffs – reports of ADAS microbiologist. In: Proceedings of the 3rd Meeting on Mycotoxins in Animal Disease, Ministry of Agriculture, Fisheries and Food, Pinner, UK

Tsubaki K (1954) *Penicillium* isolated from toxic ensilage. Trans Mycol Soc J 1:6–7

Tüller G, Armbruster G, Wiedenmann S, Hänichen T, Schams D, Bauer J (1998) Occurrence of roquefortine in silage – toxicological relevance to sheep. J Anim Physiol Anim Nutr 80:246–249

Ueno Y, Kubota K, Ito T, Nakamura Y (1978) Mutagenicity of carcinogenic mycotoxins in *Salmonella typhimurium*. Cancer Res 38:536–542

Veselý D, Veselá D, Adámková A (1981) Occurrence of the mold *Penicillium roqueforti* producing PR-toxin in maize silage. Vet Med 26:109–115

Vivier D, Rivemale M, Reverbel JP, Ratomahenina R, Galzy P (1992) Some observations on the physiology of *Penicillium roqueforti* Thom and *Penicillium cyclopium* Westling. Lait 72:277–283

Wagener RE, Davis ND, Diener UL (1980) Penitrem A and roquefortine production by *Penicillium commune*. Appl Environ Microbiol 39:882–887

Weissbach F (1996) New developments in crop conservation. In: Proceedings of the 11th International Silage Conference. University of Wales, Aberystwyth, UK, pp 11–25

Whitlow LW, Hagler WM Jr (1997) Effects of mycotoxins on the animal: the producer's perspective. In: Proceedings form the silage: field to feedbunk. Northeast Regional Agricultural Engineering Service, Hershey, Pennsylvania, pp 222–232

Whitlow LW, Hagler WM Jr (2004) The ten most frequently-asked questions about mycotoxins, cattle and dairy food products. In: Lyons TP, Jacques KA (eds) Nutritional biotechnology in the feed and food industries. Proceedings of Alltech's 20th Annual Symposium. Nottingham University Press, Nottingham, UK, pp 231–253

Wilkinson JM, Bolsen KK (1996) Production of silage and hay in Europe and North America. In: Proceedings of the 11th international silage conference, 8–11 September, Aberystwyth, UK, pp 42–43

Wilson BJ (1971) Miscellaneous *Penicillium* toxins. In: Ciegler S, Kadis S, Ajl SJ (eds) Microbial toxins. Academic, New York, pp 459–521

Woolford MK (1984) The silage fermentation. Marcel Dekker, New York

Wyllie TD (1972) Mycotoxins and mycotoxicoses: an overview. In: Proceedings of the Symposium on Mycotoxins and Mycotoxicoses, University of Missouri-Columbia, USA

Chapter 5
Aflatoxin-like Gene Clusters and How They Evolved

Kenneth C. Ehrlich and Jiujiang Yu

5.1 Polyketide Biosynthesis Overview

Almost all filamentous fungi produce one or more polyketide-derived secondary metabolites (Demain and Fang 2000). The most common secondary metabolite is 1,4-dihydroxynaphthoquinone, the precursor of melanin. Others are precursors of fungal pigments or are toxic to plants, animals, or soil microorganisms. Their toxicity often is used defensively by the fungus against competing organisms, or offensively, by facilitating invasion of plants. The best studied of the polyketides are the bis-furans known collectively as aflatoxins (AF). In this chapter, we will discuss current knowledge concerning the AF gene cluster and how such clusters evolved.

Two types of gene clusters are found in known isolates of AF-producing fungi. These differ mainly in the order of the genes (Fig. 5.1). AF-producing species such as *A. flavus* and *A. parasiticus* have the gene cluster shown in Fig. 5.1a, while the newly discovered *Aspergillus* species, *ochraceoroseus*, *astellata* and *venezuelensis*, have a gene cluster similar to that of the ST-producing species *A. nidulans* (Fig. 5.1b) (Cary and Ehrlich 2006). Two genes in the cluster (*aflR* and *aflJ*) encode proteins involved in transcriptional activation of most of the other structural genes (Chang 2003, 2004).

There are 25 enzymatic steps required for AF biosynthesis (Yu et al. 2004) which begins with a short-chain fatty acid made by a specialized hexanoylCoA synthase. HexanoylCoA acts as a starter unit for a polyketide synthase (PKS) containing a unique hexanoyl acyl transferase acceptor domain. The PKS adds seven malonylCoA units after which the resulting polyketide is cyclized to an anthraquinone. Subsequent modification steps require predominantly oxidative enzymes. First, the hexanoyl moiety is converted to a bis-furan (the genotoxic portion of the molecule) and second the anthraquinone moiety of norsolorinic acid

K.C. Ehrlich (✉)
Southern Regional Research Center, ARS, USDA, 1100 R.E. Lee Blvd., New Orleans, LA 70124, USA
e-mail: ehrlich@srrc.ars.usda.gov

M. Rai and A. Varma (eds.), *Mycotoxins in Food, Feed and Bioweapons*,
DOI: 10.1007/978-3-642-00725-5_5,

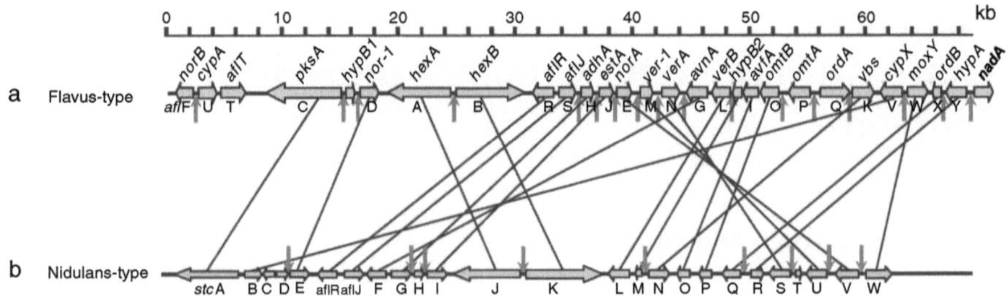

Fig. 5.1 Schematic representations of gene clusters in (**a**) section *Flavi* isolates, (**b**) section *Nidulantes* isolates. *Vertical arrows* show locations of AflR-binding sites. *Vertical lines* show gene orthologs in the section *Flavi* and section *Nidulantes* clusters. The *letter code* for naming AF-cluster genes is below the cluster, while the *descriptive gene names* are above the cluster

is converted to a xanthone. Finally, the xanthone is converted to AF. What is currently known about the AF biosynthetic steps is shown in Fig. 5.2.

5.2 *Aspergillus* Genomics

The entire genome has been sequenced for a number of fungal species including *A. flavus* and *A. oryzae*. *A. flavus* and *A. oryzae* each have eight chromosomes and similar genome sizes (36.8 Mb) (Galagan et al. 2005; Machida et al. 2005; Nierman

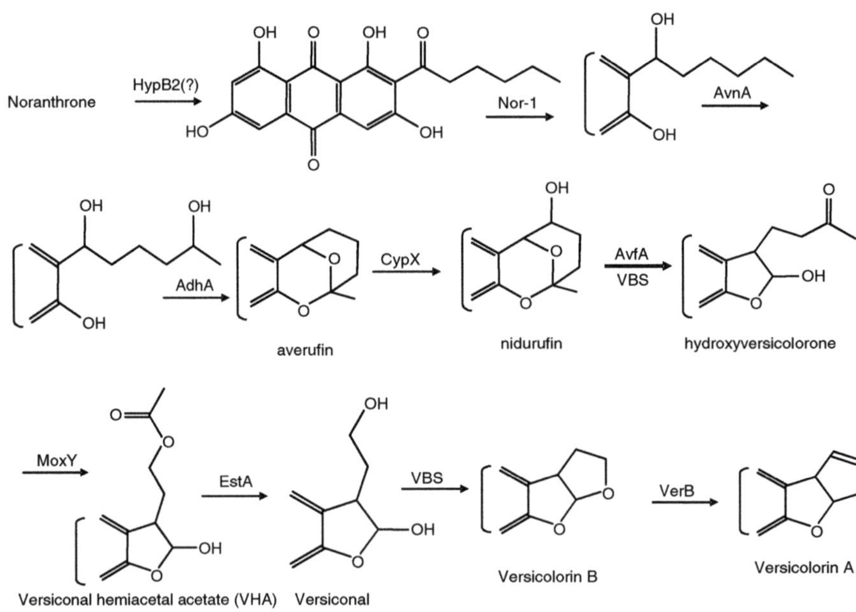

Fig. 5.2 AF biosynthesis, showing the known enzyme-catalyzed steps

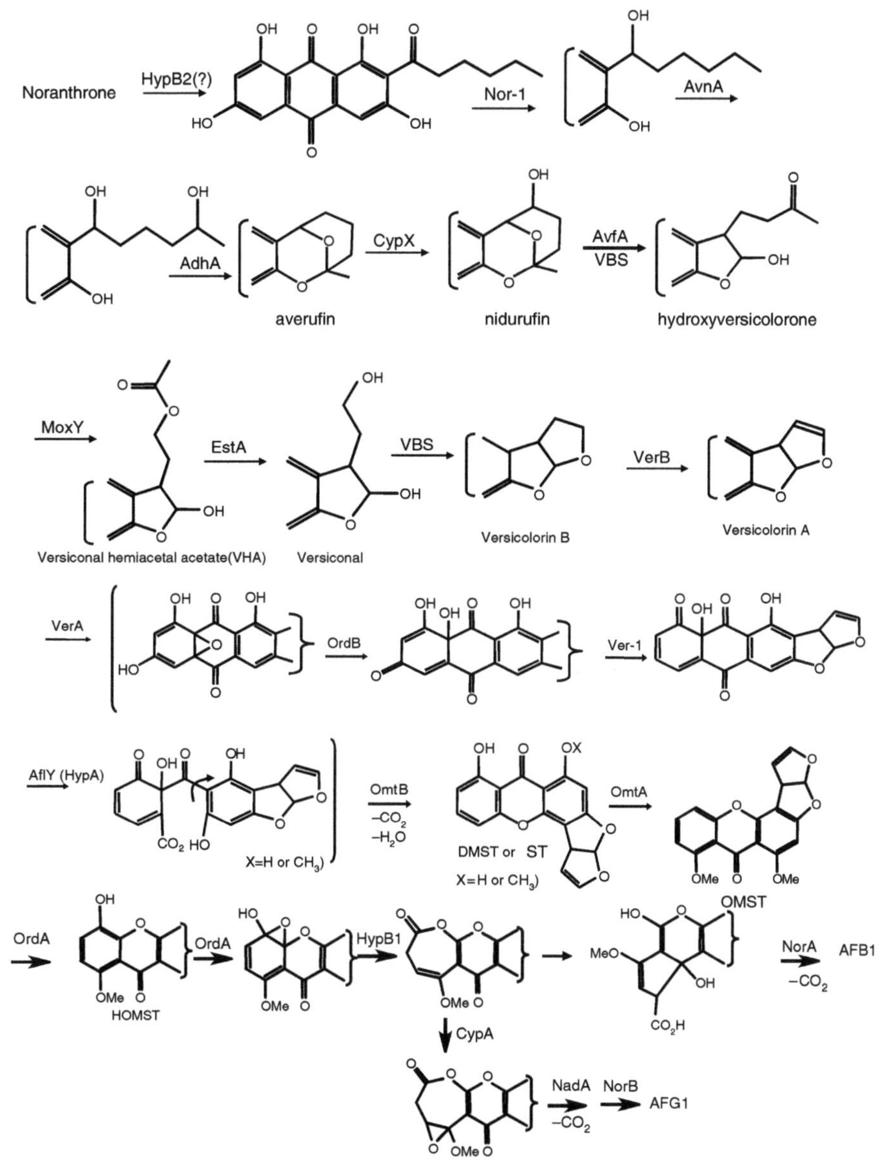

Fig. 5.2 (Continued)

et al. 2005; Payne et al. 2006; Pel et al. 2007; Ronning et al. 2005). The genome size for *A. nidulans* is 30.1 Mb (Galagan et al. 2005) and 29.4 Mb for *A. fumigatus* (Nierman et al. 2005). The *A. flavus* genome contains over 12,000 functional genes and is predicted to encode 34 PKSs, 24 non-ribosomal peptide synthases, 77 ABC-type transporters and more than 122 cytochrome P450 monooxygenases, many of which are presumed to be involved in production of secondary metabolites.

5.3 AF-Like Gene Clusters in Species Incapable of AF Production

As expected, *A. oryzae* has a gene cluster identical to that of *A. flavus* to which it is closely related. It falls into a family of non-aflatoxigenic *A. flavus* often found as isolates from agricultural soil (Cotty et al. 1994). The gene clusters of *A. oryzae* and other non-aflatoxigenic *A. flavus* have either large deletions of part or all of the genes in the cluster (Galagan et al. 2005; Lee et al. 2006) or have polymorphisms in genes that prevent formation of enzymes necessary for AF production (Chang et al. 2005; Ehrlich and Cotty 2004) (Kusumoto et al. 1998a, b; Machida et al. 2005).

Gene clusters that contain some AF/ST biosynthesis genes have been found in other related and unrelated fungi. For example, the causative agent of needle blight in a wide range of pine species, an ascomycete pathogen, *Mycosphaerella pini*, produces the red compound, dothistromin, a metabolite related to versicolorin A (Bradshaw and Zhang 2006). The genes in the dothistromin biosynthesis cluster are in three separated regions of the genome rather than in a single gene cluster. Other *Aspergillus* species, such as *A. terreus* and *A. fumigatus*, have some AF/ST biosynthesis genes. In these species the genes encoding HexA and HexB are in a separate cluster from the genes encoding the PKS and proteins similar to AflR and AflJ, OrdB and HypA. The latter two enzymes are involved in versicolorin A metabolism in AF/ST-producing *Aspergillus* species. *Coccidioides immitis*, a fungal species only distantly related to the *Aspergilli*, also contains a small cluster with a gene encoding an AF-like PKS together with orthologs of *aflR*, *aflJ*, *hexA* and *hexB*. This cluster was detected by searching the protein database with the SAT domain of PKSA. Phylogenetic analysis revealed that the PKSs known to accept hexanoyl-CoA as the starter unit have a different evolutionary history from the other PKSs.

5.4 Clusters in Species that Make B and G AFs Compared to Those that Produce Only B AFs

Comprehensive surveys of phylogenetic relationships in *Aspergillus* section *Flavi* suggest that the AFB and G-producing species, *A. nomius*, *bombycis*, and *parasiticus*, diverged prior to the divergence of the AFB-producing species, *A. pseudotamarii* and *A. flavus* (Ito et al. 2001; Peterson 1997; Peterson et al. 2001). Loss of the ability to produce G-type AFs involves different mutations of the genes encoding enzymes required for G-toxin formation. Loss of G AF production probably occurred at three separate times among extant *Aspergillus* lineages.

5.5 Origin of Genes in Secondary Metabolite Clusters

Phylogenetic evidence suggests that species with the Flavus-type gene cluster diverged from species with the Nidulans-type (Fig. 5.1) about 75 Mya (Berbee and Taylor 2001; Heckman et al. 2001; Kasuga et al. 2002). High identity between Nidulans- and Flavus-type cluster genes reflects purifying selection during divergence as measured by the ratio of rates of non-synonymous (Ka, nucleotide changes that result in amino acid changes) to synonymous nucleotide changes (Ks, nucleotide changes that result in conservation of amino acids) (Foster 2000). The Ka to Ks ratio was much lower for *ver-1* (ratio = 0.05) than for any of the 25 other Flavus-type cluster genes (ratio range = 0.2–1.04), suggesting that the most intense purifying selection was at this locus. Significantly higher Ka/Ks values in section *Flavi* compared to non-section *Flavi* species is evidence of adaptation and increased positive selection acting on genes in Flavus-type clusters with only partial gene sets. The gene *ver-1* encodes a reductase that is remarkably similar to the reductase required for melanin production. This reductase and an oxidoreductase encoded by another Flavus/Nidulans-type cluster gene, *aflX*, are similar to enzymes required for appressorium formation, a hardened mycelial structure needed for fungal virulence toward plants and insects (Henson et al. 1999; Inagaki et al. 2000). This relationship suggests that AF production and other processes geared to offensive strategies for food acquisition by fungi arose from similar genetic roots. Therefore, while extant species of AF-producing Aspergilli are entirely saprophytic, they must have had, at one time, a more aggressively invasive lifestyle.

In the Flavus-type cluster there is considerable evidence that gene duplication is involved in cluster evolution. Possible duplicated genes encode the following proteins (amino acid identity is in brackets): the cytochrome P450 monooxygenases, VerB and AvnA (39%), NorA and NorB (50%), HypB1 and HypB2 (42%), OrdB and AvfA (30%). Additional evidence for gene duplication is that after knockout of certain cluster genes, the resulting mutants still produce AF in addition to precursor intermediates. This "leaky" phenotype is found with the following knockout mutants: *nor-1*, *adhA*, *hypB1* or *hypB2*, *norA*, *norB*, and *ordB*. This suggests that gene homologs inside or outside of the cluster encode proteins that are able to substitute for the catalytic function of these Nidulans/Flavus-type cluster proteins.

Evidence was found that cluster formation involved simultaneous movement of multiple genes as modules to the cluster region (Carbone et al. 2007). At least seven gene modules (*hexA/hexB*, *aflR/aflJ*, *aflX/aflY*, *norA/norB*, *aflT/ordA*, *pksA/moxY*, and *avnA/verB*) were inferred by phylogenetic comparison of AF gene cluster duplicates across *Aspergillus* species whose entire genome sequences are available in databases. Most of the modules make sense from an evolutionary viewpoint because there is selective pressure for retention of genes that encode proteins required for overlapping functions. Although the modules are preserved, differences in cluster gene organization could arise by recombination in organisms that have maintained a sexual state, such as *A. nidulans*. Module organization is strongest in those gene pairs in which the encoded proteins act together for a particular process. For example, *hexA* and *hexB* encode subunits of the fatty acid synthase that together catalyze formation of the hexanoylCoA starter unit, and *aflR* and *aflJ* encode interacting regulatory proteins. Although *aflT* and *ordA* share no catalytic similarities, their inheritance as a module may be a function of the requirement that the OrdA-catalyzed metabolite, namely AF, would be self-toxic if the transporter encoded by *aflT* were not available.

The functions of only a few of the PKSs found in the genome are known and it is not certain that all are transcribed. Free-living organisms, such as fungi, must be able to adapt to many different ecological niches. Polyketide biosynthesis is likely to be essential to such adaptation. In this regard, fungi form symbiotic as well as competitive relationships with both insects and plants (Hoffmeister and Martin 2003; Owen and Hundley 2004). Therefore, the need to flourish and propagate must have required a diverse array of both pigmented and offensive metabolites. The large number of PKSs, whether functional or not, shows that the potential to produce bioactive metabolites is an essential part of fungal adaptive success. Phylogenetic analysis of the PKS genes shows that they evolved by the expected adaptive evolutionary processes, gene duplication and gene loss (O'Donnell et al. 1998, 2000).

5.6 Why are Gene Clusters Maintained?

Most fungal gene clusters represent regulatory islands within a region of chromatin that is at least partially heterochromatic (inactive – not able to encode proteins). It has been shown that heterologous genes inserted into the Flavus-type cluster are expressed under the same induction conditions as genes that are components of the cluster, whereas cluster genes inserted at loci removed from the cluster are severely down-regulated (Chiou et al. 2002; Keller et al. 2006). Many secondary metabolite clusters are in subtelomeric regions of the chromosome (regions generally rich in heterochromatin) (Galagan et al. 2005; Wong and Wolfe 2005). Subtelomeric regions are particularly prone to chromatin modification. They tend to have an abundance of duplicated genes as well as retrotransposons and retroposon remnants. Such elements indicate a past history of DNA deposition from other locations in the genome (Robyr et al. 2002). Subtelomeric regions contain genes which permit growth under stress conditions and on unusual nitrogen and carbon sources, but which normally are silenced (Wong and Wolfe 2005). The small *aflR*/*aflJ* cluster in *C. immitis*/*posadasii* provides evidence that cluster formation may be aided by retrotransposon-mediated gene movement, since this cluster is associated with retrotransposon elements.

A methyltransferase, LaeA, was identified that appears to specifically regulate transcription of genes in subtelomeric secondary metabolite clusters (Bok et al. 2006). It was postulated that LaeA methylates histones to turn on expression of genes within such clusters. LaeA is conserved in all fungi whose genome sequences have been obtained (Bouhired et al. 2007). LaeA's specific effect on activity of genes in secondary metabolite clusters may be mediated by variant forms of histone in these subtelomeric regions. In yeast a gene cluster that allows the yeast to grow on allantoin as the sole nitrogen source is located in a subtelomeric chromosomal region (Wong and Wolfe 2005). The chromatin associated with the allantoin gene cluster contains a variant of histone H2A (H2A.Z). This may allow the remodeling necessary for activation of genes in the cluster. LaeA and other development-specific transcription factors may also recognize a similar variant histone in fungi, and thereby activate secondary metabolite gene clusters in subtelomeric regions of the chromosome.

Induction of the active state is mediated by signaling processes that, to date, are still only partly understood, but which probably involve G-protein-mediated cell signaling (Shimizu et al. 2003; Tag et al. 2000). A G-protein signaling system has been identified that induces both secondary metabolite biosynthesis and sclerotial development (see accompanying paper by P.-K. Chang). Part of this process requires cAMP-dependent activation of protein kinase A and other kinases in the signaling cascade. Other types of signaling systems also affect AF biosynthesis. Light signaling activates some developmental processes (including AF biosynthesis) and is mediated by the transcriptional regulator VeA. Stress signaling also affects AF biosynthesis and sclerotial formation. A heat shock chaperonin, HSP30, may be involved in mediating the stress response (K. Ehrlich, unpublished results).

Expression of genes required for early stages of AF biosynthesis is modulated differently from expression of genes required for the later modification steps (Ehrlich et al. 1999; Roze et al. 2007). Roze, et al found that gene expression in the Flavus-type cluster begins at the *pksA-hypB1* locus and proceeds outward in both directions. This conclusion is drawn from the pattern of histone H_4 acetylation seen during glucose induction of AF biosynthesis in *A. parasiticus* and roughly follows the order of gene utilization in the biosynthesis steps. The different gene order in the Nidulans-type cluster may reflect utilization of global transcription factors in partnership with LaeA for recruitment of specialized histone deacetylases to render the chromatin active in the otherwise heterochromatic DNA.

After cluster formation, gene losses have occurred. The Flavus-type cluster of section *Flavi* species is missing several Nidulans-type cluster genes. These genes, *stcC, stcD, stcT*, lack promoter AflR-binding sites, and therefore are probably not regulated by AflR during ST biosynthesis. They encode proteins with unknown functions. A homolog to *stcC* is found in the dothistromin biosynthetic cluster. Such conservation suggests that this gene encodes a protein that is either no longer needed or whose function is fulfilled by a protein outside the cluster in section *Flavi* isolates. Another gene loss appears to have occurred relatively recently, perhaps within the last 1 million years. This involves a deletion of the promoter region of the cytochrome P450 monooxygenase, *cypA*, a gene necessary for formation of AFG1. Two types of deletion are found in *A. flavus*. A 1.2 kb deletion is found in *A. oryzae* and S-sclerotial morphotype *A. flavus* while a 0.8 kb deletion is found in most L morphotype *A. flavus*. It is possible that loss of G-toxin production was an adaptive change that permitted variant *A. flavus* to grow more readily on living plant tissues as opposed to being predominantly resident in the soil.

To explain retention of the genes for AF/ST/dothistromin synthesis for greater than 125 million years we hypothesize that these polyketides enabled the fungi to survive and prosper as free-living organisms in diverse environments (Bradshaw et al. 2002; Gomez and Nosanchuk 2003). This speculation may be broadened to include retention of all functional gene clusters in fungi. We assume that the reason for maintaining such gene clusters was that the precursor metabolites were critical for early fungal adaptation. Primordial fungi probably had a basal polyketide biosynthesis cluster that consisted of genes encoding only a PKS and a few enzymes capable of stabilizing the nascent polyketide. Such a basal cluster could have allowed fungi to synthesize aromatic, colorful molecules that, like plant pigments, may have attracted insects to foster spore dispersal. As mentioned above, colorful naphthoquinones and anthraquinones are found in many types of fungi, including genera that preceded divergence of the *Aspergilli*. Therefore, at this stage of evolution, rather than being toxic to insects, they may have been attractants. As the fungi dispersed into less hospitable niches, a need may have arisen for them to be competitive with associated microflora, or they may have benefited by acquiring the ability to become virulent to plants. Such competitive conditions may have selected for the duplication and adaptation of genes to allow elaboration of an even greater variety of metabolites. There is increasing evidence that AF/ST biosynthesis and conidiospore development are regulated by a shared signal transduction mechanism (Calvo et al. 2002; Hicks et al. 1997; Keller and Hohn 1997; Wilkinson et al. 2004). This finding is consistent with our suggestion that dispersal and/or maintenance of conidiospores are aided by production of attractant and protectant organic molecules.

References

Berbee ML, Taylor JW (2001) Fungal molecular evolution: gene trees and geologic time. In: McLaughlin DJ, McLaughlin E, Lemke PA (eds) The mycota. Springer, Berlin, pp 229–246

Bok JW, Noordermeer D, Kale SP, Keller NP (2006) Secondary metabolic gene cluster silencing in *Aspergillus nidulans*. Mol Microbiol 61:1636–1645

Bouhired S, Weber M, Kempf-Sontag A, Keller NP, Hoffmeister D (2007) Accurate prediction of the *Aspergillus nidulans* terrequinone gene cluster boundaries using the transcriptional regulator LaeA. Fungal Genet Biol 44:1134–1145

Bradshaw RE, Bhatnagar D, Ganley RJ, Gillman CJ, Monahan BJ, Seconi JM (2002) *Dothistroma pini*, a forest pathogen, contains homologs of aflatoxin biosynthetic pathway genes. Appl Environ Microbiol 68:2885–2892

Bradshaw RE, Zhang S (2006) Biosynthesis of dothistromin. Mycopathologia 162:201–213

Calvo AM, Wilson RA, Bok JW, Keller NP (2002) Relationship between secondary metabolism and fungal development. Microbiol Mol Biol Rev 66:447–459

Carbone I, Ramirez-Prado JH, Jakobek JL, Horn BW (2007) Gene duplication, modularity and adaptation in the evolution of the aflatoxin gene cluster. BMC Evol Biol 7:111, doi:110.1186/1471–2148–1187–1111

Cary J, Ehrlich K (2006) Aflatoxigenicity in *Aspergillus*: molecular genetics, phylogenetic relationships and evolutionary implications. Mycopathologia 162:167–177

Chang PK (2003) The *Aspergillus parasiticus* protein AFLJ interacts with the aflatoxin pathway-specific regulator AFLR. Mol Genet Genomics 268:711–719

Chang PK (2004) Lack of interaction between AFLR and AFLJ contributes to nonaflatoxigenicity of *Aspergillus sojae*. J Biotechnol 107:245–253

Chang PK, Horn BW, Dorner JW (2005) Sequence breakpoints in the aflatoxin biosynthesis gene cluster and flanking regions in nonaflatoxigenic *Aspergillus flavus* isolates. Fungal Genet Biol 42:914–923

Chiou CH, Miller M, Wilson DL, Trail F, Linz JE (2002) Chromosomal location plays a role in regulation of aflatoxin gene expression in *Aspergillus parasiticus*. Appl Environ Microbiol 68:306–315

Cotty PJ, Bayman DS, Egel DS, Elias KS (1994) Agriculture, aflatoxins and *Aspergillus*. In: Powell K (ed) The genus *Aspergillus*. Plenum, New York, pp 1–27

Demain AL, Fang A (2000) The natural functions of secondary metabolites. Adv Biochem Eng Biotechnol 69:1–39

Ehrlich KC, Cary JW, Montalbano BG (1999) Characterization of promoter elements involved in the expression of early and late aflatoxin pathway genes. In: Twentieth Fungal Genetics Conference, vol 46. Fungal Genetics Newsletter, Asilomar Conference Center, Pacific Grove, CA, p 45 (Abst 28)

Ehrlich KC, Cotty PJ (2004) An isolate of *Aspergillus flavus* used to reduce aflatoxin contamination in cottonseed has a defective polyketide synthase. Appl Microbiol Biotechnol 65:473–478

Foster PL (2000) Adaptive mutation: implications for evolution. Bioessays 22:1067–1074

Galagan JE, Calvo SE, Cuomo C, Ma LJ, Wortman JR, Batzoglou S, Lee SI, Basturkmen M, Spevak CC, Clutterbuck J, Kapitonov V, Jurka J, Scazzocchio C, Farman M, Butler J, Purcell S, Harris S, Braus GH, Draht O, Busch S, D'Enfert C, Bouchier C, Goldman GH, Bell-Pedersen D, Griffiths-Jones S, Doonan JH, Yu J, Vienken K, Pain A, Freitag M, Selker EU, Archer DB, Penalva MA, Oakley BR, Momany M, Tanaka T, Kumagai T, Asai K, Machida M, Nierman WC, Denning DW, Caddick M, Hynes M, Paoletti M, Fischer R, Miller B, Dyer P, Sachs MS, Osmani SA, Birren BW (2005) Sequencing of *Aspergillus nidulans* and comparative analysis with *A. fumigatus* and *A. oryzae*. Nature 438:1105–1115

Gomez BL, Nosanchuk JD (2003) Melanin and fungi. Curr Opin Infect Dis 16:91–96

Heckman DS, Geiser DM, Eidell BR, Stauffer RL, Kardos NL, Hedges SB (2001) Molecular evidence for the early colonization of land by fungi and plants. Science 293:1129–1133

Henson JM, Butler MJ, Day AW (1999) The dark side of the mycelium: melanins of phytopathogenic fungi. Annu Rev Phytopathol 37:447–471

Hicks JK, Yu JH, Keller NP, Adams TH (1997) *Aspergillus* sporulation and mycotoxin production both require inactivation of the FadA G alpha protein-dependent signaling pathway. EMBO J 16:4916–4923

Hoffmeister M, Martin W (2003) Interspecific evolution: microbial symbiosis, endosymbiosis and gene transfer. Environ Microbiol 5:641–649

Inagaki A, Takano Y, Kubo Y, Mise K, Furusawa A (2000) Construction of an equalized cDNA library from *Colletotrichum lagenarium* and its application to the isolation of differentially expressed genes. Can J Biochem 46:150–158

Ito Y, Peterson SW, Wicklow DT, Goto T (2001) *Aspergillus pseudotamarii,* a new aflatoxin producing species in *Aspergillus* section *Flavi*. Mycol Res 105:233–239

Kasuga T, White TJ, Taylor JW (2002) Estimation of nucleotide substitution rates in eurotiomycete fungi. Mol Biol Evol 19:2318–2324

Keller N, Bok J, Chung D, Perrin RM, Keats Shwab E (2006) LaeA, a global regulator of *Aspergillus* toxins. Med Mycol 44(Suppl):83–85

Keller NP, Hohn TM (1997) Metabolic pathway gene clusters in filamentous fungi. Fungal Genet Biol 21:17–29

Kusumoto K, Yabe K, Nogata Y, Ohta H (1998a) Transcript of a homolog of *aflR*, a regulatory gene for aflatoxin synthesis in *Aspergillus parasiticus*, was not detected in *Aspergillus oryzae* strains. FEMS Microbiol Lett 169:303–307

Kusumoto KI, Yabe K, Nogata Y, Ohta H (1998b) *Aspergillus oryzae* with and without a homolog of aflatoxin biosynthetic gene *ver-1*. Appl Microbiol Biotechnol 50:98–104

Lee YH, Tominaga M, Hayashi R, Sakamoto K, Yamada O, Akita O (2006) *Aspergillus oryzae* strains with a large deletion of the aflatoxin biosynthetic homologous gene cluster differentiated by chromosomal breakage. Appl Microbiol Biotechnol 72:339–345

Machida M, Asai K, Sano M, Tanaka T, Kumagai T, Terai G, Kusumoto K, Arima T, Akita O, Kashiwagi Y, Abe K, Gomi K, Horiuchi H, Kitamoto K, Kobayashi T, Takeuchi M, Denning DW, Galagan JE, Nierman WC, Yu J, Archer DB, Bennett JW, Bhatnagar D, Cleveland TE, Fedorova ND, Gotoh O, Horikawa H, Hosoyama A, Ichinomiya M, Igarashi R, Iwashita K, Juvvadi PR, Kato M, Kato Y, Kin T, Kokubun A, Maeda H, Maeyama N, Maruyama J, Nagasaki H, Nakajima T, Oda K, Okada K, Paulsen I, Sakamoto K, Sawano T, Takahashi M, Takase K, Terabayashi Y, Wortman JR, Yamada O, Yamagata Y, Anazawa H, Hata Y, Koide Y, Komori T, Koyama Y, Minetoki T, Suharnan S, Tanaka A, Isono K, Kuhara S, Ogasawara N, Kikuchi H (2005) Genome sequencing and analysis of *Aspergillus oryzae*. Nature 438:1157–1161

Nierman WC, Pain A, Anderson MJ, Wortman JR, Kim HS, Arroyo J, Berriman M, Abe K, Archer DB, Bermejo C, Bennett J, Bowyer P, Chen D, Collins M, Coulsen R, Davies R, Dyer PS, Farman M, Fedorova N, Feldblyum TV, Fischer R, Fosker N, Fraser A, Garcia JL, Garcia MJ, Goble A, Goldman GH, Gomi K, Griffith-Jones S, Gwilliam R, Haas B, Haas H, Harris D, Horiuchi H, Huang J, Humphray S, Jimenez J, Keller N, Khouri H, Kitamoto K, Kobayashi T, Konzack S, Kulkarni R, Kumagai T, Lafon A, Latge JP, Li W, Lord A, Lu C, Majoros WH, May GS, Miller BL, Mohamoud Y, Molina M, Monod M, Mouyna I, Mulligan S, Murphy L, O'Neil S, Paulsen I, Penalva MA, Pertea M, Price C, Pritchard BL, Quail MA, Rabbinowitsch E, Rawlins N, Rajandream MA, Reichard U, Renauld H, Robson GD, Rodriguez de Cordoba S, Rodriguez-Pena JM, Ronning CM, Rutter S, Salzberg SL, Sanchez M, Sanchez-Ferrero JC, Saunders D, Seeger K, Squares R, Squares S, Takeuchi M, Tekaia F, Turner G, Vazquez de Aldana CR, Weidman J, White O, Woodward J, Yu JH, Fraser C, Galagan JE, Asai K, Machida M, Hall N, Barrell B, Denning DW (2005) Genomic sequence of the pathogenic and allergenic filamentous fungus *Aspergillus fumigatus*. Nature 438:1151–1156.

O'Donnell K, Cigelnik E, Casper HH (1998) Molecular phylogenetic, morphological, and mycotoxin data support re-identification of the Quorn mycoprotein fungus as *Fusarium venenatum*. Fungal Genet Biol 23:57–67

O'Donnell KO, Kistler HC, Tacke BK, Casper HH (2000) Gene genealogies reveal global phylogeographical structure and reproductive isolation among lineages of *Fusarium graminearum*, the fungus causing wheat scab. Proc Natl Acad Sci U S A 97:7905–7910

Owen NL, Hundley N (2004) Endophytes—the chemical synthesizers inside plants. Sci Prog 87:79–99

Payne GA, Nierman WC, Wortman JR, Pritchard BL, Brown D, Dean RA, Bhatnagar D, Cleveland TE, Machida M, Yu J (2006) Whole genome comparison of *Aspergillus flavus* and *A. oryzae*. Med Mycol 44(Suppl):9–11

Pel HJ, de Winde JH, Archer DB, Dyer PS, Hofmann G, Schaap PJ, Turner G, de Vries RP, Albang R, Albermann K, Andersen MR, Bendtsen JD, Benen JA, van den Berg M, Breestraat S, Caddick MX, Contreras R, Cornell M, Coutinho PM, Danchin EG, Debets AJ, Dekker P, van Dijck PW, van Dijk A, Dijkhuizen L, Driessen AJ, d'Enfert C, Geysens S, Goosen C, Groot GS, de Groot PW, Guillemette T, Henrissat B, Herweijer M, van den Hombergh JP, van den Hondel CA, van der Heijden RT, van der Kaaij RM, Klis FM, Kools HJ, Kubicek CP, van Kuyk PA, Lauber J, Lu X, van der Maarel MJ, Meulenberg R, Menke H, Mortimer MA, Nielsen J, Oliver SG, Olsthoorn M, Pal K, van Peij NN, Ram AF, Rinas U, Roubos JA, Sagt CM, Schmoll M, Sun J, Ussery D, Varga J, Vervecken W, van de Vondervoort PJ, Wedler H, Wosten HA, Zeng AP, van Ooyen AJ, Visser J, Stam H (2007) Genome

sequencing and analysis of the versatile cell factory *Aspergillus niger* CBS 513.88. Nat Biotechnol 25:221–231
Peterson SW (1997) Phylogenetic relationships in *Aspergillus* based on rDNA sequence analysis. In: Samson RA, Pitt JI (eds) Classification of *Penicillium* and *Aspergillus*: integration of modern taxonomic methods. Harwood, Reading, UK, pp 323–356
Peterson SW, Ito Y, Horn BW, Goto T (2001) *Aspergillus bombycis*, a new aflatoxigenic species and genetic variation in its sibling species, *A. nomius*. Mycologia 93:689–703
Robyr D, Suka Y, Xenarios I, Kurdistani SK, Wang A, Suka N, Grunstein M (2002) Microarray deacetylation maps determine genome-wide functions for yeast histone deacetylases. Cell 109:437–446
Ronning CM, Fedorova ND, Bowyer P, Coulson R, Goldman G, Kim HS, Turner G, Wortman JR, Yu J, Anderson MJ, Denning DW, Nierman WC (2005) Genomics of *Aspergillus fumigatus*. Rev Iberoam Micol 22:223–228
Roze LV, Arthur AE, Hong SY, Chanda A, Linz JE (2007) The initiation and pattern of spread of histone H4 acetylation parallel the order of transcriptional activation of genes in the aflatoxin cluster. Mol Microbiol 66:713–726
Shimizu K, Hicks JK, Huang TP, Keller NP (2003) Pka, Ras and RGS protein interactions regulate activity of AflR, a Zn(II)2Cys6 transcription factor in *Aspergillus nidulans*. Genetics 165:1095–1104
Tag A, Hicks J, Garifullina G, Ake C, Phillips TD, Beremand M, Keller N (2000) G-protein signalling mediates differential production of toxic secondary metabolites. Mol Microbiol 38:658–665
Wilkinson H, Ramaswamy A, Sim SC, Keller NP (2004) Increased conidiation associated with progression along the sterigmatocystin biosynthetic pathway. Mycologia 96:1190–1198
Wong S, Wolfe KH (2005) Birth of a metabolic gene cluster in yeast by adaptive gene relocation. Nat Genet 37:777–782
Yu J, Chang PK, Ehrlich KC, Cary JW, Bhatnagar D, Cleveland TE, Payne GA, Linz JE, Woloshuk CP, Bennett JW (2004) Clustered pathway genes in aflatoxin biosynthesis. Appl Environ Microbiol 70:1253–1262

Chapter 6
Aflatoxin Biosynthesis and Sclerotial Development in *Aspergillus flavus* and *Aspergillus parasiticus*

Perng-Kuang Chang

6.1 Introduction

Aflatoxins are a family of fungal secondary metabolites. They are produced by species in the genus *Aspergillus*. The commonly recognized producers of aflatoxins include *A. flavus*, *A. parasiticus*, *A. nomius*, *A. tamarii*, *A. pseudotamarii*, *A. bombycis*, and *A. ochraceoroseus* (Cary et al. 2005). Aflatoxin contamination of agricultural commodities can arise from field conditions conducive to fungal growth before harvest as well as from improper storage of foods and feeds. *A. flavus* and *A. parasiticus* are the most important aflatoxigenic species. While both can produce aflatoxins, the majority of *A. flavus* isolates produce aflatoxins B_1 and B_2 although some strains have been reported to also produce aflatoxins G_1 and G_2. In contrast, *A. parasiticus* produces all four types of the aflatoxins.

The toxic, carcinogenic, mutagenic, and teratogenic effects of aflatoxin B_1 on humans and animals have been studied in great detail. Aflatoxin B_1 has been linked to a specific G to T transversion in the codon 249 in the p53 tumor suppressor gene in primary human hepatocellular carcinoma. The International Agency for Research on Cancer (IARC) has designated aflatoxin as a human liver carcinogen (Wogan 2000). To minimize potential human exposure to aflatoxins, maximum levels of aflatoxins in many commodities have been set (Wu 2004). Forty-eight countries have specific regulations limiting total aflatoxins in foodstuffs, a further 21 having regulations for aflatoxins in feedstuffs (FAO 1995). Regulatory guidelines of the U.S. Food and Drug Administration (FDA) prevent the sale of commodities if contamination by aflatoxins exceeds allowed levels. The FDA has set limits of 20 ppb total aflatoxins for interstate commerce of food and feedstuff and 0.5 ppb aflatoxin M_1 in milk. The European Commission has set the limits on

P-K. Chang
Southern Regional Research Center, Agricultural Research Service, US Department of Agriculture, 1100 Robert E. Lee Boulevard, New Orleans, LA 70124, USA
e-mail: perngkuang.chang@ars.usda.gov

M. Rai and A. Varma (eds.), *Mycotoxins in Food, Feed and Bioweapons*,
DOI: 10.1007/978-3-642-00725-5_6,

groundnuts subject to further processing at 15 ppb for total aflatoxins and 8 ppb for aflatoxin B_1, and for nuts and dried fruits subject to further processing at 10 ppb for total aflatoxins and 5 ppb for aflatoxin B_1. The aflatoxin standards for cereals, dried fruits, and nuts intended for direct human consumption are even more stringent, and the limit for total aflatoxins is 4 ppb and 2 ppb for aflatoxin B_1 (Otsuki et al. 2001).

The aflatoxin biosynthetic pathway has been well-characterized in *A. flavus* and *A. parasiticus*, and 15 stable precursors have been identified (Yabe and Nakajima 2004; Yu et al. 2004). Within the last decade, significant advances have been made in understanding the biochemistry, genetics, and gene regulation of aflatoxin biosynthesis. Many scientists have used aflatoxin biosynthesis as a model for studying fungal secondary metabolism. A number of studies have identified a genetic connection between aflatoxin biosynthesis and fungal development such as conidiation and sclerotial biogenesis (Guzman-de-Pena and Ruiz-Herrera 1997; Hicks et al. 1997; Chang et al. 2002; Bok and Keller 2004; Calvo et al. 2004). The recently available *A. flavus* genomic resources, expressed sequence tag (EST) and whole genome sequence (Payne et al. 2006; Yu et al. 2007), are having an increasing impact on aflatoxin biosynthesis and related research.

6.2 Aflatoxin Biosynthesis Gene Cluster

There are more than 25 enzymatic steps involved in aflatoxin biosynthesis and most of the genes encoding the enzymes have been cloned and characterized (Minto and Townsend 1997; Yabe and Nakajima 2004). Restriction mapping of genomic cosmid and lambda clones and chromosomal walking showed that genes involved in aflatoxin biosynthesis are clustered in the *A. parasiticus* genome spanning approximately 70 kb (Yu et al. 2004). Sequencing of the aflatoxin gene clusters of several related *Aspergillus* species including *A. flavus* isolates that produce large (L) or small (S) sclerotia, an *A. nomius* isolate, and an unnamed taxon closely related to *A. flavus* and *A. parasiticus* (Ehrlich et al. 2005b), has confirmed that the order of aflatoxin biosynthesis genes in these species is the same as that for *A. parasiticus*. A search of the 5X draft genome sequence of *A. flavus* NRRL3357 (http://www.aspergillusflavus.org/genomics/) shows that its aflatoxin gene cluster is located on the 2.3 Mb supercontig #1047283863273. Compared to the genome organization of the closely related *A. oryzae* RIB40 (http://www.bio.nite.go.jp/dogan/MicroTop?GENOME_ID = ao), *A. flavus* aflatoxin gene cluster is probably located on chromosome 3 and 87 kb from the chromosomal end although the telomeric sequence has not been identified.

In the last few years, the functions of more clustered genes involved in aflatoxin biosynthesis have been elucidated. The *cypA* gene is required for G aflatoxin formation (Ehrlich et al. 2004). The *aflY* (*hypA*) gene encodes an oxidase necessary for formation of the xanthone ring of demethylsterigmatocystin

(Ehrlich et al. 2005a). The *cypX* gene encodes a monooxygenase which converts averufin to hydroxyversicolorone (HVN) and *moxY* encodes HVN monooxygenase which catalyzes HVN to versiconal hemiacetal acetate as well as from versicolorone to versiconol acetate. The *vbs* gene encodes a cyclase which converts versiconal to versicolorin B and 5′-oxoaverantin to averufin (Sakuno et al. 2005). The *aflX* (*ordB*) gene encodes an oxidoreductase which is involved in the conversion of versicolorin A to demethylsterigmatocystin (Cary et al. 2006). Three more genes (*hypB1*, *hypB2*, and *nadA*) also have been identified in the aflatoxin gene cluster, but their functions, like *norA* and *norB*, have yet to be determined.

6.3 Instability of Aflatoxin Gene Cluster in *A. flavus*

Populations of *A. flavus* in many parts of the world vary considerably in their ability to produce aflatoxins. S-strain *A. flavus* isolates, like *A. parasiticus* and *A. nomius*, are very stable in aflatoxin production. In contrast, a significant portion of L-strain *A. flavus* isolates do not produce aflatoxins (Horn and Dorner 1999; Vaamonde et al. 2003; Pildain et al. 2004; Takahashi et al. 2004). Compared to *A. parasiticus*, *A. flavus* populations are genetically diverse and phenotypically variable. *A. parasiticus* is generally less widespread than *A. flavus* and primarily infects peanuts but not the aerial crops such as corn and cotton. *A. parasiticus* also is more adapted to soil environments and less dependent on crop infection for survival (Horn 2007). A detailed characterization of the aflatoxin gene cluster in *A. flavus* isolates collected from New Mexico to Virginia in the southern United States has revealed some interesting findings. A major portion of the *A. flavus* isolates that do not produce aflatoxins have the aflatoxin gene cluster either partially or entirely deleted, and the deletions can be characterized into seven distinct patterns (Chang et al. 2005). The subtelomeric chromosomal region including the aflatoxin gene cluster may be more susceptible to certain selective forces, as reported for *Saccharomyces cerevisiae* (Liti and Louis 2005). Genetic drift may be a driving force that is responsible for the loss of the partial or entire aflatoxin gene cluster in *A. flavus* when aflatoxins have lost their adaptive values in nature.

6.4 Transcriptional Regulation of Aflatoxin Biosynthesis

6.4.1 The *aflR* and *aflJ* Genes

Of all the clustered genes involved in aflatoxin biosynthesis, *aflR* is the only gene that encodes a Gal4-type, sequence-specific zinc-finger DNA-binding transcription factor (Payne et al. 1993; Chang et al. 1995). AflR binds to the palindromic sequence 5′-TCGN5CGR-3′ in the promoter region of aflatoxin structural genes, and is required for transcriptional activation of most, if not all, of the aflatoxin structural genes (Ehrlich et al. 1999). The promoter regions of the majority of aflatoxin biosynthesis genes have at least one binding motif within 200 nucleotides of the translation start site, and some putative binding sites have been identified further upstream. Like other Gal4-type regulatory proteins that bind to palindromic sequences, functional AflR probably is a dimer. Deletion of *aflR* in *A. parasiticus* abolished the expression of other aflatoxin pathway genes (Cary et al. 2000). Overexpression of *aflR* in *A. flavus* upregulates aflatoxin pathway gene transcription and aflatoxin accumulation (Flaherty and Payne 1997) in a fashion similar to that reported for *A. parasiticus* (Chang et al. 1995). These results demonstrate that AflR is specifically involved in the regulation of aflatoxin biosynthesis. Indeed, 23 upregulated genes identified by transcription profiling using DNA microarray assays comparing wild-type and *aflR*-deleted *A. parasiticus* strains all have the consensus AflR binding motif in their promoter regions (Price et al. 2006).

Adjacent to and divergently transcribed from *aflR* is *aflJ*. The *aflJ* gene, although not demonstrating significant homology with any other genes-encoded proteins known in databases, is necessary for expression of other genes in the aflatoxin gene cluster. Compared to *A. parasiticus aflR* transformants, the production of aflatoxin pathway intermediates was significantly enhanced in transformants that contained an additional *aflR* plus *aflJ* (Chang et al. 1995). *A. flavus aflJ*-deleted mutants failed to produce aflatoxins even though some transcripts of the aflatoxin pathway genes were still made (Meyers et al. 1998). Quantitative PCR showed that *A. parasiticus aflJ*-deleted mutants had significantly decreased transcript levels of the aflatoxin biosynthesis genes and the mutants were unable to make aflatoxin intermediates. However, deletion of *aflJ* did not have any discernible effect on *aflR* transcription, and vice versa. Overexpression of *A. flavus aflJ* did not result in elevated transcription of *ver1*, *omtA*, or *aflR*, but it appears to have some effect on *pksA*, *nor1*, *fas1*, and *fas2* (Du et al. 2007), which are required for the biosynthesis of the early aflatoxin pathway intermediate, averantin.

Regions of *A. parasiticus* AflR activating the transcription of aflatoxin pathway genes have been identified by yeast one-hybrid analyses (Chang et al. 1999). The fusion construct that encodes both the yeast Gal4p and AflR DNA-binding domains was not able to activate transcription whereas the fusion construct that encodes the Gal4p DNA-binding domain and the full-length AflR had only 15% activity of the *aflRC* construct that encodes the Gal4p DNA-binding domain and the AflR carboxy-terminal half. Deletion analysis of *aflRC* showed that the 23 amino acids (422–444) at the carboxyl terminus were important for activating the *GAL1::lacZ* gene expression. Simultaneous substitutions of Arg427, Arg429, and Arg431 with Leu decreased by 50-fold the ability to activate transcription. These results suggest that the transcriptional activation domain is located near the carboxy-terminus of AflR. Yeast two-hybrid assays showed that AflJ only bound to the carboxy-terminal region of AflR (Chang 2003). Asp436 of AflR was crucial for AflR's ability to activate transcription, but replacing it with His had little effect on the interaction of AflR with AflJ. These data demonstrate that AflJ is a transcriptional coactivator and suggest that the AflR–AflJ interacting region does not overlap with the AflR transcription activation region.

6.4.2 G-Protein Signaling Pathway and PkaA

FadA-dependent G-protein signaling pathway is an important level of control for aflatoxin biosynthesis. FadA is the alpha subunit of the heterotrimeric G-protein. When FadA is bound to GTP, it favors vegetative growth. Transforming an *A. parasiticus* strain that accumulates norsolorinic acid (the first stable intermediate in the aflatoxin pathway) with *A. nidulans* $fadA^{G42R}$ dominant active allele leads to FadA being locked in an active, GTP-bound state. This renders the transformants unable to produce aflatoxin intermediates and exhibiting a fluffy phenotype due to proliferative growth without sporulation (Hicks et al. 1997). Likewise, aflatoxin-nonproducing and aconidial phenotypes were observed for *A. flavus* transformed with $fadA^{G42R}$ (Sim and Keller, unpublished data). A primary control point of aflatoxin biosynthesis by the G-protein signaling pathway is probably at *aflR* because the *aflR* expression level in $fadA^{G42R}$-transformed *A. parasiticus* is less than 25% of the parent level (Chang et al. 2004b). *A. nidulans* produces sterigmatocystin, the penultimate precursor of aflatoxins. *A. nidulans fluG* that is involved in the synthesis of an extracellular diffusible factor and *flbA* that encodes a RGS (regulation of G-protein signaling) domain protein (Yu et al. 1996) have been demonstrated to regulate sterigmatocystin production and conidial development. In the presence of FlbA the intrinsic GTPase activity of FadA is stimulated, leading to GTP hydrolysis which inactivates the FadA-dependent signaling pathway and stimulation of sterigmatocystin production and sporulation. The *fluG* gene acts upstream of *flbA* in the regulatory circuit. FlbA/FadA signaling pathway, regulating sterigmatocystin production and conidiation in *A. nidulans*, is partially mediated by protein kinase A, PkaA (Shimizu and Keller 2001). Mutation of three PkaA phosphorylation sites in *A. nidulans* AflR restores expression of sterigmatocystin-specific genes in a *pkaA*-overexpressing strain, which suggests that phosphorylation negatively regulates AflR activity (Shimizu et al. 2003). The *sfgA* gene is a suppressor gene of *fluG*. Deletion and loss-of-function mutations of *sfgA* bypassed the need for *fluG* in conidiation and sterigmatocystin production, and overexpression of *sfgA* inhibited conidiation and delayed expression of conidiation- and sterigmatocystin-specific genes (Seo et al. 2006). Aflatoxin production is closely correlated with the conidiation process, and aconidial *A. parasiticus* mutants are unable to synthesize aflatoxins (Guzman-de-Pena and Ruiz-Herrera 1997; Wilkinson et al. 2004). Thus proteins encoded by the aforementioned genes probably regulate aflatoxin biosynthesis in *A. flavus* and *A. parasiticus* through the same mechanisms. PKA activities of *A. parasiticus* strains treated with exogenous cAMP or transformed with $fadA^{G42R}$ indicate that FadA/PKA regulate aflatoxin biosynthesis and conidiation via similar mechanisms reported for *A. nidulans*, and intracellular cAMP levels, in part, mediate a PKA-dependent regulation on aflatoxin biosynthesis and conidiation (Roze et al. 2004).

6.4.3 Calcium Signaling Pathway

Calcium is required for aflatoxin production (Maggon et al. 1977) and blocking of the calcium channel by chemicals inhibits aflatoxin production (Rao Praveen and Subramanyam 1999). Calcium, like cAMP, is a second messenger. It plays a crucial role in regulating a wide range of physiological functions of cells. The binding of many water-soluble signals to the membrane receptors stimulates an increase of calcium concentration within the cytoplasm either via a flow through calcium channels in the plasma membrane or, in most cases, from cellular stores within the endoplasmic reticulum. To turn calcium signal into a biological response, calcium ions then bind to specific effector proteins. Calmodulin (CaM) is a Ca^{2+}-binding protein that interacts with a multitude of different protein targets, thereby affecting many cellular functions. The activated Ca^{2+}–CaM complex can bind to many target proteins, including kinases, phosphatases, and other signaling proteins. Ca^{2+}–CaM has been demonstrated to be involved in various aspects of fungal development including formation of conidium and appressorium, hyphal extension and branching, and it apparently is involved in aflatoxin biosynthesis. During periods of aflatoxin production, protein phosphorylation in aflatoxigenic *A. parasiticus* was completely lacking, but protein phosphorylation was unaffected in nonaflatoxigenic *A. parasiticus*. Moreover, enhanced activity of calcineurin, a CaM-dependent protein phosphatase, was concomitant with a decreased activity of CaM-dependent protein kinase (Jayashree et al. 2000). Trifluoperazine, an antagonist of CaM, caused decreased incorporation of [^{14}C]-acetate into aflatoxin B_1. This inhibition corresponded to a decrease in acetyl-CoA carboxylase activity at periods of maximal aflatoxin production, which suggests that CaM mediates activation of acetyl-CoA carboxylase in channeling acetates for aflatoxin production (Rao Praveen and Subramanyam 2000). Most recently, putative CaM-binding domains in *A. parasiticus* proteins related to aflatoxin production have been identified by in silico analysis using the CaM target database (Rao Praveen and Subramanyam 2006). The aflatoxin-pathway-specific AflR regulator and the AflJ coactivator both contain predicted CaM-binding domains at the carboxyl termini. In addition, putative CaM-binding domains are also present in aflatoxin biosynthesis enzymes, such as Vbs, DmtA and OmtA, and in the VeA protein which regulates the expression of *aflR* and *aflJ* in *A. flavus* and *A. parasiticus* (Calvo et al. 2004; Duran et al. 2007). These findings may open new avenues for elucidating other possible mechanisms of initiation and regulation of aflatoxin biosynthesis at the posttranslational level.

6.4.4 The laeA Gene

A novel global regulatory gene, *laeA* (for lack of *aflR* expression) was first identified from *A. nidulans* (Bok and Keller 2004). It is well-conserved in fungi whose genome sequences are available. LaeA is a nuclear protein which contains an *S*-adenosylmethionine (SAM) binding motif and controls transcription of several gene clusters of secondary metabolism in the genus *Aspergillus*, including the production of sterigmatocystin and penicillin in *A. nidulans*, gliotoxin in *A. fumigatus*, and aflatoxins in *A. flavus* (Bok and Keller 2004; Bouhired et al. 2007). It also has been suggested to regulate the virulence of *A. fumigatus* (Sugui et al. 2007). A whole-genome comparison of the transcriptional profiles of wild-type and *laeA*-deleted *A. fumigatus* strains reveals that LaeA positively controls the expression of 20–40% of major classes of secondary metabolite biosynthesis (Perrin et al. 2007). The exact mechanism by which LaeA regulates secondary metabolite gene clusters is not known. Interestingly, when an unrelated gene such as *argB* was placed within the boundary of the sterigmatocystin gene cluster, it was coregulated with other clustering genes. But, when a clustering gene such as *aflR* was placed outside the boundary, its regulation was not affected by LaeA (Bok et al. 2006). One proposed possibility is that LaeA differentially methylates histone protein and it alters the chromatin structure for gene expression. The expression of *laeA* is negatively regulated by AflR via a unique feedback loop and by the signal transduction components, PKaA and RasA. Unlike the mentioned signaling factors, the primary role of LaeA is to regulate metabolic gene clusters, not sporulation, because *laeA*-deleted strains produced wild-type levels of conidia (Bok and Keller 2004). Most recent analyses of nonaflatoxigenic *A. parasiticus sec-* (for secondary metabolism negative) variants generated through serial transfer of mycelia of the *sec+* parents show that *laeA* was expressed in both *sec+* and *sec-* strain (Kale et al. 2007). This result suggests that LaeA only exerts its effect on aflatoxin biosynthesis at a certain level and is independent of other regulatory pathways that are involved in fungal development.

6.4.5 The veA Gene

The *veA* gene in *A. nidulans* (Mooney et al. 1990) is a gene initially found to be crucial for light-dependent conidiation. The light dependence is abolished by a mutation (*veA1*) which allows conidiation of *A. nidulans* to occur in darkness. VeA has been shown to positively regulate sexual reproduction in *A. nidulans* (Kim et al. 2002). Although light usually promotes higher levels of conidiation in *A. flavus* and *A. parasiticus*, abundant conidia were produced when they are grown in darkness. A comparison of the light effect on sterigmatocystin production by *A. nidulans veA+* and *veA1* strains showed that both strains produced sterigmatocystin, but the highest amount was produced by the *veA+* strain grown in darkness.

However, *veA*-deleted *A. flavus* and *A. parasiticus* strains completely lost aflatoxin production regardless of the illumination conditions (Duran et al. 2007; Stinnett et al. 2007). VeA contains a bipartite nuclear localization signal (NLS) motif and its migration to the nucleus is light-dependent and requires the importin α carrier protein. In darkness VeA is located mainly in the nucleus; under light it is located both in cytoplasm and nucleus (Stinnett et al. 2007). VeA has no recognizable DNA-binding sequences and probably exerts its effect on sterigmatocystin and aflatoxin production through protein–protein interactions with other regulatory factors. Post-translational modifications such as phosphorylation and dephosphorylation may modulate its activity. Lack of VeA production in the *veA*-deleted *A. flavus* and *A. parasiticus* strains consequently abolishes aflatoxin production because a threshold concentration of nuclear VeA might be necessary to initiate aflatoxin biosynthesis.

6.5 Oxidative Stress and Aflatoxin Production

Aflatoxins are highly oxygenated polyketide-derived metabolites. A series of oxidative steps are involved in the formation of dihydrobisfuran that leads to the formation of aflatoxins. The adaptive value of aflatoxin production is not fully understood, but synthesis of aflatoxins may act as a defense mechanism against oxidative stress. Lipoperoxidation occurs as a chain reaction initiated by free radicals, which propagates itself and can result in the formation of many equivalents of lipid peroxides. Synthetic lipoperoxides greatly stimulated aflatoxin production by *A. flavus* and *A. parasiticus* (Passi et al. 1984), but BHA, BHT and cystamine were able to block aflatoxin production induced by lipoperoxides (Fanelli and Fabbri 1989). Adding eugenol (4-allyl-2-methoxy phenol, an antioxidant) to aflatoxigenic *A. parasiticus* decreased enzyme activities involved in free radical scavenging, lipid peroxidation and maintenance of redox potential without affecting fungal growth (Jayashree and Subramanyam 1999).

Studies have demonstrated that natural phenol compounds, for example gallic acid from hydrolyzable tannins in the pellicle of walnut kernels, dramatically inhibit aflatoxin biosynthesis by *A. flavus* (Kim et al. 2006). A positive correlation has been shown between reactive oxygen species (ROS) accumulation and aflatoxin production by *A. flavus* and *A. parasiticus* (Mahoney et al. 2006; Narasaiah et al. 2006). The antioxidant enzyme activities of xanthine oxidase, superoxide dismutase, and glutathione peroxidase as well as thiobarbituric acid–reactive substances and reduced glutathione in mycelia were all increased from trophophase to idiophase in aflatoxigenic *A. parasiticus* (Jayashree and Subramanyam 2000). Further stimulation of antioxidant enzyme activities such as catalase, superoxide dismutase, and glutathione peroxidase in the same *A. parasiticus* by β-glucan (a pharmacologically active compound that stimulates antioxidant responses in animal cells) prepared from mushroom *Lentinula edodes* inhibited aflatoxin production (Reverberi et al. 2005).

6.6 Sclerotial Production by *A. flavus* and *A. parasiticus*

6.6.1 Sclerotial Morphogenesis

Besides conidiation, a relationship between sclerotial formation and aflatoxin production has long been recognized although the genes responsible for such coregulation have not been identified. Sclerotia are pigmented, compacted bodies of aggregated hyphae, which resist unfavorable environmental conditions due to their resistance to chemical and biological degradation, and are capable of remaining dormant for long periods (Willetts 1971). Sclerotia develop from localized vegetative hyphal branchings which fold back upon themselves to form bundles of parallel hyphae. Morphological stages associated with sclerotial formation consist of (1) the appearance of white tufts of thick mycelial initials, (2) the enlargement and hardening of the mycelial tufts and the appearance of exudate on the surface, (3) the continued hardening due to surface delimitation, and (4) melanin deposition in the peripheral rind cells, leaving a smooth, firm, dark surface (Littley and Rahe 1992; Willetts and Bullock 1992). Many fungi produce sclerotia, and they are considered by some researchers to be a vestige of the sexual structures, cleistothecia, which are also woven from specialized hyphae but each contains thousands of ascospores (Geiser et al. 1996).

6.6.2 A. flavus Sclerotial Morphotypes

A. flavus isolates generally can be categorized into two morphotypes based on sclerotial size, L strain and S strain [also named *A. flavus var. parvisclerotigenus* (Saito and Tsuruta 1993)]. L-strain isolates produce abundant conidia but only a few sclerotia that are usually larger than 400 μm in diameter (Cotty 1989; Horn and Dorner 1999), whereas S-strain isolates produce fewer conidia and numerous sclerotia that are usually smaller than 400 μm in diameter. The S-strain isolates typically produce higher amounts of aflatoxin than the L-strain isolates on the same media (Bayman and Cotty 1993). Despite this dichotomy, the genetic relationship between L strain and S strain is still not understood. It nonetheless suggests a positive interrelationship between sclerotial formation and aflatoxin production. Intriguingly, *A. parasiticus* do not produce distinct sclerotial types as reported for *A. flavus* strains, and the aflatoxin-producing ability of *A. parasiticus* is much more stable than that of *A. flavus* (Chang et al. 2007).

6.6.3 Secondary Metabolites in Aspergillus sclerotia

Although *A. flavus* strains generally produce only B type aflatoxins, aflatoxin G_1 was reported in sclerotia of all *A. flavus* strains but in the conidia of only one strain

(Wicklow and Shotwell 1983). The total aflatoxin levels in sclerotia of aflatoxigenic *A. flavus* and *A. parasiticus* strains vary considerably. When both aflatoxin B and G are present in sclerotia and conidia of the same strain of *A. flavus* or *A. parasiticus*, the ratios of B_1 to G_1 are nearly identical in each propagule, which implies similar distribution mechanisms or cellular activities for formation of aflatoxin B_1 and G_1. Some sclerotia contain other secondary metabolites that appear to function to deter fungivorous insects. For example, aflatrem isolated from *A. flavus* sclerotia is a toxic indole diperpene which is a potent tremorgenic compound known to cause neurological disorders (Zhang et al. 2004). Arenarins A, B and C isolated from *Aspergillus arenarius* sclerotia exhibit cytotoxicity against the dried-fruit beetle *Carpophilus hemipterus* and human tumor cell lines (Oh et al. 1998). Isokotanins B and C isolated from *Aspergillus alliaceus* sclerotia show activity against the corn earworm *Helicoverpa zea* and *C. hemipterus* (Laakso et al. 1994). Three diketopiperazine-containing metabolites 1–3 isolated from *Aspergillus ochraceus* sclerotia caused reduction in weight gain in assays against *H. zea* (de Guzman et al. 1992).

6.6.4 Nutritional Effects

A variety of nutritional and environmental factors are known to affect sclerotial biogenesis. The effects of nutrients on sclerotial production are often medium-dependent and strain-specific, and very little is known about the underlying biochemical mechanisms. The number of sclerotia produced by *A. flavus* was drastically altered by changing the amounts of nitrate and sucrose, with the optimal amounts being about 3% sucrose and 0.5% $NaNO_3$ (Hesseltine et al. 1970). Abundant sclerotia were produced in media of high concentrations of sucrose and nitrate et al. (Rudolph 1962; McAlpin 2001). Ammonium as the sole nitrogen source inhibits sclerotial production by *A. flavus* and *A. parasiticus* (Bennett et al. 1979). Sucrose is the major constituent of molasses (about 40%). Adding ammonium to molasses medium, however, significantly increased the production of sclerotia and aflatoxins by S-strain *A. flavus* isolates (Chang and Hua 2007). Some polyunsaturated fatty acids can act as morphogenic factors due to their catabolic products mimicking the action of the psi (precocious sexual inducer) factors, which affect multiple developmental processes (Calvo et al. 1999). The production of sclerotia and conidia in *A. flavus* and *A. parasiticus* was affected by linoleic acid, and linoleic acid decreased sclerotial production in the light but increased the production in darkness. The Δ12-oleic acid desaturase gene (*odeA*) is responsible for the conversion of oleic acid to linoleic acid. Sclerotial mass of *odeA*-deleted *A. parasiticus* was comparable to that of the wild-type, but sclerotial number increased significantly in the *odeA*-deleted strain. Supplementing linoleic acid to media, however, decreased sclerotial mass and number produced by the wild-type and the *odeA*-deleted strains (Chang et al. 2004a). Nutrients or C/N ratios may affect the metabolic flux favoring primary metabolism, secondary metabolism, or fungal development. Overactivation of the aflatoxin pathway apparently channels acetates for sclerotial production toward aflatoxin biosynthesis, which results in progressive decrease in sclerotial size, altered sclerotial shape, and a weakening in sclerotial structure (Chang et al. 2002).

6.6.5 Light as a Major Determinant

Among the environmental factors, light has a major effect on sclerotial biogenesis. An earlier study reported that light completely inhibited sclerotial formation of one *A. flavus* strain and largely inhibited sclerotial formation in another *A. flavus* strain (Rai et al. 1967). *A. flavus* and *A. parasiticus* strains produced sclerotia in darkness and under continuous red light, but the production was inhibited by continuous white, blue, and green light (Bennett et al. 1978). The underlying mechanism of fungal photoresponses in relation to sclerotial formation is not well understood. The *veA* gene that is crucial for light-dependent response positively regulates sexual reproduction and negatively regulates asexual conidiation in *A. nidulans*. VeA is necessary for sclerotial production of *A. flavus* and *A. parasiticus* in darkness because *veA*-deleted strains are unable to produce sclerotia (Calvo et al. 2004; Cary et al. 2007). Despite these findings, there is still a significant gap in our knowledge as to how VeA, possibly through certain signal transduction pathway(s), regulates fungal sexual and asexual development and production of secondary metabolites. Termination of sclerotial maturation has been observed to be possibly associated with a signal that controls aflatoxin biosynthesis (Cotty et al. 1994). A recent genome-wide expression study of *A. flavus* veA^{+} wild-type and its *veA*-deleted strains by DNA microarray has identified a subgroup of 15 *veA*-dependent genes that exhibit time-dependent expression profiles similar to those of aflatoxin biosynthetic genes (Cary et al. 2007). Some of these identified genes may be directly or indirectly related to sclerotial development. The global regulatory gene of secondary metabolism *laeA* also plays an important role in sclerotial production. Deletion of *laeA* in *A. flavus* NRRL 3357 led to complete loss of sclerotial formation, but reintroduction of *laeA* into the *laeA*-deleted strain restored its ability to produce sclerotia (Keller and Yu, personal communication). Further genome-wide microarray DNA analyses likely will identify candidate genes involved in sclerotial production.

6.7 Oxidative Stress and Sclerotial Biogenesis

Oxidative stress has been proposed as a major determinant for sclerotial biogenesis in fungi (Georgiou et al. 2006). The development of four types of sclerotia, that is, loose, terminal, lateral-terminal, and later-simple produced by *Rhizoctonia solani*, *Sclerotinia sclerotiorum*, *Sclerotium rolfsii*, and *Sclerotinia minor*, respectively, are directly related with the antioxidant thiol groups of glutathione and/or cysteine.

A decrease in sclerotial differentiation concurred with an increase of the thiols resulting from the use of glutathione biosynthesis modulators, *N*-acetyl-cysteine, L-oxo-thiazolidine-4-carboxylate, and L-buthionine-S,R-sulfoximine (Patsoukis and Georgiou 2007). Growing evidence has suggested that ROS play important physiological roles and a hyperoxidant state (oxidative stress) can trigger cell differentiation (Aguirre et al. 2005). Aflatoxigenic *A. parasiticus* required more oxygen than nonaflatoxigenic *A. parasiticus* at trophophase (Jayashree and Subramanyam 2000). The greater demand for oxygen may result in a higher concentration of ROS in cells, generating a hyperoxidant state. Like aflatoxin production, sclerotial production may, in part, act as a defense mechanism against oxidative stress. The types of sclerotia and amounts of aflatoxins produced by L-strain and S-strain *A. flavus* are indicative of the capacity of each to dispose of or neutralize ROS. The general inability of L strain to produce a great number of sclerotia may be compensated by the production of abundant conidia, reflecting the genetic differences of and differential mechanisms used by each sclerotial morphotype to combat ROS so as to restore fungal cells to a stable, balanced physiological state.

References

Aguirre J, Rios-Momberg M, Hewitt D, Hansberg W (2005) Reactive oxygen species and development in microbial eukaryotes. Trends Microbiol 13:111–118

Bayman P, Cotty PJ (1993) Genetic diversity in *Aspergillus flavus*: association with aflatoxin production and morphology. Can J Bot 71:23–31

Bennett JW, Fernholz FA, Lee LS (1978) Effect of light on aflatoxins, anthraquinones, and sclerotia in *Aspergillus flavus* and *A. parasiticus*. Mycologia 70:104–116

Bennett JW, Horowitz PC, Lee LS (1979) Production of sclerotia by aflatoxigenic and nonaflatoxigenic strains of *Aspergillus flavus* and *A. parasiticus*. Mycologia 71:415–422

Bok JW, Keller NP (2004) LaeA, a regulator of secondary metabolism in *Aspergillus* spp. Eukaryot Cell 3:527–535

Bok JW, Noordermeer D, Kale SP, Keller NP (2006) Secondary metabolic gene cluster silencing in *Aspergillus nidulans*. Mol Microbiol 61:1636–1645

Bouhired S, Weber M, Kempf-Sontag A, Keller NP, Hoffmeister D (2007) Accurate prediction of the *Aspergillus nidulans* terrequinone gene cluster boundaries using the transcriptional regulator LaeA. Fungal Genet Biol 44:1134–1145

Calvo AM, Hinze LL, Gardner HW, Keller NP (1999) Sporogenic effect of polyunsaturated fatty acids on development of *Aspergillus* spp. Appl Environ Microbiol 65:3668–3673

Calvo AM, Bok J, Brooks W, Keller NP (2004) *veA* is required for toxin and sclerotial production in *Aspergillus parasiticus*. Appl Environ Microbiol 70:4733–4739

Cary JW, Ehrlich KC, Wright M, Chang P-K, Bhatnagar D (2000) Generation of *aflR* disruption mutants of *A. parasiticus*. Appl Microbiol Biotechnol 53:680–684

Cary JW, Klich MA, Beltz SB (2005) Characterization of aflatoxin-producing fungi outside of *Aspergillus* section *Flavi*. Mycologia 97:425–432

Cary JW, Ehrlich KC, Bland JM, Montalbano BG (2006) The aflatoxin biosynthesis cluster gene, *aflX*, encodes an oxidoreductase involved in conversion of versicolorin A to demethylsterigmatocystin. Appl Environ Microbiol 72:1096–1101

Cary JW, O'Brian GR, Nielsen DM, Nierman W, Harris-Coward P, Yu J, Bhatnagar D, Cleveland TE, Payne GA, Calvo AM (2007) Elucidation of *veA*-dependent genes associated with aflatoxin and sclerotial production in *Aspergillus flavus* by functional genomics. Appl Microbiol Biotechnol 76:1107–1118

Chang P-K (2003) The *Aspergillus parasiticus* protein AFLJ interacts with the aflatoxin pathway-specific regulator AFLR. Mol Genet Genom 268:711–719

Chang P-K, Hua SS (2007) Molasses supplementation promotes conidiation but suppresses aflatoxin production by small sclerotial *Aspergillus flavus*. Lett Appl Microbiol 44:131–137

Chang P-K, Ehrlich KC, Yu J, Bhatnagar D, Cleveland TE (1995) Increased expression of *Aspergillus parasiticus aflR*, encoding a sequence-specific DNA-binding protein, relieves nitrate inhibition of aflatoxin biosynthesis. Appl Environ Microbiol 61:2372–2377

Chang P-K, Yu J, Bhatnagar D, Cleveland TE (1999) The carboxy-terminal portion of the aflatoxin pathway regulatory protein AFLR of *Aspergillus parasiticus* activates *GAL1::lacZ* gene expression in *Saccharomyces cerevisiae*. Appl Environ Microbiol 65:2508–2512

Chang P-K, Bennett JW, Cotty PJ (2002) Association of aflatoxin biosynthesis and sclerotial development in *Aspergillus parasiticus*. Mycopathologia 153:41–48

Chang P-K, Wilson RA, Keller NP, Cleveland TE (2004a) Deletion of the Δ-12 oleic acid desaturase gene of a nonaflatoxigenic *Aspergillus parasiticus* field isolate affects conidiation and sclerotial development. J Appl Microbiol 97:1178–1184

Chang P-K, Yu J, Yu JH (2004b) *aflT*, a MFS transporter-encoding gene located in the aflatoxin gene cluster, does not have a significant role in aflatoxin secretion. Fungal Genet Biol 41: 911–920

Chang P-K, Horn BW, Dorner JW (2005) Sequence breakpoints in the aflatoxin biosynthesis gene cluster and flanking regions in nonaflatoxigenic *Aspergillus flavus* isolates. Fungal Genet Biol 42:914–923

Chang P-K, Wilkinson JR, Horn BW, Yu J, Bhatnagar D, Cleveland TE (2007) Genes differentially expressed by *Aspergillus flavus* strains after loss of aflatoxin production by serial transfers. Appl Microbiol Biotechnol 77:917–925

Cotty PJ (1989) Virulence and cultural characteristics of two *Aspergillus flavus* strains pathogenic on cotton. Phytopathology 79:808–814

Cotty PJ, Bayman P, Egel DS, Elias KS (1994) Agriculture, aflatoxins and *Aspergillus*. In: Powell KA, Renwick A, Peberdy JF (eds) The Genus *Aspergillus*: from taxonomy and genetics to industrial application. Plenum, New York, pp 1–27

de Guzman FS, Gloer JB, Wicklow DT, Dowd PF (1992) New diketopiperazine metabolites from the sclerotia of *Aspergillus ochraceus*. J Nat Prod 55:931–939

Du W, Obrian GR, Payne GA (2007) Function and regulation of *aflJ* in the accumulation of aflatoxin early pathway intermediate in *Aspergillus flavus*. Food Addit Contam 24: 1043–1050

Duran RM, Cary JW, Calvo AM (2007) Production of cyclopiazonic acid, aflatrem, and aflatoxin by *Aspergillus flavus* is regulated by *veA*, a gene necessary for sclerotial formation. Appl Microbiol Biotechnol 73:1158–1168

Ehrlich KC, Montalbano BG, Cary JW (1999) Binding of the C6-zinc cluster protein, AFLR, to the promoters of aflatoxin pathway biosynthesis genes in *Aspergillus parasiticus*. Gene 230:249–257

Ehrlich KC, Chang P-K, Yu J, Cotty PJ (2004) Aflatoxin biosynthesis cluster gene *cypA* is required for G aflatoxin formation. Appl Environ Microbiol 70:6518–6524

Ehrlich KC, Montalbano B, Boue SM, Bhatnagar D (2005a) An aflatoxin biosynthesis cluster gene encodes a novel oxidase required for conversion of versicolorin A to sterigmatocystin. Appl Environ Microbiol 71:8963–8965

Ehrlich KC, Yu J, Cotty PJ (2005b) Aflatoxin biosynthesis gene clusters and flanking regions. J Appl Microbiol 99:518–527

Fanelli C, Fabbri AA (1989) Relationship between lipids and aflatoxin biosynthesis. Mycopathologia 107:115–120

FAO (1995) Worldwide Regulations for Mycotoxins 1995. A Compendium. FAO Food and Nutrition Paper 64, 1997. FAO, Rome, Italy

Flaherty JE, Payne GA (1997) Overexpression of *aflR* leads to upregulation of pathway gene transcription and increased aflatoxin production in *Aspergillus flavus*. Appl Environ Microbiol 63:3995–4000

Geiser DM, Timberlake WE, Arnold ML (1996) Loss of meiosis in *Aspergillus*. Mol Biol Evol 13:809–817

Georgiou CD, Patsoukis N, Papapostolou I, Zervoudakis G (2006) Sclerotial metamorphosis in filamentous fungi is induced by oxidative stress. Integr Comp Biol 46:691–712

Guzman-de-Pena D, Ruiz-Herrera J (1997) Relationship between aflatoxin biosynthesis and sporulation in *Aspergillus parasiticus*. Fungal Genet Biol 21:198–205

Hesseltine CW, Sorenson WG, Smith M (1970) Taxonomic studies of the aflatoxin-producing strains in the *Aspergillus flavus* group. Mycologia 62:123–132

Hicks JK, Yu JH, Keller NP, Adams TH (1997) *Aspergillus* sporulation and mycotoxin production both require inactivation of the FadA G alpha protein-dependent signaling pathway. EMBO J 16:4916–4923

Horn BW (2007) Biodiversity of *Aspergillus* section *Flavi* in the United States: a review. Food Addit Contam 24:1088–1101

Horn BW, Dorner JW (1999) Regional differences in production of aflatoxin B_1 and cyclopiazonic acid by soil isolates of *Aspergillus flavus* along a transect within the United States. Appl Environ Microbiol 65:1444–1449

Jayashree T, Subramanyam C (1999) Antiaflatoxigenic activity of eugenol is due to inhibition of lipid peroxidation. Lett Appl Microbiol 28:179–183

Jayashree T, Subramanyam C (2000) Oxidative stress as a prerequisite for aflatoxin production by *Aspergillus parasiticus*. Free Radic Biol Med 29:981–985

Jayashree T, Praveen Rao J, Subramanyam C (2000) Regulation of aflatoxin production by Ca^{2+}/calmodulin-dependent protein phosphorylation and dephosphorylation. FEMS Microbiol Lett 183:215–219

Kale SP, Cary JW, Hollis N, Wilkinson JR, Bhatnagar D, Yu J, Cleveland TE, Bennett JW (2007) Analysis of aflatoxin regulatory factors in serial transfer-induced non-aflatoxigenic *Aspergillus parasiticus*. Food Addit Contam 24:1061–1069

Kim H, Han K, Kim K, Han D, Jahng K, Chae K (2002) The *veA* gene activates sexual development in *Aspergillus nidulans*. Fungal Genet Biol 37:72–80

Kim JH, Campbell BC, Molyneux RJ, Mahoney NE, Chan KL, Yu J, Wilkinson JR, Cary JW, Bhatnagar D, Cleveland TE (2006) Gene targets for fungal and mycotoxin control. Mycol Res 22:3–8

Laakso JA, Narske ED, Gloer JB, Wicklow DT, Dowd PF (1994) Isokotanins A-C: new bicoumarins from the sclerotia of *Aspergillus alliaceus*. J Nat Prod 57:128–133

Liti G, Louis EJ (2005) Yeast evolution and comparative genomics. Annu Rev Microbiol 59:135–153

Littley ER, Rahe JE (1992) Sclerotial morphogenesis in *Sclerotium cepivorum* in vitro. Can J Bot 70:772–778

Maggon KK, Gupta SK, Venkitasubramanian TA (1977) Biosynthesis of aflatoxins. Bacteriol Rev 41:822–855

Mahoney N, Molyneux R, Kim JH, Chan K, Campbell BC (2006) Intensification of aflatoxigenesis by oxidative stress and its reduction by phenolic natural products. United States-Japan Cooperative Program in Natural Resources, Toxic Microorganisms Panel. November 7–9, Washington, D.C

McAlpin CE (2001) An *Aspergillus flavus* mutant producing stipitate sclerotia and synnemata. Mycologia 93:552–565

Meyers DM, Obrian G, Du WL, Bhatnagar D, Payne GA (1998) Characterization of *aflJ*, a gene required for conversion of pathway intermediates to aflatoxin. Appl Environ Microbiol 64:3713–3717

Minto RE, Townsend CA (1997) Enzymology and molecular biology of aflatoxin biosynthesis. Chem Rev 97 2537–2555

Mooney JL, Hassett DE, Yager LN (1990) Genetic analysis of suppressors of the *veA1* mutation in *Aspergillus nidulans*. Genetics 126:869–874

Narasaiah KV, Sashidhar RB, Subramanyam C (2006) Biochemical analysis of oxidative stress in the production of aflatoxin and its precursor intermediates. Mycopathologia 162:179–189

Oh H, Gloer JB, Wicklow DT, Dowd PF (1998) Arenarins A-C: new cytotoxic fungal metabolites from the sclerotia of *Aspergillus arenarius*. J Nat Prod 61:702–705

Otsuki T, Wilson JS, Sewadeh M (2001) What price precaution? European harmonization of aflatoxin regulations and African groundnuts exports. Euro Rev Agric Econ 28:263–284

Passi S, Nazzaro-Porro M, Fanelli C, Fabbri AA, Fasella P (1984) Role of lipoperoxidation in aflatoxin production. Appl Microbiol Biotechnol 19:186–190

Patsoukis N, Georgiou CD (2007) Effect of glutathione biosynthesis-related modulators on the thiol redox state enzymes and on sclerotial differentiation of filamentous phytopathogenic fungi. Mycopathologia 163:335–347

Payne GA, Nystrom GJ, Bhatnagar D, Cleveland TE, Woloshuk CP (1993) Cloning of the *afl-2* gene involved in aflatoxin biosynthesis from *Aspergillus flavus*. Appl Environ Microbiol 59:156–162

Payne GA, Nierman WC, Wortman JR, Pritchard BL, Brown D, Dean RA, Bhatnagar D, Cleveland TE, Machida M, Yu J (2006) Whole genome comparison of *Aspergillus flavus* and *A. oryzae*. Med Mycol 44(Suppl):9–11

Perrin RM, Fedorova ND, Bok JW, Cramer RA, Wortman JR, Kim HS, Nierman WC, Keller NP (2007) Transcriptional regulation of chemical diversity in *Aspergillus fumigatus* by LaeA. PLoS Pathog 3:e50

Pildain MB, Vaamonde G, Cabral D (2004) Analysis of population structure of *Aspergillus flavus* from peanut based on vegetative compatibility, geographic origin, mycotoxin and sclerotia production. Int J Food Microbiol 93:31–40

Price MS, Yu J, Nierman WC, Kim HS, Pritchard B, Jacobus CA, Bhatnagar D, Cleveland TE, Payne GA (2006) The aflatoxin pathway regulator AflR induces gene transcription inside and outside of the aflatoxin biosynthetic cluster. FEMS Microbiol Lett 255:275–279

Rai JN, Tewari JP, Sinha AK (1967) Effect of environmental conditions on sclerotia and cleistothecia production in *Aspergillus*. Mycopathol Mycol Appl 31:209–224

Rao Praveen J, Subramanyam C (1999) Requirement of Ca^{2+} for aflatoxin production: inhibitory effect of Ca^{2+} channel blockers on aflatoxin production by *Aspergillus parasiticus* NRRL 2999. Lett Appl Microbiol 28:85–88

Rao Praveen J, Subramanyam C (2000) Calmodulin mediated activation of acetyl-CoA carboxylase during aflatoxin production by *Aspergillus parasiticus*. Lett Appl Microbiol 30: 277–281

Rao Praveen J, Subramanyam C (2006) Putative calmodulin-binding domains in aflatoxin biosynthesis-regulatory proteins. Curr Microbiol 52:493–496

Reverberi M, Fabbri AA, Zjalic S, Ricelli A, Punelli F, Fanelli C (2005) Antioxidant enzymes stimulation in *Aspergillus parasiticus* by *Lentinula edodes* inhibits aflatoxin production. Appl Microbiol Biotechnol 69:207–215

Roze LV, Beaudry RM, Keller NP, Linz JE (2004) Regulation of aflatoxin synthesis by FadA/cAMP/protein kinase A signaling in *Aspergillus parasiticus*. Mycopathologia 158: 219–232

Rudolph ED (1962) The effect of some physiological and environmental factors in sclerotia aspergilli. Am J Bot 49:71–78

Saito M, Tsuruta O (1993) A new variety of *Aspergillus flavus* from tropical soil in Thailand and its aflatoxin productivity. Proc Jpn Assoc Mycotoxicol 37:31–36

Sakuno E, Wen Y, Hatabayashi H, Arai H, Aoki C, Yabe K, Nakajima H (2005) *Aspergillus parasiticus* cyclase catalyzes two dehydration steps in aflatoxin biosynthesis. Appl Environ Microbiol 71:2999–3006

Seo JA, Guan Y, Yu JH (2006) FluG-dependent asexual development in *Aspergillus nidulans* occurs via derepression. Genetics 172:1535–1544

Shimizu K, Keller NP (2001) Genetic involvement of a cAMP-dependent protein kinase in a G protein signaling pathway regulating morphological and chemical transitions in *Aspergillus nidulans*. Genetics 157:591–600

Shimizu K, Hicks JK, Huang TP, Keller NP (2003) Pka, Ras and RGS protein interactions regulate activity of AflR, a Zn(II)2Cys6 transcription factor in *Aspergillus nidulans*. Genetics 165:1095–1104

Stinnett SM, Espeso EA, Cobeno L, Araujo-Bazan L, Calvo AM (2007) *Aspergillus nidulans* VeA subcellular localization is dependent on the importin alpha carrier and on light. Mol Microbiol 63:242–255

Sugui JA, Pardo J, Chang YC, Mullbacher A, Zarember KA, Galvez EM, Brinster L, Zerfas P, Gallin JI, Simon MM, Kwon-Chung KJ (2007) Role of *laeA* in the regulation of *alb1*, *gliP*, conidial morphology, and virulence in *Aspergillus fumigatus*. Eukaryot Cell 6:1552–1561

Takahashi H, Kamimura H, Ichinoe M (2004) Distribution of aflatoxin-producing *Aspergillus flavus* and *Aspergillus parasiticus* in sugarcane fields in the southernmost islands of Japan. J Food Prot 67:90–95

Vaamonde G, Patriarca A, Fernandez Pinto V, Comerio R, Degrossi C (2003) Variability of aflatoxin and cyclopiazonic acid production by *Aspergillus* section *Flavi* from different substrates in Argentina. Int J Food Microbiol 88:79–84

Wicklow DT, Shotwell OL (1983) Intrafungal distribution of aflatoxins among conidia and sclerotia of *Aspergillus flavus* and *Aspergillus parasiticus*. Can J Microbiol 29:1–5

Wilkinson H, Ramaswamy A, Sim SC, Keller NP (2004) Increased conidiation associated with progression along the sterigmatocystin biosynthetic pathway. Mycologia 96:1190–1198

Willetts HJ (1971) The survival of fungal sclerotia under adverse environmental conditions. Bio Rev 46:387–407

Willetts HJ, Bullock S (1992) Developmental biology of sclerotia. Mycol Res 10:801–816

Wogan GN (2000) Impacts of chemicals on liver cancer risk. Semin Cancer Biol 10:201–210

Wu F (2004) Mycotoxin risk assessment for the purpose of setting international regulatory standards. Environ Sci Technol 38:4049–4055

Yabe K, Nakajima H (2004) Enzyme reactions and genes in aflatoxin biosynthesis. Appl Microbiol Biotechnol 64:745–755

Yu JH, Wieser J, Adams TH (1996) The *Aspergillus* FlbA RGS domain protein antagonizes G protein signaling to block proliferation and allow development. EMBO J 15:5184–5190

Yu J, Chang P-K, Ehrlich KC, Cary JW, Bhatnagar D, Cleveland TE, Payne GA, Linz JE, Woloshuk CP, Bennett JW (2004) Clustered pathway genes in aflatoxin biosynthesis. Appl Environ Microbiol 70:1253–1262

Yu J, Ronning CM, Wilkinson JR, Campbell BC, Payne GA, Bhatnagar D, Cleveland TE, Nierman WC (2007) Gene profiling for studying the mechanism of aflatoxin biosynthesis in *Aspergillus flavus* and *A. parasiticus*. Food Addit Contam 24:1035–1042

Zhang S, Monahan BJ, Tkacz JS, Scott B (2004) Indole-diterpene gene cluster from *Aspergillus flavus*. Appl Environ Microbiol 70:6875–6883

Chapter 7
Interaction Between Aflatoxin B_1 and Other Risk Factors in Hepatocarcinogenesis

Michael C. Kew

7.1 Introduction

Known causes of hepatocellular carcinoma (HCC) are chronic hepatitis B virus (HBV) infection; chronic hepatitis C virus (HCV) infection; dietary exposure to the fungal toxin, aflatoxin B_1 (AFB_1); cirrhosis, whatever its cause; the metobolic syndrome; hepatic iron overload and a number of rare inherited metabolic diseases. Of these, HBV and AFB_1 are the predominant risk factors in those resource-poor countries in Eastern and South-Eastern Asia, the Western Pacific islands, and sub-Saharan Africa that have the highest incidences of the tumour. Closer analysis reveals that the association between exposure to AFB_1 and the development of HCC has been reported almost exclusively in countries in which chronic HBV infection is endemic. This is a chance finding, although the possibility that AFB_1 is hepatocarcinogenic only in the presence of HBV has been mooted. Chronic HBV infection is more common in rural areas of sub-Saharan Africa and exposure to AFB_1 is virtually confined to these areas. Black patients in sub-Saharan Africa and Chinese patients in the Asian-Pacific region with HCC are appreciably younger than patients elsewhere in the world, which could be explained, at least in part, by synergism between the hepatocarcinogenic properties of the two risk factors. Such an interaction may also contribute to the very high incidence of HCC in these regions and to its higher frequency in rural than in urban dwellers in the African sub-continent.

Dietary iron overload has been documented to occur only in sub-Saharan Africa. Although the condition does occur in urban Blacks, it is far more common in rural dwellers. The geographical distribution of African iron overload thus parallels that

M.C. Kew
Department of Medicine, University of Cape Town and Groote Schuur Hospital, Cape Town and Department of Medicine, University of the Witwatersrand Medical School, Johannesburg, South Africa
e-mail: Michael.C.Kew@uct.ac.za

M. Rai and A. Varma (eds.), *Mycotoxins in Food, Feed and Bioweapons*,
DOI: 10.1007/978-3-642-00725-5_7, © Springer-Verlag Berlin Heidelberg 2010

of dietary exposure to AFB_1, as does that of the HCC that may complicate the two conditions. A synergistic interaction between these two carcinogens could be contributing to the occurrence of the tumour in older rural Black Africans.

Possible interactions between AFB_1 and other known hepatocarcinogens have received relatively little attention.

The information available on synergistic interactions between AFB_1 and other known risk factors for HCC is reviewed in this chapter.

7.2 Synergistic Hepatocarcinogenic Interaction Between Chronic Hepatitis B Virus Infection and Aflatoxin B_1 Exposure

The earliest investigations into a possible causal association between ingestion of staple foodstuffs contaminated with AFB_1 and the development of HCC, although performed in countries in which HBV infection was endemic, did not include data on the HBV status of the population studied. In later studies, the possible roles of AFB_1 exposure and HBV infection in explaining the varying frequencies of HCC in different areas of Swaziland (Peers et al. 1987) and the Guangxi Province of China (Yeh et al. 1989) were assessed. Both analyses concluded that with simultaneous exposure to AFB_1 and chronic HBV infection, exposure to the fungal toxin was a more important determinant of geographical variation in the incidence of HCC than was the viral infection, but no attempt was made to evaluate a possible synergism between the two risk factors.

The first published evidence consistent with an interaction between AFB_1 and HBV in the genesis of HCC was provided by experiments in which transgenic mice over-expressing the large envelope polypeptide of HBV were fed AFB_1. These mice developed hepatocyte dysplasia and HCC more rapidly and extensively than did unexposed littermates (Sell et al. 1991). Shortly thereafter, further experimental evidence became available of a positive hepatocarcinogenic interaction between AFB_1 and Woodchuck Hepatitis Virus (WHV) (Bannasch 1995), another member of the *Hepadnaviridae*, the family to which HBV belongs. Moreover, woodchucks infected with WHV were shown to have enhanced activation of the biologically inactive parental form of AFB_1 to the highly reactive and mutagenic metabolite, AFB_1–8,9-exo-epoxide (De Flora et al. 1989). However, later studies failed to show increased activation of AFB_1 to AFB_1–8,9-epoxide (Gemechu-Hatewu et al. 1997) or an accelerated rate of HCC formation in WHV-infected woodchucks (Tennant et al. 1990), although interpretation of the data in the latter study was hampered by the small number of surviving woodchucks (IARC Monographs on the Evaluation of Carcinogenic Risks 1994). An increased number of HCCs was reported in ducks infected with duck hepatitis virus, another member of the *Hepadnaviridae*, and exposed to AFB_1 (Cova 1990; IARC 1994), although the number of ducks studied was small.

Table 7.1 The findings in four studies comparing the risk of HBV infection alone, dietary exposure to AFB_1 alone, and the two risk factors together in the genesis of HCC

HBV alone	AFB_1 alone R.R (95% CL)[a]	HBV and AFB_1 R.R. (95% CL)	R.R. (95% CL)
Ross et al. (1992)	4.8 (1.2, 19.7)	1.9 (0.5–7.5)	60.1 (6.4–561.8)
Qian et al. (1994)	7.3 (2.2, 24.4)	3.4 (1.1–10.0)	59.4 (15.6–212)
Wang et al. (1996)	17.4 (3.6, 143.4)	0.3 (0–3.6)	70.0 (11.5–425.4)
Lunn et al. (1997)	17.0 (2.8, 103.9)	17.4 (3.4, 90.3)	67.6 (12.2, 373.2)

[a]Relative risk (95% confidence limits)

Following the introduction of laboratory methods to measure aflatoxin metabolites and aflatoxin-DNA adducts in urine and aflatoxin-albumin adducts in serum – biomarkers that were a far more accurate and reliable indicator of AFB_1 exposure than the hitherto used food sampling and dietary questionnaires – several large cohort studies were undertaken in Shanghai and Qidong counties, China and in Taiwan. All showed a synergistic interaction in hepatocarcinogenesis between exposure to AFB_1 and the HBV carrier state (Ross et al. 1992; Qian et al. 1994; Wang et al. 1996a; Lunn et al. 1997; Yu et al. 1997; Sun et al. 1999; Ming et al. 2002). In four of the investigations the relative risks of exposure to AFB_1 alone, being a HBV carrier alone, and having the two risk factors together were calculated. A greater-than-multiplicative carcinogenic effect was evident in three of the studies (Ross et al. 1992; Qian et al. 1994; Wang et al. 1996a) and a sub-multiplicative effect in the fourth (Lunn et al. 1997) (Table 7.1). A dose-response effect was shown in one of the analyses (Wang et al. 1996a). The increased relative risks for HCC in those individuals exposed to AFB_1 but not infected with HBV in these and other studies attests to the carcinogenic potential of the mycotoxin alone.

In other investigations, also performed in countries with high rates of contamination of staple foodstuffs by AFB_1, only individuals chronically infected with HBV were studied and the influence of AFB_1 exposure in further increasing their risk of HCC development was analysed. In Qidong County, China, over a 10-year prospective follow-up, the risk of HCC in male carriers of the virus with detectable urinary levels of AFB_1 metabolites was shown to be increased 3.3-fold (95% confidence limits 1.2, 8.7) (Sun et al. 1999). This result was later confirmed in a longer observation of the same cohort of carriers, when the risk of HCC was increased 3.5-fold (95% confidence limits 1.5, 8.1) (Ming et al. 2002). In addition, a dose-response relationship between urinary metabolites of AFB_1 and risk of HCC was shown in HBV carriers in Taiwan (Yu et al. 1997). Comparing high and low urinary levels of the metabolite, AFM_1, a multivariate-adjusted odds ratio of 6.0 (95% confidence limits 1.2, 29.0) was obtained. The risk was greater (odds ratio 10.0: 95% confidence limits 1.6, 60.9) when both AFM_1 and AFB_1-N^7- guanine metabolites were measured in the urine (Yu et al. 1997).

In those populations in which an interaction between the fungal toxin and HBV has been found, the viral infection is predominantly acquired in infancy or early childhood. During the early years of the infection a state of immune tolerance towards the virus exists and little if any cellular damage occurs. With loss of tolerance, chronic hepatitis with continuous or recurring cell damage supervenes. Children may be exposed to AFB_1 from an early age (Wild et al. 1993), perhaps as early as in utero (Wild et al. 1991). In China and Taiwan perinatal transmission of HBV is the predominant mode of infection, whereas in Africa, slightly later horizontal infection is the major cause. Whether chronic infection with HBV precedes exposure to AFB_1 or vice versa, and whether this is important in the aetiology and pathogenesis of HCC, is uncertain.

The effect of a synergistic interaction between AFB_1 and HBV on the age of onset of HCC was specifically addressed in a study of Taiwanese patients. HBV-infected patients in whom tumour tissue was shown to be positive for AFB_1-N^7-guanine adducts were, on average, 10 years younger than those with adduct-negative tumours (Chen et al. 1992). This observation supports the belief that co-exposure to the two risk factors is responsible, at least in part, for the significantly younger age of rural patients with HCC in sub-Saharan Africa (Kew 1992).

In some studies a positive interaction between HBV and AFB_1 seemed to depend on the presence of a polymorphism of the phase II detoxification genes, glutathione-*S*-transferase (GST) M1 and T1 and epoxide hydrolase (EPHX), whose products convert the potentially carcinogenic AFB_1-8,9-exo-epoxide to non-reactive metabolites (McGlynn et al. 1995; Chen et al. 1996; Sun et al. 2001). But no consistent pattern has emerged. In one analysis in Taiwan, the risk of HCC formation was greater in HBV carriers with the GST T1 null genotype compared with the non-null genotype (Sun et al. 2001), but in a second the risk appeared to depend on the presence of a GST M1 null genotype (Chen et al. 1996). In a study in West African and Chinese patients the risk was greater in those with null genotypes of GST M1 (McGlynn et al. 1995). A multiplicative interaction between HBV infection and mutations of the EPHX gene in the genesis of HCC was demonstrated in the third study: patients without chronic HBV infection but with at least one EPHX mutant allele had a 3.3-fold increase in HCC risk, those with HBV infection but normal EPHX alleles a 15-fold increase in risk, and in those with both HBV infection and at least one EPHX mutant allele the increased risk was 77-fold (McGlynn et al. 1995).

The presence of a guanine to thymine transversion at the third base of codon 249 of the p53 tumour suppressor gene (arginine to serine substitution; 249^{ser} p53) has been found in as many as 60% of HCCs from patients in regions with heavy dietary exposure to AFB_1 (Hsu et al. 1991; Bressac et al. 1991). A specific causative association between this inactivating mutation and dietary exposure to aflatoxins and the presence of AFB_1 biomarkers was later confirmed in epidemiological studies in regions with high or low exposure rates (Ozturk et al. 1991; Eaton and Gallacher 1994). Stemming from this observation, the presence of the 249^{ser} p53 mutation has been used as a permanent marker of dietary exposure to AFB_1 in some studies of the interactive effects between the mycotoxin and HBV. The findings have, however, been inconsistent. In an investigation of Taiwanese patients with HCC all of the 249^{ser} p53 mutations occurred in patients positive for HBsAg, giving an odds ratio of 10.0 (95% confidence limits 1.6; 175) (Lunn et al. 1997). Similarly, in a study in Qidong County, China all of the HCC patients with 249^{ser} p53 mutations showed evidence of chronic HBV infection (Ming et al. 2002), and in a second analysis in Taiwan the mutation was present in 36.3% of HBV-infected patients with HCC compared with 11.7% of those without HBV markers (Wang et al. 1996a). A study in The Gambia showed a multiplicative increase in the risk of HCC development in Black Africans when the two risk factors occurred together compared with that with either factor alone (Kirk et al. 2005). Other analyses revealed a similar but not significant trend (Sheu et al. 1992; Stern et al. 2001), and in the remaining studies from a variety of countries no association could be found (listed in Stern et al. 2001; Kimbi et al. 2005). Moreover, in a meta-analysis of 49 published studies using a method that takes into account both within-study and study-to-study variability, little evidence for HBV-AFB_1 interaction in modulating the 249^{ser} p53 mutation was found (Stern et al. 2001). Whether HBV acts as a synergistic partner or a confounder in the genesis of the p53 249^{ser} mutation thus remains unclear.

In summary, persuasive evidence is available that AFB_1 and HBV interact synergistically in the aetiology and pathogenesis of HCC. Based on the relative risks for HCC recorded in patients with exposure to AFB_1 alone, this fungal toxin is also an independent risk factor for the tumour. Whether this applies only at very high or moderately high levels of dietary exposure to AFB_1, and whether at lower levels of exposure HBV is an obligatory co-carcinogen for the mycotoxin, remains to be determined.

7.2.1 Possible Mechanisms of Interaction between Aflatoxin B_1 and Hepatitis B Virus in Hepatocarcinogenesis

Given that AFB_1 and HBV are synergistic causative agents of HCC, the possible mechanisms responsible for their interaction need to be considered. A number have been suggested. The first is that HBV infection directly or indirectly sensitises hepatocytes to the carcinogenic effects of AFB_1. One way in which this may be accomplished is that the specific cytochrome P450s that convert harmless AFB_1 to highly reactive AFB_1–8,9-exo-epoxide may be induced by either chronic hepatitis resulting from HBV infection or the presence of the virus itself. Induction of these phase I enzymes has been described in woodchucks infected with WHV (Gemechu-Hatewu et al. 1997) and in HBV transgenic mice (Chemin et al. 1999), where this effect appeared to result from hepatocyte injury rather than the presence of the virus per se (Chemin et al. 1999). The observation that Gambian and Taiwanese children and adolescents chronically infected with HBV have higher concentrations of AFB_1 adducts than uninfected individuals (Allen et al. 1992; Chen et al. 1992; Turner et al. 2000) is consistent with this mechanism although, because of the generally asymptomatic nature of HBV carriage in these children, the virus itself rather than the resulting chronic hepatitis would be favoured as the cause. On the other hand, studies in adults in China, Taiwan, and The Gambia either failed to show a significant difference in serum AFB_1-albumin adduct levels between HBsAg-positive and -negative subjects (Groopman et al. 1992; Wang et al. 1996b; Chen et al. 2002) or revealed a marginally significant difference only (Sun et al. 2002). Moreover, woodchucks with chronic WHV infection and HBV transgenic mice did not show enhanced activation of AFB_1 (Tennant et al. 1990; Kirby et al. 1994). Another way in which hepatocytes may be sensitised to the carcinogenic effects of AFB_1 is that the activity of phase II detoxification enzymes (GST and EPHX) may play a role in the genesis of HCC induced jointly by the two causative agents (Chen et al. 1996; Yu et al. 1997; Sun et al. 2001).

A second possible mechanism of synergy between the two hepatocarcinogenic agents involves the 249^{ser} p53 mutation. This mutation abrogates the normal functions of p53, including those in cell cycle control, DNA repair, and apoptosis, thereby contributing to the multistep process of hepatocarcinogenesis. The accelerated hepatocyte proliferation caused by the chronic necroinflammatory hepatic disease resulting from HBV infection increases the likelihood of both AFB_1-induced mutations, including 249^{ser} p53 mutation, being formed and the subsequent clonal expansion of hepatocytes containing these mutations (Chisari et al. 1989). The exact time at which the mutation develops is, however, not certain.

Thirdly, chronic necroinflammatory hepatic disease, including that resulting from HBV infection, results in the generation of reactive oxygen intermediates (ROI) and nitrogen species (Hussain et al. 1994; Liu et al. 1994; Ishima and Bartsch 1994). These are mutagenic. In addition, increased oxidative stress has been shown to induce the 249^{ser} p53 mutation (Hussain et al. 1994).

Finally, AFB_1-DNA adducts are normally repaired by the nucleotide excision repair pathway. Integrated sequences of the virus often include the HBV × gene. The HBV × protein interferes with the nucleotide excision repair pathway (Jia et al. 1999) and might, by this means, favour persistence of existing mutations. Furthermore, the rapid hepatocyte turnover rate in chronic hepatitis and cirrhosis may not allow sufficient time for mutated DNA to be repaired before the cell divides again, thereby "fixing" the mutation in the daughter cell DNA. HBV × protein may also contribute to uncontrolled cell proliferation. Transcription of $p21^{waf1/cip1}$, which induces cell cycle arrest at the G_1-S checkpoint, is activated by HBV × protein in a dose-dependent manner in the presence of functional p53 (Ahn et al. 2002). However, this transcription is repressed by HBV × protein when p53 is not functional or is functional at a low level (Ahn et al. 2002).

7.3 Interaction Between Aflatoxin B_1 and Chronic Hepatitis C Virus Infection in Hepatocarcinogenesis

During recent decades the incidence of chronic HCV infection has increased appreciably in many industrialised countries and in Egypt, and this virus has become a major cause of HCC in these countries. Its incidence, and that of HCV-induced HCC, continues to increase in a number of these countries.

However, in those countries in which chronic HBV infection is endemic and HBV is the predominant cause of the HCC that frequently occurs, the incidence of chronic HCV infection is considerably lower than that of HBV and this virus plays a far lesser role in the aetiology of the tumour. Thus, countries in which AFB_1 contamination of staple crops occurs have low prevalences of chronic HCV infection, and simultaneous exposure to these two risk factors would be uncommon. For this reason there has been, to my knowledge, only two studies in which a possible hepatocarcinogenic interaction between AFB_1 exposure and HCV infection was investigated. In the first study the 249^{ser} p53 mutation was used as a marker of heavy exposure to AFB_1. In the small number of patients studied no synergistic interaction between the presence of the mutation and that of HCV infection was evident (Kuang et al. 2005). In the second study aflatoxin-albumin adducts in serum were used as a measure of AFB_1 exposure, but again no evidence of a synergistic interaction between HCV infection and AFB_1 exposure was demonstrated (Chen 2007).

7.4 Synergistic Interaction Between Aflatoxin B_1 and Dietary Iron Overload in Hepatic Mutaginesis

Dietary iron overload, an important cause of HCC in the Black population of sub-Saharan Africa (Gangaidzu and Gordeuk 1995; Kew and Asare 2007), results from drinking large volumes of a traditional beer that has a high iron content as a consequence of it being home-brewed in cast iron pots or drums (Walker and Arvidsson 1953; Bothwell and Bradlow 1960). The condition (originally called Bantu visceral siderosis) occurs mainly in rural areas (Gordeuk et al. 1986): 80% of the Black population in sub-Saharan Africa lives in rural areas, more than two-thirds of adult males in these areas consume home-brewed beer, and as many as 15% of the Black adult male population may be affected. During the fermentation of sorghum or other locally-grown crops the pH of the ferment decreases to very low levels (3.7 or 3.8), leaching iron from the container into the contents (Bothwell et al. 1964) and accounting for the high iron content of the beverage (46–82 mg/l compared with <0.5 mg/l in commercial beers). This iron is in an ionised, highly bioavailable form (Bothwell and Bradlow 1960; Bothwell et al. 1964; Moyo et al. 1997; MacPhail et al. 1999). The liver is the main storage site for iron and hepatic iron concentrations comparable with those in the better known iron storage disease, hereditary haemochromatosis, result from the consumption over time of large volumes of this traditional beer. The more severe degrees of accumulated hepatic iron may be complicated by portal fibrosis or, less often, cirrhosis (Bothwell and Bradlow 1960; Bothwell and Isaacson 1962). Histological features of alcoholic liver disease are rarely evident (the alcohol content of the beer is only approximately 3%) and the amounts of iron present far exceed those that may occur in alcohol-induced liver disease (Bothwell et al. 1964; Chapman et al. 1982; Gordeuk et al. 1986;

Tavill and Qadri 2004). Both hepatocytes and macrophages are affected in dietary iron overload (Isaacson et al. 1961; Brink et al. 1976), whereas hepatocytes are predominantly affected in hereditary haemochromatosis (Valberg et al. 1975; Brink 1976).

Fibrosis and cirrhosis frequently occur in hereditary haemochromatosis, and when HCC develops in these patients it almost always does so in a cirrhotic liver (Niederau et al. 1985; Deugnier et al. 1993). Because of this almost universal association and the less frequent association in patients with African dietary iron overload, it is clear that chronic necroinflammatory hepatic disease, and cirrhosis in particular, contribute to the malignant transformation that occurs with excess hepatic iron, as they do with most other causes of this tumour (Kew and Popper 1984). During recent years, however, convincing evidence has accumulated that excess hepatic iron may also be independently hepatocarcinogenic (Kew and Asare 2007). This evidence includes an animal model for dietary iron-induced HCC in which malignant transformation occurred in the absence of fibrosis and cirrhosis (Asare et al. 2006).

Intracellular free iron is a catalyst for the formation of ROI. Iron overload thus disrupts the redox balance of the cell and generates chronic oxidative stress, which damages hepatocytes, DNA, protein, and lipids (Loeb et al. 1988; Ichiba et al. 2003; Jungst et al. 2004; Asare et al. 2006). The resulting chronic necroinflammatory hepatic disease in turn generates further ROI and additional oxidative damage. Chronic necroinflammatory hepatic disease ultimately results in cirrhosis, a pre-neoplastic condition (Ichiba et al. 2003; Jungst et al. 2004; Asare et al. 2006).

Increased lipid peroxidation (LPO) is believed to be an important contributor to hepatocarcinogenesis in iron overload (Cadenzas 1989; Esterbauer et al. 1991; Esterbauer 1993; Cheesman 1993; Benhar 2002). Oxidative stress leads to LPO of unsaturated fatty acids in membranes of cells and organelles. This results in chain breaks in fatty acids with the insertion of hydrophobic groups and cis-trans isomerization. Cytotoxic by-products such as malondialdehyde (MDA) and 4-hydroxy-2′-nonenal (4-HNE) are produced. In addition to participating in the initiation and propagation steps of LPO, iron is thought to be involved in β-cleavage of lipid hydroperoxides, producing biogenic aldehydes that interact with DNA to form exocyclic products. The cytotoxic and reactive aldehydeic by-products of LPO impair cellular function and protein synthesis (Loeb et al. 1988; Esterbauer et al. 1991; Asare et al. 2006). Furthermore, 4-HNE and MDA are capable of diffusing from their production sites to more distant sites within the cell to interact with DNA, protein, or protein nucleophiles. Both MDA and 4-HNE are cytotoxic as well as genotoxic (Cadenzas 1989; Esterbauer 1993; Cheesman 1993; Benhar et al. 2002; Hu and Tang 2004) and are implicated in carcinogenesis. 4-HNE, for example, can be further metabolised into an epoxide that can interact with DNA to form exocyclic entheno-guanine, -adenine, and -cytosine adducts. The amount of 4-HNEdG adducts significantly increases in the liver of rodents, and exocyclic guanine products have been shown to be increased in rodent models of hepatocarcinogenesis (Hagen et al. 1994; Dabbagh et al. 1994; Wacker et al. 2001).

Deoxyguanosine residues in DNA are also hydroxylated at the C8 position to form 8-hydroxy-2′-deoxyguanosine (8OHdG) (Cheng et al. 1992; Dabbagh et al. 1994; Orimo et al. 2006), the major promutagenic adduct produced by ROI and which leads to G:C to T:A transversions (Ichiba et al., 2003). 8OHdG also correlates with the rate of DNA unwinding and strand breaks in tissue (Cheng et al. 1992) and has been identified as a biomarker in HCC. An association between DNA unwinding and the risk of HCC formation has been described (Ichiba et al. 2003). It is believed that iron-induced chronic oxidative stress eventually leads to mutations in tumour suppressor genes and critical DNA repair genes. This view is supported by the increased lipid peroxidation and frequency of mutations in the tumour suppressor gene, p53, identified in non-tumourous liver samples from patients with HCC, hereditary haemochromatosis, and Wilson's disease (Vautier et al. 1999; Hussain et al. 2000; Marrogi et al. 2001).

Increased production of nitric oxide by nitric oxide synthase 2 has been demonstrated in human HCC cell lines, in hepatocytes in tissue culture, and in patients with HCC complicating hereditary haemochromatosis (Kim et al. 2000; Hon et al. 2002; Vadrot et al. 2006). Possible mechanisms of action include LPO, an effect on apoptosis, and impaired DNA repair.

Experimental evidence in animals has shown that cyclin D1, a protein involved in G_{1-2} phase of the cell cycle, is over-expressed in the iron-overloaded liver. This could contribute to cell cycle abnormalities and might play a role in iron-induced hepatocarcinogenesis (Troadec et al. 1986).

Finally, evidence suggests that excess hepatic iron induces immunologic abnormalities that may decrease immune surveillance for malignant transformation. Non-transferrin-bound iron may inhibit proliferation of lymphocytes, especially the D4 subset (Green et al. 1988). In addition, lymphocyte proliferation is inhibited by ferritin (Green et al. 1988). Tumoricidal function of mice macrophages are markedly decreased in the presence of iron and ferritin (Matzner et al. 1979).

Because contamination of certain foodstuffs by *Aspergillus* species and dietary iron overload occur together almost exclusively in rural regions of sub-Saharan Africa, the possibility of a synergistic interaction between their hepatocarcinogenic effects should be considered. No evidence in clinical practice to support such an interaction has been published, although some, as yet unconfirmed, experimental support is available (Asare et al. 2007). In Wistar albino rats fed a diet containing either AFB_1 or iron or the two together, the presence of LPO showed a significant additive effect and 8OHdG levels a significant multiplicative effect in the iron/AFB_1 group compared with the Fe and AFB_1 groups alone. Using the Ames Mutagenicity Test with the *Salmonella typhimurium* TA 100 bacteria strains, highly significantly increased levels of mutagenesis were observed in the iron/AFB_1 group (multiplicative synergy was fivefold; $p < 0.001$ compared with the iron and AFB_1 groups alone). In addition, heavy aggregates of 4-HNE and 8OHdG were observed histochemically in liver sections of the iron/AFB_1 group but not in the rats receiving iron or AFB_1 alone.

7.4.1 *Possible Mechanisms of Interaction Between Aflatoxin B_1 and Iron Overload in Hepatocarcinogenesis*

The synergism between the deleterious effects of excess tissue iron and the ingestion of AFB_1 appears to be taking place at the level of DNA damage, which would explain the increased hepatocarcinogenic potential. One possible mechanism of this interaction would be the induction of the codon 249^{ser} p53 mutation both by the dietary exposure to AFB_1 (Hsu et al. 1991; Bressac et al. 1991) and by the oxidative stress and generation of 8OHdG resulting from the increased storage iron in the liver (Cheng et al. 1992; Hussain et al. 1994; Vautier et al. 1999, Hussain et al. 2000; Marrogi et al. 2001; Ichiba et al. 2003; Orimo et al. 2006). This mutation abrogates the functions of p53, including those involved in DNA repair and apoptosis. In addition, a high level of non-heme iron in hepatocytes can inhibit nitric oxide-induced apoptosis by converting nitric oxide from a pro-apoptotic molecule to an anti-apoptotic molecule (Kim et al. 2000).

7.5 Cirrhosis as a Possible Tumour Promoter for Aflatoxin B_1-Induced Hepatocarcinogenesis

Acute exposure to very high concentrations of aflatoxins, usually as a result of eating heavily contaminated maize, causes severe liver dysfunction or frank liver failure and acute hepatic necrosis (a condition referred to as acute aflatoxicosis) (Probst 2004; Azziz-Baumgartner et al. 2004). A fatal outcome occurs in as many of 39% of those exposed. Repeated exposure to lower concentrations of aflatoxins has been reported to cause cirrhosis in humans in one study only (Kuniholm et al. 2008) and not in animals. It does, however, produce severe steatosis in mammals (Amaya-Farfan 1999). Steatosis has not been described in humans chronically exposed to aflatoxins, although steatohepatitis occurring in patients with diabetes mellitus, obesity, or the metabolic syndrome has been incriminated as a risk factor for HCC (Ratzui and Poynard 2005).

Cirrhosis, whatever its cause, acts as a tumour promoter in hepatocarcinogenesis (Kew and Popper 1984) and theoretically this could apply equally in AFB_1-induced HCC. The mechanisms by which tumour promotion by cirrhosis could be achieved include an increased hepatocyte turnover rate and the production of ROI and chronic oxidative stress. However, no reports of cirrhosis being present in patients with AFB_1-induced HCC have been published.

7.6 Alcohol as a Possible Tumour Promoter in Aflatoxin B_1-Induced Hepatocarcinogenesis

In 80–95% of patients with HCC in industrialised countries the tumour develops in a cirrhotic liver (Kew and Popper 1984; Fattovich et al. 2004). Habitual alcohol abuse or chronic HCV infection, or the two together, cause the cirrhosis in the great majority of these patients. Indeed, HCC rarely develops in alcoholics in the absence of cirrhosis (Del Olmo et al. 1998; Kuper et al. 2001), suggesting that it is the chronic necroinflammatory hepatic disease rather than the alcohol per se that is responsible for the malignant transformation. Studies in experimental animals support the belief that alcohol is not directly hepatocarcinogenic (Poschl and Seitz 2004), although it may act as a co-carcinogen or as a tumour promoter.

Increased amounts of iron accumulate in the liver in patients with alcoholic liver disease, although the amounts are considerably less than those characteristic of hereditary haemochromatosis or African dietary iron overload (Rouault 2003; Tavill and Qadri 2004). The reasons for this phenomenon have yet to be fully clarified (Tavill and Qadri 2004).

The tolerance of the liver to very high levels of storage iron in those patients with hereditary haemochromatosis or dietary iron overload but without fibrosis or cirrhosis suggests that co-factors may be necessary to activate stellate cells, the putative mediators of hepatic fibrosis (Tavill and Qadri 2004). Alcohol is likely to be one such factor because there is ample evidence that chronic alcohol abuse interacts with even slightly increased amounts of hepatic iron in the pathogenesis of fibrosis and cirrhosis (Jacobovits et al. 1979; Ioannou et al. 2000; Tsukamoto et al. 1995; Tavill and Qadri 2004). Moreover, in patients with hereditary haemochromatosis complicated by cirrhosis, chronic alcohol ingestion further increases the risk of malignant transformation (Deugnier et al. 1993).

In summary, a direct correlation between alcohol consumption per se and the development of HCC remains tenuous, although a causal association between alcohol-induced cirrhosis and HCC is undoubted.

Patients with African dietary iron overload are exposed to alcohol in parallel with excess iron. Acetaldehyde, the product of ethanol metabolism by alcohol dehydrogenase, causes cellular damage and generates free radicals that bind to numerous cellular targets, including components of cell signalling pathways and DNA (Stickel et al. 2002; Arteel 2003). Oxidative stress produces dysregulation of gene expression and cell signalling cascades, which manifest in unregulated hepatocyte proliferation (Obe and Ristow 1979). Acetaldehyde also decreases the methylation of cytosine, impairs DNA repair, and affects sister chromatid exchange, all of which may contribute to its carcinogenic potential (Lieber and DeCarli 1970; Sakamoto et al. 2006). In addition, alcohol induces cytochrome p450 2E1 (CYP2E1) (Petersen 2005). This enzyme not only facilitates the absorption of procarcinogens and their activation to tumour-promoting factors, but also generates ROI (Caro and Cederbaum 2004; McKillop and Schrum 2005). Induction of CYP2E1 may therefore result in DNA damage and mutations, or be involved in dysregulation of gene expression and cell signalling cascades, manifesting in upregulated cell proliferation (McKillop and Schrum 2005). Another result of oxidative stress is the peroxidative decomposition of membrane polyunsaturated fatty acids, such as arachidonate, and the β-cleavage of lipid hydroperoxides (Caro and Cederbaum 2004). The resulting biogenic aldehydes, particularly 4-HNE and MDA, are electrophilic and can interact with DNA or protein nucleophiles that specifically target the p53 tumour suppressor gene (Cheng et al. 1992; Dabbagh et al. 1994).

Alcohol abuse also results in interference with DNA methylation, an important epigenetic mechanism affecting the transcriptional regulation of genes involved in the development of HCC. This effect on hepatic methylation capacity is the result of its detrimental influence on the intake, absorption, and metabolism of vitamins responsible for methyl group synthesis and transfer (Murrata et al. 2004). In addition, the latter results in reduced levels of important anti-oxidants, methionine, *S*-adenosylmethionine and glutathione, which may have consequences with respect to hepatocarcinogenesis. Promoter hypermethylation also silences cyclo-oxygenase-2 and regulates growth in HCC (Austin 1991). The ingestion and metabolism of alcohol reduces hepatic retinoic acid levels and may thereby enhance cell proliferation and malignant transformation via up-regulation of activator protein-1 gene expression (Stickel et al. 2002).

A number of ways in which iron and alcohol might interact in the generation of HCC are suggested. The free radical, superoxide, is reasonably stable in aqueous solutions at pH 7.4 and its reaction rates with lipids, proteins, or DNA are slow (Kruszewsky 2003). However, its interaction with Fe^{3+} is much faster. The rapid reaction of superoxide with nitric oxide results in the formation of the very potent nitrating species, peroxynitrite. Alcohol-fed rats produce greater quantities of superoxide anion compared with those fed iron. Conversely, the iron-fed rats generated greater quantities of nitrites compared with the alcohol-fed rats. Rats fed both iron and alcohol produce lesser amounts of superoxide anion, but this might be explained by the possible combination of the excess nitrite and superoxide to form peroxynitrite. Superoxide and peroxynitrite have been shown to be involved in signalling events (Kruszewsky 2003), and they may be involved in the mobilization of iron from intracellular iron storage proteins and the formation of low molecular weight iron complexes (Chung et al. 2000).

In a study by Asare et al. (2008) total lipid hydroperoxides increased significantly in the iron-fed rats but moderately in the alcohol-fed rats. However, the two agents given together resulted in a threefold synergism in the levels of lipid hydroperoxides. Of the biogenic aldehydes, 4-HNE adducts were detected in both the iron-fed and the alcohol-fed rats but, as evidenced by the immunohistochemical staining, there was a synergistic interaction between the two agents. 4-HNE-deoxyguanosine adducts are known to be significantly increased in a rat model of hepatocarcinogenesis (Marrogi et al. 2001; Feng et al. 2004). This adduct is more potent in human cells and is preferentially formed at codon 249 of the p53 gene (Bartsch and Nair 2004), a known mutational site for HCC.

The pro-oxidant action of iron and alcohol was also seen at the level of 8-OHdG formation (Asare et al. 2008). These levels and the immunohistochemical staining were in keeping with synergistic DNA damage between iron and alcohol. Further hydroxylation of the $^{\bullet}OH$ radical occurs at position C8 of the deoxyguanine residues and misreading of this results in guanine to thymine transversion (Cadenzas 1989). Direct involvement of $^{\bullet}OH$ radicals is also seen by the formation of thymidine glycol (TG) adducts. These adducts are derived from the hydroxyl free radical attack on the carbon 5 and 6 ethylenic bond of the thymine moiety of TG. The heavy deposits of TG adducts, as seen immunohistochemically, supports a synergistic DNA-damaging effect produced by the •OH free radical (Asare et al. 2008). Although levels of TG adducts were not measured in this study, in a study of another second generation DNA adduct, ε-DNA, increased hepatic levels were detected in patients with hereditary haemochromatosis (Deugnier 2003).

These observations are consistent with previous reports on the possible synergy of iron and alcohol in hepatocarcinogenesis (Deugnier 2003; Fletcher and Powell 2003; Tavill and Qadri 2004; Bartsch and Nair 2004; Petersen 2005).

7.7 Interaction Between Aflatoxin $B_1$and Polycyclic Aromatic Hydrocarbons in Hepatocarcinogenesis

Polycyclic aromatic hydrocarbons (PAH) are ubiquitous environmental pollutants produced during combustion of organic materials and are found in polluted air, cigarette smoke, and the diet (Chen et al. 2002). Exposure to high levels of these pollutants increases the risk of developing a number of human cancers, including breast and lung cancer (Chen et al. 2002).

Increased levels of PAH-DNA adducts have recently been demonstrated in liver tissue adjacent to HCC (Chen et al. 2002). In a follow-up study a possible synergistic interaction between PAH-albumin adducts, AFB_1 exposure, and chronic HBV infection in causing HCC was investigated (Wu et al. 2007). A trend towards an interaction between PAH-albumin adducts and exposure to AFB_1 was shown, with a definite interaction between exposure to PAH and the combined presence of exposure to AFB_1 and the presence of chronic HBV infection (odds ratio 8.2: 95% confidence intervals 3.6–19). However, because interaction between AFB_1 and HBV would itself increase the risk of HCC formation this study is incomplete.

The possible mechanism of PAH-induced hepatocarcinogenesis is not fully understood, although the generation of ROI intermediates has been demonstrated (Autrup et al. 1999). Oxidative stress is also produced by chronic HBV infection and is a possible explanation for the malignant transformation in the patients reported.

References

Ahn JI, Jung EY, Kwun HJ, Lee CW, Sung YC, Jang KL (2002) Dual effects of hepatitis B virus X protein on the regulation of cell cycle depending on the ststus of cellular p53. J Gen Virol 83:2765–2772

Allen SJ, Wild CP, Wheeler JG, Riley EM, Montesano M, Bennett S Whittle HC, Hall AJ, Greenwood BM (1992) Aflatoxin exposure, malaria and hepatitis B infection in rural Gambian children. Trans Roy Soc Trop Med Hyg 86:426–430

Amaya-Farfan J (1999) Aflatoxin B1-induced hepatic steatosis: role of carbonyl compounds and active diols on steatogenesis. Lancet 353:747–748

Arteel GE (2003) Oxidants and antioxidants in alcohol-induced liver disease. Gastroenterology 124:778–790

Asare GA, Bronze M, Naidoo V, Kew MC (2007) Interaction between aflatoxin B_1 and dietary iron overload in hepatic mutagenesis. Toxicology 234:157–166

Asare GA, Mossanda KS, Kew MC, Paterson AC, Kahler-Venter CP, Siziba K (2006) Hepatocellular carcinoma caused by iron overload: a possible mechanism of direct hepatocarcinogenicity. Toxicology 219:41–52

Asare GA, Bronze M, Naidoo V, Kew MC (2008) Synergistic interaction between excess hepatic iron and alcohol ingestion in hepatic mutagenesis. Toxicology 254:11–18

Austin A (1991) The role of tobacco use and alcohol consumption in the etiology of hepatocellular carcinoma. In: Tabor E, DiBiscegie AM, Purcell R, (eds) Etiology, pathology, and treatment of hepatocellular carcinoma in North America. Gulf Publishing Co, Houston, pp 57–76

Autrup H, Daneshvar LO, Dragsted M, Gamborgt M, Hansen S, Loft H (1999) Biomarkers of exposure to ambient air pollution — comparison of carcinogen-free DNA adduct levels and other exposure markers and markers for oxidative stress. Environ Health Perspect 107:233–238

Azziz-Baumgartner E, Lindblade L, Gieseker K, Schurz Rogers H, Kieszak S, Njapau H Schleicher R, McCoy LF, Misore A, DeCock K, Rubin C, Slutsker L (2005) Aflatoxin Investigative Group: case-control study of an acute aflatoxicosis outbreak, Kenya, 2004. Environ Health Perspect 113:1779–1783

Bannasch P, Khoshkou NI, Hacker HJ, Raedeva S, Mrzek M, Zillmann U, Kopp-Schneider A, Haberkom U, Elgas M, Tolle T (1995) Synergistic hepatocarcinogenic effect of hepadnaviral infection and dietary aflatoxin B_1 in woodchucks. Cancer Res 55:3318–3330

Bartsch H, Nair J (2004) Oxidative stress and lipid peroxidation-derived DNA lesions in inflammatory-driven carcinogenesis. Cancer Detection Prev 28, 385–391.

Benhar M, Engelberg D, Levitzki A (2002) Reactive oxygen stress, stress-activated kinases and stress signalling in cancer, EMBO Rep 3:420–425

Bothwell TH, Bradlow BA (1960) Siderosis in the Bantu: a combined histopathological and chemical study. Arch Path 70:279–292

Bothwell TH, Isaacson C (1962) Siderosis in the Bantu: a comparison of incidence in males and females. Br Med J 1:522–524

Bothwell TH, Seftel H, Jacobs P, Torrance JD, Baumslag N (1964) Iron overload in Bantu subjects. Studies on the availability of iron in Bantu beer. Am J Clin Nutrit 14:47–51

Bressac B, Kew MC, Wands JR, Ozturk M (1991) Selective G to T mutations of p53 gene in hepatocellular carcinoma from southern Africa. Nature 350:429–430

Brink B, Disler P, Lynch SR, Jacobs P, Charlton RW, Bothwell TH (1976) Patterns of iron storage in dietary iron overload and idiopathic haemochromatosis. J Lab Clin Med 88:725–731

Cadenzas E (1989) Biochemistry of oxygen toxicity. Ann Rev Biochem 58:79–110

Caro AA., Cederbaum A (2004) Oxidative stress, toxicology, and pharmacology of CYP2E1. Ann Rev Pharmacol Toxicol 44:9–18

Chapman RW, Morgan MY, Laulicht M, Hoffbrand AV, Sherlock S (1982) Hepatic iron stores and markers of iron overload in alcoholics and patients with idiopathic haemochromatosis. Dig Dis Sci 27:909–916

Cheeseman KH (1993) Lipid peroxidation in cancer. In: Halliwell B, Arouma AI, (eds) DNA and Free Radicals. Ellis Horwood, London, pp 109–144

Chemin I, Ohgaki H, Chisari FV, Wild CP (1999) Altered expression of hepatic carcinogen metabolizing enzymes with liver injury in HBV transgenic mouse lineages expressing various amounts of hepatitis B surface antigen. Liver 19:81–87

Chen C-J, Yu M-W, Liaw Y-F, Wang LW, Chiampresert S, Martin F, Hirvonen A, Bell DA, Santella RM (1996) Chronic hepatitis B carriers with null genotypes glutathione-S-transferase M1 and T1 polymorphisms who are exposed to aflatoxin are at increased risk of hepatocellular carcinoma. Am J Human Genet 59:128–134

Chen SY, Wang LY, Lun WY, Tsai PH, Lee CS, Ahsan YJ, Zhang VJ, Chen CJ, Santella RM (2002) Polycyclic aromatic hydrocarbon-DNA adducts in the liver tissues of hepatocellular carcinoma patients and controls. Internat J Cancer 99:14–21
Chen CJ, Zhang YJ, Lu SN, Santella RM (1992) Aflatoxin-B_1-adducts in smeared liver tissue of hepatocellular carcinoma patients. Hepatology 16: 1150–1155.
Chen CH, Wang MH, Wang JH, Hung CH, Hu TH, Lee SC, Tung HD, Lee CM, Changchien CS, Chen PF, Hsu MC, Lu SN (2007) Aflatoxin exposure and hepatitis C virus in advanced liver disease in a hepatitis C virus endemic area in Taiwan. Am Trop Med Hyg 77:674–752
Cheng KC, Cahill DS, Kasai H, Nishimura S, Loeb LA (1992) 8-hydroxyl guanine, an abundant form of oxidative DNA damage, causes G-T and A-C substitutions. J Biol Chem 267: 166–172
Chisari FV, Klopchin K, Moriyama T, Pasquinelli C, Dunsford HA, Sell S, Prinkert CA, Brinster RL, Palmiter RD (1989) Molecular pathogenesis of hepatocellular carcinoma in hepatitis B transgenic mice. Cell 59:1145–1156
Chung FL, Nath RG, Ocando J, Nishkawa A, Zhang L (2000) Deoxyguanine adducts of *t*-4-hydroxy-2-nonenal are endogenous DNA lesions in rodents and humans: detection and potential sources. Cancer Res 60:1507–1511
Cova L, Wild CP, Mehrotra R, Tursov V, Shirai T, Lambert V, Jaquet C, Tomatis L, Trepo C, Montesano R (1990) Contribution of aflatoxin B_1 and duck hepatitis B virus in the induction of liver tumors in ducks. Cancer Res 50:2156–2163
Dabbagh AJ, Mannion T, Lynch SM, Frei B (1994) The effect of iron overload on rat plasma and liver oxidant status *in vivo*. Biochem J 300:799–803
De Flora S, Hietane E, Bartsch H, Camoirano A, Izzotti A, Bagnasco M, Millman I (1989) Enhanced metabolic activation of chemical hepatocarcinogenesis in woodchucks infected with woodchuck hepatitis virus. Carcinogenesis 10:1099–1106
Del Olmo JA, Serra MA, Rodriguex F, Escudero A, Gilabert S, Rodrigo JM (1998) Incidence and risk fractors for hepatocellular carcinoma in 967 patients with cirrhosis. J Cancer Res Clin Oncol 124:560–564
Deugnier Y (2003) Iron and liver cancer. Alcohol 30:145–150
Deugnier YM, Guyader D, Crantock L, Lopez JM, Turlin B, Yaouang J, Jouanolle H, Campion JP, Launois B, Halliday JW (1993) Primary liver cancer in genetic haemochromatosis: a clinical, pathological and pathogenetic study of 54 cases. Gastroenterology 104:228–234
Eaton DL, Gallagher EP (1994) Mechanisms of aflatoxin carcinogenesis. Annu Rev Pharmacol Toxicol 34:135–172
Esterbauer H (1993) Cytotoxicity and genotoxicity of lipid-oxidation products. Am J Clin Nutr 57 (suppl):779–860
Esterbauer H, Schaur RJ, Zollner H (1991) Chemistry and biochemistry of 4-hydroxynonenal, malondialdehyde and related aldehydes. Free Radic Biol Med 11:81–128
Fattovich G, Stroffolini T, Zagni I, Donato F (2004) Hepatocellular carcinoma in cirrhosis: incidence and risk factors. Gastroenterology 127:S35–S50
Feng Z, Hu W, Tang MS (2004) Trans-4-hydroxy-2-nonenal inhibits nucleotide excision repair in human cells: A possible mechanism for lipid peroxidation-induced carcinogenesis. Proc Natl Acad Sci USA 101:8598–9602
Fletcher LM, Powell LW (2003) Hemochromatosis and alcoholic liver disease. Alcohol 30: 131–136
Gangaidzo IT, Gordeuk VR (1995) Hepatocellular carcinoma and African iron overload. Gut 37:727–730
Gemechu-Hatewu M, Platt K-L, Oesch F, Hacker HJ, Bannasch P, Steinberg P (1997) Metabolic activation of aflatoxin B_1 to aflatoxin B_1_8,9-epoxide in woodchucks undergoing chronic active hepatitis. Int J Cancer 73:587–591
Gordeuk VR, Boyd RD, Brittenham GM 1986 Dietary iron overload persists in rural sub-Saharan Africa. Lancet 1:1310–1313

Green R, Esparza I, Schreiber R 1988 Iron inhibits the non-specific tumoricidal activity of macrophages. A possible contributory mechanism for neoplasia in hemochromatosis. Ann N Y Acad Sci 526:301–309

Groopman JD, Hall AJ, Whittle H, Hudson GJ, Wogan JN, Montesano R (1992) Molecular dosimetry of aflatoxin-N^7 - guanine in human urine obtained in The Gambia. Cancer Epidemiol Biomarkers Prev 1:221–227

Hagen TM, Huang S, Curnutte J, Fowler P, Martinez V, Wehr CM, Ames BM, Chisari FV (1994) Extensive oxidative DNA damage in hepatocytes of transgenic mice with chronic active hepatitis destined to develop hepatocellular carcinoma. Proc Natl Acd Sci USA 91:12808–12812

Hon WM, Lee KH, Khoo HE (2002) Nitric oxide in liver diseases: friend, foe, or just passer-by? Ann NY Acad Sci 962:275–295

Hsu IC, Metcalf RA, Sun T, Welsh JA, Wang NJ, Harris CC (1991) Mutational hotspot in the p53 gene in human hepatocellular carcinomas. Nature (Lond) 350:427–428

IARC Monographs on the Evaluation of Carcinogenic Risks to Humans (1994) Hepatitis B Virus. IARC, Lyon 59:45–164

Hu W, Tang MS (2004) Trans-4-hydroxy-2-nonenal inhibits nucleotide excision repair in human cells: A possible mechanism for lipid peroxidation-induced carcinogenesis. Proc Natl Acad Sci USA 101:8598–9602

Hussain SP, Aquilar F, Amstad P, Cerutti P (1994) Oxy-radical induced mutagenesis of hotspot codons 248 and 249 of the human p53 gene. Oncogene 9:2277–2281

Hussain SP, Raja K, Amstad PA, Sawyer M, Trudel TJ, Wogan GN, Hofseth LJ, Shields PG, Billiar TR, Trautwein C, Hohler T, Galle PR, Phillips DH, Markin R, Morraji AJ, Harris CC (2000) Increased p53 mutation load in non-tumorous human liver of Wilson disease and haemochromatosis: Oxyradical overload disease. Proc Natl Acad Sci U S A 97:12770–12775

Ichiba M, Maeta Y, Mukoyama T, Saeki T, Yasui S, Kanbe T, Okano J, Tanabe Y, Hirooka Y, Yamada S, Kurimasa A, Murawaki Y, Shiota G (2003) Expression of 8-hydroxy-2′-deoxyguanosine in chronic liver disease and hepatocellular carcinoma. Liver Internat 23:1781–1789

Ioannou GN (2000) The effect of alcohol consumption on the prevalence of iron overload, iron deficiency, and iron deficiency anemia. Gastroenterology 126:1293–1301

Isaacson C, Seftel HC, Keeley KJ, Bothwell TH (1961) Siderosis in the Bantu: the relationship between iron overload and cirrhosis. J Lab Clin Med 58:845–853

Ishima H, Bartsch H (1994) Chronic infection and inflammatory processes as cancer risk factors: possible role for nitric oxide in carcinogenesis. Mutation Res 305:253–264

Jacobovits A, Morgan MY, Sherlock S (1979) Hepatic siderosis in alcoholics. Dig Dis Sci 24: 305–310

Jia L, Wang XW, Harris CC (1999) Hepatitis B virus x protein inhibits nucleotide excision repair. Int J Cancer 80:875–879

Jungst C, Cheng B, Gehrke R, Schmitz V, Nischalke HD, Ramakers J, Schramel P, Schirmacher P, Sauerbruch C, Caselmaan WH (2004) Oxidative damage is increased in human liver tissue adjacent to hepatocellular carcinoma. Hepatology 39:1663–1672

Kew MC (1992) Chronic hepatitis-B virus infection and hepatocellular carcinoma in Africa. S Afr J Sci 88:524–528

Kew MC, Asare GA (2007) Dietary iron overload in the African and hepatocellular carcinoma. Liv Internat 27:735–741

Kew MC, Popper H (1984) Relationship between hepatocellular carcinoma and cirrhosis. Semin Liv Dis 4:136–146

Kim YM, Chung HT, Simmons RL, Billia TR (2000) Cellular non-heme iron content is a determinant of nitric oxide-mediated apoptosis, necrosis, and caspase inhibition. J Biol Chem 275:10954–10961

Kimbi GA, Kew MC, Yu MC, Arakawa K, Hodkinson J (2005) 249 serine p53 mutation in the serum of black southern Africans with hepatocellular carcinoma. J Gastro Hepatol 20: 1185–1190

Kirby G, Chemin I, Montesano R, Chisari FV, Lang MA, Wild CP (1994) Induction of specific cyotochrome p-450s involved in aflatoxin B_1 metabolism in hepatitis B transgenic mice. Mol Carcinogenesis 11:74–80

Kirk GD, Lesia OA, Mendy M, Szymanska K, Whittle H, Goedert JT, Hainaut P, Montesano R (2005) 249 ser TP53 mutation in plasma, hepatitis B viral infection, and risk of hepatocellular carcinoma. Oncogene 24:5858–5867

Kruszewsky M (2003) Labile iron pool: the main determinant of cellular response to oxidative stress. Mutat Res 531:81–92

Kuang SY, Lekawanvijit S, Maneenkarn N, Thongswat S, Brodovicz K, Nelson K, Groopman JD (2005) Hepatitis B 1762T/1764A mutations, hepatitis C infection and codon 249 p53 mutations in hepatocellular carcinomas in Thailand. Cancer Epidemiol Biomarkers Prev 14: 380–384

Kuniholm MH, Lesis OA, Mendy M, Akano AO, Sam O, Hall AJ, Whittle H, Bah E, Goedert JJ, Hainaut P, Kirk GD (2008) Aflatoxin exposure and viral hepatitis in the etiology of cirrhosis in The Gambia, West Africa. Environ Health Perspect 116:1553–1557

Kuper H, Ye W, Broome U, Romelsjo A, Mucci LA, Ekbom A, Adami HO (2001) The risk of liver and bile duct cancer in patients with chronic viral hepatitis, alcoholism, and cirrhosis. Hepatology 34:714–718

Lieber CS, DeCarli LM (1970) Hepatic microsomal ethanol-oxidizing systems: in vitro characteristics and adaptive properties. J Biol Chem 245:2505–2512

Liu RH, Jacob JR, Hotchkiss JH, Cote PJ, Gerin JL, Tennant BC (1994) Woodchuck hepatitis virus surface antigen induces nitric oxide synthesis in hepatocytes: possible role in hepatocarcinogenesis. Carcinogenesis 15:2875–2877

Loeb LA, James EA, Waltersdorph AM, Klebanorf SJ (1988) Mutagenesis by the autoxidation of iron with isolated DNA. Proc Natl Acad Sci U S A 85:3918–32

Lunn RM, Zhang Y-J, Wang L-Y, Chen CJ, Lee PH, Lee CS, Tsai WY,Santella RM (1997) Mutations, chronic hepatitis B virus infection, and aflatoxin exposure in hepatocellular carcinoma in Taiwan. Cancer Res 57:3471–3477

McGlynn KA, Rosvold EA, Lustbader ED, Hu Y, Clapper ML, Zhou T, Wild CP, Xia XL, Baffoe-Bonnei A, Ofori-Adjei D (1995) Susceptibility to hepatocellular carcinoma is associated with genetic variation in the enzymatic detoxification of aflatoxin B_1. Proc Natl Acad Sci USA 92:2384–2387

MacPhail AP, Mandishona E, Bloom P, Paterson AC, Rouault TA, Gordeuk VR (1999) Measurements of iron status and survival in African iron overload. S Afr Med J 89:966–972

Marrogi AJ, Khan MA, van Gijssel HE, Welsh JA, Rahim H, Demetris AJ, Kowdley V, Hussain SP, Nair J, Bartsch H, Okby N, Poirier MC, Ishak KG, Harris CC (2001) Oxidative stress and p53 mutations in the carcinogenesis of iron-overload-associated hepatocellular carcinoma. J Natl Cancer Inst 93:1652–1655

Matzner Y, Hershko C, Polliak K, Konijn A, Izak G (1979) Suppressive effect of ferritin on in vitro lymphocyte function. Br J Hematol 42:443–448

McKillop IH, Schrum LW (2005) Alcohol and liver cancer. Alcohol 35:195–203

Ming L, Thorgeirsson SS, Gail MH, Lu P, Harris CC, Wang L, Shao Y, Wu Z, Wang X, Sun Z (2002) Dominant role of hepatitis B virus and co-factor role of aflatoxin in hepatocarcinogenesis in Qidong, China. Hepatology 36:1214–1220

Moyo VM, Gangaidzo IT, Gomo ZA, Khumalo H, Saungweme T, Kiire CF, Roualt T, Gordeuk VR (1997) Traditional beer consumption and the iron status of spouse pairs from a rural community in Zimbabwe. Blood 89:2159–2166

Murata H, Tsuji S, Tsuji, M, Sakaguchi Y, Fu HY, Kawano S, Hori M (2004) Promoter hypermethylation silences cyclo-oxygenase-2 (cox-2) and regulates growth of human hepatocellular carcinoma cells. Lab Invest 84:1050–1059

Niederau C, Fisher R, Sonnenberg A, Stremmel W, Trampisch HAJ, Stromeyer MD (1985) Survival and causes of death in cirrhotic and non-cirrhotic patients with primary haemochromatosis. N Engl J Med 313:11256 1262

Obe G, Ristow H (1979) Mutagenic, cancerogenic and teratogenic effects of alcohol. Mutat Res 65:229–259
Orimo H, Toura Y, Hino R, Kasai H (2006) Formation of 8-hydroxy-2′-deoxyguanosine in the DNA of cultured keratinocytes by clinically used doses of narrowband and broadband B and psoralen plus ultraviolet light. Cancer Sci 97:99–105
Ozturk M (1991) A p53 mutational hotspot in primary liver cancer is geographically localised to high aflatoxin areas of the world. Lancet 338:260–265
Peers F, Bosch X, Kaldor J, Linsell A, Pleijmen M (1987) Aflatoxin exposure, hepatitis B virus infection and liver cancer in Swaziland. Int J Cancer 39:545–553
Petersen DR (2005) Alcohol, iron-associated oxidative stress, and cancer. Alcohol 35:2 43–249
Poschl G, Seitz HK (2004) Alcohol and cancer. Alcohol 39:155 –165
Probst C, Njapau H, Cotty PJ (2004) Outbreak of acute aflatoxicosis in Kenya in 2004: identification of the causal agent. Appl Environ Microbiol 73:2762–2764
Qian GS, Ross RK, Yu MC, Yan JM, Gao YT, Henderson BE, Wogan GN, Groopman JD (1994) A follow-up study of urinary markers of aflatoxin exposure and liver cancer risk in Shanghai, People's Republic of China. Cancer Epidemiol Biomarkers Prev 3:3–10
Ratziu V, Poynard T (2005) Hepatocellular carcinoma in NAFLD. In: Farrell GC, George J, de la Hall P, McCullough AJ, (eds) Fatty liver disease: NASH and related disorders. Blackwell, Oxford, 276–288
Rouault TA (2003) Hepatic iron overload in alcoholic liver disease: why does it occur and what is its role in pathogenesis? Alcohol 30:103 – 106
Ross RK, Yuan JM, Yu MC, Wogan GN, Qian GS, Tu JT, Groopman JD, Gao YT, Henderson BE (1992) Urinary aflatoxin biomarkers and the risk of hepatocellular carcinoma. Lancet 339: 943–946
Sakamoto T, Hara M, Higaki Y, Ichiba M, Horita M, Mizuta T, Eguchi Y, Yasatake T, Osaki I, Yamamoto K, Onohara S, Kawasoe S, Shigematsu H, Tanaka K (2006) Influence of alcohol consumption and gene polymorphisms of ADH2 and ALDH2 on hepatocellular carcinoma in a Japanese population. Int J Cancer 118:1501–1507
Sell S, Hunt JM, Dunsford HA, Chisari FV (1991) Synergy between hepatitis B virus expression and chemical hepatocarcinogenesis in transgenic mice. Cancer Res 51:1278–1285
Sheu J-C, Huang G-T, Lee P-H, Chung JC, Chou HC, Lai My, Wang JT, Lee HS, Shih LN, Yang PM (1992) Mutations of p53 in hepatocellular carcinomas in Taiwan. Cancer Res 52: 6098–6100
Stern MC, Umbach DM, Yu MC, London SJ, Zhang Z-Q, Taylor JA (2001) Hepatitis B, aflatoxin B_1, and p53 codon 249 mutation in hepatocellular carcinoma from Guangxi, People's Republic of China, and a meta-analysis of existing studies. Cancer Epidemiol Biomark Prev 10:617–625
Stickel F, Schuppan D, Hahn EG, Seitz HK (2002) Carcinogenic effects of alcohol in hepatocrcinogenesis. Gut 1:132–139
Sun C-A, Wang L-Y, Chen C-J, Lu SN, You SL, Wang LW, Wang O, Wu DM, Santella RM (2001) Genetic polymorphisms of glutathione-S-transferases M1 and T1 associated with susceptibility to aflatoxin-related carcinogenesis among chronic hepatitis B carriers: a nested case-control study in Taiwan. Carcinogenesis 22:1289–1294
Sun CA, Wu DM, Wang LY, Chen CJ, You SL, Santella RM (2002) Determinants of formation of aflatoxin-albumin adducts: a seven township study in Taiwan. Br J Cancer 87:966–970
Sun Z, Lu P, Gail MH, Pee D, Zhang O, Ming L, Wang J, Wu Y, Liu G, Wu Y, Zhu Y (1999) Increased risk of hepatocellular carcinoma in male hepatitis B surface antigen carriers with chronic hepatitis who have detectable urinary aflatoxin metabolite M1. Hepatology 30: 379–383
Tavill AS, Qadri AM (2004) Alcohol and iron. Semin Liver Dis 24:318–325
Tennant BC, Hornbuckle WE, Yeager AE, Cote PJ, Baldwin BH, Korba BE (1990) Effects of aflatoxin B_1 on experimental woodchuck hepatitis virus infection and hepatocellular carcino-

mas. In: ' Hollinger FB, Lemon SM, Margolis H (eds) Viral Hepatitis and Liver Disease. Williams and Wilkins, Baltimore, pp 599–600

Troadec M-B, Courselaud B, Detwaud L, Haziza-Pigeon C, Leroyer P, Brissot P, Loreal O (2006) Iron overload promotes cyclin D1 expression and alters cell cycle in mouse hepatocytes. J Hepatol 44:391–399

TsuTsukamoto H, Horne W, Kamimura S, Niemela O, Parkkila S, Yla-Herttuala S, Brittenham GM (1995) Experimental liver cirrhosis induced by alcohol and iron. J Clin Invest 96:620–630

Turner PC, Mendy M, Whittle H, Fortuin M, Hall AJ, Wild CP (2000) Hepatitis B infection and aflatoxin biomarker levels in Gambian children. Trop Med Int Health 5:837–841

Vadrot N, Legrand A, Nello E, Bringuier AF, Guillot R, Feldman G (2006) Inducible nitric oxide synthase could be responsible for resistance or sensitivity to IFN-gamma-induced apoptosis in several human hepatoma cell lines. J Interferon Cytokine Res 26:901–913

Valberg LS, Simon JB, Manley PM, Corbett WE, Ludwig J (1975) Distribution of storage iron as body stores expand in patients with haemochromatosis. J Lab Clin Med 86:479–489

Vautier G, Bomford AB, Portmann BC, Metivier E, Williams R (1999) p53 mutations in British patients with hepatocellular carcinoma: clustering in genetic hemochromatosis. Gastroenterology 117:154–160

Wacker M, Wanek P, Eder E (2001) Detection of 1,N2-propanodeoxyguanosine adducts of trans-4-hydroxy-2-nonenal after gavage of trans-4-hydroxy-2-nonenal or induction of lipid peroxidation with carbon tetrachloride in F344 rats. Chem Biol Interact 137:269–83

Walker ARP, Arvidsson UB (1953) Iron overload in the South African Bantu. Tr Roy Soc Trop Med Hyg 47:536–548

Wang L-Y, Hatch M, Chen C-J, Levin B, You SL, Lu SN, Wu MH, Wu MP, Wang LW, Wang Q, Huang GT, Yang PM, lee HS, Santella RM (1996a) Aflatoxin exposure and risk of hepatocellular carcinoma in Taiwan. Br J Cancer 67:620–630

Wang JS, Qian GS, Zarba A, He X, Zhu YR, Zhang BC, Jacobson L, Gange SL, Munoz A, Kensler TW (1996b) Temporal patterns of aflatoxin-albumin adducts in hepatitis B surface antigen-positive and -negative residents of Daxin, Qidong County, People's Republic of China. Cancer Epidemiol Biomarkers Prev 5:253–261

Wild CP, Fortuin M, Donato F, Whittle HC, Hall AJ, Wolf CR, Montesana R (1993) Aflatoxin, liver enzymes and hepatitis B virus infection in Gambian children. Cancer Epidemiol Biomarkers Prev 2:555–561

Wild CP, Rasheed FN, Jawla MF, Hall AJ, Jansen LA, Montesano R (1991) In-utero exposure to aflatoxin in West Africa. Lancet 337:1602

Wu HC, Wang Q, Wang LW, Yang HI, Ahsan H, Tsai WY, Tsai MW, Wanh LY, Chen SY, Chen SJ, Santella R (2007) Polycyclic aromatic hydrocarbons and aflatoxin;albumin adducts, hepatitis B virus infection and hepatocellular carcinoma in Taiwan. Cancer Lett 252:104–114

Yeh F-S, Yu MC, Mo C-C, Luo S, Tong MJ, Henderson BE (1989) Hepatitis B virus, aflatoxins, and hepatocellular carcinoma in southern Guangxi. China Cancer Res 49:2506–2509

Yu M-W, Lien J-P, Chiu Y-H, Santella RM, Liaw Y-F, Chen C-J (1997) Effect of aflatoxin metabolism and DNA adduct formation on hepatocellular carcinoma among chronic hepatitis B carriers in Taiwan. J Hepatol 27:120–130

Chapter 8
Zearalenone and its Derivatives: Known Toxins in New Aspects

Piotr Golinski, Agnieszka Waskiewicz, and Karolina Gromadzka

8.1 Introduction

Mycotoxins are not a new problem in foods and feeds, and some suggest that they were first observed after the Flood reported in the Bible. However, a correlation of disease symptoms in humans and/or animals with the occurrence of pathogenic fungi and their metabolites was reported for the first time at the beginning of the twentieth century. For example, Svend Larsen (1928), a Danish veterinary inspector in a slaughterhouse, found that the macroscopic changes and abnormalities in the kidneys of slaughtered pigs, which were the effect of mouldy feed, probably containing toxic metabolites. After about 40 years, ochratoxin A related to mycotoxic porcine nephropathy was found in feeds. The metabolite was extracted, purified and the chemical structure of the compound was determined. The role of ochratoxin A in the etiology of the disease was proved and confirmed in biological experiments.

For many years studies on mycotoxins were not successful because the metabolites are present in the matrix (usually at ppb or ppm levels) at much lower concentrations than other biologically active compounds recorded in samples. A necessary condition of mycotoxin biosynthesis is the presence of toxigenic fungi in the environment. Very important factors facilitating mycotoxin formation include substrate, temperature and humidity. Under ambient temperature and high air humidity, rapid growth and development of the fungus on the host tissue may easily be observed. A fungus having a new source of energy (organic matter) develops rapidly, using and transforming host tissue into energy, simultaneously forming mycotoxins. One of the theories explaining why biosynthesis of these toxic compounds by fungi is observed is that an increasing

P. Golinski (✉)
Department of Chemistry, University of Life Sciences, Wojska Polskiego 75, 60-625, Poznan, Poland
e-mail: piotrg@up.poznan.pl

M. Rai and A. Varma (eds.), *Mycotoxins in Food, Feed and Bioweapons*,
DOI: 10.1007/978-3-642-00725-5_8,

concentration of by-products of primary metabolism (delivering energy) such as acetates, malonates and propionates — according to Le Chatelier's principle — stops reactions of primary metabolism. Since energy is necessary for the life of the fungus (for continuation of metabolism), secondary metabolism with the biosynthesis of mycotoxins is initiated, originating from the above-mentioned anionic residues as the reaction substrate. That is why so many toxic metabolites may be formed by fungi, with some repeating similarities in their chemical structures, as in the case of trichothecenes group A and B, aflatoxins, ochratoxins, and fumonisins.

Global changes in climate conditions, together with increasing exchange and trade of goods, are the factors responsible for worldwide distribution of fungi previously typical for certain zones only. For example in Poland at the beginning of the 1970s, positive samples naturally contaminated with ochratoxin A were estimated at about 7%, and toxigenic fungi of *Aspergillus* and *Penicillium* at about 20% of the total amount of fungi present in cereals (Szebiotko et al. 1981). At the turn of the 1970s and 1980s (Chelkowski and Golinski 1983) the number of positive samples (contaminated with ochratoxin A) increased. We believe that this was due to a change in the distribution of fungi present in stored cereals because of the development of sensitive analytical methods and tools in mycotoxin chemical determination. As a result, taking into consideration also weather and climate conditions (high air humidity and moisture content of stored cereals), symptoms of mycotoxic porcine nephropathy were observed in pig herds (Golinski et al., 1984, 1985). During this time ochratoxin A was recognized as an economically significant problem (deterioration of stored crops) and the most important in the climate conditions of Poland.

In contrast, the problem of zearalenone was considered then as marginal. In fact, the percentage of toxigenic isolates of *Fusarium* (ZON producers) increased significantly in comparison to the early 1970s and was estimated at about 31% of fungi present in cereals, but ZON-positive cereal samples were still below 0.5%, with low levels of toxin concentrations (Chelkowski et al. 1983). During the following years, an increasing percentage of both toxigenic *Fusarium* isolates (ZON producers) and ZON-positive feed samples and their cereal components were observed.

In the last decade, especially after the introduction of extrusion in feed and feed component preparation (many feed producers still believe that extrusion is a panacea for all food/feed contaminants, including mycotoxins), the situation started to become worse. Since cereals (including corn) were introduced (to replace meat) to pelleted pet food for species belonging to *Carnivora* (cats and dogs), zearalenone has been found in pet food and in animal tissue (Waskiewicz 2006). The reason is that cereals used in the production of such feeds, especially corn, are very susceptible to infection with pathogens (*Fusarium culmorum* and *F. graminearum*), followed by *Fusarium* cob blight and zearalenone biosynthesis, which in consequence results in toxin (ZON) residues in animal tissue (Waskiewicz 2006).

8.2 Chemical Structure, Properties and Derivatives of Zearalenone

The interest in mycotoxins had already started when aflatoxins were found to be carcinogens widespread in foodstuffs and feedstuffs. Today, mycotoxins and mouldy feedstuffs are known causes of animal diseases. Symptoms are often subtle and there are contributing factors, e.g. environmental stress, exposure to multiple mycotoxins and infectious agents, and nutrient/vitamin deficiencies. Often it is difficult to find cause-effect relationships with contaminated feedstuffs.

In the late 1920s it was observed that feeding of swine with mouldy corn resulted in the development of hyperoestrogenic symptoms and metritis in animals. Other reports of genital changes in swine following the consumption of mouldy corn were published (Koen and Smith 1945). Stob et al. (1962) were the first to isolate a uterotrophic compound from mouldy corn, while in 1965 Christensen et al. isolated a compound (called preliminary F-2) from corn inoculated with *Fusarium*. In 1966 Urry with co-workers isolated the same compound from contaminated corn and they called it zearalenone (ZON). The metabolite is considered as a natural contaminant of food with significant *Fusarium* infection of cereal origin components. Biosynthesis of the compound was observed in cereals (kernels) such as corn, rice and wheat, infected by several species of *Fusarium*, including *F. graminearum*, *F. culmorum*, *F. crookwellense*, *F. equiseti* and *F. semitectum* (Betina 1989).

The concentration of accumulated ZON in cereals depends on many factors such as substrate, temperature, duration of *Fusarium* growth and strain of fungal species. Moreover, a humid tropical climate promotes microbial proliferation on food and feedstuffs and finally mycotoxins biosynthesis (Nuryono et al. 2005).

The compound name zearalenone was derived from *Gibberella zeae*, but chemically it is [6-(10-hydroxy-6-oxo-trans-1-undecenyl)b-resorcylic-acid-lactone]. ZON is soluble in alkaline solutions, ether, benzene, acetonitrile, methyl chloride, chloroform, acetone and alcohols, while it is virtually insoluble in water. It is heat stable, which makes it difficult to remove and/or decompose from food (Hidy et al. 1977; Kuiper-Goodman et al. 1998)

Zearalanone, α-, β-zearalenol, and α-, β-zearalanol are the most frequently observed derivatives of ZON (Golinski et al. 1988; Tiemann et al. 2003). According to Richardson et al. (1985), only α-zearalenol is observed in naturally contaminated cereals. However, Golinski et al. (1988) described the biosynthesis of α- and β-zearalenols by an *F. crookwellense* strain isolated from dry rotted potato tubers. Moreover, Plasencia and Mirocha (1991) reported that zearalenone-4-sulfate was extracted from cultures of four different *Fusarium* species grown on rice substrate. Zearalenone-4-glucoside is also naturally occurring in *Fusarium* contaminated cereals (Berthiller et al. 2007).

Toxicity of ZON and its metabolites described as oestrogenic properties are related to the chemical structure of the mycotoxins, a structure similar to the naturally occurring oestrogens, i.e. oestradiol, oestrone and oestriol. Interaction of such compounds with human oestrogen receptors in competition with 17 β-oestradiol was also reported (Miksicek 1994). The oestrogenic potency of zearalenone has been shown to be several times higher than that of other environmental oestrogens in various test assays (Shier et al. 2001).

8.3 Occurrence of Zearalenone in Foods and Feeds

The worldwide contamination of food and feeds with mycotoxins is a significant health and economic problem (Hussein and Brasel 2001; Bennett and Klich 2003). *Fusarium* spp. are commonly present in the agricultural environment. Species dominating in cereals vary from season to season, depending on the geographical location and climate conditions, and are a significant factor in food/feed component contamination with zearalenone and other mycotoxins (Chelkowski et al. 2001).

Zearalenone and its derivatives have been observed in many important crops such as corn, wheat, sorghum, barley, oats, sesame seed, hay and corn silage (D'Mello et al. 1999). Several studies carried out in Europe and a number of transcontinental countries have reported a high incidence of ZON in cereals and feeds (De Saeger et al. 2003; Scudamore and Patel 2000). Many factors such as temperature, duration of growth, substrate and strain of fungal species influence the amounts of accumulated ZON in crops (Jimenez and Mateo 1997). In the last few years, in the Central Europe climatic zone, as studies show, mycotoxins produced by fungi of the *Fusarium* genus, especially zearalenone, play a dominant role in food/feed deterioration (Čonková et al. 2003). The highest amounts of zearalenone formed by *Fusarium* were observed below a temperature of 25°C, at a high amplitude of daily temperature and at 16% humidity (Zwierzchowski et al. 2005).

Available data in Europe indicate that wheat and maize are cereals with a high incidence and high levels of contamination with ZON; however oat, as well as barley, has been found to be contaminated occasionally with this toxin.

Germany seems to be the European country where more data can be found on ZON in cereals than elsewhere in Europe. Surveys of cereals and derivatives for several years confirmed their contamination with ZON (Schollenberger et al. 2006). In Yugoslavia, ZON was found at high levels (up to 10.0 mg kg^{-1}) in corn (Balzer et al. 1977). In Hungary, Fazekas et al. (1996) reported the contamination with ZON of stored mouldy corn at a range between 0.01 and 11.8 mg kg^{-1}. In Scotland, according to Gross and Robb (1975), high contamination of barley (stored for 3 months to about 1 year) with ZON was detected and the levels ranged between 2.1 and 26.5 mg kg^{-1}.

Concerning recent data on human exposure to ZON in Europe, the occurrence of the toxin was reported in 32% of mixed cereal samples (n = 4,918) from nine European countries. The distribution showed that much of this contamination was in maize kernels and wheat grain. A high incidence of ZON was found in samples of oat from Finland (47% of samples containing >0.2 mg kg^{-1} with the highest level of 1.3 mg kg^{-1} being reported) and a high incidence of ZON in wheat from France

(16% of samples above 0.2 mg ZON per kg, the highest being 1.8). Raw maize was the food commodity with the highest levels of ZON (14% of maize with levels >0.2 mg kg^{-1}, the highest level of 6.5 mg kg^{-1}), reported in a sample of maize from Italy (SCOOP 2003).

Data from both Americas indicate that the highest contamination of cereals with ZON was observed in the USA and Argentina. High ZON levels in maize and sorghum samples were reported by Bennett et al. (1985) and Bagneris et al. (1986), respectively. In Argentina, ZON was found in corn-based foods (Resnik et al. 1996) and poultry feeds (Dalcero et al. 1998). Recent data from Argentina reported the contamination of cattle feeds with ZON at levels that ranged between 1.2 and 3.1 mg kg^{-1} (Cavaglieri et al. 2005).

Among Asiatic countries, only in Japan were high zearalenone concentrations in barley (11.0–15.0 mg kg^{-1}) observed (Yoshizawa 1997); however, in other countries the highest level of cereal contamination with the toxin did not exceed 1.4 mg kg^{-1} in wheat (Li et al. 2002; Park et al. 2005).

Even though most African countries have a climate characterized by high humidity and high temperature, which favour the growth of moulds, little information is available on the occurrence of *Fusarium* toxins, particularly ZON in foods and feeds. High contaminations of the raw material are a constant problem. Legal regulations are not prepared in the field of food exhibition and retailing, and mycotoxin problems have already been associated with some food contamination in some areas in Africa (Zinedine et al. 2007). In Egypt, several commodities (especially maize, wheat and rice) were reported to contain ZON (Abd Alla 1997). Maize from Egypt was also found to be contaminated with high levels of zearalenone, ranging between 9.8 and 38.4 mg kg^{-1} (El-Maghraby et al. 1995). A high level of ZON (17.5 mg kg^{-1}) was also found in maize from Nigeria (Gbodi et al. 1986). In North African countries, no information is available on the occurrence of *Fusarium* toxins in foods and feeds. The first publication from Morocco reported the co-occurrence of ZON with fumonisin B_1 and ochratoxin A (OTA) in corn (Zinedine et al. 2006). In New Zealand, ZON was detected in maize at high levels (up to 10.0 mg kg^{-1}) (Lauren et al. 1996), while in Australia ZON concentration did not exceed 1.0 mg kg^{-1} (Blaney et al. 1987).

8.4 Guidelines

As a result of increasing awareness of the human health risk posed by mycotoxins, guidelines and/or regulatory limits for mycotoxins were introduced in 35% of the world's countries in the period of 1987–1997 (Food and Agriculture Organization of the United Nations (FAO)).

In the European Union the ZON tolerance level in foodstuffs is as follows: (1) up to 100 μg kg^{-1} in unprocessed cereals other than maize, up to 350 μg kg^{-1} in unprocessed maize, (2) up to 75 μg kg^{-1} in cereals intended for direct human consumption, cereal flour, germ, and bran sold as the end-product for direct human consumption, (3) up to 100 μg kg^{-1} in maize for direct human consumption, maize flour, maize meal, maize grits, maize germ and refined maize oil, (4) up to 50 μg kg^{-1} in bread (including small bakery wares), pastries, biscuits, cereal snacks and breakfast cereals, with the exception of maize snacks and maize-based breakfast cereals, and (5) up to 20 μg kg^{-1} in processed maize- and cereal-based foods for infants and young children (EC 2007).

8.5 Toxic Effects on Animals of Experimental Feeds Contaminated with ZON and Natural Occurrence of the Toxins in Animal Tissue

ZON is a phyto-estrogen demonstrating, due to the presence of a phenolic ring in its chemical structure, activity with and affinity to both oestrogen receptors, ERa and ERb, found in mammalian tissues (Shier et al. 2001). By its action it causes several functional changes in the reproductive system, similar to naturally occurring oestrogens (Gora et al. 2004).

Maaroufi et al. (1996), in their studies on haematotoxic properties of ZON in rats, observed dysfunction of blood coagulation accompanied by changes of some blood parameters (MCV, haematocrit, platelet count and WBC). In *in vivo* tests they also observed changes in levels of such biochemical markers as aspartate aminotransferase (AST), alanine aminotransferase (ALT), alkaline phosphatase (ALP), serum creatinine and bilirubin, which indicates the metabolite's hepatotoxicity. Čonková et al. (2001) examined the influence of low and high doses of ZON on the activity of selected blood serum enzymes of rabbits. A significant increase in the activity of ALP was recorded only in the experimental group of 10 μg ZON per kg of body weight, while that of AST, ALT, ALP, GGT and LD was recorded in the group with the toxin level of 100 μg kg^{-1} body weight, which indicates possible liver toxicity of ZON due to chronic effects of the toxin.

Recently, Nikaido et al. (2004) studied the effects of prenatal ZON on the postnatal development of the reproductive tract and mammary glands of mouse offspring. In their study four daily subcutaneous injections of 0.5 or 10.0 mg kg^{-1} per day were given, beginning on GD 15, and the offspring were killed at 4, 8, 12, and 16 weeks of age. A lack of corpora lutea and vaginal cornification was observed, and mammary growth decreased in the 10 mg kg^{-1} experimental group. The above results are in agreement with a report by Williams et al. (1989) on neonatal offspring exposed to zearalenone.

High concentrations of ZON (50–100 μg kg^{-1}) in swine diets have been reported to adversely affect cycling, conception, ovulation and implantation. Placental membrane and fetal development may also be disrupted, resulting in decreased litter size and diminished viability of neonates. The toxin caused embryonic death, inhibition of fetal development and decreased numbers of fetuses in exposed swine. A majority of cases of ZON intoxication occurred after animals were fed cereal grains contaminated with the toxin. It was demonstrated in New Zealand and Europe that sheep and cattle grazing on pastures in which grass is contaminated with ZON can subsequently have reproductive problems (Task Force Report 2003).

Additionally, the introduction of ready-to-eat dry and moist dog and cat food has caused considerable demand on the part of consumers. Poor storage conditions of dry food may result in deterioration of their nutritive value as well as development of bacterial and fungal flora even within its shelf life (Goliński and Nowak 2004). Both maintenance and therapeutic ready-to-eat balanced dog and cat food contain as much as 80% plant origin components (cereals, vegetables, pulses). Thus, they may constitute a potential source of disadvantageous substances, including zearalenone (Skorska-Wyszynska et al. 2004), while the technological manufacturing process may not be completely effective in preventing the development of mould fungi (Popiel et al. 2004).

Poisoning with long-acting, low-threshold values, i.e. those encountered most frequently in the case of human consumption or animal feeding, seems to be especially dangerous (McEvoy et al. 2001).

In females of domestic animals (bitches, female cats) during routine medical examination changes were observed resulting, e.g. from oestrogenization caused by mycotoxins contained in feed (Fritsche and Steinhart 1999). Symptoms of hyperoestrogenism observable in those animals include swelling and enlargement of the vulva, balding and excessive pigmentation of skin in the perineum, sides and the abdomen, secondary seborrhoea and ceruminal otitis externa (Tomaszewski et al. 1998). In turn, changes detected during clinical examinations may be divided into three categories:

- Disturbed sexual cycle, including sterility
- Disturbances in the physiology of pregnancy and parturition, including miscarriages
- Lesions within female reproductive organs and the mammary glands.

Moreover, in the course of intra-operative examination of surgical specimens and histopathological examination, ovarian cysts (Fig. 8.1) and hormonally active gonadal tumours are observed, determining the development of colpitis. What is more, the effect of oestrogenizing factors on the formation of ovarian, uterine, vaginal and breast cancer may not be excluded (Fig. 8.2). Differences in the sensitivity to zearalenone and its derivatives among domestic animals and pets relate to different metabolic profiles characteristic of individual species.

Golinski and Nowak (2004) suggested that recently introduced, new recipes of pelleted feeds for dogs and cats, with increasing amounts of cereals replacing meat, constitute a real possibility and danger of ZON consumption by these animals. Since ovarian cysts were detected surgically, which is recognized as the first stage of the endometrial pyometra complex found in about 30% of females, the authors assumed that ZON present in a dog's diet might possibly cause the pathological aberrations mentioned above.

Fig. 8.1 Ovarian cyst in a bitch and follicular cysts on ovaries: ZON level in the tissue — 1.8 ng g^{-1} (photo by Nowak T., "My Pet" Animal Clinic)

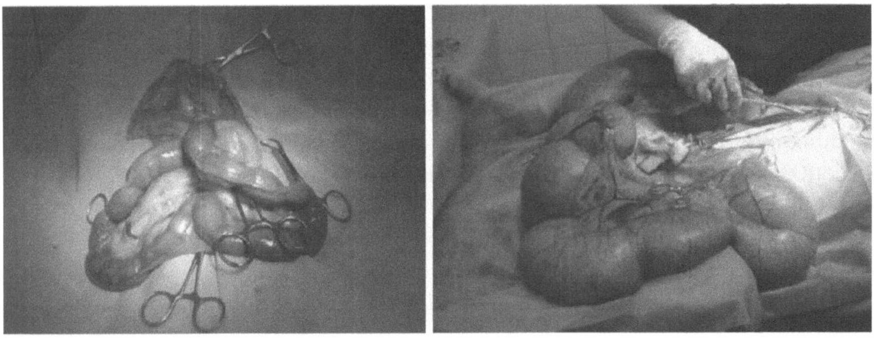

Fig. 8.2 Chronic follicular endometriti: ZON level in the tissue — 2.9 ng g^{-1} (photo by Nowak T., "My Pet" Animal Clinic)

Studies conducted so far indicate prompt ZON absorption from the digestive tract, since as early as 30 minutes later zearalenone and its metabolites, α- or β-zearalenol as well as α- and β-zearalanol, were detected in the blood (Berek et al. 2001; Gora et al. 2004). It needs to be stressed that naturally occurring zearalenone derivatives also exhibit oestrogenic action (Tiemann et al. 2003, Seeling et al. 2005).

So far few articles concerning the natural occurrence of ZON in animal tissue have been published. Curtui et al. (2001) examined ZON concentration in blood serum, kidney, liver and muscles of swine. ZON was detected only in serum samples with the highest concentration of 0.96 ng ml^{-1}.

Studies concerning ZON incidence in tissues and blood of domestic animals were conducted in Poland in 2005 (Waskiewicz 2006). ZON was detected in pathologically changed tissues of reproductive organs and blood of pets (dogs, cats) at a level of 0.5–2.9 ng ml^{-1} and 0.5–2.8 ng g^{-1} respectively. Examinations showed high frequency of this toxin in 65% of tissues and 90% of blood samples. Additionally, it was observed that in 55% of cases this toxin was present both in blood and tissues of examined animals. Further studies conducted in Poland concerned changes within ovaries in bitches (Skorska-Wyszynska et al. 2004). Animals were given ZON at 200.0 μg kg^{-1} b.w. for a week, with lesions detected during histopathological examinations in ovaries with numerous congestions in the medullary part and damage to the membrana granulosa of ovarian follicles.

According to Aurich et al. (2006), the toxin concentration detected in naturally contaminated oats does not have relevant effects on the release of reproductive hormones, cycle length and uterine histology in mares. The shortened cycle was most likely caused by the collection of endometrial biopsies on day 10 of the cycle. In contrast to pigs (Edwards et al. 1987) and ruminants (Mirocha et al. 1968; Bloomquist et al. 1982), horses seem to be less sensitive to the above mycotoxins with respect to reproductive functions.

According to D'Mello et al. (1999), the presence of ZON in a woman's body, administered above a certain threshold, is particularly undesirable during pregnancy, as this toxin both reduces embryonic survival and decreases fetal weight. Additionally, ZON may influence the uterus by decreasing LH and progesterone secretion, and by altering the morphology of uterine tissues. It was observed that in male pigs the toxin may reduce serum testosterone, testis weight and spermatogenesis, as well as induce feminization and suppress libido. ZON contributes to alterations in the reproductive tract of laboratory and domestic animals (mice, rats, guinea pigs, hamsters, rabbits). During laboratory studies several oestrogenic effects were observed (decreased fertility, increased embryolethal resorptions, reduced litter size, changed weight of adrenal, thyroid and pituitary glands, and changes in serum levels of progesterone and oestradiol); however, no teratogenic effects were found in mice, rats, guinea pigs or rabbits (Zinedine et al. 2007).

To study the potential oestrogenic effects of this compound in the brain, Turcotte et al. (2005) examined the effects of ZON on the expression of neuronal progestin receptors and feminine sexual behaviour in female rats. Ovariectomized rats were treated daily for 3 days with ZON (0.2, 1.0, and 2.0 mg), oestradiol benzoate, or vehicle (propylene glycol with 10% ethanol). They were then either perfused and had their progestin receptors visualized by immunocytochemistry, or were injected with progesterone and tested for sexual receptivity with male rats. Progestin receptor-containing cells were counted in the medial preoptic area and ventromedial hypothalamus. The two highest doses of zearalenone increased the concentration of neuronal progestin receptors, as did 10 μg of oestradiol. The highest dose of zearalenone (2 mg) also induced progestin receptor staining density comparable to that of 10 μg of oestradiol benzoate. In behavioural tests ovariectomized animals treated with 2 mg of ZON followed by progesterone showed levels of sexual receptivity comparable to females treated daily with oestradiol benzoate (2 μg) followed by progesterone. These studies suggest that, although structurally distinct and less potent than oestradiol, ZON can act as an oestrogen agonist in the rat brain.

Vlata et al. (2006) investigated the in vitro cytopathic effects of ZON on freshly isolated human peripheral blood mononuclear cells (PBMC) in relation to the proliferation and cell death patterns of untreated and mitogen-activated cells. At concentrations higher than 30 μg ml^{-1}, ZON totally inhibited T and B lymphocyte proliferation from stimulation with phytohaemagglutinin and pokeweed mitogen. The inhibitory effects of ZON were further related to cell necrosis/apoptosis.

The presence of zearalenone in foodstuffs may also cause hyperoestrogenism in women, whose organisms are highly sensitive to the action of oestrogenic hormones. Too high levels of them may result in numerous systemic disorders, manifested, e.g. by a loss of libido, ovarian and uterine dysfunctions, ovulation, infertility and neoplastic changes in the entire reproductive system (Shier et al. 2001; Tiemann et al. 2003).

Recent studies indicate the effect of ZON on the growth of human breast cancer cells containing oestrogen response receptors. Oestrogenic effects of high ZON concentrations in food with an increasing incidence of breast cancer indicates that possibly exposure to ZON may contribute to breast cancer development (Yu et al. 2005).

In Poland, in order to detect the presence of zearalenone in blood serum of women with neoplastic changes in the reproductive system, examinations were conducted in the course of which the toxin was recorded in 13.5% of cases with a positive diagnosis indicating the presence of this toxin together with its derivative α-zearalenol in the tested samples (Gajecki et al. 2004).

8.6 Methods of ZON Detoxication

To avoid the presence of mycotoxins in food, feeds and human/animal tissue, several strategies have been investigated and suggested (Doyle et al. 1982, Ramos and Hernandez 1997), which may be divided into pre- and post-harvest technologies and into biological, chemical, and physical methods.

Prevention of fungal development and mycotoxin biosynthesis seems to be the best way to reduce risk (Miedaner and Reinbrecht 1999), e.g. by harvesting grain at maturity and low moisture content and storing it under cool and dry conditions, which obviously is difficult to do in countries with a warm and humid climate. Furthermore, the growth and development of fungi followed by mycotoxin formation could be reduced by the use of propionic acid or ammonium isobutyrate. Feed additives such as antioxidants, sulphur-containing amino acids, vitamins and trace elements are also suggested as detoxicants (Nahm 1995).

Biological methods are especially popular in the case of zearalenone. According to Takahashi-Ando et al. (2002), ZON is converted into a compound with much lower oestrogenic properties by incubation with *Clonostachys rosea* IFO 7063. Alkaline hydrolase, responsible for detoxification, was purified and the entire coding region of the gene was cloned by PCR techniques. The authors recorded very good activity of the enzyme in ZON degradation in heterologous hosts carrying the cloned gene, which could be a promising genetic resource for in planta detoxification of the mycotoxins in important crops.

Utermark and Karlovsky (2007) showed a strong inhibitory effect of ZON on ascomycete fungi. Zearalenone inhibited the growth of *Sordaria fimicola* at a concentration as low as 2 μg L^{-1}, and of isolates of *Epicoccum purpurascens*, *Cladosporium herbarum* and *Alternaria alternata* at a concentration of 20 μg L^{-1}. The above-mentioned authors found a gene encoding zearalenone-specific lactonase in *Gliocladium roseum* and postulated that this enzyme may be one of the reasons why the growth of *Gliocladium* spp. was found not to be inhibited by zearalenone.

According to Gromadzka et al. (2008), selected non-toxigenic *Trichoderma* and *Gliocladium* isolates were found to be able to significantly reduce (by 95–100%) the biosynthesis of zearalenone by *F. graminearum* and *F. culmorum* on solid substrate.

Chemical methods seem to be a promising solution, but because of their side effects (deterioration of alimentary valuable compounds), a solution with a limited range of applications. Some mycotoxins may be destroyed with calcium hydroxide monoethylamine, ozone or ammonia (McKenzie et al. 1997; Lemke et al. 1999). In the case of zearalenone in particular, ozonation yields good results of ZON detoxication. After a 15-s ozone application McKenzie et al. (1997) observed high efficiency of ZON degradation, with no detectable by-products. Additionally, the toxicity of this compound was significantly reduced after ozone treatment.

Physical methods are focused on the removal of mycotoxins by different adsorbents added to diets contaminated with mycotoxins (Ramos et al. 1996).

Many effective mineral adsorbents based on clay — montmorillonite and the zeolite clinoptilolite — adsorbed mycotoxins with polar groups, such as aflatoxins (Phillips et al. 1995). Less polar mycotoxins, i.e. zearalenone and ochratoxin A, are not effectively adsorbed on hydrophilic negatively charged surfaces of these unmodified minerals (Santin et al. 2002). Surface properties of the above minerals may be modified and their hydrophobicity may be controlled by natural charge-balance cations (Na, K, Ca or Mg) being replaced by high molecular weight quaternary ammonium ions. Amines exchange these cations only on the external surfaces of zeolites, while in bentonite all exchangeable positions are equally available for quaternary ammonium ions (Bowman et al. 1995). Therefore, to achieve the same surface properties for zeolites a lower number of quaternary ammonium ions is needed than for bentonites (Dakovic et al. 2001). This is important since the practical use of these chemically modified mineral adsorbents in animal feed is limited by the type and concentration of organic modifiers required. According to Tomasevic-Canovic et al. (2003), organo-zeolites with high efficiency adsorb aflatoxin B_1, zearalenone, ochratoxin A and ergopeptine alkaloids. The method of preparation of organo-zeolites has little influence on adsorption of mycotoxins. Organic modification of zeolitic tuff by both processes (wet and dry) changed its surface properties and improved its binding properties for less polar mycotoxins such as zearalenone and ochratoxin A. The presence of long chain organic cations on the zeolitic surface significantly improved the adsorption of ochratoxin A and zearalenone, suggesting that an increase in hydrophobicity of the zeolitic surface probably influences adsorption of these mycotoxins. According to Huwig et al. (2001), aluminosilicates are the preferred adsorbents, followed by activated charcoal and special type polymers. The efficiency of mycotoxin binders depended mainly on the chemical structure of both the adsorbent and the toxin. As far as the applicability of aluminosilicates is concerned, it may be concluded that they are very effective in preventing aflatoxicosis, but their efficacy against zearalenone, ochratoxin, and trichothecenes is limited. In addition to the narrow binding range for different mycotoxins, aluminosilicates have the disadvantage resulting from their inclusion rates for vitamins and minerals. Ramos et al. (1996) examined another adsorbent, crospovidone (polyvinylpyrrolidone), a highly polar amphoteric polymer, the in vitro adsorption of which was measured as 0.3 mg g^{-1} for zearalenone. Sabater-Vilar et al. (2007) described the adsorption capacity of a variety of potential binders, including compounds that have not been evaluated before, such as humic acids. All compounds were tested at realistic inclusion levels for their capacity to bind ZON, using an in vitro method that resembles different pH conditions in the gastrointestinal tract of pigs. Those authors showed that some of the selected smectite clays, humic substances and yeast-wall-derived products efficiently adsorbed ZON (>70%). Binding efficiency was indirectly confirmed by the reduction of toxicity in in vitro bioassays. Avantaggiato et al. (2003) examined the intestinal absorption of mycotoxins by using a laboratory model that mimics the metabolic processes of the gastrointestinal (GI) tract of healthy pigs. This model was used to evaluate the small-intestinal absorption of zearalenone from contaminated wheat (4.1 mg kg^{-1}) and the effectiveness of activated carbon and cholestyramine at four inclusion levels (0.25, 0.5, 1 and 2%) in reducing toxin absorption. Approximately 32% of ZON intake (247 mg) was released from the food matrix during 6 h of digestion and was efficiently absorbed at the intestinal level. A significant reduction of intestinal ZON absorption was found after inclusion of activated carbon or cholestyramine, even at the lowest dose of adsorbents, with a more pronounced effect exhibited by activated carbon. In particular, when 2% of activated carbon or cholestyramine was added to the meal the ZON intestinal absorption was lowered from a 32% ZON intake to 5 and 16%, respectively.

8.7 Conclusion

Today's awareness of food safety has led to an increasing interest in mycotoxins. Due to modern laboratory methods and a growing interest in this field of research, more than 300 different mycotoxins have been differentiated so far. However, for a practical consideration in the feed-manufacturing process only a small number of toxins are of relevance, with aflatoxins, trichothecenes, ochratoxins, fumonisins and zearalenone being of particular interest. ZON is among the most widespread *Fusarium* mycotoxins in agricultural commodities. Unexpectedly high concentration of mycotoxins can be found in relation to the severity of the *Fusarium* infection. Surveys on zearalenone occurrence in several commodities are routinely published: ZON, varying within a wide concentration range, is mainly influenced by the considered matrix, the timing of the harvest season, climate and location. It is acknowledged that ZON is of relatively low toxicity but its role as a mammalian endocrine disrupter is being recognized, with effects in both males and females of different species.

It is easy to conclude that nature should not be considered as a static, but rather a dynamically changing medium — what was true 20–30 years ago is not necessarily true nowadays: old, solved problems are replaced with new ones and in the near future quite a number of topics have to be elucidated, concerning even metabolites known for many years, such as, for example, zearalenone. Fungi (similar to other living organisms), competing for energy (organic matter), in changing global climate conditions, change the distribution of mycoflora in our environment. As often happens, metabolites known for many years emerge as new problems. Thus, it is advisable to consider screening analyses of both toxigenic fungi present in agricultural products and mycotoxins contaminating foods and feeds that are of prime concern in protecting human health and in reducing of economic losses in the agricultural sector.

References

Abd Alla ES (1997) Zearalenone: toxigenic fungi and chemical decontamination in Egyptian cereals. Nahrung 41:362–365

Aurich JE, Hoppen HO, Trampler R, Zentek J, Boehm J, Razzazi-Fazeli E, Aurich C (2006) Effects of mycotoxins on reproductive function in mares. Anim Reprod Sci 94:238–241

Avantaggiato G, Havenaar R, Visconti A (2003) Assessing the zearalenone-binding activity of adsorbent materials during passage through a dynamic in vitro gastrointestinal model. Food Chem Toxicol 41:1283–1290

Bagneris RW, Gaul JA, Ware GM (1986) Liquid chromatographic determination of zearalenone and zearalenol in animal feeds and grains, using fluorescence detection. J Assoc Off Anal Chem 69:894–898

Balzer I, Bogdanic C, Muzic S (1977) Natural contamination of corn (*Zea mays*) with mycotoxins in Yugoslavia. Ann Nutr Aliment 31:425–430

Bennett JW, Klich M (2003) Mycotoxins. Clin Microbiol Rev 16(3):497–516

Bennett GA, Shotwell OL, Kwolek WF (1985) Liquid chromatographic determination of alpha-zearalenol and zearalenone in corn: Collaborative study. J Assoc Off Anal Chem 68:958–961

Berek L, Petri IB, Mesterhazy A, Teren J, Molnhr J (2001) Effects of mycotoxins on human immune functions in vitro. Toxicol In Vitro 15:25–30

Berthiller F, Lemmens M, Werner U, Krska R, Hauser MT, Adam G, Schuhmacher R (2007) Short review: Metabolism of the *Fusarium* mycotoxins deoxynivalenol and zearalenone in plants. Mycotoxin Res 23:68–72

Betina V (1989) Mycotoxins. In: Betina V (ed) Bioactive molecules. Elsevier, Amsterdam, p 271

Blaney BJ, Moore CJ, Tyler A L (1987) The mycotoxins dash 4-deoxynivalenol, zearalenone and aflatoxin dash in weather-damaged wheat harvested 1983–1985 in south-eastern Queensland. Aust J Agric Res 38:993–1000

Bloomquist C, Davidson JN, Pearson EG (1982) Zearalenone toxicosis in prepubertal dairy heifers. J Am Vet Med Assoc 180:164–165

Bowman RS, Haggerty GM, Hundleston RG, Neel D, Flunn MM (1995) In: Sabadini DA, Knox RC, Harwell JH (eds) Surfactant enchased subsurface remediation, Proc. ACS symposium, Series 594, San Diego, California, 13–17 March 1994. American Chemical Society, Washington, p 54

Cavaglieri LR, Gonzalez Pereyra LM, Pereyra CM, Magnoli CE, Chulze SN, Dalcero AM (2005) Fungal and mycotoxin contamination of cow feeding stuffs in Argentina. In: International Conference. Fifth Framework Program. European Union Myco-Globe Project. European Commission, 13–16 September, Accra, Ghana

Chelkowski J, Golinski P (1983) Mycotoxins in cereal grain. Part VII. A simple method to assay mycotoxins potential of cereal grain and cereal products. Nahrung 27(4):305–310

Chelkowski J, Golinski P, Manka M, Trojanowska K, Wiewiorowska M, Szebiotko K (1983) Mycotoxins in cereal grain. Part IX. Zearalenone and Fusaria in wheat, barley, rye and corn kernels. Nahrung 27(6):525–531

Chelkowski J, Perkowski J, Grabarkiewicz-Szczesna J, Kostecki M, Golinski P (2001) Toxigenic fungi and mycotoxins in cereal grains and feeds in Poland. In: Logrieco A (ed) Occurrence of toxigenic fungi and mycotoxins in plants, food and feed in Europe. COST Action 835, European Commission, pp 111–130

Christensen CM, Nelson GH, Mirocha CJ (1965) Effect on the white rat uterus of a toxic substance isolated from *Fusarium*. Appl Microbiol 13:653

Čonková E, Laciakova A, Pastorová B, Seidel H, Kovác G (2001) The effect of zearalenone on some enzymatic parameters in rabbits. Toxicol Lett 121:145–149

Čonková E, Laciakova A, Kovac G, Seidel H (2003) Fusarial toxins and their role in animal diseases. Vet J 165:214–220

Curtui VG, Gareis M, Usleber E, Märtlbauer E (2001) Survey of Romanian slaughtered pigs for the occurrence of mycotoxins ochratoxins A and B and zearalenone. Food Add Contam 18: 730–738

Dakovic AS, Tomasevic-Canovic MR, Dondur VT, Stojsic DG, Rottinghaus GE (2001) Zeolites and mesoporous materials at the dawn of the 21st century, Stud Surf Sci Catal. In: Galarneau A, Di Renzo F, Fajula F, Vedrina J (eds) Proc 13th International Zeolite Conference, Montpellier, France, 8–13 July, vol 135. Elsevier, Amsterdam, p 5276

Dalcero A, Magnoli C, Luna M, Ancasi G, Reynoso MM, Chiacchiera S, Miazzo R, Palacio G (1998) Mycoflora and naturally occurring mycotoxins in poultry feeds in Argentina. Mycopathologia 141:37–43

De Saeger S, Sibanda L, Van Peteghem C (2003) Analysis of zearalenone and α-zearalenol in animal feed using high-performance liquid chromatography. Anal Chem Acta 487:137–143

Doyle MP, Applebaum RS, Brackett RE, Marth EH (1982) Physical, chemical and biological degradation of mycotoxins in foods and agricultural commodities. J Food Protect 45:946–971

Edwards S, Cantley TC, Rottinghaus GE, Osweiler GD, Day BN (1987) The effect of zearalenone on reproduction in swine. The relationship between ingested zearalenone dose and anestrous in non-pregnant, sexually mature gilts. Theriogenology 28:43–49

El-Maghraby OM, El-Kady I A, Soliman S (1995) Mycoflora and *Fusarium* toxins of three types of corn grains in Egypt with special reference to production of trichothecene-toxins. Microbiol Res 150:225–232

European Commission (EC) (2007) Commission Regulation (EC) no 1126/2007 of 28 September 2007 amending Regulation (EC) no 1881/2006 setting maximum levels for certain contaminants in foodstuffs as regards *Fusarium* toxins in maize and maize products. Off J Eur Union L255:14–17

Fazekas B, Kis M, Hajdu ET (1996) Data on the contamination of maize with fumonisin B1 and other fusariotoxins in Hungary. Acta Vet Hung 44:25–37

Food and Agriculture Organization of the United Nations (FAO) (1997) Worldwide regulations for mycotoxins in food and feed

Fritsche S, Steinhart H (1999) Occurrence of hormonally active compounds in food: a review. Eur Food Technol 209:153–179

Gajecki M, Przybylowicz M, Obremski K, Zielonka Ł, Zwierzchowski W, Skorka-Wyszynska E, Gajecka M, Polak M, Jakimiuk E (2004) Preliminary results of monitoring research over zearalenone presence in blond of women with neoplastic lesions in reproductive system. Polish J Vet Sci 7:153–156

Gbodi TA, Nwude N, Aliu YO, Ikediobi CO (1986) The mycoflora and some mycotoxins found in maize (*Zea mays*) in the Plateau State of Nigeria. Vet Hum Toxicol 28:1–5

Golinski P, Hult K, Grabarkiewicz-Szczesna J, Chelkowski J, Kneblewski P, Szebiotko K (1984) Mycotoxic porcie neprhropathy and spontaneous occurrence of ochratoxin A residues in kidneys and blond of Polish swine. Appl Environ Microbiol 47:1210–1212

Golinski P, Hult K. Grabarkiewicz-Szczesna J, Chelkowski J, Szebiotko K (1985) Spontaneous occurrence of ochratoxin A residues in porcine kidney and serum samples in Poland. Appl Environ Microbiol 49(4):1014–1015

Golinski P, Nowak T (2004) Dietary origin of mycotoxins with estrogenic potential and possible health implications to female dogs. Polish J Vet Sci 4:337–341

Golinski P, Vesonder RF, Latus-Zietkiewicz D, Perkowski J (1988) Formation of fusarenone X, nivalenol, zearalenone, α-trans-zearalenol, β-trans-zearalenol, and fusarin C by *Fusarium crookwellense*. Appl Environ Microbiol 54:2147–2148

Gora M, Luczynski MK, Smoczynski L, Obremski K, Polak M, Swist M, Zielonka L, Gajecki M (2004) Modification of zearalenone structure in model and natural conditions. Polish J Vet Sci 7(3):181–185

Gross VJ, Robb J (1975) Zearalenone production in barley. Ann Appl Biol 80:211–216

Gromadzka K, Buśko M, Waśkiewicz A, Kostecki M, Chełkowski J, Popiel D (2008) Impact of *Trichoderma* and *Glocladium* on *Fusarium* mycotoxins production in solid substrate. Cereal Res Commun 36(Suppl B):707–710

Hidy PH, Baldwin RS, Greasham RL, Keithy CL, McMuller JR (1977) Zearalenone and some derivatives. Adv Appl Microbiol 22:59–82

Hussein HS, Brasel JM (2001) Toxicity, metabolism, and impact of mycotoxins on humans and animals. Toxicology 167:101–134

Huwig A, Freimund S, Kappeli O, Dutler H (2001) Mycotoxin detoxication of animal feed by different adsorbents. Toxicol Lett 122:179–188

Jimenez M, Mateo R (1997) Determination of mycotoxins produced by *Fusarium* isolates from banana fruits by capillary gas chromatography and high-performance liquid chromatography. J Chromatogr A 778:363–372

Koen JS, Smith HC (1945) An unusual case of genital involvement in swine associated with eating moldy corn. Vet Med 40:101–124.

Kuiper GJ, Lemmen JG, Carlsson B, Corton C, Safe SH, van der Saag PT (1998) Interaction of esteogenic chemicals and phytoestrogens with estrogen receptor b. Endocrinology 139(10):4252–4263

Larsen S (1928) On chronic degeneration of the kidneys caused by mouldy rye. Maanegsskr Dyrl 40:259–284, 289–300

Lauren DR, Jensen DJ, Smith WA, Dow BW, Sayer ST (1996) Mycotoxins in New Zealand maize: a study of some factors influencing contamination levels in grain New Zealand. NZ. J Crop Hortic Sci 24:13–20

Lemke SL, Mayura K, Ottinger SE, McKenzie KS., Wang N, Fickey C, Kubena LF, Phillips TD (1999) Assessment of the estrogenic effects of zearalenone after treatment with ozone utilizing the mouse uterine weight bioassay. J Toxicol Environ Health A 56:283–295

Li FQ, Li YW, Luo XY, Yoshizawa T (2002) *Fusarium* toxins in wheat from an area in Henan Province, PR China, with a previous human red mould intoxication episode. Food Addit Contam 19:163–167

Maaroufi K, Chekir L, Creppy EE, Ellouz F, Bacha H (1996) Zearalenone induces modifications in haematological and biochemical parameters in rats. Toxicon 34:534–540

D'Mello JPF, Placinta CM, McDonald AMC (1999) *Fusarium* mycotoxins: a review of global implications for animal health, welfare and productivity. Sci Technol 80:183–205

McEvoy TG, Robinson JJ, Ashworth JA, Rooke JA, Sinclair KD (2001) Feed and forage toxicants affecting embryo survival and fetal development. Theriogenology 55:113–129

McKenzie K.S, Sarr AB, Mayura K, Bailey RH, Miller DR, Rogers TD, Norred WP, Voss KA, Plattner RD, Kubena LF, Phillips TD (1997) Oxidative degradation and detoxification of mycotoxins using a novel source of ozone. Food Chem Toxicol 35:807–820

Miedaner T, Reinbrecht C (1999) Fusarien in Getreide — Bedeutung von Pflanzenbau und Resistenzzüchtung zur Verminderung von Ertragsverlusten und einer Kontamination mit Mycotoxinen. Getreide Mehl Brot 53:135–140

Miksicek RJ (1994) Interaction of naturally occurring non-steroidal estrogens with expressed recombinant human estrogen receptor. J Steroid Biochem Mol Biol 49:153–160

Mirocha CJ, Harrison J, Nichols AA, McClintock M (1968) Detection of a fungal estrogen in hay associated with infertility in dairy cattle. Appl Microbiol 16:797–798

CAST (2003) Mycotoxins: risk in plant, animal and human systems, Task Force Report No. 139. Council for Agricultural Science and Technology – AST, Ames, IA, USA

Nahm KH (1995) Possibilities for preventing mycotoxicosis in domestic fowl. World Poult Sci J 51:177–185

Nikaido Y, Yoshizawa K, Danbara N, Tsujita-Kyutoku M, Yuri T, Uehara N, Tsubura A (2004) Effects of maternal xenoestrogen exposure on development of the reproductive tract and mammary gland in female CD-1 mouse offspring. Reprod Toxicol 18:803–811

Nuryono N, Noviandi CT, Böhm J, Razzazi-Fazeli E (2005) A limited survey of zearalenone in Indonesian maize-based food and feed by ELISA and high performance liquid chromatography. Food Contr 16:65–71

Park JW, Shoi SY, Hwang HJ, Kim YB (2005) Fungal mycoflora and mycotoxins in Korean polished rice destined for humans. Int J Food Microbiol 103:305–314

Phillips TD, Sarr BA, Grant PG (1995) Selective chemisorption and detoxification of aflatoxins by phylosilicate clay. Nat Toxins 3:204

Plasencia J, Mirocha CJ (1991) Isolation and characterization of zearalenone sulfate produced by *Fusarium* spp. Appl Environ Microbiol 57:146–150

Popiel J, Nicpon J, Sapikowski G (2004) Effect of the dry feed contaminated with moulds on the prevalence of alimentary tract disorders in cats. Polish J Vet Sci 7:149–151

Ramos AJ, Hernandez E (1997) Prevention of aflatoxicosis in farm animals by means of hydrated sodium calcium aluminosilicate addition to feedstuffs. A review. Anim Feed Sci Technol 65:197–206

Ramos AJ, Hernandez E, Pla-Delfina JM, Merino M (1996) Intestinal absorption of zearalenone and in-vitro study of non-nutritive sorbent materials. Int J Pharm 128:129–137

Resnik S, Neira S, Pacin A, Martinez E, Apro N, Latreite S (1996) A survey of the natural occurrence of aflatoxins and zearalenone in Argentine field maize: 1983–1994. Food Addit Contam 13:115–120

Richardson KE, Hagler WM, Mirocha CJ (1985) Production of zearalenone a- and b-zearalenol and a- and b-zearalanol by *Fusarium* spp. in rice culture. J Agric Food Chem 33:862–866

Sabater-Vilar M, Malekinejad H, Selman MHJ, van der Doelen MAM, Fink-Gremmels J (2007) In vitro assessment of adsorbents aiming to prevent deoxynivalenol and zearalenone mycotoxicoses. Mycopathologia 163:81–90

Santin E, Paulillo AC, Maiorka PC, Alessi AC, Krabbe EL, Maiorka A (2002) The effects of ochratoxin/aluminosilicate interaction on the tissues and humoral immune response of broilers. Avian Pathol 31:73–79

Schollenberger M, Muller HM, Rufle M, Suchy S, Planck S, Drochner W (2006) Natural occurrence of 16 *Fusarium* toxins in grains and feedstuffs of plant origin from Germany. Mycopathologia 161:43–52

SCOOP (2003) Collection of occurrence data of *Fusarium* toxins in food and assessment of dietary intake by the population of EU Member States. In: Vidnes A, Bergsten C, Paulsen B (eds) Subtask zearalenone. SCOOP European Project. Task 3.2.10, pp 241–481

Scudamore KA, Patel S (2000) Survey for aflatoxins, ochratoxin A, zearalenone and fumonisins in maize imported into the United Kingdom. Food Addit Contam 17(5):407–416

Seeling K, Dänicke S, Ueberschär KH, Lebzien P, Flachowsky G (2005) On the effects of *Fusarium* toxin-contaminated wheat and the feed intake level on the metabolism and carry over of zearalenone in dairy cows. Food Addit Contam 22(9):847–855

Shier WT, Shier AC, Xie W, Mirocha CJ (2001) Structure-activity relationships for human estrogenic activity in zearalenone mycotoxins. Toxicon 39:1435–1438

Skorska-Wyszynska E, Jakimiuk E, Gajecka M, Mlynarczyk J, Obremski K, Gajecki M (2004) Preliminary evaluation of influence of zearalenone on cocultures of granulosa and internal heca cells of ovarian follicles in bitches *in vitro* culture. Polish J Vet Sci 7:305–309

Stob M, Baldwin RS, Tuite J, Anders FN, Gillette KG (1962) Isolation of an anabolic, utertrophic compound from corn infected with *Gibberella zeae*. Nature 196:1318

Szebiotko K, Chełkowski J, Dopierala G, Godlewska B, Radomyska W (1981) Mycotoxins in cereal grain. Part I. Ochratoxin, citrinin, sterigmatocystin, penicillic acid and toxigenic fungi in cereal grain. Nahrung 25(5):415–421

Takahashi-Ando N, Kimura M, Kakeya H, Osada H, Yamaguchi I (2002) A novel lactonohydrolase responsible for the detoxification of zearalenone: enzyme purification and gene cloning. Biochem J 365:1–6

Tiemann U, Tomek W, Schneide F, Vanselow J (2003) Effects of the mycotoxins α- and β-zearalenol on regulation of progesterone synthesis in cultured granulosa cells from porcine ovaries. Reprod Toxicol 17:673–681

Tomasevic-Canovic M, Dakovic A, Rottinghaus G, Matijasecic S, Duricic M (2003) Surfactant modified zeolites – ew efficient adsorbents for mycotoxins. Micropor Mesopor Mat 61: 173–180

Tomaszewski J, Miturski R, Semczuk A, Kotarski J, Jakowicki J (1998) Tissue zearalenone concentration in normal, hyperplastic human endometrium. Ginekol Pol 69(5):363–366

Turcotte JC, Hunt PJB, Blaustein JD (2005) Estrogenic effects of zearalenone on the expression of progestin receptors and sexual behavior in female rats. Horm Behav 47:178–184

Urry WH, Wehrmeister HL, Hodge EB, Hidy PH (1966) The structure of zearalenone. Tetrahedron Lett 27:3109–3114

Utermark J, Karlovsky P (2007) Role of zearalenone lactonase in protection of *Glocladium roseum* from fungitoxic effects of the mycotoxins zearalenone. Appl Environ Microbiol 73:637–642

Vlata Z, Porichis F, Tzanakakis G, Tsatsakis A, Krambovitis E (2006) A study of zearalenone cytotoxicity on human peripheral blood mononuclear cells. Toxicol Lett 165:274–281

Waskiewicz A (2006) Selected mycotoxins in human and animal tissue as a result food and feed contamination. PhD Thesis, Agricultural University, Poznan, Poland

Williams BA, Mills KT, Burroughs CD, Bern HA (1989) Reproductive alterations in female C57BL/Cgl mice exposed neonatally to zearalenone, an estrogenic mycotoxin. Cancer Lett 46:225–230

Yoshizawa T (1997) Geographic difference in trichothecene occurrence in Japanese wheat and barley. Bull Inst Compr Agric Sci Kinki Univ 5:23–30

Yu Z, Zhang L, Wu D, Liu F (2005) Anti-apoptotic action of zearalenone in MCF-7 cells. Ecotoxicol Environ Safe 62:441–446

Zinedine A, Brera C, Elakhdari S, Catano C, Debegnach F, Angelini S, De Santis B, Faid M, Benlemlih M, Minardi V, Miraglia M (2006) Natural occurrence of mycotoxins in cereals and spices commercialized in Morocco. Food Contr 17:868–874

Zinedine A, Soriano JM, Molttŏbreve; JC, Mañes J (2007) Review on the toxicity, occurrence, metabolism, detoxification, regulations and intake of zearalenone: an oestrogenic mycotoxin. Food Chem Toxicol 45:1–18

Zwierzchowski W, Przybylowicz M, Obremski K, Zielonka Ł, Skorska-Wyszynska E, Gajecka M, Polak M, Jakimiuk E, Jana B, Rybarczyk L, Gajecki M (2005) Level of zearalenone in blood serum and lesions in ovarian follicles of sexually immature gilts in the course of zearalenone micotoxicosis. Polish J Vet Sci 8(3):209–218

Chapter 9
Zearalenone: Undesirable Substance

Maciej Gajęcki, Magdalena Gajęcka, Ewa Jakimiuk, Łukasz Zielonka, and Kazimierz Obremski

9.1 Introduction

The presence of toxic or undesired substances in plant material can disturb homeostasis in both humans and animals (Massart et al. 2008; Shephard 2008b). These toxins enter the organism via the gastrointestinal tract, resulting in numerous adverse consequences (Cavret and Lecoeur 2006; Sergent et al. 2008). These substances include: (a) plant protective agents (Trucksess and Scott 2008); (b) industrial pollutants emitted to the environment (Crain et al. 2008; Shephard 2008a); (c) residues of veterinary medicines used in therapy; and (d) active substances naturally produced by plants, e.g., mycotoxins (Minervini and Dell'Aquilla 2008). The latter can cause millions of dollars of loss by unfavourably influencing human and animal health and can reduce the health qualities of plant products (Blandino et al. 2008). They have significant economic influence by diminishing livestock populations or causing problems in their growth and breeding (Egmond et al. 2007). According to legal regulations already in force concerning some mycotoxins (EU Commission Decree Nr 123/2005, 856/2005 and 1126/2007), their presence in edible goods results in national and international trade bans. Regarding feeding stuffs, only Commission Recommendation 576/2006 is in force in Europe. The above-mentioned mycotoxins are secondary metabolites of moulds mostly from the *Penicillium*, *Aspergillus* and *Fusarium* genera, which are very toxic and have mutagenic properties (aflatoxins, fumonizins, ochratoxin A, luteoscirin, toxin T-2), teratogenic properties (ochratoxin A, patulin, aflatoxin B_1, toxin T-2) and/or estrogenic properties (zearalenone) (Jarvis and Miller 2005).

M. Gajęcki (✉)
Department of Veterinary Prevention and Feed Hygiene, Faculty of Veterinary Medicine, University of Warmia and Mazury in Olsztyn, St. M.Oczapowskiego 13, 10-719, Olsztyn, Poland
e-mail: gajecki@uwm.edu.pl

M. Rai and A. Varma (eds.), *Mycotoxins in Food, Feed and Bioweapons*,
DOI: 10.1007/978-3-642-00725-5_9, © Springer-Verlag Berlin Heidelberg 2010

In this chapter, we present an initial analysis of the influence of zearalenone on animals, especially with regard to the immune and hormonal systems (Liska et al. 2006).

9.2 Moulds and Their Products

Moulds are eukaryotic organisms that lack chlorophyll. While most moulds are parasites, many species are also saprophytes. Their presence in materials intended for human or animal foodstuffs should be treated as a threat to health (Čonková et al. 2003). Unfavourable effects on the health of mammals depend both on the degree of mould contamination, encompassing the presence of proteins on the surface of the mould and in their spores, and on the number of secondary metabolites (mycotoxins) present in the substrate (Kuhn and Ghannoum 2003). Very little is known about either the allergenic proteins related to moulds found in fungal tissues or about the allergenic potency of mycotoxins which are carried in organisms by protein carriers (Jarvis and Miller 2005). We can anticipate pathological effects arising from the moulds and the mycotoxins that they introduce to humans or to animals. However, it is difficult to recognize clinical changes deriving from a single acute intoxication, and even less is known about the effects of mixed intoxications.

Fusarium, Aspergillus, and *Penicillium* genera are very common moulds in the plant world. Although more than 100 species of moulds have been described, only 300 mycotoxins have been identified due to the analytical difficulties in assaying many mycotoxins in plant material (Gutleb et al. 2002). The effects of only a few identified mycotoxins have been studied with respect to the metabolism of humans and animals and these results have been only fragmentary (Shephard 2008b). Some of the mechanisms of their pathogenic activity have been partially elucidated. Problems arising from interactions between mycotoxins are still uncharacterized (Speijers and Speijers 2004).

The toxic activities of mycotoxins can differ (Cetin and Bullerman 2005). Acute toxic effects are observed only exceptionally. Rather, long-term exposure to low concentrations of particular mycotoxins can result in chronic disease states, liver or kidneys neoplasms, or other diseases together with allergies (Jarvis and Miller 2005).

Some mycotoxins show mutagenic, teratogenic, and estrogenic activities (zearalenone) (Cavaliere et al. 2005; Jarvis and Miller 2005). At the initial phase of their activity, some mycotoxins interfere with protein synthesis, causing skin hypersensitivity or necrotic lesions or, ultimately, they bring about decreased levels of antibodies, as in the case of zearalenone (Atroshi et al. 2002; Gajęcka et al. 2004). Other recognized mycotoxins are neurotoxins. In low concentration, these cause seizures in animals; a small increase in their concentration can result in brain lesions or even death (Pitt 2000).

Mycotoxins are aromatic hydrocarbons or, in some cases, aliphatic hydrocarbons of low molecular weight. These chemical properties determine their resistance to environmental factors and account for their low or absent immunogenic properties (Speijers and Speijers 2004; Cavaliere et al. 2005), while still modulating immune system functions.

Recent investigations have examined the integration of the effects of specific mycotoxins or groups of mycotoxins with vital cell functions, with the goal of elucidating the specific pathological changes in particular organs, tissues, or cells. Some mycotoxins serve as triggering factors (Cavaliere et al. 2005), while other mycotoxins have opposing activity which results in negative reactions.For example, patulin exhibits the beneficial effect of protecting fats from oxygenation (Riley and Norred 1996; Riley 1998). In contrast, trichothecenes suppress the immune systems of animals (Riley 1998; Zielonka et al. 2003, 2004; Pestka et al. 2005). From other examples, it appears that the kidney is the organ where the activity of the mycotoxins is strongest, but systemic reactions such as oedema or allergic states can also be present (Fischer and Dott 2003; Jarvis and Miller 2005).

At present, we most often deal with fusariotoxins, secondary metabolites produced by moulds from the genus *Fusarium*. They occur during both the growth and development phases of the *Fusarium* life cycle (Olsen and Kiessling 1983; Cavret and Lecoeur 2006). Moulds of this kind can produce one or many mycotoxins (e.g., deoxynivalenol, nivalenol, T-2 or HT-2 toxins and many others from the trichothecenes group, and, beyond this group, zearalenone or fumonisin B), but not all of these fungi are toxigenic. Weather anomalies present worldwide encourage mould growth and mycotoxin production by some species of mould during the vegetative stages of plants, during storage of unprocessed and processed farm products, and also in the end products, e.g., feeding stuffs (Hollinger and Ekperigin 1999). Fusariotoxins enter humans and animals in very low doses, mostly *per os* (Rhyn and Zoller 2003; Gareis et al. 2003; Jarvis and Miller 2005), and, apart from systemic toxicity, they also exhibit carcinogenic (Isaacson 2005), mutagenic, and teratogenic features (Hussein and Brasel 2001) and cause gastrointestinal tract disturbances in animals (Sweeney 2002; Gajęcka 2006) and humans (Kuciel-Lisieska 2006). Additionally, man can be indirectly exposed to mycotoxins through contaminated edible goods of animal origin (Cavret and Lecoeur 2006).

Corn contamination with fusariotoxins is very common despite different preventive measures. Fusariotoxins are compounds of great chemical persistence and until now there has been no method of effective grain decontamination (Polak 2007). The degree of fusariotoxin contamination of grain has been related to human and animal health status and to economic losses in animal production (Čonková et al. 2003).

Currently, strategies for preventive and monitoring activities are being directed towards grain humidity control in the field and during storage (Binder 2007).

9.3 Phytoestrogens

Phytoestrogens are present in plants and are similar in structure and hormonal activity to estrogens (Morgavi and Riley 2007). They are generally classified as flavonoids, isoflavonoids, lignans, coumestans, mycotoxins and stilbenes (Stopper et al. 2005). Flavonoids, isoflavonoids, and stilbenes most commonly show immunomodulating potential.

There are lower incidences of hypertrophic and chronic inflammatory diseases in human and animal populations ingesting diets rich in flavonoids and isoflavones, indicating that their presence in foods favours biological balance of immune processes (McClain et al. 2005). Simultaneously, numerous in vitro and *in vivo* studies in laboratory animals confirm selective influences of these diverse compounds on the functioning of many immunological cells.

Probably the polytrophic effect of phytoestrogens activity is related not only to receptor specificity, but, depending on the (estrogen receptors — ERs) ERα and ERβ receptor (Benassayag et al. 2002) interactions of these compounds (Rosselli et al. 2000), also to other properties of phytoestrogens, which directly influence signal transduction pathways in immune cells (Stopper et al. 2005).

From the existing literature, phytoestrogens appear to be modulators of the proliferative reaction of lymphocytes, due to their estrogen-dependent and -independent modes of action.

9.4 Zearalenone

Zearalenone is a non-steroidal mycotoxin with estrogenic activity and also a specific hormone regulating sexual reproduction of *Fusarium* (sexual state of *Gibberella zeae*) (Voigt et al. 2007). In favourable climate conditions, different genes of *Fusarium* produce ~150 zearalenone derivatives. The mycotoxins most frequently isolated from animal and plant tissues are fusaric mycotoxins (Gajęcki 2002; Zinedine et al. 2007). Zearalenone [6-(-10-hydroxy-6-oxo-*E*-1-undecenyl)-β-resorcylic acid lactone] has a stable chemical structure and does not undergo degradation during storage, grinding, technical processing with high temperatures, or during the milling of moist grain, which favours zearalenone formation in high concentration in the gluten fractions. Zearalenone is soluble in alkaline solutions, ether, benzene, acetonitrile, ethylene alcohol, but is practically insoluble in water. In mammals, the C-6 ketone group can be reduced to the stereo isomeric metabolites of zearalenol (α- and β-isomers); these reduction products are also synthesized by moulds, but in much lower concentrations then the zearalenone parent compound. These metabolites have been found in grain, rye, and silages, and in grain by-products or soy meal.

Mould growth and zearalenone production depend mostly on microclimate conditions. The largest amounts of zearalenone are produced at temperature below 25°C, with elevated daily ambient temperature (5–25°C) and at ~16% humidity. Its biosynthetic yield also depends on the mould species, e.g., *F. roseum* produces around 3,000–15,000 ppm zearalenone, while *F. moniliforme* yields only 1–19 ppm (Goliński et al. 2002).

The intoxication of female mammals with zearalenone during pregnancy results in foetal death and/or neonatal low body weight (Minervini and Dell'Aquilla 2008).

Zearalenone influences uterus morphology, reducing luteinizing hormone and progesterone release. In different animal species, particularly in sexually immature gilts, clinical oestrus signs have been observed due to hormonal disturbances provoked by hyperestrogenism, but without tolerance reaction (present even in sterilized animals). In boars, hyperestrogenism can result in low libido, lower testicular weight, slower spermatogenesis, and lower testosterone concentrations. In cattle, zearalenone intoxication results in infertility, lower milk production, and hyperestrogenism. The most sensitive species are pigs and sheep. Human intoxication with zearalenone can induce activation of proliferative processes in estrogenic cells in uterus, ovarian and mammary tissues neoplasms (Kuciel-Lisieska 2006), duodenum, or large intestines. These events probably result from the saturation of nuclear estrogenic receptors (ERα and ERβ) being saturated with zearalenone and its metabolites (Chen et al. 2005). Consequently, this mycotoxin can modify the transcription process (Minervini and Dell'Aquilla 2008). ER receptors, from group IV (depending on DNA-binding domain structure) in the ligand-free form, are localized in the cell cytoplasm, creating complexes with chaperone proteins. The binding of a ligand (an endogenic hormone, phytoestrogen, or xenobiotic) results in a receptor conformational change, followed by separation from the chaperone protein and movement of the receptor with bound ligand to the cellular nucleus, where they can be homodimerised or heterodimerised. After dimerisation, the receptor-ligand complex binds to specific sites of endogenic hormone reaction. However, structurally diverse natural and synthetic compounds can also serve as ligands (including phytoestrogens or xenobiotics), which can bind to regulatory parts of the target genes. This process is followed by an increase in the transcription of specific genes or by competitive blocking of the access of the natural hormone to the ER by phytoestrogen (e.g., zearalenone) or xenobiotic. Endogenic hormones usually inhibit proliferation, stimulate differentiation, and promote apoptosis. Phytoestrogens usually act similarly. In low doses, zearalenone acts as a chemopreventive and in higher doses it promotes proliferative processes (Mueller 2002). In pathological processes, these hormones or estrogen-like compounds (hormonmimetics) can produce a "functional imbalance" that stimulates development of existing hormonally sensitive neoplasms or by provoking intensive proliferative processes in hormone-reactive cells (Vlata et al. 2006).

9.4.1 Zearalenone Absorption

Gastrointestinal absorption determines the transition of mycotoxins from the lumen of the gastrointestinal tract into the blood, and ultimately guides their distribution in the organism (Zinedine et al. 2007). With knowledge of the rate and the mechanisms of the absorption processes in the gastrointestinal tract, one can theoretically evaluate fusariotoxin absorption routes (Sergent et al. 2006). In comparison with other fusaric mycotoxins, zearalenone appears in blood much sooner (by 30 min) in concentrations similar to those found in the gastrointestinal tract. The zearalenone absorption process is so effective that after a single administration, 85% of the initial zearalenone dose is present in blood.

Little is known about how fusariotoxins cross the intestinal barrier. Endocytosis seems unlikely. The possibility of intercellular transport is small, because only small hydrophilic particles can be transported by this process, and only in the apical part of the cell. Mycotoxin transport through passive or active diffusion seems more probable. According to Ramos et al. (1996) zearalenone intestinal absorption occurs by passive diffusion, which is the spontaneous passage of zearalenone and/or its metabolites from chyme through the intestinal wall to blood. The effectiveness of this process depends upon changing the physicochemical properties of the mycotoxin (weight, number of particles, lipophility, or conformation ability), which allow passage through apical or basal lipophilic membranes. The rapid appearance of most of the fusariotoxins in blood and serum shows that the absorption process begins at the level of the stomach and small intestines. Avantaggiato et al. (2003) demonstrated, with in vitro studies using a dynamic phantom model of the gastrointestinal tract of swine, that the absorption of zearalenone and some of the trichothecenes took place mainly in the small intestines lumen, particularly in jejunum (from 70 to 85%) and ileum (from 30 to 15%).

9.4.2 Zearalenone Biotransformation

Biotransformation is a complex of defensive mechanisms that occur before or after absorption of dangerous or undesirable substances by an organism (Liska et al. 2006). Humans and animals have different enzymatic systems for transforming xenobiotics, naturally occurring biologically active substances, and drugs to hydrophilic metabolites, which allows for fast elimination from the organism via bile or urine. Biotransformation reactions take place mainly in the liver (with numerous enzymatic biotransformation systems). To a lesser degree, they also occur in lungs, kidneys, intestines, or granular cells of ovarian follicles (Malekinejad et al. 2006a).

Identification of the enzymes that metabolize in fusariotoxins, including zearalenone, as well as knowledge about factors accompanying this process, are necessary to understand the distribution of mycotoxins in the organism and the transformation processes influencing the modification of their physical and chemical properties (Cavret and Lecoeur 2006).

Based on the current state of knowledge, studies by Nobel et al. (2001) prove that enzymatic activation of non-active hormones or their precursors to biologically active compounds or, conversely, the alteration of active hormones to non-active forms in specific cells are important mechanisms that regulate the affinity for specific endogenic receptors, a new form of hormonal regulation. Active hormones have increased affinity for their receptors, while the non-active hormones have lower affinity. Enzymes that mediate the interchange between active and non-active forms of hormones and the enzymes that immediately change activity at the molecular level are termed enzymatic pre-receptoral regulators (Penning 2003). Specific enzymes regulate the concentration of active steroid hormones at the pre-receptoral level in specific tissues. Enzymatic regulation at the pre-receptoral level concerns the second phase of biotransformation, the conjugation phase. Enzymes that function in the conjugation phase include cytochrome P-450, enzymes of aldo-keto reductase (AKR) family and the hydroxysteroid dehydrogenase (HSD) enzymes, most of which belong to the short-chained dehydrogenase/reductase (SDR) family (Mindnich et al. 2004) and function in the regulatory processes at the pre-receptoral level.

The publications of Rosselli et al. (2000) indicate that phytoestrogens have multiple actions. They can be agonists and antagonists for estrogen receptors (ERα and/or ERβ) or other receptors. Importantly, at the pre-receptoral level, they can modulate the activity of enzymes taking part in estrogen biosynthesis, such as aromatase, sulphatase, sulphtransferase, 3β-HSD and 17β-HSD (Tiemann et al. 2003).

Zearalenone biotransformation to α-zearalenol and β-zearalenol is performed by ketol group reduction in the 6′ position. This reaction is very similar to the processes occurring during metabolism of steroids catalyzed by HSDs. Malekinejad et al. (2006b) examined zearalenone biotransformation in granulose cells with or without the endogenic substrates for 3α- and 3β-HSD. These results document, that α-zearalenol production is reduced, because of substrate concentration increase, which is for 3β-HSD to be used or to reduce oxydase 3b-HSD 3β-HSD, activity reduction as a result of substrate gathering. Tiemann et al. (2003) stated that — depending on the α-zearalenol and β-zearalenol dose, thanks to reduction of P450scc and 3β-HSD activity reduction of progesterone synthesis in granulose cell cultures from swine is present.

In other studies, Malekinejad et al. (2006a) confirmed that during zearalenone biotransformation in the liver of swine mostly α-zearalenol is produced, but in cattle β-zearalenol is the predominant form. In chickens, as a result of liver biotransformation, β-zearalenol is the main product in both the isolated microsomal and postmitochondrial fractions, which is consistent with previous studies on poultry hepatocytes. Biotransformation of zearalenone in the liver of rats produces mainly β-zearalenol, which is in contrast to the results presented by Ueno and Tashiro (1981), who suggested that α-zearalenol is the only transformation product. These results confirmed earlier suggestions regarding the varied effects of zearalenone biotransformation in different animal species, which can reflect differences in the distribution of HSDs in cells. The rate of the glucuronidation process depends on uridine diphosphate glucuronyltransferase (UDPGT) activity in the endoplasmic reticulum and the accessibility of uridine diphosphate glucuronic acid (UDPGA). The efficiency of the glucuronidation process towards zearalenone has been determined for only a few animal species. In swine, total glucuronidation of zearalenone occurs in liver samples at low zearalenone concentrations (10 μM), whereas in other species, at comparable concentrations, zearalenone glucuronate conjugation is low. In swine, the dominant metabolite is α-zearalenol, which correlates with higher sensitivity of these animals to zearalenone as an estrogenic factor. Currently, there are basically two zearalenone biotransformation processes in organisms: first zearalenone is reduced to α-zearalenol and β-zearalenol, a reaction catalyzed by 3α-HSD and 3β-HSD; second, zearalenone and its reduced metabolites are conjugated with glucuronic acid, a transformation catalysed by UDPGT.

9.4.3 Zearalenone Hormonal Properties

Zearalenone is a phytoestrogen belonging to the class of naturally occurring biologically active substances with estrogenic properties, but effective only towards peripheral reproductive tissues, e.g., causing uterus weight increase and other minor indicators of estrogen activity. Clinical states of intoxication by this mycotoxin are difficult to differentiate from the phases of the reproductive cycle in, for instance, swine (Minervini and Dell'Aquilla 2008). Intoxication gives signs similar to typical oestrus signs in swine. Pigs, particularly gilts before first oestrus, are the species most susceptible to zearalenone intoxication. Typical signs of intoxication are oedema and reddening of the vulva, uterus weight increase, and in extreme cases, vaginal and anal prolapse. Even low concentrations of this mycotoxin in animal feed (20–40 $\mu g\ kg^{-1}$) can cause reproductive disturbances, including first oestrus delay, false pregnancy, early abortion, small litter size, or stillbirths. The most commonly observed microscopic changes are oedema, uterus cells hyperplasia, and metaplasia of mammary gland ducts and swine cervical and vaginal cells in bitches (Gajęcka et al. 2007). These metabolic changes in different reproductive organ tissues, particularly in endometrium, are attributed to the elevated concentrations of either estradiol, which intensifies proliferative processes, or progesterone, which decreases proliferative processes (Kustritz 2005).

In in vitro studies, zearalenone and its metabolites show affinity for many estrogen receptors (ERs) (Scippo et al. 2002). These specific receptors are observed in uterus, mammary gland, liver, brain, hypothalamus, and even in malignant neoplasms of different reproductive tissues of different animal species. The number of ERs in target tissues or cells with affinity towards estradiol or zearalenone is very diverse. The affinity of zearalenone for estrogen receptor types ERα and ERβ is similar to that for progesterone receptors in rats (Lephart et al. 2005; Minervini and Dell'Aquilla 2008).

9.5 Environmental Estrogens

Ansar-Ahmed (2000) suggests two common terms from endocrinology should be reconsidered. First, sexual hormones can not be treated only as "reproductive" hormones, but also as compounds influencing "non-reproductive" functions of many tissues, particularly in the immune system. Secondly, estrogens exist not only as naturally occurring compounds in an organism, but also as environmental xenobiotics. These currently recognized substances, not necessarily contaminants, called environmental estrogens, create a group of compounds called endocrine disrupters (EDs), commonly found in the environment, in soil, air, water, and edible goods (Yurino et al. 2004). The EDs naturally present in the environment are phytoestrogens (genistein, cumestrol) and micoestrogens, product of moulds from *Fusarium* spp. family (zearalenone) (Gajęcki 2002).

There is no clear correlation between environmental estrogens and the ability to modulate endocrine systems (Sweeney 2002) and their eventual negative effects on human and animal organisms. Research performed on wild, farm, and companion animals forces us to consider updating the methods of determining the harmfulness of substances that can modulate the endocrine system (Gajęcka 2006). Rather, EDs should be divided into categories according to their characteristics and to the duration of their effects, i.e., whether they are reversible or irreversible.

This division should take into account the double action of EDs on the immune system, which has often been recognized (Yurino et al. 2004). Reproductive hormones influence the morphology of the thymus and other lymphatic organs. Estrogens administration causes thymus involution. On the other hand, thymus involution, particularly after parturition, can cause changes in ovarian, testicular, or thyroid endocrine tissues. Estrogens regulate the synthesis of serum and uterine IgM, IgA, and IgG immunoglobulins (Gajęcka et al. 2004), increasing the production of both specific and non-specific antibodies (Kurup et al. 2000). Hormones and cytokines probably play an important role in transferring information between the reproductive and the immune systems. The close correlation between these two systems shows that the influence of environmental estrogens on reproductive tissues can also affect the immune system. Whether this effect is caused by direct or indirect activity of environmental estrogens is still open to question. Future studies should examine whether the immune system is really a target for EDs in general or for environmental estrogens in particular.

Estrogen hormones probably participate in pathological states resulting from autoimmune or allergic diseases. These illnesses or pathological states are more common in females then in males. Reproductive hormones are usually a primary cause of allergic disease states, such as asthma, nasal mucosal inflammation or skin inflammation in people and animals (Stubner et al. 1999).

9.6 Cooperation of the Immunological System with Steroid Hormones

In mammals, general systemic homeostasis is maintained by the nervous system (Jana and Majewski 2007), the endocrinal system (Chen et al. 2005), and the immune system (Inadera 2006). Even small disturbances in functioning and activity of one of these systems caused by, for instance, mycotoxins can cause both reversible and irreversible disease states. Unwanted or dangerous substances present in the environment, edible goods, or feeding stuffs can greatly influence the incidence and dynamics of immune-mediated disease progression, e.g., atopic allergies or anaphylaxis (Inadera 2006). Parallel observations from laboratory research and clinical observation in both humans and animals indicate that the incidence of autoimmune diseases is increased when pathological immune states occur in the gastrointestinal tract (Acheson 2004).

The immune system plays an important role in integrating the functions of diverse organ systems, thanks to its specific organization and pleiotropic action (Jana and Majewski 2007). The organized structures, represented by primary and secondary lymphatic organs, and circulating cells within this system assure constant contact between immunological cells and other tissues. The actions of effectory leukocytes not only protect an organism from foreign agents, but also directly influencs reactions of the central nervous system and the endocrine glands. The most important effectory mechanisms of immune cells acting directly on nervous and endocrinal cells are the production of cytokines and growth factors.

Interactions of immune system mediators and steroid hormones with the peripheral nervous system are important, in our view. These interactions have direct characteristics, because receptors of high estrogen and progesterone affinity are present in lymphocytes and, simultaneously, cytokines originating from leukocytes participate in the functioning of the gonads, which is controlled by the peripheral nervous system (Jana and Majewski 2007).

An interaction between the immune system and the activity of the gonads is well documented in females, because it seems that estrogens have the greatest influence on immune reactions. Lower perinatal mortality of female neonates, higher concentrations of circulating immunoglobulin, and higher incidence of autoimmune diseases in females show, indirectly, that estrogens are modulate immunity levels (Lang 2004). In this context, natural and xenobiotic phytoestrogens may be considered as having the potential to modulate immune and endocrinal activity.

Experimental investigations have documented that estrogens take part in proliferation, apoptosis, and processes of lymphocytes differentiation. The range of lymphocytes proliferation can be a result of direct estrogens interference in cell cycle process. Cells exposed to estrogens are subjected not to division, but apoptosis. This phenomenon is accompanied by lower transcription and translation of bcl-2 protein, an important regulator of proliferation and apoptosis control. Estrogens inhibit T lymphocytes proliferation, which is connected to lower interleukin-2 (IL-2) synthesis together with reduced IL-2 receptor expression. IL-2 synthesis depends on transcription factor activity, and studies have shown that estrogens inhibit binding of these factors to nuclear DNA, which can alter IL-2 expression. Furthermore, estrogens can decrease lymphocyte B cells total numbers and, at the same time, increase the number of plasma cells producing antibodies. Estrogens are also able to modulate effector functions of immune cells.

The molecular basis for the above-mentioned activity of estrogen on the immune system is probably the presence of ERα in lymphocytes and macrophages and ERβin monocytes and macrophages (Lang 2004). In this respect, phytoestrogens may be assumed to be agonists and/or antagonists of these receptors, and modulation of their activity can influence proliferation, apoptosis and immune cells differentiation.

Only a few studies have investigated the potential of phytoestrogens to treat autoimmune diseases. Nevertheless, these substances are used in chronic inflammatory diseases, for instance rheumatoid arthritis (Verbeek et al. 2004). Pleiotropic influence of phytoestrogens on cells and structures of the immune system shows promise for their use in prevention of autoimmune diseases and hypersensitivity reactions.

9.7 Conclusion

After considering the work of different authors and our own research concerning the presence of zearalenone in organisms of different animal species, the following conclusions can be drawn:

- Although zearalenone does not have a steroid structure, it is a substrate for both 3α- and 3β-HSD.
- High concentration and/or prolonged zearalenone administration results in endocrine effects from the dual activity with estrogen receptors, both with respect to the interaction with synthesis enzymes and with inactivation processes mediated by endogenous steroid hormones.
- Certain animal species have individual susceptibility to zearalenone caused by different biotransformation of this mycotoxin.
- Based on various research results, zearalenone biotransformation in the liver into α-zearalenol may be considered as a bioactivation reaction and zearalenone transformation into β-zearalenol may be considered as a "detoxication" reaction.
- Despite extensive research on wild and farm or companion animals, we still question the safety of the presence of zearalenone and its influence on immune systems in humans. The number of research studies in humans is too small to have clear conclusions. Therefore, there is a critical need for investigations into the influence of zearalenone on immune systems in animals. It would then be advisable to determine the pathological effects of low doses of zearalenone on the immune system over an extended period of time, and to eventually focus on the examination of allergic or autoimmune effects.

References

Acheson DWK (2004) Mucosal immune responses. Best Pract Res Cl Ga 18:387–404

Ansar Ahmed S (2000) The immune system as a potential target for environmental estrogens (endocrine disrupters): a new emerging field. Toxicology 150:191–206

Atroshi F, Rizzo A, Westermarck T, Ali-Vehmas T (2002) Antioxidant nutrients and mycotoxins. Toxicology 180:151–167

Avantaggiato G, Havenaar R, Visconti A (2003) Assessing the zearalenone binding activity of adsorbent materials during passage through a dynamic in vitro gastrointestinal model. Food Chem Toxicol 41:1283–1290

Blandino M, Reyneri A, Vanara F (2008) Effect of plant density on toxigenic fungal infection and mycotoxin contamination of maize kernels. Field Crop Res 106:234–241

Benassayag C, Perrot-Applanat M, Ferre F (2002) Phytoestrogens as modulators of steroid action in target cells. J Chromatogr B 777:233–248

Binder EM (2007) Managing the risk of mycotoxins in modern feed production. Anim Feed Sci Tech 133:49–166

Cavaliere C, D'Ascenzo G, Foglia P, Pastorini E, Samperi R, Laganà A (2005) Determination of type B trichothecenes and macrocyclic lactone mycotoxins in field contaminated maize. Food Chem 92:559–568

Cavret S, Lecoeur S (2006) Fusariotoxin transfer in animals. Food Chem Toxicol 44:444–453

Chen J-Q, Yager JD, Russo J (2005) Regulation of mitochondrial respiratory chain structure and function by estrogens/estrogen receptors and potential physiological/pathophysiological implications. Biochim Biophys Acta 1746:1–17

Cetin Y, Bullerman LB (2005) Cytotoxicity of Fusarium mycotoxins to mammalian cell cultures as determined by the MTT bioassay. Food Chem Toxicol 43:755–764

Crain DA, Janssen SJ, Edwards TM, Heindel J, Ho S, Hunt P, Iguchi T, Juul A, McLachlan JA, Schwartz J, Skakkebaek N, Soto AM, Swan S, Walker C, Woodruff TK, Woodruff TJ Giudice LC, Guillette LJ (2008) Female reproductive disorders: the roles of endocrinedisrupting compounds and developmental timing. Fertil Steril 90:911–940

Čonková E, Laciaková A, Kováč G, Seidel H (2003) Fusarial toxins and their role in animal diseases. Vet J 165:214–220

Egmond HP, Schothorst RC, Jonker MA (2007) Regulations relating to mycotoxins in food perspectives in a global and European context. Anal Bioanal Chem 389:147–157

Fischer G, Dott W (2003) Relevance of airborne fungi and their secondary metabolites for environmental, occupational and indoor hygiene. Arch Microbiol 179:75–82

Gajęcki M (2002) Zearalenone — undesirable substances in feed. Pol J Vet Sci 5:117–122

Gajęcka M (2006) The impact of experimental zearalenon micotoxicosis on the reproductive tract of bitches [Doctoral thesis]. University of Warmia and Mazury in Olsztyn, Poland (in Polish)

Gajęcka M, Jakimiuk E, Skorska-Wyszyńska E, Zielonka Ł, Polak M, Paluszewski A, Rybarczyk L, Gajęcki M (2004) Influence of zearalenone micotoxicosis on selected immunological, hematological and biochemical indexes of blood plasma in bitches. Pol J Vet Sci 7:177–180

Gajęcka M, Janowski T, Jakimiuk E, Polak M, Podhalicz-Dzięgielewska M, Rotkiewicz T, Otrocka-Domagała I, Obremski K, Zielonka Ł, Gajęcki M (2007) Histopathological and immunohistochemical examination of the uterus in bitches following zearalenon mycotoxicosis. Pol J Vet Sci 10:143–151

Gareis M, Zimmermann C, Schothorst R, Paulsch W, Vidnes A, Bergsten C, Paulsen B, Brera C, Miraglia M, Grossi S, Debegnach F (2003) Reports on tasks for scientific cooperation — Collection of occurrence date of Fusarium toxins in food and assessment of dietary intake by the population of EU Member States, Directorate-General Health and Consumer Protection. Report of experts participating in Task 3.2.10 – April 2003

Goliński PK, Kaczmarek A, Kiecana I, Wiśniewska H, Kaptur P, Kostecki M, Chełkowski J (2002) Fusarium head blight of common Polish winter wheat cultivars – omparison of effects *Fusarium avenaceum* and *F. culmorum* on yield components. J Phytopathol 150:135–141

Gutleb AC, Morrison E, Murek AJ (2002) Cytotoxicity assays for mycotoxins produced by Fusarium strains: a review. Environ Toxicol Phar 11:309–320

Hollinger K, Ekperigin HE (1999) Micotoxicosis in food producing animals. Vet Clin N Am-Food A 15:133–165

Hussein HS, Brasel JM (2001) Toxicity, metabolism, and impact of mycotoxins on humans and animals. Toxicology 167:101–134

Inadera H (2006) The immune system as a target for environmental chemicals: Xenoestrogens and other compounds. Toxicol Lett 164:191–206

Isaacson C (2005) The change of the staple diet of black South Africans from sorghum to maize (corn) is cause of the epidemic of squamous carcinoma of the oesophagus. Med Hypotheses 64:658–660

Jana B, Majewski M (2007) Influence of peripheral nervous system on ovarian function. Med Weter 63:1163–1167 (in Polish)

Jarvis BB, Miller JD (2005) Mycotoxins as harmful indoor air contaminants. Appl Microbiol Biot 66:367–372

Kuciel-Lisieska G (2006) The presence of zearalenone in women with neoplastic lesions in mammary gland [Doctoral thesis]. II Faculty of Medicine, Medical Academy in Warsaw, Poland (in Polish)

Kuhn DM, Ghannoum MA (2003) Indoor mold, toxigenic fungi, and *Stachybotrys chartarum*: infectious disease perspective. Clin Microbiol Rev 16:144–172

Kurup VP, Shen HD, Banerjee B (2000) Respiratory fungal allergy. Microbes Infect 2:1101–1110

Kustritz MVR (2005) Reproductive behavior of small animals. Theriogenology 64:734–746

Lang TJ (2004) Estrogen as immunomodulator. Clin Immunol 113:224–230

Lephart ED, Setchell KDR, Lund TD (2005) Phytoestrogens: hormonal action and brain plasticity. Brain Res Bull 65:193–198

Liska D, Lyon M, Jones DS (2006) Detoxification and biotransformational imbalances. Explore-NY 2:122–140

Malekinejad H, Maas-Bakker R, Fink-Gremmels J (2006a) Species differences in the hepatic biotransformation of zearalenone. Vet J 172:96–102

Malekinejad H, Van Tol HTA, Fink-Gremmels J (2006b) Expression of 3a- and 3b-hydroxy steroid dehydrogenase mRNA in COCs and granulosa cells determines Zearalenone biotransformation. Toxicol In Vitro 20:458–463

Massart F, Meucci V, Saggese G, Soldani G (2008) High growth rate of girls with precocious puberty exposed to estrogenic mycotoxins. J Pediatr 152:690–695

McClain RM, Wolz E, Davidovich A, Pfannkuch F, Bausch J (2005) Subchronic and chronic safety studies with genistein in dogs. Food Chem Toxicol 43:1461–1482

Mindnich R, Moeller G, Adamski J (2004) The role of 17beta- hydroxysteroid dehydrogenases. Mol Cell Endocrinol 218:7–20

Minervini F, Dell'Aquilla ME (2008) Zearalenone and reproductive function in farm animals. Int J Mol Sci 9:2570–2584

Morgavi DP, Riley RT (2007) An historical overview of field disease outbreaks known or suspected to be caused by consumption of feeds contaminated with *Fusarium* toxins. Anim Feed Sci Tech 137:201–212

Mueller SO (2002) Overview of in vitro tools to assess the estrogenic and antiestrogenic activity of phytoestrogens. J Chromatogr B 777:155–165

Nobel S, Abrahmsen L, Oppermann U (2001) Metabolic conversion as a pre-receptor control mechanism for lipophilic hormones. Eur J Biochem 268:4113–4125

Olsen M, Kiessling KH (1983) Species differences in zearalenone reducing activity in subcellular fractions of liver from female domestic animals. Acta Pharmacol Toxicol (Copenhagen) 52:287–291

Penning TM (2003) Hydroxysteroid dehydrogenases and pre-receptor regulation of steroid hormone action. Hum Reprod Update 9:193–205

Pestka JJ, Uzarski RL, Islam Z (2005) Induction of apoptosis and cytokine production in the Jurcat human T cells by deoxynivalenoldeoxynivalenol: role of mitogen-activated protein kinases and comparision to other 8-ketotrichothecenes. Toxicology 206:207–219

Pitt JI (2000) Toxigenic fungi and mycotoxins. Brit Med Bull 56:184–192

Polak M (2007) The evaluation of the efficacy of sodium carbonate as zearalenone destructor in the feed destined for gilts feeding [Doctoral thesis]. University of Warmia and Mazury in Olsztyn, Poland (in Polish)

Ramos AJ, Hernandez E, Pla-Delfin JM, Merino M (1996) Intestinal absorption of zearalenone and in vitro study of non-nutritive sorbent materials. Int J Pharm 128:129–137

Rhyn P, Zoller O (2003) Zearalenone in cereals for human nutrition: relevant data for this Swiss population. Eur Food Res Technol 216:319–322

Riley RT (1998) Mechanistic interactions of mycotoxins: theoretical consideration. In: Sinha KK, Bhatanagar D (eds) Mycotoxins in agriculture and food safety. Marcel Dekker, Inc., Basel, pp 227–254

Riley RT, Norred WP (1996) Mechanisms of mycotoxicity. In: Howard DH, Miller JD (eds) The mycota, vol VI. Springer, Berlin, pp 194–195

Rosselli M, Reinhart K, Imthurn B, Keller PJ, Dubey RK (2000) Cellular and biochemical mechanisms by which environmental oestrogens influence reproductive function. Hum Reprod Update 6:332–350

Scippo ML, Van De Weerdt C, Willemsen P, Françoise RD, Muller M, Matrial JA, Maghuin-Rogister G (2002) Detection of illegal growth promoters in biological samples using receptor binding assays. Anal Chim Acta 473:135–141

Sergent T, Parys M, Garsou S, Pussemier L, Schneider YJ, Larondelle Y (2006) Deoxynivalenol transport across human intestinal Caco-2 cells and its effects on cellular metabolism at realistic intestinal concentration. Toxicol Lett 164:167–176

Sergent T, Ribonnet L, Kolosova A, Garsou S, Schaut A, De Saeger S, Van Peteghem C, Larondelle Y, Pussemier L, Schneider YJ (2008) Molecular and cellular effects of food contaminants and secondary plant components and their plausible interactions at the intestinal level. Food Chem Toxicol 46:813–841

Shephard GS (2008a) Committee on natural toxins on food allergens. J AOAC Int 91:1B–16B

Shephard GS (2008b) Impact of mycotoxins on human health in developing countries. Food Addit Contam 25:146–151

Speijers GJA, Speijers MHM (2004) Combined toxic effects of mycotoxins. Toxicol Lett 153:91–98

Stopper H, Schmitt E, Kobras K (2005) Genotoxicity of phytoestrogens. Mutat Res-Fund Mol M 574:139–155

Stubner UP, Gruber D, Berger UE, Toth J, Marks B, Huber J, Horek F (1999) The influence of female sex hormones on nasal reactivity in seasonal allergic rhinitis. Allergy 54:865–871

Sweeney T (2002) Is exposure to endocrine disrupting compounds during fetal/post-natal development affecting the reproductive potential of farm animals? Domest Anim Endocrin 23:203–209

Tiemann U, Tomek W, Schneider F, Vanselow J (2003) Effects of the mycotoxins α- and β-zearalenol on regulation of progesterone synthesis in cultured granulose cells from porcine ovaries. Reprod Toxicol 17:673–681

Trucksess MW, Scott PM (2008) Mycotoxins in botanicals and dried fruits: A review. Food Addit Contam 25:181–192

Ueno Y, Tashiro F (1981) α-zearalenol, a major hepatic metabolite in rats of zearalenone, an oestrogenic mycotoxin of Fusarium species. J Biochem 89:563–571

Verbeek R, Plomp AC, van Tol EAF, van Noor JM (2004) The flavones luteolin and apigenin inhibit in vitro antigen-specific proliferation and interferon-gamma production by murine and human autoimmune T cells. Biochem Pharmacol 68:621–629

Vlata Z, Porichis F, Tzanakakis G, Tsatsakis A, Krambovitis EA (2006) Study of zearalenone cytotoxicity on human peripheral blood mononuclear cells. Toxicol Lett 165:274–281

Voigt CA, von Scheidt B, Gácser A, Kassner H, Lieberei R, Schäfer W, Salomon S (2007) Enhanced mycotoxin production of a lipase-deficient Fusarium graminearum mutant correlates to toxin-related gene expression. Eur J Plant Pathol 117:1–12

Yurino H, Ishikawa S, Sato T, Akadegawa K, Ito T, Ueha S, Inadera H, Matsushima K (2004) Endocrine disrupters (environmental estrogens) enhance autoantibody production by B1 cells. Toxicol Sci 81:139–147

Zielonka Ł, Gajęcki M, Obremski K, Zwierzchowski W (2003) Influence of low doses of deoxynivalenol applied per os on chosen indexes of immune response in swine. Pol J Vet Sci 6:74–77

Zielonka Ł, Gajęcki M, Obremski K, Zwierzchowski W (2004) Influence of low doses of deoxynivalenole on the level of this mycotoxin in pig serum. Med Weter 60:534–536 (in Polish)

Zinedine A, Soriano JM, Moltó JC, Mañes J (2007) Review on the toxicity, occurrence, metabolism, detoxification, regulations and intake of zearalenone: An oestrogenic mycotoxin. Food Chem Toxicol 45:1–18

Part II
Agriculture and Food

Chapter 10
Mycotoxins in Lithuanian Cereals and Grain Products

Audrone Mankeviciene

10.1 Introduction

The central goal of grain cultivation is the production of high-quality food-related raw materials for the processing industry. Natural toxins, such as mycotoxins, have emerged as a significant factor affecting the safety image of cereal grains as a raw material for the food and feed industry. Previous studies in Lithuania (Bakutis and Januškevičienė 1997; Keblys et al. 2000; Baliukonienė et al. 2003; Gaurilčikienė et al. 2005; Semaškienė et al. 2005; Mankevičienė et al. 2006, 2007) and other European countries (Park et al. 1996; Muller et al. 1998; Döll et al. 2002; Schollenberger et al. 2002) as well as globally (Webley and Jackson 1998; Bennett and Klich 2003) have shown that there is reason to focus on *Fusarium* toxins and their appearance. A planned European Union directive will specify the maximum limits for trichothecenes such as deoxynivalenol (DON), T-2 toxin and HT-2 toxin and for zearalenone (ZEN) and fumonisins. DON is the most frequently found contaminant of barley, wheat, oats and corn throughout the world (Scott 1989; Eskola 2002). In addition to DON, T-2 toxin, HT-2 toxin and ZEN frequently occur in cereal crops cultivated in northern temperate regions (Muller et al. 1998; Grabarkiewicz-Szczesna et al. 2001; Rasmussen et al. 2003; Edwards 2004; Hietaniemi et al. 2004). Fumonisins cause more extensive problems in the Southern Hemisphere than in the Northern (Shephard et al. 1996). The most frequently isolated *Fusarium* species are *Fusarium graminearum, Fusarium culmorum, Fusarium moniliforme, Fusarium poae, Fusarium eqiseti* and *Fusarium proliferatum* (Eriksen and Alexander 1998; Eskola et al. 2001; Creppy 2002).

The ochratoxin A (OTA) frequently contaminates cereal grains, beans coffee, nuts, olives (Wood et al. 1996; Fazekas et al. 2002). In Northern Europe, OTA is

A. Mankeviciene
Department of Plant Pathology and Protection, Lithuanian Institute of Agriculture, LT-58344 Akademija, Kėdainiai District, Lithuania
e-mail: audre@lzi.lt

M. Rai and A. Varma (eds.), *Mycotoxins in Food, Feed and Bioweapons*,
DOI: 10.1007/978-3-642-00725-5_10,

mainly produced by the molds *Penicillium verrucosum* and *Aspergillus ochraceus* during storage of cereal grain (Krogh 1987; Kuiper-Goodman and Scott 1989; Frisvad and Samson 1991; Creppy 2002; Lund and Frisvad 2003). OTA contamination has mainly been associated with post-harvest conditions (Abramson et al. 1990; Mills 1990). Species of *Penicillium* genus have been found to be responsible for OTA-contaminated grain in colder areas (Scandinavia and Canada). *A. ochraceus* was isolated in warmer climatic zones (Yugoslavia, Australia) and in coffee bean-producing countries (Frisvad and Samson 1991). The experimental evidence on ochratoxin A occurrence in Lithuanian-grown cereals is rather limited, and hence more comprehensive research has been done into the fungi producing this mycotoxin (Lugauskas et al. 2004).

Aflatoxins can contaminate agricultural commodities including corn, wheat, rice, peanuts and many other crops (Sinha and Sinha 1991; Aly 2002). Aflatoxins, each of which is a group of closely related mycotoxins, may be produced by *Aspergillus flavus* and *A. parasiticus*. Four different aflatoxins, B1, B2, G1 and G2, have been identified with B1 being the most toxic, carcinogenic and prevalent. Monitoring of aflatoxins and their producers in Lithuania is mainly done for imported commodities; however, the data on Lithuania-grown grain contamination with this toxin are insufficient.

Temperature and moisture conditions during the growing season and insect infestations are critical factors affecting fungal infection and toxin synthesis (Cromey et al. 2001). More mycotoxins were produced during the warm, dry summers than in rainy and cool summers (Grabarkiewicz-Szczesna et al. 2001).

The aim of the present study was to investigate the occurrence of mycotoxins DON, T-2, ZEN, ochratoxin A, aflatoxin (total) in the grain of different cereal species grown for food and feed in Lithuania during 2003–2007.

10.2 Experimental

10.2.1 Samples

Grain samples of winter and spring wheat (*Triticum aestivum* L.), spring barley (*Hordeum distichon* L.), winter triticale (x*Triticosecale* Wittm.), winter rye (*Secale*

Table 10.1 The number of cereal grain samples tested for mycotoxin contamination during 2003–2007

Mycotoxin	Number of samples						
	Winter cereals			Spring cereals			Total
	Wheat	Rye	Triticale	Wheat	Barley	Oats	
DON	333	43	29	98	173	24	700
ZEN	280	35	26	76	169	17	603
T-2	184	15	12	52	93	15	371
Ochratoxin (A + B)	76	20	4	20	51	–	171
Aflatoxin (B1 + B2 + G1 + G2)	46	18	–	18	96	–	178

cereale L.) and spring oats (*Avena sativa* L.) were collected at harvest during 2003–2007 from the Lithuanian Institute of Agriculture in Dotnuva and analyzed for contamination by DON, ZEN, T-2 toxin, aflatoxin (B1 + B2 + G1 + G2) and ochratoxin (A + B). The number of samples analyzed is given in Table 10.1.

Part of each sample was subjected to mycotoxicological contamination, and another part (about 50 g) was air-dried, milled in a mill IKA A11 Basic and kept at −20°C until analysis.

Grain, bran and flour tests were done on winter wheat grain grown under conventional production conditions in 2006. A laboratory mill, the Brabender "Quadrumat junior", was used for flour fractionation and bran separation. The distribution of mycotoxins DON, ZEN, ochratoxin (A + B), aflatoxin (B1 + B2 + G1 + G2) and T-2 toxin in grain, bran and flour was determined.

10.2.2 Analysis of Mycotoxins

The wheat, rye, oats and barley samples were analysed by the ELISA (enzyme-linked immunosorbent assay) method (Wilkinson et al. 1992; Bennett et al. 1994). The method is based on the antibody-antigen interaction, provides sensitive, rapid and accurate monitoring of mycotoxins and is suitable for screening large numbers of samples. The Veratox test kits (Neogen Corporation, Scotland), approved by the AOAC Research Institute (Certificate N 950702) were used for the analysis. Mycotoxin extraction and tests were performed according to the manufacturer's instructions. The optical densities of samples and controls from standard curve were estimated by a multichannel programmable photometer, Multiskan MS (Labsystems,

Table 10.2 The highest permissible mycotoxin concentrations in grain

EU document	Cereals and cereal products	Mycotoxin	Maximum levels ($\mu g\ kg^{-1}$) (ppb)
Commission regulation (EC) no. 856/2005	Unprocessed cereals (including durum wheat, oats and maize) durum wheat, oats and maize	Deoxynivalenol (DON)	1,250 1,750
Commission regulation (EC) no. 856/2005	Unprocessed cereals (including maize) maize	Zearalenone (ZEN)	100 200
Commission regulation (EC) no. 123/2005	Unprocessed cereals and cereal products	Ochratoxin	3–5
Commission regulation (EC) no. 2174/2003	Unprocessed cereals and cereal products	Aflatoxin ($B_1 + B_2 + G_1 + G_2$)	4
Global research recommendations (Eriksen and Alexander 1998)	Unprocessed cereals and cereal products	T-2 toxin	100

Finland), using a 650-nm filter and the calculation mode Point to Point. Measured absorbances were automatically converted to the mycotoxin concentration units – μg kg^{-1}. The results were estimated taking into account the lowest calibration curve's mycotoxin concentration value (LOD – limit of detection), which is for:

DON: 100 μg kg^{-1} (ppb)
ZEN: 10 μg kg^{-1} (ppb)
T-2 toxin: 7.5 μg kg^{-1} (ppb)
Ochratoxin: 1 μg kg^{-1} (ppb)
Aflatoxin: 1 μg kg^{-1} (ppb).

While assessing our data with regard to food and forage safety, we referred to the EU document no. 856/2005 for deoxynivalenol, zearalenone, no. 123/2005 for ochratoxin, no. 466/2001 for aflatoxins and global research recommendations for T-2 toxin (Table 10.2).

10.3 Results

10.3.1 Deoxynivalenol (DON)

The most widespread and most comprehensively researched trichothecene, and the one causing the greatest problems in animal production, is deoxynivalenol (DON). Although this toxin is not especially toxic, its low or moderate doses continuously getting into a human or animal organism through food or feed can cause health damage and disturb the immune system. DON-contaminated grain is found in Europe, North and South America, South Africa, Asia and New Zealand. The concentrations of this mycotoxin range between 1 and 67,000 μg kg^{-1} (Placinta et al. 1999). In Europe, during the period 1999–2002 the highest DON contamination was identified for maize and wheat grain (Rafai et al. 2000).

Results of DON in the grain samples of winter and spring cereals in 2003–2007 are reported in Table 10.3. DON was found to be the most widespread mycotoxin in Lithuanian-grown cereal grain; it was present in 88.4–98.3% of grain samples tested. DON was more prevalent in spring cereals compared with winter cereals (94.7–98.3 and 88.4–93.3%). The greatest content of contaminated grain with higher DON concentrations was determined in spring wheat samples. In most spring wheat samples, DON concentration was higher than in winter cereal samples; however, the highest DON contents were identified in separate samples of winter wheat (987.0 μg kg^{-1}) and winter rye (691.0 μg kg^{-1}) grain. Although DON was the most prevalent of all mycotoxins tested in Lithuania, its contents identified at harvesting were not high and did not exceed the allowable levels. According to the Commission of the European Communities regulation for mycotoxins, which came into effect on July 2006, maximum level of DON in unprocessed cereals is 1,250 μg kg^{-1}. In other countries also DON contamination at grain harvesting

Table 10.3 DON contents in the grain of different cereal species and seasonal types in 2003–2007

Species	Total number of samples	Positive		Distribution of positive samples according to content of DON			Average positive concentration ($\mu g\ kg^{-1}$)	Max concentration ($\mu g\ kg^{-1}$)
		Samples	%	<100 $\mu g\ kg^{-1}$	100–300 $\mu g\ kg^{-1}$	>300 $\mu g\ kg^{-1}$		
Winter cereals								
Wheat	333	295	88.6	171	120	4	112.6	987.0
Rye	43	38	88.4	17	19	2	140.6	691.0
Triticale	30	28	93.3	11	14	3	151.7	427.0
Spring cereals								
Wheat	99	96	97.0	28	56	12	174.0	847.0
Barley	178	175	98.3	74	99	2	153.0	375.0
Oats	19	18	94.7	7	8	3	186.0	314.6

ranges from 60 to 100%, but the levels identified are very dependent on the weather conditions at cereal flowering and harvesting (Scott 1997; Schollenberger et al. 2002; Tutelyan 2004).

10.3.2 Zearalenone (ZEN)

Zearalenone poses the greatest problems for pig production, since it causes hyperestrogenism which results in premature development of secondary sexual attributes in gilts, swelling of mammary glands, atrophy of testicles, and deterioration of semen quality (Mankevičienė 2002). Various ZEN contents are identified in almost every agricultural product, forage raw materials and prepared forage. Ninety percent of wheat and maize grain grown in Austria, Germany, and France is contaminated with zearalenone (Eriksen and Alexander 1998). In Lithuania, ZEN contamination in the grain of spring cereals was higher than that in the grain of winter cereals (Table 10.4). Spring wheat, barley and oats contamination ranged from 67.5 to 82.9%. In one spring barley grain sample the concentration of ZEN exceeded the allowable level almost twice (193.4 μg kg^{-1}). According to the Commission of the European Communities regulation for mycotoxins mentioned above, maximum level of ZEN in unprocessed cereals is 100 μg kg^{-1}. Experimental evidence suggests that the incidence of *Fusarium* spp. is generally higher on spring cereal grain than on winter cereal grain (Semaškienė et al. 2005), therefore mycotoxin concentrations may be higher. Winter wheat, triticale, and rye were less contaminated (42.3–58.6%). Similar results were obtained by researchers of other countries investigating ZEN in winter wheat grain (Lepschy-v Gleissenthal et al. 1989).

10.3.3 T-2 Toxin

The data on the distribution of T-2 toxin belonging to the trichothecene mycotoxin group in agricultural produce are scarce. Bearing in mind its harmful effects on animals (it causes inflammations of digestive tract, and exhibits embryotoxic and teratogenic effects), reliable scientific data on the variation of its concentrations are insufficient. As a result, our task was to do as many tests as possible with various cereal grain.

T-2 toxin concentration in positive-tested grain samples of various cereals varied between 10.8 and 32.1 μg kg^{-1} (Table 10.5); however, there are reports that exceeded the allowable level (Eriksen and Alexander 1998). Lower concentrations of this mycotoxin were found in spring and winter wheat samples, although the contamination level ranged from 64.7% to 76.9%. In the larger part of the tested samples, T-2 toxin concentrations varied between 7.5 and 30 μg kg^{-1}. Spring barley, winter rye and oats were contaminated more heavily.

Table 10.4 ZEN contents in the grain of different cereal species and seasonal types in 2003–2007

Species	Total number of samples	Positive		Distribution of positive samples according to content of ZEN			Average positive concentration (μg kg^{-1})	Max concentration (μg kg^{-1})
		Samples	%	<10 μg kg^{-1}	10–30 μg kg^{-1}	>30 μg kg^{-1}		
Winter cereals								
Wheat	280	164	58.6	87	65	12	16.5	75.6
Rye	35	20	57.1	8	10	2	17.8	52.2
Triticale	26	11	42.3	9	2	0	<10	14.1
Spring cereals								
Wheat	76	63	82.9	18	34	11	17.9	95.6
Barley	169	114	67.5	52	48	14	16.9	193.4
Oats	17	14	82.4	1	12	1	20.6	50.4

Table 10.5 T-2 toxin contents in the grain of different cereal species and seasonal types in 2003–2007

Species	Total number of samples	Positive		Distribution of positive samples according to content of T-2 toxins			Average positive concentration ($\mu g\ kg^{-1}$)	Max concentration ($\mu g\ kg^{-1}$)
		Samples	%	<7.5 $\mu g\ kg^{-1}$	7.5–30 $\mu g\ kg^{-1}$	>30 $\mu g\ kg^{-1}$		
Winter cereals								
Wheat	184	119	64.7	49	69	1	10.8	32.9
Rye	15	14	93.3	4	7	3	28.5	153.6
Triticale	12	9	75.0	4	1	4	25.9	59.7
Spring cereals								
Wheat	52	40	76.9	4	36	0	16.9	23.0
Barley	93	79	84.9	14	42	23	29.7	316.0
Oats	15	15	100	0	8	7	32.1	122.0

In one spring barley grain sample, the content of T-2 toxin was three times the recommended level (Eriksen and Alexander 1998), and was as high as 316.0 μg kg^{-1}. Higher than allowable concentrations were found in winter rye (153.6 μg kg^{-1}) and oat grain samples (122.0 μg kg^{1}). The data on T-2 toxin occurrence are limited, so it is necessary to conduct comprehensive research into the factors that determine its concentration in grain and grain products, since under Lithuania's conditions T-2 toxin producers *Fusarium poae* and *F. sporotrichioides* are rather common on grain, especially on oats. The frequency of identification of these fungi in oats was 43 and 5%, respectively. *F. graminearum*, which is the main producer of DON and ZEN in oats was identified in low concentrations (Mankeviciene et al. 2006).

The occurrence of the *Fusarium* species in Lithuania in the samples of various cereal species suggests that in wheat and barley grain, *F. poae, F. avenaceum, F. sporotrichioides*, which are not producers of DON and ZEN, were more frequent; therefore, the concentrations of these toxins were low in the samples tested (Mankeviciene et al. 2006).

The trends of DON, ZEN, T-2 toxin variation in cereals grain might have been determined not only by the weather conditions but also by other factors: soil peculiarities in different districts, application of plant protection products, choice of varieties and others (Cromey et al. 2001; Edwards 2004; Hietaniemi et al. 2004; Heier et al. 2005).

10.3.4 Ochratoxin (A + B)

Ochratoxins are found more frequently and in higher concentrations in Northern and Eastern European regions (Denmark, Germany, Poland). Ochratoxin A, which is the most toxic, is the most often mentioned of all ochratoxins. Especially conducive conditions for the production of this mycotoxin occur in wet summers if grain is stored improperly. High ochratoxin doses cause nefropathy in piglets, which is common in European countries, and occurs especially frequently in Denmark (Hald 1991). In Lithuanian-grown cereals, higher ochratoxin contents were identified in spring wheat and barley samples (Table 10.6). All the barley samples tested were 52.9%-contaminated with ochratoxin, and in 22.2% of these samples the concentrations identified exceeded 3.0 μg kg^{-1}. Barley grain samples were found in which the contents of this toxin exceeded the allowable limits by five times (49.6 μg kg^{-1}). Quite high ochratoxin contamination (80%) was identified in winter rye grain samples, but the concentrations determined were low. These results indicate that the problem of ochratoxin in Lithuanian-grown barley grain is undoubtedly relevant. This is corroborated by previously conducted tests suggesting that barley grain, especially during storage is heavily infested with the fungi of *Penicillium* genus, which are capable of producing ochratoxin (Lugauskas et al. 2004).

Table 10.6 Ochratoxin (A + B) contents in different cereal species and seasonal types in 2003–2007

Species	Total number of samples	Positive		Distribution of positive samples according to content of Ochratoxin A + B			Average positive concentration ($\mu g\ kg^{-1}$)	Max concentration ($\mu g\ kg^{-1}$)
		Samples	%	<1 $\mu g\ kg^{-1}$	1–3 $\mu g\ kg^{-1}$	>3 $\mu g\ kg^{-1}$		
Winter cereals								
Wheat	76	25	32.9	20	5	1	<1	3.9
Rye	20	16	80.0	11	5	0	<1	1.8
Spring cereals								
Wheat	20	8	40.0	4	3	1	4.0	26.1
Barley	51	27	52.9	15	6	6	5.0	49.6

Table 10.7 Aflatoxin (B1 + B2 + G1 + G2) contents in different cereal species and seasonal types in 2003–2007

Species	Total number of samples	Positive		Distribution of positive samples according to content of aflatoxin (B1 + B2 + G1 + G2)			Average positive concentration ($\mu g\ kg^{-1}$)	Max concentration ($\mu g\ kg^{-1}$)
		Samples	%	<1 $\mu g\ kg^{-1}$	1–3 $\mu g\ kg^{-1}$	>3 $\mu g\ kg^{-1}$		
Winter cereals								
Wheat	46	36	78.3	5	21	10	3.0	7.4
Rye	18	12	66.7	1	8	3	2.4	5.1
Spring cereals								
Wheat	18	12	66.7	0	8	4	3.1	6.7
Barley	96	57	59.4	17	37	3	1.5	4.8

10.3.5 Aflatoxin (B1 + B2 + G1 + G2)

Lithuanian climate conditions are thought to be unfavorable for the occurrence of this mycotoxin. Higher concentrations of aflatoxin are found in imported produce. However, during various technological processes (forage production, drying, storage) conditions occur for the spread of aflatoxin. Control of aflatoxin as well as that of other toxins is performed in Lithuanian-grown plant produce.

From 59.4% to 78.3% of Lithuanian-grown cereal grain is contaminated with aflatoxin (B1 + B2 + G1 + G2). The most frequent concentrations both in winter and spring cereal grain vary within the 1–3 $\mu g\ kg^{-1}$ range (Table 10.7). Of the winter wheat grain samples grown during the period 2003–2007, 78.3% were contaminated with aflatoxin (B1 + B2 + G1 + G2). The highest concentration determined was 7.4 $\mu g\ kg^{-1}$. Of 96 spring barley samples tested, 57 (59.4%) were found to be contaminated with aflatoxin (B1 + B2 + G1 + G2). The highest concentration reached 4.8 $\mu g\ kg^{-1}$. Regardless of the low aflatoxin contamination level, the previous research shows that *A. flavus*, which is the chief producer of aflatoxin, is quite frequent in barley grain (Lugauskas et al. 2004).

10.3.6 Mycotoxins in Grain and Grain Products

Various grain products, including wheat, are the basis of a healthy nutrition pyramid. Wheat bran as a dietry product is often recommended for improvement of digestion. However, analyses of mycotoxins in grain, and bran and flour produced from this grain, showed that the highest concentrations of mycotoxins (DON, ZEN, ochratoxin (A + B), aflatoxin (B1 + B2 + G1 + G2) were found in grain bran (Table 10.8). DON-

Table 10.8 Mycotoxin contents in food grain, bran and flour

Contamination level and concentrations	Mycotoxin ($\mu g\ kg^{-1}$)				
	DON	ZEN	T-2	Ochratoxin (A + B)	Aflatoxin (B1 + B2 + G1 + G2)
Grain					
Percentage of positive samples	100	100	57.1	57.1	100
Number of samples with higher than allowable concentration (%)	0	0	0	0	14.3
Mean concentration of all samples ($\mu g\ kg^{-1}$)	103.3	10.7	<7.5	<1.0	2.7
Maximum concentration ($\mu g\ kg^{-1}$)	167.0	17.2	<7.5	<1.0	4.4
Bran					
Percentage of positive samples	85.7	100	100	100	100
Number of samples with higher than allowable concentration (%)	0	0	0	0	100
Mean concentration of all samples ($\mu g\ kg^{-1}$)	108.3	<10.0	<7.5	1.0	4.6
Maximum concentration ($\mu g\ kg^{-1}$)	205.0	10.0	12.8	1.3	4.9
Flour					
Percentage of positive samples	100	28.6	42.9	71.4	57.1
Number of samples with higher than allowable concentration (%)	0	0	0	0	71.4
Mean concentration of all samples ($\mu g\ kg^{-1}$)	<100	<10	<7.5	<1.0	2.8
Maximum concentration ($\mu g\ kg^{-1}$)	101	<10	<7.5	<1.0	4.3

contaminated samples accounted for 85.7% of all bran samples tested, while the other mycotoxins were found in 100% of bran samples tested. The levels of DON, ZEN, T-2 toxin, and ochratoxin (A + B) were found to be low; however, the concentrations of aflatoxin (B1 + B2 + G1 + G2) reached the allowable limits. As a result, it is dangerous to use such bran as a dietary food product. Flour samples were less contaminated with mycotoxins compared with grain and bran. Mycotoxin concentrations in them were also lower. Flour was found to be more contaminated with DON than bran (100 and 85.7%). Heavier contamination of bran with ZEN, T-2 toxin, ochratoxin, and aflatoxin shows the highest mycotoxin localisation in separate grain fractions. Flour was less contaminated with mycotoxins.

10.4 Future Perspectives

It is impossible to grow plant produce completely free from mycotoxin contamination; however, efforts can be made to minimise their concentrations by using good agricultural practice. It is very important for agricultural producers and processors to realise that good agricultural practice is the key factor helping to control grain contamination with *Fusarium* toxins. Another important factor is good production practice and its implementation while working with grain intended for human food and animal feed, its storage, processing and distribution.

The data from research done globally as well as in Lithuania indicate that the *Fusarium* spp. are detected most frequently on grains, which can produce two or more toxins in different amount. (Fink-Gremmels 1999; Bennett and Klich 2003).

The EU commission's recommendation 2006/57/EB, adopted on 17 August 2006, specifies that grain and forage samples should be tested for the incidence of DON, ZEN, ochratoxin A, T-2 toxin, and fumonisin, since it is very important to estimate to what extent these toxins occur together. There is little experimental research dealing with this possible risk factor, either in Lithuania or globally.

The tests revealed the trend that one sample could contain several mycotoxins. This leads to the assumption that the toxic metabolites DON, ZEN, T-2 toxin, ochratoxin A, and aflatoxin, though in low concentrations but occurring together, can exert a stronger toxic effect than one toxin occurring at higher than allowable levels. Currently, the data from tests in this area of products intended for humans and animals are very limited.

10.5 Conclusions

Mycotoxin DON is the most frequent toxin produced by the *Fusarium* genus found in Lithuanian-grown cereal grain. It was identified in 88.4–98.3% of cereal grain samples assayed during 2003–2007. Spring cereals were found to be more heavily infested by this toxin than winter cereals.

ZEN was more prevalent in the grain of spring wheat, barley and oats (82.9, 67.5 and 82.4%, respectively). Higher risk of zearalenon contamination was revealed in spring barley grain samples.

T-2 toxin contaminated 64.7–100% of spring and winter cereal samples. Higher T-2 toxin contamination level was identified in the samples of rye (93.3%), spring barley (84.9%), and oats (100%). T-2 toxin control in Lithuanian-grown grain is very important, because climate conditions are conducive to the development of its producers.

Ochratoxin (A + B) contaminated 25–80% of grain. In the grain of winter cereals the concentrations of this mycotoxin were lower than those determined in spring cereals. The highest ochratoxin concentration in spring barley grain reached 49.6 $\mu g\ kg^{-1}$.

From 59.4% to 78.3% of Lithuanian-grown cereal grain was found to be contaminated with aflatoxin (B1 + B2 + G1 + G2) but the contents determined in most cases do not exceed 3 $\mu g\ kg^{-1}$ limit. Higher contents of alfatoxin (B1 + B2 + G1 + G2) exceeding the allowable limits are found only in separate cases.

From the tests on grain, bran and flour contamination with mycotoxins, it was revealed that the highest contents of DON, ZEN, ochratoxin (A + B) and aflatoxin (B1 + B2 + G1 + G2) accumulated in bran. Flour was found to be less contaminated with mycotoxins.

Acknowledgements Separate stages of this study were supported by the Lithuanian State Science and Studies Foundation's and the Lithuanian Ministry of Agriculture's Programme "Kenksmeda" (contract no. P-39/07 and 8P-636).

References

Abramson D, Sinha RN, Mills JT (1990) Mycotoxin formation in HY-320 wheat during granary storage at 15 and 19% moisture content. Mycopathologia 111:181–189

Aly SE (2002) Distribution of aflatoxins in product and by-products during glucose production from contaminated corn. Nahrung 46:341–344

Bakutis B, Januškevičienė G (1997) Pašaruose vyraujančių mikotoksinų poveikis gyvulių sveikatai. Žemės ūkio mokslai 4:86–88

Baliukoniene V, Bakutis B, Stankevicius H (2003) Mycological and mycotoxicological evaluation of grain. Anna Agr Environ Med 10:223–227

Bennett JW, Klich M (2003) Mycotoxins. Clin Microbiol Rev 16(3):497–516

Bennett GA, Nelsen TC, Miller BM (1994) Enzyme-linked immunosorbent assay for detection of zearalenone in corn, wheat and pig feed: collaborative study. J AOAC Int 77:1500–1509

Creppy EE (2002) Update of survey, regulation and toxic effects of mycotoxins in Europe. Toxicol Lett 127:19–28

Cromey MG, Lauren DR, Parkes RA Sinclair KI, Shorter SC, Wallace AR (2001) Control of Fusarium head blight of wheat with fungicides. Austral Plant Pathol 30(4):301–308

Döll S, Valenta H, Dänicke S, Flachowsky G (2002) *Fusarium* mycotoxins in conventionally and organically grown grain from Thuringia/Germany. Landbauforsch Völk 2(52):91–96

Edwards SG (2004) Investigation of *Fusarium* mycotoxins in UK wheat production. In: Canty SM, Boring T, Wardwell J, Ward RW (eds) Proceedings of the 2nd International Symposium on

Fusarium Head Blight; The 8th European Fusarium Seminar, Michigan State University, East Lansing, MI, pp 398–400

Eriksen GS, Alexander J (1998) Fusarium toxins in cereals – a risk assessment. TemaNord report 502. Nordic Council of Ministers, Copenhagen, pp 10–165

Eskola M (2002) Study on trichothecenes, zearalenone and ochratoxin A in Finnish cereals: occurrence and analytical techniques. National Veterinary and Food Research Institute (EELA) and University of Helsinki. (Academic dissertation)

Eskola M, Parikka P, Rizzo A (2001) Trichothecenes, ochratoxin A and zearalenone contamination and Fusarium infection in Finnish cereal samples in 1998. Food Addit Contam 18 (8):707–718

Fazekas B, Tar AK, Zomborszky-Kovacs M (2002) Ochratoxin A contamination of cereal grains and coffee in Hungary in the year 2001. Acta Vet Hung 50:177–188

Fink-Gremmels J (1999) Mycotoxins: their impact on human and animal health. Vet Quart 21:115–120

Frisvad JC, Samson RA (1991) Mycotoxins produced by species of Penicillium and Aspergillus occurring in cereals. In: Chelkovski J (ed) Cereal grain. Mycotoxins, fungi and quality in drying and storage. Developments in food science. Elsevier Science Pub, Amsterdam, pp 441–476

Gaurilčikienė I, Mankevičienė A, Dabkevičius Z (2005) Impact of triazole and strobilurin fungicides on the incidence of toxic fungi and mycotoxins on winter wheat grain. Bot Lithuanica 7:27–35

Grabarkiewicz-Szczesna J, Kostecki M, Golinski P, Kiecana I (2001) Fusariotoxins in kernels of winter wheat cultivars field samples collected during 1993 in Poland. Nahrung 45(1):28–30

Hald B (1991) Porcine nephropathy in Europe. In: Castegnaro M, Plestina R, Dirheimer G et al. (eds) Mycotoxins, endemic nephropathy and urinary tract tumours. IARC Scientific Publications, vol 115. International Agency for Research on Cancer, Lyon, pp 49–56

Heier T, Jain SK, Kogel K-H, Pons-Kühneman J (2005) Influence of N-fertilization and fungicide strategies on Fusarium head blight severity and mycotoxin content in winter wheat. J Phytopathol 153(9):551–557

Hietaniemi V, Kontturi M, Rämö S, Eurola M, Kangas A, Niskanen M, Saastamoinen M (2004) Contents of trichothecenes in oats during official variety, organic cultivation and nitrogen fertilization trials in Finland. Agr Food Sci 13(1–2):54–67

Keblys M, Flåøyen A, Langseth W (2000) The occurrence of type A and B trichothecenes in Lithuanian cereals. Acta Agric Scand (B) Soil Plant Sci 50(3):155–160

Krogh P (1987) Ochratoxins in food. In: Krogh P (ed) Mycotoxins in food. Academic Press, London, pp 97–121

Kuiper-Goodman T, Scott PM (1989) Review: Risk assessment of the mycotoxin ochratoxin A. Biomed Environ Sci 2:179–248

Lepschy-v Gleissenthal J, Dietrich R, Martlbauer E, Schuster M, Süβ A, Terplan G (1989) Survey on the occurrence of *Fusarium* mycotoxins in Bavarian cereals from the 1987 harvest. Z Lebensm Unters For 188(6):521–526

Lugauskas A, Krasauskas A, Repečkienė J (2004) Ekologiniai veiksniai, lemiantys mikromicetu paplitimą ant javu grūdų ir sojų sėklų. Ekologija 2:21–32

Lund F, Frisvad JC (2003) Penicillium verrucosum in wheat and barley indicates presence of achratoxin A. J Appl Microbiol 95:1117–1123

Mankevičienė A (2002) The effects of mycotoxin zearalenone on reproductive performance of boars and measures for inactivation of zearalenone. Proceedings of Latvia University of Agriculture, Raksti 7(302):48–58

Mankevičienė A, Gaurilčikienė I, Dabkevičius Z Semaškienė R, Mačkinaitė R, Supronienė S (2006) Mycotoxin contamination of Lithuania-grown cereal grain and factors determining it. Ekologija 3:21–27

Mankevičienė A, Butkutė B, Dabkevičius Z, Supronienė S (2007) *Fusarium* mycotoxins in Lithuanian cereals from the 2004–2005 harvests. Ann Agr Environ Med 14(1):103–107

Mills JT (1990) Mycotoxins and toxigenic fungi on cereal grains in western Canada. Can J Physiol Pharmacol 68:982–986
Muller HM, Reimann J, Schumacher U, Schwadorf K (1998) Natural occurrence of *Fusarium* toxins in oats harvested during five years in an area of southwest Germany. Food Addit Contam 15(7):801–806
Park JJ, Smalley EB, Chu FS (1996) Natural occurrence of Fusarium mycotoxins in field samples from the 1992 Wisconsin corn crop. Appl Environ Microbiol 62:1642–1648
Placinta C, D'Mello J, Macdonald S (1999) A review of worldwide contamination of cereal grains and animals feed with Fusarium mycotoxins. Animal Feed Sci Technol 78:21–37
Rafai P, Bata AL, Jakab L, Vanyi A (2000) Evaluation of mycotoxin-contaminated cereals for their use in animal feeds in Hungary. Food Addit Contamin 17(9):799–808
Rasmussen PH, Ghorbani F, Berg T (2003) Deoxynivalenol and other Fusarium toxins in wheat and rye flours on the Danish market. Food Addit Contamin 20(4):396–404
Schollenberger M, Jara HT, Suchy S, Drochner W, Muller HM (2002) Fusarium toxins in wheat flour collected in an area in southwest Germany. Int J Food Microbiol 72(1–2):85–89
Scott PM (1989) The natural occurrence of trichothecenes. In: Beasley VR (ed) Trichothecene mycotoxicosis: pathophysiologic effects, vol. 1. CRC Press, Boca Raton, pp 1–26
Scott PM (1997) Multi-year monitoring of Canadian grains and grain-based foods for trichothecenes and zearalenone. Food Addit Contam 14(4):333–339
Semaškienė R, Mankevičienė A, Dabkevičius Z, Leistrumaitė A (2005) Toxic fungi infection and mycotoxin level in organic grain. Bot Lithuanica 7:17–25
Shephard GS, Thiei PG, Stockenström S, Sydenham EW (1996) Worldwide survey of fumonisin contamination of corn and corn-based products. J AOAC Int 79(3):671–687
Sinha KK, Sinha AK (1991) Monitoring and identification of aflatoxins in wheat, grain and maize flours in Bihar state (India). Food Addit Contam 8:453–457
Tutelyan VA (2004) Deoxynivalenol in cereals in Russia. Toxicol Lett 153:173–179
Webley DJ, Jackson KL (1998) Mycotoxins in cereals – a comparison between North America, Europe and Australia. In: Banks HJ, Wright EJ, Damcevski KA (eds) Proceedings of the Australian Post-harvest Technical Conference. CSIRO, Canberra, Australia, pp 63–66
Wilkinson AP, Ward CM, Morgan MRA (1992) Immunological analysis of mycotoxins. In: Linskens HF, Jackson JF (eds) Plant toxin analysis. Springer, Berlin, pp 185–225
Wood GM, Patel S, Entwisle AS, Boenke A (1996) Ochratoxin A in wheat: a second intercomparison of procedures. Food Addit Contam 13:519–539

Chapter 11
Control of Mycotoxin Contamination in Cereals by Breeding

Ákos Mesterházy

11.1 Introduction

The mycotoxin contamination of cereal commodities is a much older problem than its detection and characterization. Molded grains often caused different animal and human health problems long before the causal agents were known. The first mycotoxin detected is the aflatoxin B1 in the 1960s and since then hundreds of toxins have been described (D'Mello and McDonald 1997; Bartók et al. 2006). They represent chemically very different groups, very far from each other in physiological actions; some, such as aflatoxins or fumonisins, are highly carcinogenic. The most numerous group of toxins is at present the trichothecenes with about 100 toxins and toxic degradation products. At least 90 fumonisin toxins are known (Bartók et al. 2008), most of them without physiological and toxic characterization. For now, we have a rather comprehensive view on the occurrence of the different mycotoxins in cereal grains; most of the synthesis and decomposition mechanisms of the major mycotoxins are known. It is very probable that new toxins, maybe toxic groups, will be characterized in the future. Most of the synergism between toxins is not known. It could be high, as the same severity of swine toxicoses from natural deoxynivalenol (DON) could be reached by the addition of 5–10 times more chemically clean DON (Miller and Trenholm 1994). Increasing knowledge of the toxins has verified the food and feed safety risks. Now many countries, such as EU member states (Anonymous 2005), have binding toxin regulations or suggested toxin levels, as in the USA, Canada, etc. As the EU is a significant market, EU regulations generate stricter toxin regulations in the source countries. Many accept the view that the bioethanol industry can utilize toxin-contaminated commodities. However, the industry has shown (Stepanik et al. 2007) that the ethanol yield from toxin-contaminated grain is lower, and the

Á. Mesterházy
Cereal Research non-profit Company, H-6701 Szeged, P.O. Box 391, Hungary
e-mail: akos.mesterhazy@gabonakutato.hu

M. Rai and A. Varma (eds.), *Mycotoxins in Food, Feed and Bioweapons*,
DOI: 10.1007/978-3-642-00725-5_11,

toxin-containing remains (dried distillery grain with solubles) provides large problems in economic utilization, because a costly detoxification is needed. This means also that non-food uses might need healthy grains.

The most important toxins are *aflatoxin B1* of *Aspergillus flavus*, *ochratoxin A* of *Aspergillus ochreaceus*, *DON*, *nivalenol*, *T-2* toxin, *zearalenone* and *fumonisins* from the genus *Fusarium*. In this chapter, we review the *Fusarium* head blight in wheat and the *Aspergillus* and *Fusarium* ear rot of corn, as here we have the most experience of breeding and toxicology.

For a long time toxicoses were considered to result from storage problems. Indeed, high amounts of toxic feeds and foods were identified from wrongly stored commodities where microbial activity was high and toxic metabolites could also be synthesized in large amounts.

About 20 years ago, significant aflatoxin contamination was found in freshly harvested grains in US, where maize fields were exposed to very high temperatures and drought. *Aspergillus flavus* was also found, so it becomes clear that "typical" storage fungi can really cause field infection and field-originated aflatoxin contamination. Therefore, we conclude that the strict differentiation between field and storage fungi is out of date, and we have to expect their activity under both conditions. We believe that in tropical fruits and nuts containing high ratio of protein and fatty acid we may face the same problem of lack of resistance.

In the past decades, considerable experience has been gained in breeding for resistance or toxin resistance, which can inhibit toxin accumulation in the field. Toxin contamination can also be reduced by fungicide application. This problem needs intensive research investment (Mesterházy et al 2003, 2004; Mesterházy 2003b), but will not be treated in this chapter.

11.2 Legal Toxin Regulations and Their Consequences

Toxin regulations always represent a compromise. Food and feed safety need zero tolerance; for example, trichothecenes such as DON inhibit human T-cell activity by 80% at 100 ppb concentration, being important in immune response (Berek et al. 2001). The limit is, however, 1,250 ppb (856/2005 EU regulation). Even under non-epidemic conditions, 200–400 ppb regularly occurs. This would indicate that 60–70% of the grain harvested should be withdrawn from human consumption. This would cause an economic crisis for the cereal grower, and masses of people would die of hunger. We think that at present the susceptibility level of cultivars, plant protection technology, and the present toxin limits represent a good compromise. This encourages the breeders to pay attention to this trait which they have neglected until now; it also encourages those in animal husbandry to avoid economic losses from toxicoses by buying only healthy food, and food companies will also be aware not to put products on the shelf with toxin contamination higher than that allowed. We can forecast without much risk that toxin limits will be lowered in line with improving resistance and plant protection. So breeding for resistance against toxigenic fungi will become an integral part of the process. Methodology of the resistance tests is crucial. They should be repeatable, and resistance in artificial inoculation should result in resistance under natural conditions (Dill-Macky 2003; Mesterházy et al. 2006, 2007).

11.3 Resistance to Disease and Toxin

Disease and toxin resistance are often used as synonyms (Clements and White 2004; Williams 2006; Menkir et al. 2006). Other papers deal with resistance to toxin, and seek to identify genetic factors that might inhibit toxin accumulation or degrade toxins which have been synthesized. There are papers that deal with disease resistance and measure the toxin contamination as a mean of expressing resistance. We should state, however, that disease and toxin resistance are not synonymous expressions.

Resistance to diseases and resistance to toxin are normally governed by different mechanisms. Full agreement was found only in pathotoxins, where the toxin itself causes the disease. Classic examples are the *Helminthosporium* (*Bipolaris maydis* race T HT toxin) and *H. carbonum* HC toxin. In these cases, the resistance to toxins is equal to resistance to disease (Hoppe 1997). However, in most cases there is no proof that mycotoxins would be pathotoxins, even where in some cases transitional signs can be found. As an example of the latter, DON has a role in the aggressiveness. After disruption of the Tri5 gene (Proctor et al. 1995) that is important for the synthesis of the trichothecene skeleton, the new line lost both DON production ability and aggressiveness compared to the wild type strain to wheat. Jansen et al. (2005) opined that trichothecenes are not virulence factors during infection through the fruit coat. However, specific resistance to DON exists. Miller et al. (1985, 1994) and Miller and Arnison (1986) described a mechanism which decomposes DON much more intensively in resistant plants than in a susceptible one. Poppenberger et al. (2003) detected a DON-glycoside transferase in *Arabidopsis*. Lemmens et al. (2005) found that DON was detoxicated by putative DON-glycosyl transferases in wheat, and the amount of the non-toxic product could be measured.

The fact that aggressiveness of the DON-producing fungi correlates well with aggressiveness was clearly stated (Mesterházy 1995, 2002; Mesterházy et al. 1999, 2005; Lemmens et al. 2003; Miedaner et al. 2003). The results indicated that we had here a pathotoxin. This gave the idea that resistance to this toxin could ensure resistance to disease. Chinese breeders were able to select plants with higher resistance from toxin resistance-screened calluses (Chen et al. 1990), supporting the idea. Somaclones were divided into two parts against T-2 toxin and against a toxic culture filtrate (Ahmed et al. 1992, 1996a,b); half were exposed to toxin and the other half were regenerated without selective agent. Than the two populations in R1–R3 generations were compared for seedling Fusarium resistance. In the toxin-selected population a rather large variability was found, from susceptible to high resistance. If the toxin resistance meant disease resistance, all plants should have been resistant. The control plants, however, showed the same variation without any toxin mediation. This means that the reason for the appearance of disease-resistant plants was due not to toxin selection, but to the existing variability in the somaclone population. So the toxin selection did not perform a real selection pressure in the direction of disease resistance. The Chinese study did not have a control group, so the authors could suppose that the higher disease resistance came through selection via toxin-bridge. However, the contrary seems to be true: that resistance to resistance to DON does not automatically mean resistance to the disease. Similar work was done by Snijders in The Netherlands and Lemmens in Austria (pers. comm. 1995 and 2003 respectively) with the same result, so the selection against toxins was ceased.

From the breeding side, resistance behavior of wheat genotypes was tested against different *Fusarium* spp. The very early data (Mesterházy 1978, 1983) supported the idea of common resistance to *F. graminearum* and *F. culmorum* As both are DON producers, the common disease resistance could also be explained by the hypothesis of toxin resistance. However, *F. avenaceum* and *F. sporotrichioides* were also tested and the finding was the same (Mesterházy 1977). Therefore, the hypothesis of the common resistance was extended to other *Fusarium* spp. (Mesterházy 2002; Mesterházy et al. 2005; Tóth et al. 2008). The conclusion is that *Fusarium* resistance is not race-specific and is also species-non-specific. This means that a plant having resistance to *F. graminearum* will also have resistance to other *Fusarium* spp. such as *F. sporotrichioides*, *F. avenaceum*, *F. poae* and *F. verticillioides*, etc. The connection is not toxin resistance, but an unknown common mechanism.

11.4 Resistance and Toxin Relation in the Wheat *Fusarium* Head Blight Syndrome

As a conclusion, resistance to *Fusarium* in small grains is not specific. This is very important as it makes breeding easier (Mesterházy 2002; Mesterházy et al. 2005; Tóth et al. 2008). Additionally, numerous morphological traits influencing disease and resistance expression have been described (Mesterházy 2003a). So far, five physiological resistance types have been described (Schroeder and Christensen 1963; Mesterházy 1995). The fact that experiments are often poorly repeatable is not merely due to a genotype x environmental interaction; rather, problems with method, and heterogeneity of the populations tested, are responsible for the discouraging results (Mesterházy et al. 2006). Genetically the resistance is characterized by QTLs, but no one resistance gene has been identified until now. The QTL fhb1 on 3BS from Sumai 3 is one of the mostly studied, but its function is not yet clear. So the genetic characterization is undetermined; therefore, the relationship between resistance types and genetic regulation is not yet understood. Additionally, only three well-characterized large effect QTLs are known (3BS and 5A from Sumai 3, 3BSc from W114); many other QTLs with small and medium effect have high instability. Their interactions are only partially known and they might be very important as determinants of the native resistance (Brown-Guedira et al. 2008). Transgressive segregation might also be important. For example, the two parents of Sumai 3 are the medium susceptible Funo and Taiwanxiaomai (Liu and Wang 1991) and their cooperating QTLs gave an excellent result. In other cases this might also be possible.

For these reasons, in this section, we will analyze results from practical work concerning the suitability of different traits for measuring resistance. In addition, DON data will be analyzed, as this is the most important trait we face; the commodities will be qualified according to the level of the toxin contamination and not according to FDK or FHB severity.

Table 11.1 presents the results of a 4-year investigation of FHB resistance in wheat. This was the first report that analyzed not only visual head symptoms, but also FDK, yield loss and DON contamination (Table 11.1).

Data clearly show that genotypes differ significantly according to FHB severity. The repeatability of the tests was good. The correlation coefficient between FDK

Table 11.1 FHB resistance of wheat genotypes against FHB. General means across eight *F. graminearum* and *F. culmorum* isolates, 1990–1993 (Mesterházy 1995)

Genotype	FHB (%)	FDK (%)	Yield loss (%)	DON (mg kg^{-1})
SVP 72017	7.18	11.79	12.21	0.85
Arina	8.66	5.66	10.35	0.34
SVP 75059–28	10.66	10.08	17.62	1.44
81F349	10.88	11.42	12.60	0.46
85.92	12.29	18.67	18.12	2.35
82F328	12.47	15.45	22.37	0.42
RC103	13.04	18.81	19.81	2.07
Copain	15.06	18.33	18.46	1.65
Kri-Mon	16.93	12.23	25.56	0.74
Szõke	17.37	15.21	24.31	2.55
Sgv/GT	17.44	20.73	26.57	2.58
Bence	17.91	23.55	20.11	1.19
Rechsler	18.47	28.35	29.79	2.87
SVP72059–32	21.10	25.04	29.42	2.81
Ok-Sp	21.45	22.64	20.33	2.07
Bty-Mo*Kr[a]	23.70	17.36	33.11	3.05
Ni-Kr*Dol[a]	24.01	17.32	28.10	1.28
Mon-Ar	24.88	16.90	23.76	1.37
SVP 72005	27.23	34.51	30.68	1.50
Zombor	28.64	40.11	35.49	6.67
Mean	17.47	19.21	22.94	1.91
LSD (5%)	3.09	2.38	1.32	0.54
LSD (1%)	4.07	3.13	4.08	0.71

	FHB (%)	FDK (%)	Yield loss (%)
FDK (%)	0.7430**		
Yield loss (%)	0.8594**	0.7266**	
DON (mg kg^{-1})	0.5748***	0.7322**	0.6746**

**P = 0.1%, ***P = 1%

A * means that a single cross was crossed with the cultivars designated with sympols Kr and Mon

Table 11.2 Mean reactions of wheat genotypes to FHB. Summary of 1994–1996 data across eight *F. graminearum* and *F. culmorum* isolates (Mesterházy et al. 1999)

Trait	DON (mg kg^{-1})	FHB (%)	FDK (%)	Yield loss (%)
Maximum	42.25	38.92	50.43	48.21
Minimum	0.32	2.61	0.92	3.68
Variation width	41.93	36.31	49.51	44.53
Mean	13.37	21.70	21.48	28.23
LSD 5%	8.58	0.98	2.70	3.59
Correlation analysis				
	DON (mg kg^{-1})	FHB (%)	FDK (%)	
FHB (%)	0.8671***			
Kernel inf (%)	0.8199***	0.8357***		
Yield loss (%)	0.8693***	0.9526***	0.8294***	

***P = 0.1%

and DON was closer than between FHB severity and DON. It seems that DON is located mostly in the visually infected grains. So resistance level seems to be more important in DON than specific effects. In several cases, however, when FHB data are very similar, DON data differ significantly, as in SVP 72005 and Zombor; SVP 72005 seems to have an additional DON-regulating mechanism.

In the next experiment, the genotype set was changed and genotypes from exotic spring wheat resistance sources were included (Table 11.2). The most resistant genotypes showed only sporadic infection; they had DON lower than 1.25 mg kg^{-1}, whereas the most susceptible genotypes produced 30–40 mg kg^{-1} DON and up to 50% kernel infection. The differences across 3 years and eight isolates were highly significant, indicating that repeatability of the tests was good. The correlations between traits were very close; the DON contamination correlated with the other traits between $r = 0.82$ and $r = 0.86$. It seems that for Sgv-GT and Zombor we can suppose additional DON-resistance factors. However, it seems again that the resistance is the most important DON-reducing agent, and not the additional toxin-regulating mechanisms.

The next experiment (Mesterházy et al. 2005) was set up with 20 isolates of eight *Fusarium* spp. with and without DON-producing ability. Highly resistant spring wheat sources, their selected progenies and also highly susceptible western European wheat varieties and lines were included. The low FDK values result in mostly low DON data, but there are several genotypes where low FDK results in high DON contamination, such as the line RSt//MM/Nobeoka Bozu. In this test, Sumai 3 has the same resistance level as Nobeoka Bozu; in some years, however, NB used had somewhat higher values. On the susceptible side, however, we have a number of genotypes which have about the same FDK values at about 60–70%, but in which DON data strikingly differ. Öthalom, the susceptible Hungarian cultivar, or Zugoly have only 30–50 mg kg^{-1} DON, but several late western European wheat genotypes show 2–5 times higher data. The background is unknown. Therefore, not only is resistance to DON important, but susceptibility factors allowing extreme high DON contamination should also be investigated carefully. It is clear that this

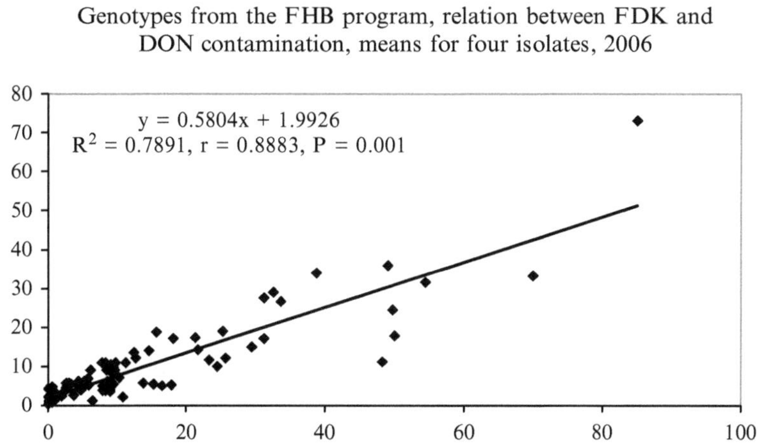

Fig. 11.1 Materials from the resistance program, relation between FDK and DON contamination, means for four isolates, 2006 ($n = 83$)

cultivar type has extraordinary food and safety risk. A clear explanation does not yet exist.

The next data come from advanced breeding material, tested in 2006. The correlation between FDK and DON contamination is very high; additional DON resistance or susceptibility can be supposed where the genotypes have significant distance from the regression line (Fig. 11.1). In 2004, the same phenomenon was recorded. Again, very close correlation was registered between FDK and DON contamination ($y = 0.0293x^2 - 0.1258x + 8.6469$, $R^2 = 0.8136$, $n = 96$). Other European literature sources support the close correlations between FDK and DON contamination (Miedaner et al. 2004; Paul et al. 2005; Wilde and Miedaner 2006; Polisenska and Tvaruzek 2007; Wilde et al. 2007).

It seems, therefore, that at least two distinct resistance mechanisms operate in wheat. The first is a non-specialized resistance background that inhibits fully or partially disease development. Toxin contamination is often used as a resistance trait; however, it has not been determined whether the lower toxin contamination is a result of less disease or specific toxin degrading activity of the host plant. The other one acts specifically on DON (maybe similar mechanisms will be found in other toxins), transforms it to a non-toxic product and so helps to keep infection severity lower through reducing the aggressiveness of the DON-producing isolates (Lemmens et al. 2005). Experimental results show (Mesterházy et al. 2007) that when the two distinct mechanisms co-operate, the chances of less toxin are higher. For this reason, the terms resistance to disease and resistance to toxin are not interchangeable. DON and disease severity measured by FDK have very close correlations, in well-planned and executed experiments normally above $r = 0.80$ or higher. This shows that the toxin content is strongly proportional to the intensity of disease; i.e., the resistance level is the most important DON-regulating agent. Other traits were also found that additionally modified DON by down- and up-regulating it, but at present little is known of them. For this reason, breeding should concentrate on high disease resistance.

11.5 Resistance and Toxin Resistance Relations Against *Aspergillus flavus* and *Fusarium* Ear Blight in Maize

Aflatoxin contamination is a chronic problem in the southern US, in the Midwestern states, and in other countries like in Africa where very high temperatures accompanied by drought stress occur (Anderson et al. 1975; Zuber et al. 1978). The disease is caused by *A. flavus* and *A. parasiticus*. The initial disease development is good in humid and warm weather, but aflatoxin B1 production is at extreme high temperatures in mid summer. (Paul et al. 2003). A survey of literature reveals that nearly 400 articles deal with different aspects of the problem. Artificial ear inoculation methods were developed (Zummo and Scott 1989). At present the genetics and regulation of the aflatoxin are moderately understood, but the knowledge is growing rapidly (Bhatnagar et al. 2006).

Tubajika et al. (2000) found that most commercial hybrids are highly susceptible to *Aspergillus* ear rot. The most susceptible hybrids (DK 683 and Myogen) contained 25 and 26 ng g^{-1} aflatoxin B1, the most resistant only 3–5 μg kg^{-1}. The authors evaluated the differences as resistance deviation. The GT-MAS:gk maize genotype is resistant to *A. flavus* (Russin et al. 1997) as its wax layer has a component that reduces colonization of the kernels. However, this component was not found in any of the tested commercial hybrids. So this compound can be responsible for *A. flavus* resistance in that genotype. Brown et al. (1999) and Guo et al. (1996) also confirmed the significance of this same wax component, and a fungal growth inhibitor was found to be associated with resistance to disease and aflatoxin resistance. From the data it is clear that the reduced aflatoxin from the resistant material was due to inhibited fungal growth, i.e., disease resistance was the major cause and not a direct aflatoxin-degrading or -inhibiting activity. Another study (Windham and Williams 1998) found significant differences in kernel infection following artificial inoculation. The resistant hybrids contained significantly less aflatoxin than the susceptible ones. The differences in aflatoxin content were significant during the whole harvest time from 49 to 63 days after midsilk, but visual symptom differences in infection severity were observed after only 42 days. Hamblin and White (2000) stated that in one selection cycle against *A. flavus* ear rot severity decreased by 8.5% and the aflatoxin contamination by 19 ng g^{-1}. So the decrease in aflatoxin contamination was the consequence of the higher resistance. Brown et al. (2001) screened African maize genotypes and compared them to the GT-MAS:gk and M182. Half of them provided similar level of growth reduction to *A. flavus*. They also found several resistant lines that had rather an inhibitory effect on aflatoxin synthesis, which led directly to fungal infection. Menkir et al. (2006) developed a kernel-screening assay. The correlation between symptoms and aflatoxin contamination was good, so the low aflatoxin contamination was mostly due to lower disease severity. Consequently, Williams et al. (2003) speak about resistance of aflatoxin, but they consider it as a consequence of preharvest kernel infection in the field, i.e., disease resistance is the focus rather than specific toxin resistance. Gardner et al. (2007) measured aflatoxin contamination from artificially inoculated ears as simple disease trait. Therefore, their resistance to aflatoxin contamination is actually a result of disease resistance, and no specific genetic system regulating aflatoxin contamination was found. Betran and Isakeit (2004) tested a number of hybrids for *A. flavus* resistance and aflatoxin contamination. The most resistant hybrid had 176 ng g^{-1} and the most susceptible more than 2,000 in one site; on another site, 780 and 3,500 were the lowest and highest values. They detected that hybrids with good husk coverage (husk leaves overgrow the cob tip by at least 2.5 cm) are important, as average aflatoxin content was here about 500 ng g^{-1}; however, the hybrids where cobs significantly overgrew husk leaves had 2000 ng g^{-1} aflatoxin and more. It seems that such a simple trait can significantly influence colonization and so aflatoxin contamination; most early hybrids belonged to this group. As *Aspergillus* kernel infection may not be determined exactly by visual scoring, Scott and Zummo (1995) used grinded kernels. Disease-infected particles showed as green under "black light," and the number of particles was proportional to the aflatoxin concentration.

QTLs for *A. flavus* resistance have also been detected, but they were of low effect, with LOD values just above 2.0. They were not stable; their appearance depended on year and location (Paul et al. 2003). So the development of resistant lines with many small-effect QTLs will not be easy; although the first attempts were made to pyramid QTLs (Widstrom et al. 2003).

Interestingly, for Fusarium ear rot caused by *F. graminearum* resistance and DON contamination, only a few articles were found. Reid and co-workers (Reid and Sinha 1998, Reid et al. 2000) found a significant correlation between ear rot and DON levels, indicating the role of resistance in controlling toxin contamination. More papers deal with *F. verticillioides* and fumonisin contamination. Schaafsma et al. (2006) found significant correlation between Gibberella ear rot and DON, as well as *F. verticillioides* ear rot and fumonisin B1 concentration respectively. Ear rot correlated with DON and fumonisin ($r = 0.67$ and $r = 0.69$ respectively), indicating a close correlation between disease severity and toxin contamination (Toldi et al. 2008). Robertson et al. (2006) found close ($r = 0.64$–0.96) correlations between disease severity and fumonisin data; the authors found that it is much cheaper to select for resistance than to check fumonisin contamination. Pasquale et al. (2002) came to the same conclusion. There is another fact we should consider. Bacon (1996) found that *Fusarium* infection is not always visible, even though fumonisin contamination might be present.

Most of the authors differentiated between disease and toxin resistance. Numerous reports dealt with aflatoxin resistance; however, resistance to disease was mostly initiated by artificial inoculation. In some cases no visual evaluation was made; only toxin contamination was measured. This means that resistance to toxin contamination was considered equal to disease resistance. Most data indicate close correlations between severity of disease symptoms and toxin contamination, and this verifies the major role of resistance in combatting the aflatoxin problem. In some cases, specific toxin-resistance mechanisms can also be present. Studies on the direct aflatoxin resistance of different corn plants by using pure toxin have not been found up to now, so although aflatoxin resistance might be supposed, it has not yet been demonstrated.

11.6 Conclusions and Outlook

The experimental data and the literature sources clearly show that the increase in resistance is the key to decreasing risks from toxin contamination. In most cases, the cited resistance to toxin should be understood as resistance to disease. Of course, resistance to toxins also exists, but this is rather of secondary importance. Therefore, screening for specific toxin resistance seems to be an ineffective diversion. Breeding is a slow process for replacing the cultivars in a country by more resistant ones. The first step is to identify more resistant genotypes among the candidates in the registration process; here results can be achieved within 4–5 years from the start of official tests. By this simple resistance test we can improve food or/and feed safety by about 50%. Additional improvement will come when breeding for resistance is included in the breeding process; this takes over 10 years.

We have analyzed two host-pathogen relationships more carefully. We think that the lessons could be utilized for other diseases not analyzed here in detail.

In wheat we have the necessary resistance, but the combination of FHB resistance with the other traits such as yielding ability, quality, resistance to other diseases, etc. is even now a worldwide problem. We have gained the basic knowledge that *Fusarium* resistance in wheat is race-non-specific and also species-non-specific, indicating that resistance to *F. graminearum* secures resistance to all other *Fusarium* spp. tested until now; and the probability is very high that this is valid for others which have not yet been tested. For this reason, we do not forecast the maximum level of resistance in commercial cultivars, but a strong medium level of resistance as has been achieved in China (Ma et al. 2008). At higher FHB resistance level, the agronomical parameters may not be competitive.

In maize the *Aspergillus* resistance is the key to decreased aflatoxin contamination. In spite of the work invested, progress is slow. In Fusarium ear rot caused by different *Fusarium spp.* there are also strong indications that resistance basically determines the toxin contamination. We have preliminary data that a common resistance to different *Fusarium* spp. may also be the case in corn. We think that resistance to Fusarium ear rot, as the key factor in feed safety, will enjoy much more attention in the future.

For tropical fruits, dried fruits, peanuts and tree nuts, we have a large amount of toxicological data, but according to Adams (1996) and Hamid (1981) rice, wheat, groundnuts, black pepper, maize, oil seed and even wine contained mostly aflatoxin. All available cultivars of the above seeds and oil seed were experimentally tested by *A. flavus* to find whether there was any variety of them resistant to aflatoxin contamination. No item was observed to be resistant to either fungal growth or production of aflatoxin. It was concluded that under favorable environmental conditions (moisture and temperature) there were chances for fungal growth and accumulation of aflatoxins. Muir et al. (2003) demonstrated differences in resistance to aflatoxin produced by *Aspergillus flavus* in walnut cultivars (gallic acid overproduction). Anderson et al. (1996) report that out of 831 peanut accessions, 70 had significantly lower aflatoxin than Florunner, a widely produced cultivar in Georgia US. Nagarajan and Bhat (1973) found significant resistance differences between peanut cultivars in India. Of course, prevention by more resistant cultivars has a priority; storage and trade conditions are also very important for keeping the commodity healthy from farm to fork.

References

Adams S (1996) Genetic resistance a key to controlling aflatoxin — prevention of aflatoxin infection on peanuts. Agricultural Research, July, 1996 (findarticles.com/p/articles/mi_m3741/is_n7_v44/ai_18598626-30k)

Ahmed KZ, Mesterházy Á, Sági F (1992) *In vitro* techniques for selecting wheat (*Triticum aestivum* L.) for Fusarium-resistance. I. Double-layer culture technique. Euphytica 57: 251–257

Ahmed KZ, Mesterházy Á, Bartók T, Sági F (1996a) II. Culture filtrate technique and inheritance of Fusarium resistance in the somaclones. Euphytica 91:341–349

Ahmed KZ, Mesterházy Á, Sági F (1996b) *In vitro* production of *Fusarium* resistant wheat plants. In: Bajaj YPS (ed) Biotechnology in agriculture and forestry, vol 36. Somaclonal variation in crop improvement II. Springer, Heidelberg, pp 3–19

Anderson HW, Nehring EW, Wichser WR (1975) Aflatoxin contamination of corn in the field. J Agric Food Chem 23:775–782

Anderson WF, Holbrook CC, Wilson DM (1996) Development of greenhouse screening for resistance to *Aspergillus parasiticus* infection and preharvest aflatoxin contamination in peanut. Mycopathologia 135(2):115–118

Anonymous (2005) Commission regulation (EC) No 856/2005 of 6 June 2005, Official Journal of European Union 7.6.2005, L 143/3

Bacon CW (1996) Symptomless endophytic colonization of maize by *Fusarium moniliforme*. Can J Bot 74:1195–1202

Bartók T, Szécsi Á, Szekeres A, Mesterházy Á, Bartók M (2006) Detection of new fumonisins mycotoxins and fumonisin-like compounds by reversed phase high-performance liquid chromatography/electrospray ionization — ion-trap mass spectrometry. Rapid Commun Mass Spectrom 20:1–17

Bartók T, Szekeres A, Szécsi Á, Bartók M, Mesterházy Á (2008) New type of fumonisin series appears on the scene of food and feed safety. Cereal Res Commun 36(Suppl B):315–320

Berek L, Perti IB, Mesterházy Á, Téren J, Molnár J (2001) Effect of mycotoxins on human immune functions *in vitro*. Toxicol In Vitro 15:25–30

Betran FJ, Isakeit T (2004) Aflatoxin accumulation in maize hybrids of different maturities. Agron J 96:565–570

Bhatnagar D, Cary JW, Ehrlich K, Yu J, Cleveland TE (2006) Understanding the genetics of regulation of aflatoxin production and *Aspergillus flavus* development. Mycopathologia 162:155–166

Brown RL, Chen ZY, Cleveland TE, Russin JS (1999) Advances in the development of host resistance in corn to aflatoxin contamination by *Aspergillus flavus*. Phytopathology 89:113–117

Brown RL, Chen ZY, Menkir A, Cleveland TE, Cardwell K, Kling J, White DG (2001) Resistance to aflatoxin accumulation in kernels of maize inbreds selected for ear rot resistance in West and Central Africa. J Food Prot 64:396–400

Brown-Guedira G, Griffey C, Kolf F, McKendry A, Murphy JP, Van Sanford D (2008) Improved scab resistance in China: sources of resistance and problems. In: Breeding FHB resistant soft red wheat: progress and prospects. Cereal Res Comm 36(Suppl B):31–36

Chen HG, Wang YW, Wang YZ (1990) Identification of resistance to *Gibberella zeae* in wheat cultivars by *Fusarium* toxin-etiolated wheat coleoptile assay. Jiangshu Agr Sci Nr 6, 23–24

Clements MJ, White DG (2004) Identifying sources of resistance to aflatoxin and fumonisin contamination in corn grain. J Toxicol Toxin Rev 23:381–396

D'Mello JPF, MacDonald AMC (1997) Mycotoxins. Anim Feed Sci Technol 69:155–166

Dill-Macky R (2003) Inoculation methods and evaluation of Fusarium head blight resistance in wheat. In: Leonard KJ, Bushnell WR (eds) Fusarium head blight of wheat and barley. APS Press, St. Paul, MN, USA, pp 184–210

Gardner HD, Williams WP, Windham GL (2007) Diallele analysis of aflatoxin accumulation in maize. Field Crops Res 102:60–63

Guo BZ, Russin JS, Brown RL, Cleveland TE, Widstrom NW (1996) Resistance to aflatoxin contamination in corn as influenced by relative humidity and kernel germination. J Food Prot 59:276–281

Hamblin AM, White DG (2000) Inheritance of resistance to *Aspergillus* ear rot and aflatoxin production of corn from Tex 6. Phytopathology 90:292–296

Hamid A (1981) Mycotoxins in grains. MS Dissertation, University of Punjab, Lahore, p 250

Hoppe H-H (1997) Fungal phytotoxins. In: Hartleb H, Heitefuss R, Hoppe H-H (eds) Resistance of crop plants against fungi. Gustav Fischer, Stuttgart, pp 58–83

Jansen C, von Wettstein D, Schaefer W, Kogel K-H, Felk A, Frank J, Maier FJ (2005) Infection patterns in barley and wheat spikes inoculated with wild-type and trichodiene synthase gene disrupted *Fusarium graminearum*. Proc Natl Acad Sci U S A 102:16892–16897

Lemmens M, Krska R, Buerstmayr H, Josephs R, Schuhmacher R, Grausgruber H, Ruckenbauer P (2003) Fusarium head blight reactions and accumulation of deoxynivalenol, moniliformin and zearalenone in wheat grains. Cereal Res Commun 31:407–414

Lemmens M, Scholz U, Berthiller F, Dall'Asta C, Koutnik A, Schuhmacher R, Adam G, Buerstmayr H, Mesterházy A, Krska R, Ruckenbauer P (2005) The ability to detoxify the mycotoxin deoxynivalenol co-localizes with a major QTL for Fusarium head blight resistance in wheat. Mol Plant Microbe Interact 18:1318–1324

Liu ZZ, Wang ZY (1991) Improved scab resistance in China: Sources of resistance and problems. In: Saunders DA (ed) Wheat for the nontraditional warm areas. Proc Int Conf CIMMYT, DF, Mexico, pp. 178–188. Publ. CIMMYT El-Batán, Mexico D. F., Mexico

Ma H, Yao J, Zhou M, Zhang X, Ren L, Yu G, Lu W (2008) Molecular breeding for wheat Fusarium head blight resistance in China. Cereal Res Commun 36(Suppl B):203–212

Menkir A, Brown RL, Bandyopadhyay R, Chen Z-Y, Cleveland T (2006) A USA–Africa collaborative strategy for identifying, characterizing, and developing maize germplasm with resistance to aflatoxin contamination. Mycopathologia 162:225–232

Mesterházy Á (1977) Reaction of winter wheat varieties to four *Fusarium* species. Phytopath Z 90:104–112

Mesterházy Á (1978) Comparative analysis of artificial inoculation methods with *Fusarium* spp. on winter wheat varieties. Phytopath Z 93:12–25

Mesterházy Á (1983) Relationship between resistance to stalk rot and ear rot of corn influenced by rind resistance, premature death and the rate of drying of the ear. Maydica 28:425–437

Mesterházy Á (1995) Types and components of resistance against Fusarium head blight of wheat. Plant Breed 114:377–386

Mesterházy Á (2002) Role of deoxynivalenol in aggressiveness of *Fusarium graminearum* and *F. culmorum* and in resistance to Fusarium head blight. Eur J Plant Pathol 108:675–684
Mesterházy Á (2003a) Breeding wheat for *Fusarium* head blight resistance in Europe. In: Leonard K, Bushnell W (eds) *Fusarium* head blight of wheat and barley. APS Press, St. Paul, MN, USA, pp 211–240
Mesterházy Á (2003b) Control of Fusarium head blight of wheat by fungicides. In: Leonard KJ, Bushnell WR (eds) *Fusarium* head blight of wheat and barley. APS Press, St. Paul, MN, USA, pp 363–380
Mesterházy Á, Bartók T, Mirocha CM, Komoróczy R (1999) Nature of resistance of wheat to Fusarium head blight and deoxynivalenol contamination and their consequences for breeding. Plant Breed 118:97–110
Mesterházy Á, Bartók T, Lamper CS (2003). Influence of cultivar resistance, epidemic severity, and *Fusarium* species on the efficacy of fungicide control of *Fusarium* head blight in wheat and deoxynivalenol (DON) contamination of grain. Plant Dis 87:1107–1115
Mesterházy Á, Kászonyi G, Tóth B, Bartók T, Varga M (2004) Prothioconazole fungicides against FHB in Wheat, 2003/2004 results. In: 2nd international symposium on fusarium head blight. Orlando, 11–15 December 2004, pp 355–359
Mesterházy Á, Bartók T, Kászonyi G, Varga M, Tóth B, Varga J (2005). Common resistance to different *Fusarium* spp. causing Fusarium head blight in wheat. Eur J Plant Pathol 112: 267–281
Mesterházy Á, Tóth B, Kászonyi G (2006) Sources of "environmental interactions" in phenotyping and resistance evaluation ways to neutralize them. In: Ban T, Lewis JM, Phipps EE (eds) The global *Fusarium* initiative for international collaboration – strategic planning workshop held at CIMMYT, El Batán, Mexico, 14–17 March 2006, pp 84–92
Mesterházy Á, Buerstmayr H, Tóth B, Lehoczki-Krsjak SZ, Szabó-Hevér Á, Lemmens M (2007) An improved strategy for breeding FHB-resistant wheat must include Type I resistance. Proceedings of the 5th Canadian workshop on Fusarium head blight, Delta Winnipeg, pp 51–66
Miedaner T, Schneider B, Geiger HH (2003) Deoxynivalenol (DON) content and Fusarium head blight resistance in segregating populations of winter rye and winter wheat. Crop Sci 43:519–526
Miedaner T, Heinrich N, Schneider B, Oettler G, Rohde S, Rabenstein F (2004) Estimation of deoxynivalenol (DON) content by symptom rating and exoantigen content for resistance selection in wheat and triticale. Euphytica 139:123–132
Miller JD (1994) Epidemiology of *Fusarium* diseases of cereals. In: Miller JD, Trenholm HL (eds) Mycotoxins in grain compounds other than Aflatoxin. Eagan Press, St. Paul, MN, USA, pp 19–36
Miller JD, Arnison PG (1986) Degradation of deoxynivalenol by suspension cultures of the *Fusarium* head blight resistant wheat cultivar Frontana. Can J Plant Pathol 8:147–150
Miller JD, Trenholm HL (eds) (1994) Mycotoxins in grain – compounds other than aflatoxin. Eagan Press, St. Paul, MN, USA, p 552
Miller JD, Young JC, Sampson RD (1985) Deoxynivalenol and *Fusarium* head blight resistance in spring cereals. Phytopath Z 113:359–367
Muir R, Dandekar A, McGranahan G, Vail P, Molyneux R, Mahoney N, Leslie CH, Uratsu S, Tebbets S (2003) Genetic engineering and breeding of walnuts for control of aflatoxin. Walnut Res. ucdavis.edu/2003/2003_407.pdf
Nagarajan V, Bhat RV (1973) Aflatoxin production in peanut varieties by *Aspergillus flavus* link and *Aspergillus parasiticus* speare. Appl Microbiol 25:319–321
Pasquale M, Visconti A, Chelkowski J (2002) Ear rot susceptibility and mycotoxin contamination of maize hybrids inoculated with *Fusarium* species under field conditions. Eur J Plant Pathol 108(7):645–651
Paul PA, Lipps PE, Madden LV (2005) Relationship between visual estimates of Fusarium head blight intensity and deoxynivalenol accumulation in harvested wheat grain: a meta-analysis. Phytopathology 95:1225–1236

Paul C, Naidoo G, Forbes A, Mikkilineni V, White D, Rocheford T (2003) Quantitative trait loci for low aflatoxin production in two related maize populations. Theor Appl Gen 107:263–270
Polisenska I, Tvaruzek L (2007) Relationships between deoxynivalenol content, presence of kernels infected by *Fusarium* spp. pathogens and visually scabby kernels in Czech wheat in 2003–2005. Cereal Res Commun 35:1437–1448
Poppenberger P, Berthiller F, Lucyshyn D, Sieberer T, Schuhmache R, Krska R, Kuchler K, Glössl J, Luschnig CH, Adam G (2003) Detoxification of the Fusarium mycotoxin deoxynivalenol by a UDP-glucosyltransferase from *Arabidopsis thaliana*. J Biol Chem 278:(48) 47905–47914
Proctor RH, Hohn TM, McCormick SP (1995) Reduced virulence of *Gibberella zeae* caused by disruption of a trichothecene toxin biosynthetic gene. Mol Plant Microbe Interact 8:593–601
Reid LM, Sinha RC (1998) Maize maturity and the development of Gibberella ear rot symptoms and deoxynivalenol after inoculation. Eur J Plant Pathol 104(2):147–154
Reid LM, Zhu X, Savard ME, Sinha RC, Vigier B (2000) Pre-harvest accumulation of deoxynivalenol in sweet corn ears inoculated with *Fusarium graminearum*. Food Addit Contam 17:689–701
Robertson LA, Kleinschmidt CE, White DG, White DG, Payne GA, Maragos CM, Holland JB (2006) Heritabilities and correlations of Fusarium ear rot resistance and fumonisin contamination resistance in two maize populations. Crop Sci 46(1):353–361
Russin JS, Guo BZ, Tubajika KM, Brown RL, Cleveland TE, Widstrom NW (1997) Comparison of kernel wax from corn genotypes resistant or susceptible to *Aspergillus flavus*. Phytopathology 87:529–533
Schaafsma AW, Tamburic-Illincic L, Reid LM (2006) Fumonisin B-1 accumulation and severity of Fusarium ear rot and Gibberella ear rot in food-grade corn hybrids in Ontario after inoculation according to two methods. Can J Plant Pathol 28(4):548–557
Schroeder HW, Christensen JJ (1963) Factors affecting resistance of wheat to scab by *Gibberella zeae*. Phytopathology 53:831–838
Scott GE, Zummo N (1995) Size of maize sample needed to determine percent kernel infection by *Aspergillus flavus*. Plant Dis 79(8):861–864
Stepanik T, Kost D, Nowicki T, Gaba D (2007) Effects of electron beam irradiation on deoxynivalenol in distillers dried grain and soluble and production intermediates. In: 5th Canadian Workshop on Fusarium Head Blight. Delta Winnipeg, Winnipeg, p 96. (Abstract)
Toldi É, Bartók T, Varga M, Szekeres A, Tóth B, Mesterházy Á (2008) The role of breeding in reducing mycotoxin contamination in maize. Cereal Res Commun 36(Suppl B):175–178
Tóth B, Kászonyi G, Bartók T, Varga J, Mesterházy Á (2008) Common resistance of wheat to members of the *Fusarium graminearum* species complex and *F. culmorum*. Plant Breed 127:1–8
Tubajika KM, Mascagni HJ, Damann KE, Russin JS (2000) Susceptibility of commercial corn hybrids to aflatoxin contamination in Louisiana. Cereal Res Commun 28:463–467
Widstrom NW, Butron A, Guo BZ, Wilson DM, Snook ME, Cleveland TE, Lynch RE (2003) Control of preharvest aflatoxin contamination in maize by pyramiding QTL involved in resistance to ear-feeding insects and invasion by *Aspergillus* spp. Eur J Plant Pathol 19(4):563–572
Wilde F, Miedaner T (2006) Selection for Fusarium head blight resistance in early generations reduces the deoxynivalenol (DON) content in grain of winter and spring wheat. Plant Breed 125:96–98
Wilde F, Korzun V, Ebmeyer E, Geiger HH, Miedaner T (2007) Comparison of phenotypic and marker-based selection for Fusarium head blight resistance and DON content in spring wheat. Mol Breed 19:357–370
Williams WP (2006) Breeding for resistance to aflatoxin accumulation in maize. Mycotoxin Res 22:27–32
Williams WP, Windham GL, Buckley PM (2003) Enhancing maize germplasm with resistance to aflatoxin contamination. J Toxicol Toxin Rev 22:175–193

Windham GL, Williams WP (1998) *Aspergillus flavus* infection and aflatoxin accumulation in resistant and susceptible maize hybrids. Plant Dis 82:281–284

Zuber MS, Calvert OH, Kwolek WF, Lillehoj EB, Kang MS (1978) Aflatoxin B1 production in an eight-line diallele of *Zea mays* infected with *Aspergillus flavus*. Phytopathology 68: 1346–1349

Zummo N, Scott GE (1989) Evaluation of field inoculation techniques for screening maize genotypes against kernel infection by *Aspergillus flavus* in Mississippi. Plant Dis 73: 313–316

Chapter 12
Fusarium and Fumonisins in Maize in South America

Adriana M. Torres, Maria L. Ramirez, and Sofia N. Chulze

12.1 Introduction

South America, with its considerable North-South extent, is subject to climates that vary from tropical, subtropical, and warm temperate to temperate. This situation makes the area suitable for the production of a wide range of agricultural products, including cereals, oil seeds, beans, fruits, and nuts.

Maize (*Zea mays* L.) is one of the world major cereal grains, with a world annual average production (2004–2006) of 678 million metric tons (Mt), representing 32% of the 2.15 billion metric tons of world cereal grain production. South America produces 64 million Mt, 10% of the world maize production. This cereal is the staple food source for humans, and also the feed source for domesticated animals. The end-use products can be foods such as breakfast cereals, indigenous foods such as tortillas, tamales, tacos, enchiladas and porridge, snack foods, and feeds; 67% of the world maize production is intended for animal feed, or industrial uses, depending on the regions and their cultural preferences. Also several countries base their economies on this cereal marketing; the main exporting nations are USA (53%), China (19%), and Argentina (17%) (Chung et al. 2007). Both the climatic conditions and the amount of production in South America can be favorable for fungal contamination and mycotoxin occurrence. Maize can be infected with several fungal species, which are potential mycotoxin producers. Among these species, *Fusarium verticillioides* (Sacc.) Nirenberg and *Fusarium proliferatum* (Matsushima) Nirenberg are relevant, since they can infect maize worldwide and produce fumonisins. Fumonisin B_1 is the best known and studied of the fumonisins, but other derivatives are known to occur naturally as well (Plattner et al. 1996; Sewram et al. 2005).

S.N. Chulze (✉)
Mycology Section, Department of Microbiology and Immunology, National University of Río Cuarto, Ruta 36 km. 601 (5800), Río Cuarto, Córdoba, Argentina
e-mail: schulze@exa.unrc.edu.ar

M. Rai and A. Varma (eds.), *Mycotoxins in Food, Feed and Bioweapons*,
DOI: 10.1007/978-3-642-00725-5_12, © Springer-Verlag Berlin Heidelberg 2010

Dietary exposure to fumonisin causes adverse effects in farm and laboratory animals; these toxins have been associated with leukoencephalomalacia in horses (Ross et al. 1992), pulmonary edema syndrome in pigs (Harrison et al. 1990), liver and kidney toxicity in rats (Voss et al. 1988), and apoptosis in many types of cells (Jones et al. 2001). Fumonisins are phytotoxic, but their role in plant diseases caused by *F. verticillioides*, if any, has not been clearly defined (Lamprecht et al. 1994). Epidemiological studies have shown some evidence of a connection between intake of fumonisins and esophageal cancer in Africa, Brazil, China, and Italy. In addition, fumonisin B_1 reduces the folate uptake in cell lines, and fumonisin intake has been related to neural tube defects (Marasas et al. 2004).

Since mycotoxins are unavoidable contaminants in food and feed chains, their presence need to be reduced in order to avoid their effects on human and animal health and to diminish the annual market loss through rejected maize. Previous reports on the situation of mycotoxins in South America covering the periods 1995–2000 and 2000–2004 showed that the major problems appeared to be fumonisins in maize and maize products (Rodriguez Amaya and Sabino 2002; Chulze 2004).

The impact of mycotoxins in South America is significant from the point of view of:

(1) *The effects on public health.* The staple diet of the population sometimes is based on a commodity such as maize, which is susceptible to contamination with more than one mycotoxin, for example fumonisins and aflatoxins. The level of contamination of the agricultural products and by-products is sometimes of great concern; and taking into account the consumption of the contaminated food by the population, the daily intake of mycotoxins could sometimes be higher than the maximum tolerable daily intake established by the Joint FAO/WHO Expert Committee on Food Additives (JECFA) (FAO/WHO 2002).

Using a biomarker approach, exposure to fumonisins at individual level has been determined in populations from Brazil and Argentina (Solfrizzo et al. 2004).

(2) *The losses in animal production.* The loss is due to diversification of products, with high contamination levels of mycotoxins to animal feeding.
(3) *The problems in the trading of agricultural commodities.* Mycotoxins limits and regulations established by the European Community and Mercosur (regional trading block) can cause problems in trading agricultural commodities.

12.2 Fumonisin Producers *Fusarium* Species

Surveys carried out to evaluate the contamination with *Fusarium* species in maize harvested in different countries have shown that *F. verticillioides* (Sacc.) Nirenberg, (*Giberrella fujikuroi* mating population A) teleomorph *G. moniliformis* was the dominant species isolated, while other *Fusarium* species including *F. proliferatum* (Matsushima) Nirenberg (*G. fujikuroi* mating population D), teleomoph *G. intermedia* and *F. subglutinans* (*G. fujikuroi* mating population E) teleomorph *G. subglutinans* were also isolated in significant frequencies. In some areas of central and northwestern Argentina, *F. subglutinans* was isolated in higher frequency than *F. verticillioides* (Chulze et al. 2000; Torres et al. 2001).

F. verticillioides is similar to *F. proliferatum*, but the latter species is distinguished by its ability to form chains of microconidia from polyphialides (Leslie and Summerels 2006). *F. verticilliodes* can enter a maize plant systemically from seed, through wounds in the plant or through infections of the silks (Oren et al. 2003). Studies carried out in Argentina and Brazil showed that the fungus usually appears in the kernels during the early stages of development and continues to increase until harvest (physiological maturity), when it may represent 90% or more of the *Fusarium* isolates from the grain (Chulze et al. 1996; Almeida et al. 2002).

Based on vegetative compatibility groups (VCG) studies and using molecular markers such as AFLPs, high genetic diversity was observed among members of the mating populations A isolated from maize in Argentina and Uruguay (Chulze et al. 2000; Silva et al. 2000). It was also observed that multiple genetically distinct isolates of *F. verticillioides* may be recovered from a single maize plant (Kedera et al. 1994). Fumonisins are produced by *F. proliferatum* at high levels, and some atypical strains, producers of higher levels of FB_2 than of FB_1, were isolated from maize in Argentina (Chulze et al. 1998).

F. subglutinans is relatively easily separated from other major species in the *G. fujikuroi* species complex. *F. subglutinans* produce microconidia only in false heads, which distinguishes it from both *F. proliferatum* and *F. verticillioides*, which both produce microconidia in chains.

F. subglutinas was isolated in higher frequency in some areas of Argentina; the frequency of isolation and the ability of the strains to produce beauvericin were related to climatic conditions of the area (Torres et al. 2001). *F. subglutinans* was isolated in colder regions; a similar situation was observed in other surveys carried out in other cool areas in Europe (Lew et al. 1991; Logrieco et al. 1993).

Coproduction of fumonisins, fusaproliferin and beauvericin also among members of *G. intermedia* (*Gibberella fujikuroi* mating population D) and *G. subglutinans* (*Gibberella fujikuroi* mating population E) isolated from maize in Argentina was observed by Reynoso et al. (2004).

Others species described as fumonisin producers are *F. dlamini* Marasas, Nelson and Toussoun, isolated from plant debris present in soil from maize fields in South Africa but also described in maize in Argentina (Magnoli et al. 1999), and *F. nygamai* Burges and Trimboli, sexual stage *G. nygamai*; some strains produce high levels of fumonisin and carry the FUM biosynthetic genes cluster (Proctor et al. 2004). *F. nygamai* has been isolated from maize in Argentina (Etcheverry et al. 1999).

12.3 Fumonisin Biosynthesis

The fumonisin biosynthetic pathway in *Fusarium* species begins with formation of a linear dimethylated polyketide and condensation of the polyketide with alanine, followed by a carbonyl reduction, oxygenations and esterification with two propane -1, 2,3 tricarboxylic acids. The entire fumonisin gene cluster has 42 kb; fumonisin polyketide synthase (FUM1) was the first gene cloned, and is the anchor of a cluster of 15 co-regulated fumonisin biosynthetic genes (Desjardins and Proctor 2007). It has been observed that the fumonisin biosynthetic gene clusters in *F. proliferatum* and *F. verticillioides* have a high degree of synteny, but the flanking regions are completely different (Waalwijk et al. 2004).

12.4 Ecophysiology of Fungal Species

Fungal growth and mycotoxin production result from the complex interaction of several factors, and therefore an understanding of each factor involved is essential to understand the overall process, and to predict and prevent mycotoxin development (Charmley et al. 1994). Environmental conditions have a major impact on the fungal growth, and play a critical role in mycotoxicosis epidemiology. In addition, mycotoxin production is genetically regulated in response to environmental conditions (Holliger and Ekperigin 1999). Temperature and water availability are the primary environmental factors that influence growth and mycotoxin production by several *Fusarium* species (Magan and Lacey 1984; Marín et al. 1995a, b, 1996; Hope et al. 2005).

It has been demonstrated that within the same species in the same culture conditions, toxin production by *Fusarium* strains may vary sharply; some strains produce large amounts of mycotoxins, whereas others produce small or undetectable amounts of mycotoxins (Bakan et al. 2001). However, it is necessary to carry out a wide study to know whether the geographical origin of the strain can determine the optimal values for a_W and temperature for growth and mycotoxin production.

Prevention of mycotoxin contamination of food raw materials is now considered more important than subsequent cure. Thus, hazard analysis critical control point (HACCP) approaches are being developed to examine the critical control points at which mycotoxicogenic moulds and mycotoxins may enter a range of food chains. Accurate information is therefore needed on the impact of key environmental factors, such as water availability and temperature, and regarding which are marginal and which are optimum for growth and toxin production (Sanchis and Magan 2004).

Studies carried out in Argentina have been focused on the impact of different abiotic factors ($a_{W,}$ temperature and preservatives) and biotic factors on *Fusarium* Section *Liseola* species, particularly *F. verticillioides* and *F. proliferatum*, and on growth rates and fumonisins production.

Production of fumonisins B_1, B_2, and B_3 by *F. verticillioides* has been evaluated on irradiated corn kernels inoculated with different spore concentrations (10, 10^2, 10^3, 10^5, and 10^6), a water activity of 0.97, and a temperature of 25°C. There was a direct relationship between the level of toxin produced and inoculum size. The highest levels of total fumonisin produced after 35 days of incubation were 5,028 and 9,063 ng g^{-1} at 10^5 and 10^6 spores per ml respectively. The pattern of fumonisin production ($FB_1 > FB_2 > FB_3$) in cultures growing from different inocula was not affected during the 35 days of incubation. The ratio between FB_2 and FB_1 varied from 0.15 to 0.42, whereas the ratio between FB_3 and FB_1 varied from 0.34 to 0.87 (Chulze et al. 1999).

Different strategies have been examined to diminish the entry of fumonisins into the human and animal food chains. Efficient drying of maize immediately after harvest is the most effective way to try and limit fumonisin contamination of maize grains. However, maize is often harvested slightly wet, and if subsequent drying is delayed, fumonisin contamination can significantly increase. Postharvest treatment with antimicrobial agents has been examined to determine whether control can be achieved. Recently, Etcheverry et al. (2002) examined the effect of single food-grade antioxidants such as butylated hydroxyanisole (BHA), butylated hydroxytoluene (BHT), trihydroxybutyrophenone (THBP) and propylparaben (PP) (at concentrations of 1–20 mmol l^{-1}) on growth of and fumonisin production by Argentinian strains of *F. verticillioides* and *F. proliferatum.* Studies on lag phases prior to growth, relative growth rates and fumonisin concentrations were carried out *in vitro* in relation to water activity (0.995–0.93 a_W) and temperature (18 and 25°C) on a maize meal agar. Overall, PP was the antioxidant which was most effective at inhibiting strains of both species. The lag phase prior to growth and growth rates were significantly decreased by PP and BHA at 10 and 20 mmol l^{-1}, regardless of the temperature or a_W level tested. Total fumonisin production was higher at 0.98 a_W and decreased by about 45–50% at 0.995 and 0.95 a_W. Overall, BHT only inhibited fumonisin production at 0.95 a_W at 10 and 20 mmol l^{-1}, while BHA was effective at most a_W levels tested at 10 and 20 mmol l^{-1}. PP completely inhibited fumonisin production by both *F. verticillioides* and *F. proliferatum* at >1 mmol l^{-1}, regardless of the temperature or a_W level. Small interstrain differences in the levels of inhibition by the antioxidants were observed for three *F. verticillioides* and four *F. proliferatum* strains at 0.995, 0.98 and 0.95 a_W. PP and BHA completely inhibited the growth of both species at the concentrations evaluated, regardless of the a_W level. In conclusion, two antioxidants (PP and BHA) combined with lowered a_W and temperature could help to control *F. verticillioides* and *F. proliferatum* growth and fumonisin production *in vitro.*

In another study, Torres et al. (2003) evaluated the effect of the two best food-grade antioxidants BHA and PP (100, 200, 500 μg g^{-1}) and a_W (0.995, 0.98, 0.95) of irradiated maize (*in situ* study) on the lag phase prior to growth, growth rate and fumonisin production by *F. verticillioides* and *F. proliferatum*; the evaluation was carried out at 25°C. Both antioxidants had an effect on growth characteristics, and fumonisin production. However, this was dependent on the dose used and the a_W treatment. Both antioxidants significantly reduced the production of fumonisin by both *Fusarium* species, especially at 0.98 and 0.95 a_W. These results suggest that these antioxidants have potential for treatment of maize grain for controlling growth of these mycotoxigenic species and for prevention of fumonisin accumulation. The potential also exists for using mixtures of these antioxidants to try to obtain synergistic effects on the control of growth and fumonisins by these important pathogens of maize.

However, few studies have examined the efficacy of combinations of antioxidants to determine whether additive or synergistic effects on growth inhibition or control of fumonisins might be achieved. Studies carried out by Magan (1993) and Marin et al. (1998) suggested that the activity of certain hydrolytic enzymes was critical for colonization of cereal grains by spoilage fungi. However, the effect of antifungal treatments on such hydrolytic enzymes has not been studied in detail. Reynoso et al. (2002) evaluated the effect of single or mixtures of antioxidants on the lag phase prior to growth, growth rate, hydrolytic enzyme production (*N*-Acetyl β-glucosaminidase, β-D-glucosidase and α-D-galactosidase) and fumonisin production by *F. verticillioides* and *F. proliferatum* on maize-based media at 25°C, and under different water activity (a_w) conditions. An increase in the lag phase (h) was observed for both *F. verticillioides* and *F. proliferatum,* especially with PP + BHA treatments, at all a_W levels tested. For both species, PP alone or in combination with BHA at concentrations of 0.5 and 1 mM, reduced the growth rates by >85% at the three a_W levels tested (0.995, 0.98 and 0.95). PP + BHT or THBP were less effective in controlling growth, regardless of a_W level. Combinations of PP + BHA reduced the fumonisin concentrations produced by *F. verticillioides* and *F. proliferatum* at 0.995 and 0.98 a_W significantly. However, at low concentrations of antioxidants (0.5 mM) some stimulation in fumonisin production was observed with some treatments. The efficacy of the treatments was reflected in the impact on enzyme production. In the untreated control, the highest total enzyme activity of three hydrolytic enzymes was observed at 0.995 a_W after 96 h. All the antioxidant treatments alone or combined resulted in a significant reduction ($P < 0.001$) of the total enzyme activity at the a_W levels tested. During this study a synergistic effect of the mixture PP-BHA was observed, which is important, as this resulted in a higher reduction in the level of fumonisin produced at both 0.995 and 0.98 a_W. The use of combinations of antioxidants could be a good strategy to diminish the entry of fumonisin into the animal feed and human food chains.

Further work has been carried out in order to examine the efficacy of mixtures of these antioxidants (BHA and PP) in controlling *Fusarium* species in natural maize grain, and to examine the effect on growth and mycotoxin production. Farnochi et al. (2005) examined the temporal effect of the antioxidants BHA and PP at doses of 500 and 1,000 μg g^{-1} on the growth of *F. verticillioides* and *F. proliferatum* inoculated on natural maize grain, in the presence of competing mycoflora and fumonisin production at 0.98 and 0.95 water activity (a_W) over a 28-day storage period. The reduction in the colony-forming units (CFU) of *Penicillium, Aspergillus* and *Fusarium* populations was 10- to 100-fold, depending on dose of BHA or PP, a_W and time. However, the populations of all three groups were higher at 0.98 a_W than 0.95 a_W. BHA at 500 μg g^{-1} and 0.95 a_W reduced the fumonisin content by

82% after 7–14 days incubation, but at the end of the experimental period the reduction was only 32%. A higher reduction in the level of fumonisin produced (77%) was achieved with BHA at 1,000 μg g^{-1} after 28 days. PP at 500 and 1,000 μg g^{-1} decreased fumonisin production throughout the incubation period in the drier treatment, but at 0.98 a_W, control of toxin production was only achieved after 7–14 days. The reduction in the fumonisin levels could be due to the combined effect of antioxidants, and the competing mycoflora, mainly *Aspergillus* and *Penicillium* species.

12.5 Occurrence of Fumonisins

In Venezuela, rice and maize are main components of the diet of the population, and co-occurrence of mycotoxins like fumonisins and aflatoxins has been observed in these staple foods (Martinez et al. 2004).

In Chile, *Fusarium* spp and *Alternaria* spp are the dominant species of maize. Among the *Fusarium* species, fumonisin producers *F. verticillioides* and *F. proliferatum* have been isolated. The fumonisin levels found ranged from 0.32 to 37.5 mg kg^{-1} (Vega et al. 2000).

In Brazil, the tropical and subtropical climates favor fungal growth and mycotoxin contamination. Fumonisins in maize and maize-based foods have been reported (Ono et al. 1999; Rodriguez Amaya and Sabino 2002).

Another aspect that needs to be considered is the climate difference between the crop production areas within each country, which could render different levels of contamination with toxigenic fungi and mycotoxins. For example, Ono et al. (2001) found different levels of fumonisins in maize from Central Western and Central Southern Parana State, Brazil.

Also, a survey in different areas of Parana State, Brazil, during the 1999/2000 harvest season showed differences in the percentages of contamination with *F. verticillioides*, ranging from 10% to 18%, and in the fumonisin levels detected, which ranged from 1.7 to 3.4 mg kg^{-1} (Van der Westhuizen et al. 2003) (Table 12.1).

A survey carried out on ten different corn-based food products (208 samples) sold in the Federal District of Brazil and produced in the southern and south eastern regions of Brazil, which are the main maize-producing areas of the country, showed that 80.7% and 71.6% respectively of the samples had quantifiable levels of FB_1 and FB_2. Mean levels of total fumonisin (FB_1 and FB_2) ranged from 0.127 mg kg^{-1} for corn flakes to 2.04 for corn meal, and 0.664 for pop corn. No fumonisins was detected in sweet maize and fresh maize on the cobs (Caldas and Silva 2007).

In addition, the co-occurrence of aflatoxin, fumonisins and zearalenone, or ochratoxin, zearalenone and aflatoxins has been observed in maize for human consumption in Brazil (Machinski and Soares 2000; Vargas et al. 2001) (Table 12.2). Results of exposure assessment conducted in Brazil have shown that certain members of the population, such those consuming large amounts of maize products,

Table 12.1 Fumonisin levels and fungal contamination in maize samples collected from different areas of Brazil

Maize samples	Fumonisins (FB1 + FB2) (mg kg^{-1})	Fungal contamination[a] (%)	References
Paraná State		NA	Ono et al. (2001)
Central Southern	1.14		
Central Western	5.08		
Northern	9.85		
Santa Catarina State (1990–2000)			
Human consumption		14	Van der Westhuizen et al. (2003)
Western (Mountainous areas)	3.2	11	
Northern	3.4	18	
Southern (Prairies)	1.7	10	
Animal Feed			
Southern	1.5		

[a]*F. verticillioides* percentage

Table 12.2 Co-occurrence of fumonisins and other mycotoxins in maize from Brazil

Sample	Positive samples/ no. of samples	Toxin	Positive (%)	Range (μg kg^{-1})	Reference
Maize	82/214	AFB1	38.3	0.2–129	Vargas et al. (2001)
		ZEA	30.4	37–719	
		FUM	99.1	200–6,100	

AFB1 + FUM = 100% of the samples
AFB1 + ZEA + FUM = 18 samples
AFB1 + AFB2 + FUM = 43 samples

Table 12.3 Surveys of fumonisin contamination in Brazilian maize based foods and maize based infant foods

Sample	Fumonisins	Range (mg kg^{-1})	Mean (mg kg^{-1})	References
Corn-based food for infant	FBs	<0.02–8.03	2.2	Castro et al. (2004)
Instant corn-based baby food	FBs	<0.02–1.0	0.4	
Infant cereal	FBs	<0.02–1.7	0.6	
Maize-based infant food	FB1	0.030–6.1		Sewram et al. (2005)
	FB2	0.053–1.7		
Corn flour	FB1	0.47–7.20	2.11	Bittencourt et al. (2005)
	FB2	0.12–1.76	0.67	
Corn meal	FB1	1.08–15.29	5.17	
	FB2	0.23–3.94	1.00	
Polenta precooked corn flakes	FBs	0.52–4.07		Costa et al. (2004)
Popcorn	FBs	0.66–3.98	1.31	

and children, might be at greater risk when consuming maize products in Brazil (Table 12.3). Maize samples from Argentina showed different levels of contamination according to the region from which samples were collected. A survey of samples collected from the central region showed levels of FB_1 ranging from

Table 12.4 Fumonisin occurrence in maize samples from Argentina

Samples	Fumonisin levels (mg kg^{-1})		
	Range	Mean	References
Maize Central Region	FB1 = 0.06–11.1	FB1 = 3.18	Hennigen et al.
	FB2 = 0.02–3.52	FB2 = 0.99	(2000)
	FB1 = 0.43–2.93	FB1 = 1.44	Reynoso (2002)
	FB2 = 0.12–2.23	FB2 = 0.42	
	FB3 = 0.05–0.47	FB3 = 0.19	
Maize North-Eastern Region	FB1 = 0.024–9.2	FB1 = 1.4	Pacin et al. (2001)
	FB2 = 0.018–0.29	FB2 = 0.1	
	FB3 = 0.02–0.15	FB3 = 0.06	
Pop corn	FB1 = 0.03–0.53	FB1 = 0.11	Pacin et al. (2002)
	FB2 = 0.07–0.22	FB2 = 0.05	
	FB3 = 0.02–0.10	FB3 = 0.018	
Maize and maize-byproducts			Torres et al. (2001)
North-Western Region	FB1 + FB2 = 0.60–1.88	FB1 = 0.97	
		FB2 = 0.48	

0.06 to 11.1 mg kg^{-1}, mean 3.18, and of FB_2 from 0.02 to 3.52, mean 0.99 (Hennigen et al. 2000; Pacin et al. 2001; Torres et al. 2001) (Table 12.4).

From a survey of 1,711 samples collected from different agro-climatic zones in Argentina in the period 1999–2005, it was found that the median of total fumonisin contamination (FB_1 and FB_2) was 1,712 μg kg^{-1}. Even though contamination for all zones varied by year, there did not seem to followed a trend, median values ranging between 343 and 4,814 μg kg^{-1} (Pacin et al. 2007).

12.6 Stability of Fumonisin During Processing

In general, mycotoxins are stable compounds, and fumonisins that occur commonly in maize and maize-based products can remain after food processing operations such as sorting, milling, baking, frying, canning, roasting, flaking, nixtamalization and extrusion.

12.6.1 Sorting and Cleaning

Sorting and cleaning have shown some effect in reducing the levels of mycotoxins by removal of the contaminated material; however, these operations do not destroy mycotoxins. Since high levels of fumonisins can be found in asymptomatic maize grains, these operations have only a limited effect on the reduction of fumonisins.

In Argentina, studies using sieves for cleaning maize at the storage entrance showed reduction in the levels of fumonisins. It was observed that sieves of 6.5 mm size clean almost twice the amount cleaned by a 7-mm sieve (Resnik 2006).

12.6.2 Milling

During the dry-milling of maize, fumonisin B_1 was found in the highest amounts in the bran fraction, followed by the germ fraction; both are used for animal feeds. In flaking grits and flour, lower amounts were found (Katta et al. 1997; Brera et al. 2004). During wet milling of maize, fumonisin can be dissolved into the water or distributed among the byproducts, but not destroyed. During wet-milling, fumonisin was dissolved in the steep water or distributed to the gluten, fiber and germ fractions; the starch showed no detectable amounts of the toxin (Bennett et al. 1996).

A study conducted in Argentina on the analysis of corn meal, corn grits, corn flour, germ and brand originating from dry-milled maize naturally contaminated with 9.7 $\mu g\ g^{-1}$ of fumonisins showed that the levels of fumonisin in germ and bran were 3-fold higher than in the whole maize, 13-fold higher than in the maize flour, and 29-fold higher than in the maize meal and maize grits (Resnik 2006).

12.6.3 Brewing

Maize as grits or syrup can be used during beer production; reduction of fumonisin B_1 (3–28%) and fumonisin B_2 (9–17%) has been observed during brewing fermentation (Scott et al. 1995).

12.6.4 Thermal Processing and Extrusion

Fumonisin B_1 is a head stable compound; *F. verticillioides* culture material was stable after being boiled in water for 30 min and dried at 60°C during 24 h (Alberts et al. 1990). Higher temperatures such as those used for baking, roasting, frying or canning showed reduction in the fumonisin levels (Jackson et al. 1997; Castelo et al. 1998). Production of tortillas by alkaline cooking and steeping, followed by further processing into tortillas chips and corn chips, produces a reduction in the fumonisin levels (Jackson et al. 1997). The greatest reduction of fumonisins occurs at extrusion temperatures of 160°C or higher in presence of glucose. Extrusion of maize grits contaminated with fumonisin to which glucose 10% was added resulted in 75–85% reduction of the toxin. During extrusion, some products of fumonisin degradation are formed, such as hydrolyzed fumonisin B_1 and N-carboxymethyl fumonisin B_1 (Castelo et al. 2001). Also, Bullerman and Bianchini (2007) showed reduction in fumonisin B_1 toxicity after a maize grits extrusion process to which glucose was added. The type of screw, screw speed, and temperature also have effect on the reduction of fumonisins, Extrusion cooking resulted in more apparent loss of fumonisin B_1 with mixing screws than nonmixing ones (Castelo et al. 1998).

Maize grits spiked with fumonisin B_1 and extruded in a co-rotating twin screw extruder at different temperatures and screw speeds showed reductions ranging from 34% to 95%, depending on the extrusion parameters used (Katta et al. 1999).

12.6.5 Corn Flake Process

During the flaking process, a reduction in the level of fumonisins of 60–70% was observed by De Girolamo et al. (2001). Another study showed reduction in FB_1 of 53.5% and 48.7% after cooking and toasting respectively, while the addition of glucose to the process gave the highest reduction of fumonisin B_1 (86–89%). Although reductions in the levels of fumonisins have been observed during the cornflake proces,s the presence of protein-binding fumonisin (hidden fumonisins) have been detected in commercial cornflakes (Kim et al. 2003).

12.7 Methods of Detection of Fumonisins

Screening of fumonisins can be done by TLC methods with normal phase (silica) and reverse phase (C18) chromatography (Gelderblom et al. 1988; Rotthinghaus et al. 1992). However, the selected method for fumonisin analysis is HPLC, which is carried out after appropriate clean-up using C18 cartridges, SAX columns or IAC columns (Bennet and Richard 1994). The absence of chromophore UV in the fumonisin B structure makes necessary a chemical modification to be detected. Various precolumn derivatization techniques involving reactions involving the primary amino group to produce a fluorescent derivative have been reported (Shephard et al. 1990; Syndenham et al. 1996; Visconti et al. 2001). In surveys conducted in South America, ELISA and HPLC methods have been used to evaluate fumonisin occurrence in maize and maize-based foods and feeds (Castelo et al. 1998; Caldas and Silva 2007).

12.8 Biological Effect

Fumonisins are of concern in animal and human health safety, since they have been associated with many animal diseases such as leukoencephalomalacia (hole in the head syndrome) in equines (Marasas et al. 1988) and rabbits (Bucci et al. 1996), pulmonary edema and hydrothorax in swine (Harrison et al. 1990), and hepatotoxic and carcinogenic effects (Gelderblom et al. 1996) and apoptosis in the liver of rats (Pozzi et al. 2000). With respect to humans, studies on the prevalence of esophageal cancer in regions of South Africa, China, Italy, and Iran revealed an association between this disease and the consumption of maize contaminated by *Fusarium* spp

(Franceschi et al. 1990; Rheeder et al. 1992; Chu and Li 1994; Marasas 1996; Wang et al. 2000). Finally, fumonisins can cause neural tube defects in experimental animals, and thus may also have a role in human cases. It has been hypothesized that a cluster of anencephaly and spine bifida cases in southern Texas may have been related to fumonisins in corn products (Hendricks et al. 1999; Missmer et al. 2000). The International Agency for Research on Cancer (IARC) evaluated in 1992 the toxins derived from *F. verticillioides* as possibly carcinogenic to humans (IARC 1993). More recently, based on the research results obtained so far, FB_1 has been evaluated as possibly carcinogenic to humans (class 2B) (IARC 2002).

Bouhet and Oswald (2007) in a review have summarized the data dealing with the impact of FB_1 on the intestine. Although FB_1 is poorly absorbed and metabolized in the intestine, it induces intestinal disturbances (abdominal pain or diarrhea) The main toxicological effect of $FB(_1)$ reported *in vitro* and *in vivo* is the accumulation of sphingoid bases associated with the depletion of complex sphingolipids (Merril et al. 2001). This disturbance of the sphingolipid biosynthesis pathway could explain the other observed toxicological effects, such as an alteration in intestinal epithelial cell viability and proliferation, a modification of cytokine production, and a modulation of intestinal physical barrier function.

Theumer et al. (2002) investigated the immunobiological effects of subchronic dietary exposure to FB_1 in rats. The authors found histopathologic changes consisting of histiocytic perivascular infiltrate and an increased number of Kupffer cells in the liver, necrosis and apoptosis of tubular epithelial cells in the kidney, and increased mitotic figures and lymphocytic infiltrate in the small intestine. Serum enzyme alkaline phosphatase was significantly elevated in rats fed FB_1, while triglyceride levels decreased compared to controls. Treatment with FB_1 *in vivo* or *in vitro* did not have a significant effect on mitogen-induced proliferation of spleen mononuclear cells (SMC). However, increased levels of interleukin-4 (IL-4) and decreased levels of IL-10 were released by these cells in culture compared to controls. FB_1 *in vivo* or *in vitro* decreased the hydrogen peroxide released by peritoneal macrophages, while no changes in levels of superoxide anion produced by total peritoneal cells were detected. The results demonstrated that subchronic FB_1 intake could affect the small intestine, and alter the interleukin profile and some main functions of macrophages in antitumor activity.

Posterior studies were carried out with maize co-contaminated with aflatoxin B_1 (AFB_1) and fumonisin B_1 (FB_1). In *in vitro* studies, lower proliferations of SMC preexposed to the mixture of toxins were detected. The SMC of animals fed with both toxins produced higher levels of IL-4, lower levels of IL-10 and equal levels of IL-2 than the control. The peritoneal macrophages of animals fed with the mixture of toxins produced higher levels of H_2O_2 (Theumer et al. 2003). The co-exposure to fumonisins and AFB_1 produced a higher liver toxicity, with respect to their individual administration, inducing apoptosis and mitotic hepatocytes. There was an inversion of the typical Sa/So ratio in rats. Therefore, the mixture of fumonisins and AFB1 induced toxic responses which could not be considered a sum of the effects caused individually by these mycotoxins (Theumer et al. 2008).

12.9 Control of Fumonisin Production

The prevention of contamination of corn by fungi and their mycotoxins can be dealt with using strategies targeted at the pre- or postharvest periods. Prevention of mycotoxin contamination of cereal grains is better than detoxification, as the latter is not always fully effective (Moss 1998; Placinta et al. 1999). However, prevention is often difficult, as many of the conditions in the field cannot be manipulated by man, and the fumonisin producer fungi are ubiquitous in their geographical distribution.

Several approaches have been applied to try and diminish the human and animal exposure to fumonisins. In the field, strategies that have been reported include breeding for resistance to fungal and insect invasion (D'Mello and Macdonald 1997; CAST 2003), genetic engineering to confer resistance to fungal and insect invasion (Munkvold and Desjardins 1997; Dowd 2001; Bakan et al. 2002), and application of insecticides and fungicides (D'Mello and Macdonald 1997; Placinta et al. 1999).

Fusarium ear rot severity and fumonisin concentration in grain of 16 Argentinian maize hybrids were assessed after inoculation with *F. verticillioides*. Differences among hybrids for both traits indicate that field contamination with fumonisins might be reduced by sowing the most resistant genotypes currently grown in Argentina (Presello et al. 2007).

Maize breeding programs in Argentina have included resistance to ear rots as a selection trait, and most current hybrids are moderately resistant to these diseases. However, concentration of fumonisins in grain may be above guidelines whenever outbreaks occur. To improve resistance, it might be useful to employ new sources of resistance. Local landraces are potential sources of resistance to *Fusarium* spp. A study was conducted with the aim of characterize the potential of sources of resistance. Results from this research suggest that the sources of resistance evaluated might be used to enhance resistance to field fumonisin accumulation (Iglesias et al. 2007).

Munkvold and Desjardins (1997) stated that the engineering of plants to produce antifungal proteins or to detoxify mycotoxins in planta were feasible approaches to minimizing the risk they pose. Transgenic corn has mostly been genetically manipulated to include the gene from *Bacillus thuringiensis* (*Bt*) responsible for the production of the protein *Bt* cry1Ab, known to be toxic to insects. This corn is now widely known as *Bt* corn and has been demonstrated to have higher yields and lower levels of insect damage and infection by *Fusarium* compared to nontransgenic corn (Bakan et al. 2002; Gatch and Munkvold 2002). In addition, it has been observed that total fumonisins were reduced in some cases by as much as 30–50 times in *Bt* corn hybrids compared to the levels in non *Bt* hybrids (Dowd 2001; Clements et al. 2003; Hammond et al. 2004). In a study carried out in seven localities from the Argentinean maize-growing area, Barros et al. (2009) found similar results in a *Bt* hybrid derived from event MON 810.

Novel preharvest strategies evaluated to control fungal growth in the field include the use of endophytic bacterium such as *Bacillus mojavensis* and *Bacillus subtilis* as a biological control agents (Bacon et al. 2001), and the introduction of nonmycotoxigenic strains (Plattner et al. 2000).

In Argentina, Cavaglieri et al. (2004) evaluated the correlation between different screening methods, and suggested an adequate procedure that could be used to select bacterial agents with potential biocontrol against *F. verticillioides* in the maize rhizosphere. The authors carried out the assays with *Arthrobacter* spp., *Azotobacter* spp., *Pseudomonas* spp., and *Bacillus* spp. paired with 13 *F.verticillioides* strains isolated from maize endorhizosphere.

Pereyra et al. (2007) tested the ability of *Bacillus amyloliquefaciens* and *Microbacterium oleovorans* to reduce *F. verticillioides* populations and fumonisin accumulation in the maize agroecosystem. The impact of releasing these biocontrol agents on rhizospheric bacterial and fungal groups was also evaluated through isolation and identification of culturable microorganisms. When applied as seed coatings, both agents were effective in reducing *F. verticillioides* counts and FB_1 and B_2 content from maize grains. Richness and diversity indexes calculated for bacteria and fungi inhabiting the rhizosphere of maize remained unchanged following the addition of both biocontrol agents to seeds.

Effective preharvest inhibition of fungal growth ensures a lower fungal inoculum level enters storage. Thereafter, good agricultural/management practices at harvest and during storage should effectively ensure the prevention of fungal invasion and related mycotoxin production (Bhat and Miller 1991). Unlike during the preharvest period, most of the conditions during the postharvest period can be controlled. Good agricultural practices that have been found to limit mycotoxin production from the point of harvesting and during storage include rapid drying to moisture contents of about 13–14% (Munkvold and Desjardins 1997; Hussein and Brasel 2001), storage in dry cool rooms with adequate aeration, and application of effective pesticides and fungicides (D'Mello and Macdonald 1997; Placinta et al. 1999).

Several other potential methods to control fumonisin production on corn during storage have also been proposed, including the application of compounds with antifungal effects such as essential oils and their components (Velluti et al. 2003; Nguefack et al. 2004; López et al. 2004; Dambolena et al. 2007), weak organic acids and their salts (Marin et al. 2000), natural phenolic compounds (Beekrum et al. 2003; Samapundo et al. 2007a), or the use of modified atmospheres (Samapundo et al. 2007b, c). Varying degrees of efficacy have been achieved, which have not necessarily resulted in commercial success. Moreover, most of the studies have been carried out on artificial media, and their effects would still need to be validated on corn.

Studies by Thompson (1993, 1994) showed that the germination and growth of some mycotoxigenic *Fusarium* and *Penicillium* spp., including one moniliformin-producing strain of *F. verticillioides* (*F. moniliforme*), could be inhibited by some antioxidants. Most of these experiments were carried out using nutritionally rich media and without any consideration of the effect of key environmental factors, particularly water availability and temperature. Based on these findings, studies were carried out to determine the efficacy of a range of food-grade antioxidants to control the growth and fumonisin production by *Fusarium* spp. in Argentina (Etcheverry et al. 2002; Reynoso et al. 2002; Torres et al. 2003; Farnochi et al. 2005).

12.10 Regulations for Fumonisins

Regulations for fumonisins in foods and feeds have been recently established. Guidelines in USA for total fumonisins ($FB_1 + FB_2 + FB_3$) are 2 mg kg^{-1} for demerged, dry-milled maize products, and 3 mg kg^{-1} for popcorn grain (FDA 2001). In the EU, the maximum limit ($FB_1 + FB_2$) is 1 mg kg^{-1} for maize flour, grits, semolina, germ and oil, 0.4 mg kg^{-1} for maize-based products ready for consumption, and 0.2 mg kg^{-1} for maize based products for babies and children (Commision Regulation EC 2006). In South America, there are currently no regulations for fumonisins in foods and feeds.

12.11 Conclusions and Future Perspectives

Although differences exist between countries in the level of well-trained human resources, laboratory facilities, and financial resources, progress to diminish the impact of fumonisins is ongoing in the area.

The multidisciplinary efforts by geneticists, breeders, growers, handlers, industry, and cereal chemist/food scientists should be focused to ensure a total quality-based marketing system (considering fumonisin) to serve customers and consumers in the domestic and export markets.

Different strategies are ongoing in the area to diminish the entry of fumonisins into the food and feed chains, including looking for maize-resistant germplasms, instituting chemical and biological control and using *Bt* maize.

More knowledge regarding the biodiversity, ecophysiological and genetic characteristics of the *Fusarium* fumonisin producer species will be necessary in order to improve the management of maize to reduce fumonisin impact.

Regional differences in maize consumption, *Fusarium* species isolation and fumonisin incidence, mainly related to climatic conditions, make it necessary to establish regulatory levels for fumonisins in food in the area.

Acknowledgments The authors are member of the Research Carier of Consejo Nacional de Investigaciones Científicasy Técnicas (CONICET). The authors are grateful to the ANPCyT, SECyT-UNRC and CONICET for the financial support to carry out part of the results reported in the present chapter.

References

Alberts JF, Gelderblom WCA, Thiel PG, Marasas WFO, Van Schalkwyk DJ, Behrend Y (1990). Effects of temperature and incubation period on production of fumonisin B_1 by *Fusarium moniliforme*. Appl Environ Microbiol 56(6):1729–1733

Almeida AP, Fonseca H, Fancelli AL, Direito GM, Ortega EM, Correa B (2002). Mycoflora and fumonisin contamination in Brazilian corn from sowing to harvest. J Agric Food Chem 50(13):3877–3882

Bacon CW, Yates I, Hinton DM, Meredith F (2001) Biological control of *Fusarium moniliforme* in maize. Environ Health Perspect 109(Suppl 2):325–327
Bakan B, Pinson L, Cahagnier B, Melcion D, Semon E, Richard-Molard D (2001) Toxigenic potential of *Fusarium culmorum* strains isolated from French wheat. Food Addit Contam 18:998–1003
Bakan B, Melcion D, Richard-Molard D, Cahagnier B (2002) Fungal growth and *Fusarium* mycotoxin content in isogenic traditional maize and genetically modified maize grown in France and Spain. J Agric Food Chem 50(4):728–731
Barros G, Magnoli C, Reynoso MM, Ramirez ML, Farnochi MC, Torres A, Dalcero AM, Sequeira J, Rubinstein C, Chulze S (2009) Fungal and mycotoxin contamination in Bt maize and non-Bt maize grown in Argentina. World Mycotoxin Journal 2:53–60
Bennett GA, Richard JL (1994) Liquid chromatographic method for analysis of the naphthalene dicarboxaldehyde derivative of fumonisin. J AOAC Int 77:501–506
Bennett GA, Richard JL, Eckhoff SR (1996) Distribution of fumonisins in food and feed products prepared from contaminated corn. Adv Exp Med Biol 392:317–322
Bhat RV, Miller JD (1991) Mycotoxins and food supply. Food Nutr Agric 1:27–31
Bittencourt ABF, Oliveira CAV, Dilkin P, Correa B (2005) Mycotoxin occurrence in corn meal and flour traded in Sao Paulo, Brazilocurrence, fumonisinsBrazil. Food Control 16:117–120
Bouhet S, Oswald IP (2007) The intestine as a possible target for fumonisin toxicity. Mol Nutr Food Res 51(8):925–931
Brera C, Debegnach F, Grossi S, Miraglia M (2004) Effect of industrial processing on the distribution of fumonisin B_1 in dry milling corn fractions. J Food Prot 67(6):1261–1266
Bucci T, Hansen DK, LaBorde JB (1996) Leucoencephalomalacia and hemorrhage in the brain of rabbits gavaged with mycotoxin fumonisin B_1. Nat Toxins 4(1):51–52
Bullerman LB, Bianchini A (2007) Stability of mycotoxins during food processing. Int J Food Microbiol 119(1–2):140–146
Caldas ED, Silva ACS (2007) Mycotoxins in corn-based food products consumed in Brazil: an exposure assessment for fumonisins. J Agric Food Chem 55(19):7974–7980
Castelo MM, Summer SS, Bullerman LB (1998) Stability of fumonisins in thermally processed corn products. J Food Prot 61(8):1030–1033
Castelo MM, Jackson LS, Hanna MA, Reynolds BH, Bullerman LB (2001) Loss of fumonisin B_1 in extruded and baked corn-based foods with sugars. J Food Sci 66:416–421
Castro MF, Shephard GS, Sewram V, Vicente E, Mendoca TA, Jordan AC (2004) Fumonisins in Brazilian corn based foods for infant consumption. Food Addit Contam 21(7):693–699
Cavaglieri LR, Passone A, Etcheverry M (2004) Correlation between screening procedures to select root endophytes for biological control of *Fusarium verticillioides* in *Zea mays* L. Biol Control 31(3):259–267
Chamley LL, Rosenberg A, Trenholm HL (1994) Factors responsible for economics losses due to *Fusarium* mycotoxin contamination of grains, food and feedstuffs. In: Miller JD, Trenholm HL (eds) Mycotoxins in grain compounds other than aflatoxin. Eagan Press, St. Paul, MN, USA
Chu FS, Li GO (1994) Simultaneous occurrence of fumonisin B_1 and other mycotoxins in mouldy corn collected from the People's Republic of China in regions with high incidence of oesophageal canceroesophageal cancer. Appl Environ Microbiol 60(3):847–852
Chulze S (2004). Mycotoxins and toxigenic fungi in South America. Myco-Globe Conference, Brussels, October 2004
Chulze S, Ramírez ML, Farnochi MC, Pascale M, Visconti A, March G (1996) *Fusarium* and fumonisins occurrence in Argentinean corn at different ear maturity stages. J Agric Food Chem 44:2797–2801
Chulze S, Ramirez ML, Pascale M, Visconti A (1998) Fumonisins production by, and mating populations of, *Fusarium* Section *Liseola* isolates from maize in Argentina. Mycol Res 102:141–144
Chulze S, Etcheverry M, Lecumberry S, Magnoli C, Dalcero A, Ramírez ML, Pascale M, Rodriguez MI (1999) Fumonisin production on irradiated corn kernels: effect of inoculum size. J Food Prot 62:814–817

Chulze SN, Ramírez ML, Torres A, Leslie JF (2000) Genetic variation in *Fusarium* Section *Liseola* from no-till maize in Argentina. Appl Environ Microbiol 66:5312–5315

Chung OK, Park SH, Kim Y (2007) Concerted efforts in cereal grain quality improvement. ICC International Conference on cereals and cereal products quality and safety, Rosario, Argentina, September 2007

Clements MJ, Campbell KW, Pilcher C, Headrick JM, Pataky JK, White DG (2003) Influence of cry1Ab protein and hybrid genotype on fumonisin contamination and *Fusarium* ear rotmaize-Fusarium ear rot of corn. Crop Sci 43:1283–1293

Commission Regulation (EC) (2006) Setting maximum levels for certain contamiants in foods. Official J Eur Commission L364:5–24

Costa LLF, Peres M, Scussel VM (2004) Fumonisins and aflatoxins in corn-based extruded snacks and other food products for children. XI International IUPAC Symposium on Mycotoxins and Phycotoxins, Bethesda, MD, USA, May 2004

Council for Agricultural Science and Technology (CAST) (2003) Mycotoxins: risks in Plant, Animal, and Human Systems. Task Force Report No.139. Ames, IA, USA

D'Mello JPF, Macdonald AMC (1997) Mycotoxins. Anim Feed Sci Technol 69:155–166

Dambolena JS, López AG, Cánepa MC, Theumer MG, Zygadlo JA, Rubinstein HR (2007) Inhibitory effect of cyclic terpenes (limonene, mentho, menthone and thymol) on *Fusarium verticillioides* MRC 826 growth and fuminsin B_1 biosynthesis. Toxicon 51(1):37–44

De Girolamo A, Solfrizzo M, Visconti A (2001) Effect of processing on fumonisin concentration in corn flakes. J Food Prot 64(5):701–705

Desjardins AE, Proctor RH (2007) Molecular biology of *Fusarium* mycotoxins. Int J Food Microbiol 119 (1–2):47–50

Dowd F (2001) Biotic and abiotic factors limiting efficacy of Bt corn in indirectly reducing mycotoxin levels in commercial fields. J Econ Entomol 94:1067–1074

Etcheverry M, Nesci A, Barros G, Torres A, Chulze S (1999) Occurrence of *Aspergillus* section *Flavi* and aflatoxin B_1 in corn genotypes and corn meal in Argentina. Mycopathologia 147:37–41

Etcheverry M, Torres A, Ramirez ML, Chulze S, Magan N (2002) *In vitro* control of growth and fumonisin production by *Fusarium verticillioides* and *F. proliferatum* using antioxidants under different water availability and temperature regimes. J Appl Microbiol 92(4):624–632

Farnochi C, Torres AM, Magan N, Chulze S (2005) Effect of antioxidants and competing mycoflora on *Fusarium verticillioides* and *F. proliferatum* populations and fumonisin production on maize grain. J Stored Prod Res 41:211–219

Food and Drug Administration (FDA) (2001) Fumonisins levels in human foods and animal feeds. Guidance for industry, final guidance. US Food and Drug Administration Center for Food Food Safety and Applied Nutrition, Rockville, MD, USA, November 2001

Franceschi S, Bidoldi E, Baron AE, La Vecchia C (1990) Maize and risk of cancers of the oral cavity, pharynx, and oesophagus in Northeastern Italy. J Natl Cancer Inst 82(17):1407–1411

Gatch EW, Munkvold GP (2002) Fungal species composition in maize stalks in relation to European corn borer injury and transgenic insect protection. Plant Dis 86:1156–1162

Gelderblom WC, Smuts CM, Abel S, Snyman SD, Cawood ME, van der Westhuizen L, Swanevelder S (1996) Effect of fumonisin B_1 on protein and lipid synthesis in primary rat hepatocytes. Food Chem Toxicol 34(4):361–369

Hammond BG, Campbell KW, Pilcher CD, Degooyer TA, Robinson AE, McMillen BL, Spangler SM, Riordan SG, Rice LG, Richard JL (2004) Lower fumonisin mycotoxin levels in the grain of *Bt* corn grown in the United States in 2000–2002. J Agric Food Chem 52(5):1390–1397

Harrison LR, Colvin BM, Greene JT, Newman LE, Cole JR Jr (1990) Pulmonary edema and hydrothorax in swine produced by fumonisin B_1 a toxic metabolite of *Fusarium moniliforme*. J Vet Diagn Invest 2(3):217–221

Hendricks KA, Simpson JS, Larsen RD (1999) Neural tube defects along the Texas-Mexico border, 1993–1995. Am J Epidemiol 149L:1119–1127

Hennigen MR, Valente Soarez LM, Sanchez S, Migdal Di Benedetto N, Longhi A, Eyherabide G, Torroba J (2000) Fumonisins in maize genotypes grown in Argentina during two consecutives seasons. X International IUPAC Symposium on Mycotoxins and Phycotoxins, Guaruja, Brazil, May 2000

Holliger K, Ekperigin HE (1999) Mycotoxins in food-producing animals. Vet Clin North Am 15:133–165

Hope R, Aldred D, Magan N (2005) Comparison of environmental profiles for growth and deoxynivalenol production by *Fusarium culmorum* and *F. graminearum* on wheat grain. Lett Appl Microbiol 40(4):295–300

Hussein HS, Brasel JM (2001) Toxicity, metabolism, and impact of mycotoxins on humans and animals. Toxicology 167(2):101–134

IARC (1993) Monograph on the evaluation of carcinogenic risk to humans. 56:257–263

Iglesias J, Presello D, Díaz C, Botta G, Gómez LG (2007) Potential of Argentinean germoplasm to improve an elite single cross of maize for field accumulation of fumonisins in grain. 1st Latin American ICC International Conference ARGENTINA 2007 "Cereals and cereal products: quality and safety-new challenges of world demand", Rosario, Argentina, September 2007

International Agency for Research on Cancer (IARC) (2002) Some traditional herbal medicines, some mycotoxins, naphthalene and styrene. IARC Monogr Eval Carcinog Risks Hum 82:171–301

Jackson LS, Katta SK, Fingerhut DD, De Vries JW, Bullerman LB (1997) Effects of baking and frying on the fumonisin B_1 content of corn based foods. J Agric Food Chem 45:4800–4805

Jones C, Ciacci-Zanella JR, Zhang Y, Henderson G, Dickman M (2001) Analysis of fumonisin B_1 induced apoptosis. Environ Health Perspect 109(2):315–320

Katta SK, Cagampang AE, Jackson LS, Bullerman LB (1997) Distribution of *Fusarium* molds and Fumonisins in dry-milled corn fractions. Cereal Chem 74:858–863

Katta SK, Jackson LS, Summer SS, Hanna MA, Bullerman LB (1999) Effect of temperature and screw speed on stability of fumonisin B_1 in extrusion cooked corn grits. Cereal Chemistry 76:16–20

Kedera CJ, Leslie JF, Claffin LE (1994) Genetic diversity of *Fusarium* section *Liseola* (*Gibberella fujikuroi*) in individual maize plants. Phytopathology 84:603–607

Kim EK, Scott PM, Lau BPY (2003) Hidden fumonisins in corn flakes. Food Addit Contam 20(2):161–169

Lamprecht SC, Marasas WFO, Alberts JF, Cawood ME, Gelderblom WCA, Shephard GS, Thiel PG, Calitz FJ (1994) Phytotoxicity of fumonisins and TA toxin to corn and tomato. Phytopathology 84:383–391

Leslie JF, Summerell BA (2006) Species descriptions. In: The *Fusarium* laboratory manual. Blackwell, London

Lew H, Adler A, Edinger W (1991) Moniliformin and the European corn borer. Mycotoxin Res 7:71–76

Logrieco A, Moretti A, Altomare C, Bottalico A, Carbonell Torres E (1993) Occurrence and toxicity of *Fusarium subglutinans* from Peruvian maize. Mycopathologia 122(3):185–190

Lopez AG, Theumer MG, Zygadlo JA, Rubinstein HR (2004) Aromatic plants essential oils activity on *Fusarium verticillioides* fumonsin B_1 production in corn grain. Mycopathologia 158(3):343–349

Machinski M, Soares LMV (2000) Fumonisins B_1 and B_2 in Brazilian corn-based food products. Food Addit Contam 17:875–879

Magan N, Lacey J (1984) Water relations of some *Fusarium* species from infected wheat ears and grain. Trans Br Mycol Soc 83:281–285

Magnoli CE, Saenz MA, Ferrero S, Chiacchera SM, Dalcero AM (1999) Natural occurrence of *Fusarium* species and fumonisins-production by toxigenic strains isolated from poultry feeds in Argentina. Mycopathologia 145(1):35–41

Marasas WFO (1996) Fumonisins: history, world-wide occurrence and impact. In: Jackson LS, DeVries JW, Bullerman LB (eds) Fumonisins in food. Plenum, New York, pp 1–17

Marasas WFO, Kellerman TS, Gelderblom WCA, Coetzer JAW, Thiel PG, Van Der Lugt JJ (1988) Leukoencephalomalacia in horse induced by fumonisin B_1 isolated from *Fusarium moniliforme*. Onderstepoort J Vet Res 55(4):197–203

Marasas WFO, Riley RT, Hendricks KA, Stevens VL, Sadler TW, Gelineau van Waes J, Missmer SA, Cabrera J, Torres O, Gelderblom WCA, Allegood J, Martinez C, Maddox J, Miller JD, Starr L, Sullards MC, Roamn AV, Voss KA, Wang E, Merrill AH Jr (2004) Fumonisins disrupt sphingolipid metabolism, folate transport, and neural tube development in embryo culture and *in vivo*: a potential risk factor for human neural tube defects among populations consuming fumonisin-contaminated maize. J Nutr 134(4):711–716

Marín S, Sanchis V, Viñas I, Canela R, Magan N (1995a) Effect of water activity and temperature on growth and fumonisin B_1 and B_2 production by *Fusarium proliferatum* and *F. moniliforme* on maize grain. Lett Appl Microbiol 21(5):298–301

Marín S, Sanchis V, Magan N (1995b) Water activity, temperature and pH effects on growth of *Fusarium moniliforme* and *Fusarium proliferatum* isolates from maize. Can J Microbiol 41 (12):1063–1070

Marín S, Sanchis V, Teixido R, Saenz AJ, Ramos AJ, Magan N (1996) Water activity, temperature relationships and microconidial germination of *Fusarium moniliforme* and *Fusarium proliferatum* from maize. Can J Microbiol 42(10):1045–1050

Marín S, Sanchis V, Ramos AJ, Magan N (1998) Effect of water activity on hidrolitic enzyme production by *Fusarium moniliforme* and *Fusarium proliferatum* during colonization of maize. Int J Food Microbiol 42(3):185–194

Marin S, Magan N, Abellana M, Canela R, Ramos AJ, Sanchis V (2000) Selective effect of propionates and water activity on maize mycoflora and impact on fumonisin B_1 accumulation. J Stored Prod Res 36:203–214

Martinez A, Raybaudi R, Alvarado R, Gomez N, Escobar E, Casanova R (2004) Occurrence of aflatoxins and fumonisins in corn and rice from Venezuela and its mycoflora. XI International IUPAC Symposium on Mycotoxins and Phycotoxins, Bethesda, MD, USA, May 2004

Merrill AH Jr, Sullards MC, Wang E, Voss KA, Riley RT (2001) Sphingolipid metabolism: roles in signal transduction and disruption by fumonisins. Environ Health Perspect 109(2):283–289

Missmer S, Hendricks KA, Suarez L, Larsen RD, Rothman IJ (2000) Fumonisins and neural tube defects. Epidemiology 11:183–184

Moss MO (1998) Recent studies of mycotoxins. J Appl Microbiol Symp Suppl 84:62S–76S

Munkvold GP, Desjardins AE (1997) Fumonisins in maize. Can we reduce their occurrence? Plant Dis 81:556–565

Nguefack J, Leth V, Amvam Zollo PH, Mathur SB (2004) Evaluation of five essential oils from aromatic plants of Cameroon for controlling food spoilage and mycotoxin producing fungi. Int J Food Microbiol 94:329–334

Ono EY, Sugiura Y, Homechin M, Kamogae M, Vizzoni E, Ueno Y, Hirooka EY (1999) Effect of climatic conditions on natural microflora in fumonsisins freshly harvested corn of the State of Parana, Brazil. Mycopathologia 147:139–148

Ono EY, Ono MA, Funo FY, Medinal AE, Oliveira TC, Kawamura O, Ueno Y, Hirooka EY (2001) Evaluation of fumonisin–aflatoxin occurrence in Brazilian corn hybrids by ELISA. Food Addit Contam 18:719–729

Oren L, Ezrati S, Cohen D, Sharon A (2003) Early events in the *Fusarium verticillioides* maize interaction characterized by using a green fluorescent protein expressing transgenic isolate. Appl Environ Microbiol 69:1695–1701

Pacin AM, Broggi LE, Resnik SL, Gonzalez HHL (2001) Mycoflora and mycotoxins natural occurrence in corn from Entre Rios Province, Argentina. Mycotox Res 17:31–38

Pacin AM, Boca RT, Gonzalez HHL, Resnik SL, Burak R, Broccoli AM, de Souza JC (2002) Natural occurrence of mycotoxins and mycoflora of Argentinean popcorn. Mycotox Res 18:90–96

Pacin A, Whitechurch E, Cano G, Taglieri D, Ciancio Bovier E (2007) Mycotoxins in Argentinean maize in preharvest from 1999 to 2005. ICC International Conference on cereals and cereal products quality and safety, Rosario, Argentina, September 2007

Pereira P, Nesci A, Etcheverry M (2007) Effects of biocontrol agents on *Fusarium verticillioides* count and fumonisin content in the maize agroecosystem: impact on rhizospheric bacterial and fungal groups. Biol Control 42(3):281–287

Placinta CM, D'Mello JPF, Macdonald AMC (1999) A review of the worldwide contamination of cereal grains and animal feed with *Fusarium* mycotoxins. Anim Feed Sci Technol 78:21–37

Plattner RD, Desjardins AE, Leslie JF, Nelson PE (1996) Identification and characterization of strains of *Gibberella fujikuroi* mating populations A (*Fusarium moniliforme*) with rare fumonisin phenotypes. Mycologia 87:416–424

Plattner R, Proctor RH, Brown DW, Desjardins AE (2000) Potential for minimisation of fumonisin levels in corn using fumonisin non-producing strains. Aflatoxin/Fumonisin Workshop, October 25–27 2000, Tenaya Lodge, Fish Camp, Yosemite, CA, USA, pp 35–37

Pozzi CR, Correa B, Xavier JG, Direito GM, Orsi RB, Matarazzo SV (2000) Effects of prolonged oral administration of fumonisin B_1 and aflatoxin B_1 in rats. Mycopathologia 151:21–27

Presello DA, Iglesias J, Botta G, Eyhérabide GH (2007) Severity of *Fusarium* ear rot and concentration of fumonisin in grain of Argentinian maize hybrids. Crop Prot 26(6): 852–855

Proctor RH, Plattner RD, Brown DW, Seo JA, Lee YW (2004) Discontinuous distribution of fumonisin biosynthetic genes in the *Gibberella fujikuroi* species complex Mycol Res 108:815–822

Resnik SL (2006) Food processing to reduce the entry of mycotoxins to the food and feed chains. Conference on "Advances in research on toxigenic fungi and mycotoxins in South America-maize ensuring food and feed safety in a mycoglobe context" Carlos Paz, Cordoba, Argentina, March 2006

Reynoso MM (2002) Especies biológicas dentro del complejo *Gibberella fujikuroi* aisladas de maíz en Argentina. Aspectos genéticos y ecofisiológicos. PhD Thesis, Universidad Nacional de Río Cuarto

Reynoso MM, Torres AM, Ramirez ML, Rodriguez MI, Chulze S, Magan N (2002) Efficacy of antioxidant mixtures on growth, fumonisin production and hydrolytic enzyme production by *Fusarium verticillioides* and *F. proliferatum in vitro* on maize-based media. Mycol Res 106:1093–1099

Reynoso MM, Torres, AM, Chulze SN (2004) Fusaproliferin, beauvericin and fumonisin production by different mating populations among the Gibberella fujikuroi complex isolated from maize Mycol Res 108(2):154–160

Rheeder JP, Marasas WFO, Thiel PG, Sydenham EW, Shephard GS, Van Schalkwyk DJ (1992) *Fusarium moniliforme* and fumonisins in corn in relation to human oesophageal canceroesophageal cancer in Transkei. Phytopathology 82:353–357

Rodriguez Amaya DB, Sabino M (2002) Mycotoxin research in Brazil: the last decade in review. Braz J Microbiol 33:1–11

Ross PF, Rice LG, Osweiler GD, Nelson HA, Richard JL, Wilson TM (1992) A review and up-date of animal toxicoses associated with fumonisin-contaminated feeds and production of fumonisins by *Fusarium* isolates. Mycopathologia 117:109–114

Rottinghaus GE, Coatney CE, Minor HC (1992) A rapid, sensitive thin layer chromatographic procedure for the detection of fumonisin B_1 and B_2. J Vet Diagn Invest 4:326–329

Samapundo S, De Meulenaer B, Osei-Nimoh D, Lamboni Y, Debevere J, Devlieghere F (2007a) Can phenolic compounds be used for the protection of corn from fungal invasion and mycotoxin contamination during storage? Food Microbiol 24(5):465–473

Samapundo S, De Meulenaer B, Atukwase A, Debevere J, Devlieghere F (2007b) The influence of modified atmospheres and their interaction with water activity on the radial growth and fumonisin B(1) production of *Fusarium verticillioides* and *F. proliferatum* on corn. Part I: the effect of initial headspace carbon dioxide concentration. Int J Food Microbiol 114 (2):160–167

Samapundo S, De Meulenaer B, Atukwase A, Debevere J, Devlieghere F (2007c) The influence of modified atmospheres and their interaction with water activity on the radial growth and fumonisin B(1) production of *Fusarium verticillioides* and *F. proliferatum* on corn. Part II: the effect of initial headspace oxygen concentration. Int J Food Microbiol 113(3):339–345

Sanchis V, Magan N (2004) Environmental conditions affecting mycotoxins. In: Magan N, Olsen M (eds) Mycotoxins in food: detection and control. Woodhead, Oxford

Scott PM, Kanhere SR, Lawrence GA, Daley EF, Farber JM (1995) Fermentation of wort containing added ochratoxin A and fumonisinsB_1 and B_2. Food Addit Contam 12:31–40

Sewram V, Mshicileli N, Shephard GS, Vismer HF, Rheeder JP, Lee YW, Leslie JF, Marasas WFO (2005) Production of fumonisin B and C analogs by several *Fusarium* species. J Agric Food Chem 53:4861–4866

Shephard GS, Sydenham EW, Thiel PG, Gelderblom WCA (1990) Quantitative determination of fumonisins B1 and B2 by high-performance liquid chromatography with fluorescence detection. J Liq Chromatogr 13:2077–2080

Silva GE, Piñeiro MS, Zeller KA, Leslie JF (2000) In: Sabino M, Rodriguez-Amaya D, Correa B (eds) Fumonisins and *Gibberella fujikuroi* from maize in Uruguay. X International IUPAC Symposium on Mycotoxins and Phycotoxins, Guaruya, Brasil

Solfrizzo M, Chulze SN, Mallmann C, Visconti A, De Girolamo A, Rojo F, Torres A (2004) Comparison of urinary sphingolipids in human populations with high and low maize consumption as a possible biomarker of fumonisin dietary exposure. Food Addit Contam 21(11):1090–1095

Sydenham EW, Shephard GS, Thiel PG, Stochenstrom S, Snijman PW, van Schalkwyk DJ (1996) Liquid chromatographic determination of fumonisin B1, B2 and B3 in corn: AOAC IUPAC collaborative study. J AOAC Int 79:688–696

Theumer MG, López AG, Maíz DT, Chulze SN, Rubinstein, HR (2002) Immunobiological effects of fumonisin B1 in experimental subchronic mycotoxicoses in rats. Clin Diagn Lab Immunol 9(1):149–155

Theumer MG, López AG, Maíz DT, Chulze SN, Rubinstein HR (2003) Imunobiological effects of AFB1 and AFB1 + FB1 mixture in experimental subchronic mycotoxicoses in rats. Toxicology 186(1–2):159–170

Theumer MG, López AG, Aoki MP, Cánepa MC, Rubinstein HR (2008) Subchronic mycotoxicoses in rats. Histopathological changes and modulation of the sphinganine to sphingosine (Sa/So) ratio imbalance induced by *Fusarium verticillioides* culture material, due to the coexistence of aflatoxin B1 in the diet. Food Chem Toxicol 46(3):967–977

Thompson DP (1994) Minimum inhibitory concentrations of esters of p-hydroxybenzoic acid (paraben) combinations against toxigenic fungi. J Food Prot 57:133–135

Thompson DP, Metevia L, Vessel T (1993) Influence of pH alone and in combination with phenolic antioxidants on growth and germination of mycotoxigenic species of *Fusarium* and *Penicillium*. J Food Prot 56:134–138

Torres A, Reynoso MM, Rojo F, Ramírez ML, Chulze S (2001) Fungal and mycotoxin contamination in home grown maize harvested in the north area of Argentina. Food Addit Contam 18:836–843

Torres A, Ramirez ML, Arroyo M, Chulze S, Magan N (2003) Potential for control of growth and fumonisin production by *Fusarium verticillioides* and *F. proliferatum* on irradiated maize grain using anti-oxidants. Int J Food Microbiol 83:319–324

Van der Westhuizen L, Shephard GS, Scussel VM, Costa LL, Vismer HF, Rheder JP, Marasas WF (2003) Fumonisin contamination and *Fusarium* incidence in corn from Santa Catarina, Brazil. J Agric Food Chem 51:5574–5578

Vargas EA, Preis RA, Castro L, Silva CMG (2001) Co –occurrence of aflatoxins B_1, B_2, G_1, G_2, zearalenone and fumonisin B_1 in Brazilian corn. Food Addit Contam 18:981–986

Vega HM, Saelzer G, Rios E, Herlitz C, Bastias M, Veja M (2000) Contaminación por micotoxinas de alimentos y piensos en Chile: Resultados del triênio 1997–2000. III Congreso Latinoamericano de Micotoxicologia, Los Cocos, Cordoba, Argentina

Velluti A, Sanchis V, Ramos AJ, Egido J, Marin S (2003) Inhibitory effect of cinnamon, clove, lemongrass, oregano and palmarose essential oils on growth and fumonisin B1 production by *Fusarium proliferatum* on maize grain. Int J Food Microbiol 59:59–66

Visconti A, Solfrizzo M, De Girolamo A (2001) Determination of fumonisins B1 and B2 in corn and corn flakes by liquid chromatography with immunoaffinity column cleanup: collaborative study. J AOAC Int 84(6):1828–1837

Voss KA, Plattner RD, Riley RT, Meredith FI, Norred WP (1988) *In vivo* effects of fumonisin B_1 producing and fumonisin B_1 non-producing *Fusarium moniliforme* isolates are similar: Fumonisins B_2 and B_3 cause hepato- and nephrotoxicity in rats. Mycopathologia 141:45–58

Waalwijk C, van der Lee Theo, de Vries I, Hesselink T, Arts J, Kema GHJ (2004) Synteny in toxigenic *Fusarium* species: the fumonisin gene cluster and the mating type region as examples. Eur J Plant Pathol 110:533–544

Wang H, Wei H, Ma J, Luo X (2000) The fumonisin B1 content in corn from North China, a high-risk area of esophageal cancer. J Environ Pathol Toxicol Oncol 19(1–2):139–141

World Health Organization (2002) Evaluation of certain mycotoxins in food. Fifty-sixth Report of the Joint FAO/WHO Expert Committee on Food Additives (JECFA). WHO Tech Rep 906:16–27

Chapter 13
Prevention and Management of Mycotoxins in Food and Feed

Bulent Kabak

13.1 Introduction

The contamination of food and animal feeds with mycotoxins is a worldwide problem, while mycotoxins occur more frequently under tropical conditions as a result of environmental conditions in the field coupled with improper harvesting and bad storage. The FAO has estimated that up to 25% of the world's food crops are significantly contaminated with mycotoxins. The global volume of agricultural products such as maize, groundnuts, copra, palm nuts and oilseed cake, which are high-risk commodities, is about 100 million tonnes — 20 million tonnes of which come from the developing countries (FAO 1996). Several management strategies have been developed to help prevent the growth of mycotoxigenic fungi as well as to decontaminate and/or detoxify mycotoxin contaminated foods and animal feeds (Bata and Lásztity 1999). These strategies include:

- Prevention of mycotoxin contamination at pre-harvest, during harvesting and in the post-harvest stage
- Decontamination/detoxification of mycotoxins present in food and feed
- Inhibition of mycotoxin absorption in the gastrointestinal tract

13.2 Prevention of Mycotoxin Contamination

Fungal infection and subsequently mycotoxin formation can occur in the field, during harvesting or during storage period and processing. Therefore, strategies for managing mycotoxin accumulation in agricultural products may be divided into three categories, pre-harvest, harvesting and post-harvest strategies.

B. Kabak
Department of Food Engineering, Faculty of Engineering, Hitit University, Corum, Turkey
e-mail: bkabak@cu.edu.tr, bulentkabak@hitit.edu.tr

M. Rai and A. Varma (eds.), *Mycotoxins in Food, Feed and Bioweapons*,
DOI: 10.1007/978-3-642-00725-5_13,

13.2.1 Pre-Harvest Management

Prevention through pre-harvest management is the first step for the control of mycotoxin contamination (Lopez-Garcia et al. 1999). Many *Aspergillus* species infect nuts such as groundnuts, peanuts and pistachio, and cause decay of the kernels before harvest (Pitt et al. 1991). The main problem associated with pistachios is "early split" during pre-harvest season. The shells of most pistachio nuts split naturally in the orchard prior to harvest. In this case, the hull covering the shell usually remains intact, protecting the kernel from invasion by moulds and invasion. Nevertheless, in a small percentage of the pistachios in an orchard have the hull attached to the shell, so that the hull splits with the shell, exposing the kernel to mould and insects (Boutrif 1998).

However, the main mycotoxin hazards associated with small grain cereals, particularly wheat pre-harvest, are the toxins produced by *Fusarium* species, *F. culmorum* and *F. graminearum* in the growing crop. These two species can produce ZEA and trichothecenes including NIV, DON and T-2 toxin. *Fusarium* species can also cause a serious disease called *Fusarium* head (ear) blight (FHB), also referred to as scab, which can lead to significant losses in crop yield and quality (Aldred and Magan 2004). A variety of strategies are possible for controlling fungal infection and mycotoxin formation in field crops, and it is likely that eventual success may involve a combination of several approaches. These are critically reviewed below.

13.2.1.1 Environmental Conditions

While many factors are known to influence the production of mycotoxins in the field, of these, drought stress during plant growth is among the most important, with reports indicating that drought stress during groundnut maturation increases the susceptibility of aflatoxigenic mould invasion and aflatoxin production (Moss 1992).

Infection of the ear occurs at a critical period during anthesis, and is favoured by cool temperatures and high humidity (EC Report 1999). Previous work by Lacey et al. (1999) observed that *Fusarium* infection in the UK is exacerbated by wet periods at a critical time in early flowering in the summer, which is the optimum window for susceptibility.

13.2.1.2 Resistant Varieties

With respect to genetic resistance to *Aspergillus* infection, since the early 1970s much work has been done to identify genetically resistant crop genotypes in both laboratory- and field-based experiments to help control aflatoxigenic mould growth and aflatoxin biosynthesis (Cleveland et al. 2003).

Resistance in cereals to FHB using conventional breeding and through transgenic approaches has been reviewed extensively by Dahleen et al. (2001) and Snijders (2004). Resistance is quantitatively inherited, with significant genetic variation among breeding materials, and under the different environmental and agronomic conditions in which crops are cultivated (Edwards 2004). In a previous work, Langesth and Rundberget (1999) observed that oats had higher levels of DON than barley and wheat in Norway from 1996 to 1999, whereas DON levels in wheat, barley and oats were similar when grown under the same field conditions in Western Canada in 2001 (Tekauz 2002). In a very extensive work by Meidaner et al. (2001), although rye genotypes were, on average, affected by *Fusarium* infections much the same as wheat genotypes, wheat accumulated twice as much DON as rye.

13.2.1.3 Field Management

Field management practices involving crop rotation, tillage practices, irrigation and fertilization approaches are preventive actions that control the production of mycotoxins in the field. Crop rotation is important and is intended to break the chain of production of infectious material, for example by using wheat/legume rotations (Aldred and Magan 2004). In maize-dominated crop rotations, maize stubble bears a high infection risk for the following cereal crop (Schmidt and Nitzsche 2004). In another study, Schaafsma et al. (2001) observed that the average DON level in fields where maize and wheat were grown 2 years before wheat was at least 57% greater than in fields where soybean was the crop 2 years previous to wheat.

Several techniques are used for soil cultivation. One is ploughing, where the top 10–30 cm of soil is inverted; the other technique is minimum tillage, where the crop debris is mixed with the top 10–20 cm of soil; and no-till, where seed is directly drilled into the previous crop stubble with minimum changes of the soil structure. Field results from Germany showed that ploughing and twice power-harrowing (15 cm deep) both reduced the *Fusarium* infection of wheat, whereas direct sowing yielded high mycotoxin content in wheat kernels (Schmidt and Nitzsche, 2004). In another study, Schaafsma et al. (2001) reported that from 1996 to 1999 the average DON levels of wheat grain in minimum tillage systems were always higher than in no-till or conventional tillage systems. Irrigation is also a valuable method of reducing plant stress in some growing situations.

Fertilisers have been found to be a contributing factor in the control of *Fusarium* spp. contamination of crops, by altering the rate of residue decomposition, by acting on the rate of plant growth, and by changing the soil structure and its microbial activity (Jouany 2007). The incidence of *Fusarium* infection in wheat, barley and triticale was significantly enhanced by increasing N from 70 to 170 kg ha^{-1} (Martin et al. 1991). In addition, Blandino et al. (2008) reported that the higher fertilization dose (400 kg ha^{-1}) led to a significant increase in ZEA and OTA in 2002 and DON in 2004, compared with plots with lower N fertilization doses or without any N fertiliser application.

13.2.1.4 Use of Chemical and Biological Agents

Many species of insects can facilitate the entry of mycotoxin-producing fungi to several major crops such as cotton seed, maize, peanuts and tree nuts. The mycotoxins most commonly associated with insect damage are aflatoxin and fumonisin. It is possible for insecticides to be of value in indirectly controlling mycotoxigenic fungi and mycotoxins (Dowd 2003). Additionally, fungicides are known to have an important effect on controlling FHB infection and mycotoxin synthesis. Tebuconazole has been reported as an effective fungicide for selectively controlling *Fusarium* infection of ears and DON formation (Ioos et al. 2005).

Biocontrol using non-toxigenic biocompetitive microorganisms is also a potentially useful technology that may reduce mycotoxin contamination. The application of non-aflatoxigenic strains of *A. flavus*/*A. parasiticus* has clearly been observed to reduce aflatoxin contamination of agricultural products including peanuts (Dorner et al. 2003), maize (Dorner et al. 1999) and cottonseed (Cotty 1994) mainly through competition for substrate and through the production of inhibitory metabolites. It has been also reported that some microorganisms have the ability to decrease the growth of Fusaria as well as DON production (Palazzini et al. 2007).

13.2.2 Harvest Management

Timing of harvesting is one of the important factors in reducing mycotoxin contamination. Field crops should be harvested as soon as the crop is fully grown and the crop cycle is completed (Lopez-Garcia et al. 1999). For example, higher levels of aflatoxin contamination have been reported in over-mature or unmature groundnuts. Secondly, during the harvesting process it is important that every effort is made to avoid physical damage to the agricultural commodities (Heathcote and Hibbert 1978), with crops which have been physically damaged being more susceptible to fungal growth.

13.2.3 Post-Harvest Management

Post-harvest strategies include improving of drying and storage conditions, together with the use of chemical and natural agents, and irradiation.

13.2.3.1 Improving of Drying and Storage Conditions

Storage, whether on the farm, at the manufacturing premises or in the grocery store, is the most critical post-harvest phase in the food handling. An inappropriate storage facility, improper packaging and/or the state of the foods can cause mycotoxin contamination during storage (Park et al. 1999). The water availability, which may be expressed as the moisture content, is one of the most important factors in the prevention of fungal growth and mycotoxin production (Abramson 1998). It has been observed that grain stored at a moisture content equivalent to $0.70a_w$ (<14.5% moisture by weight) or less will not be subject to spoilage and mycotoxin formation (Aldred and Magan 2004). The moisture content of wheat at this relative humidity is about 13%, but it is lower for seeds containing more oil, approximately 7% and 10% for peanuts and cottonseeds respectively (Heathcote and Hibbert 1978). As well as humidity, the temperature influences fungal contamination during storage.

In addition, modified atmospheres or alternative gases such as carbon dioxide (Cairns-Fuller et al. 2005), nitrogen, carbon monoxide and sulphur dioxide could potentially be employed for the protection of cereal grain from fungal spoilage and mycotoxin contamination during the post-harvest period.

13.2.3.2 Use of Chemical and Natural Agents

Since fungal infection of grain, nuts and fruit is often preceded by physical damage caused by insect invasion, much effort has been expended on the potential of insecticides to reduce infestation, infection and, therefore, mycotoxin contamination from "storage" fungi such as the Aspergilli. It has been reported that insecticides including dichlorvos, landrin, malathion and diazinon are very effective in the inhibition of AFB_1 production by *A. parasiticus* (D'Mello et al. 1998).

Natural phytochemicals may be an alternative to synthetic chemicals for controlling fungal growth and mycotoxin production in stored products. Nesci et al. (2007) showed that the natural phytochemicals trans-cinnamic acid and ferulic acid could be considered as effective fungitoxicants for *A. flavus* and *A. parasiticus* in maize in the a_w range 0.99–0.93. Similarly, application of phenolic compounds (vanillic and caffeic acid) to maize during storage has been reported to significantly reduce FB_1 and AFB_1 production by *Fusarium* and *Aspergillus* isolates respectively (Samapundo et al. 2007).

A more promising approach, and a possible alternative to fungicide treatment in the prevention of mycotoxin formation, particularly post-harvest, is the potential use of antagonistic bacteria, fungi and yeast. Up to now, many studies have reported on the antifungal properties of various microorganisms, but most are conducted in laboratory conditions. On the other hand, there are many technological and economic hurdles associated with the application of natural products from bacteria, yeast and fungi, and essential oils from plants.

13.2.3.3 Irradiation

Recent studies showed that radiation technology can be used for the elimination of toxigenic mould growth and subsequent mycotoxin formation. It must be noted that fungal strain, condition of storage, humidity, inoculum size, irradiation dose levels, storage periods and composition of the substrate play a great role in mould growth and subsequent mycotoxin formation (Aziz et al. 2002).

13.3 Decontamination/Detoxification of Mycotoxins

When contamination can not be prevented at pre-harvest or during the post-harvest stage, decontamination/detoxification procedures play an important role in helping prevent exposure to the toxic and carcinogenic effect of mycotoxins through the physical separation and physical, chemical and biological inactivation and/or removal of the toxin (Kabak et al. 2006). Any detoxification procedure to reduce the toxic and economic impact of mycotoxins needs the following basic criteria (Jemmali 1979):

- It must destroy, inactivate or remove the mycotoxins in foods and feeds.
- It must not produce or leave toxic and/or carcinogenic residues in the final products.
- It should not alter significantly the nutritional and technological properties of the product.
- It must be capable of destroying fungal spores and mycelia in order to avoiding new toxin forming under favourable conditions.
- It has to be technically and economically feasible.

13.3.1 Removal of Mycotoxins from Contaminated Commodities

Several methods have been reported for removal of mycotoxins from contaminated commodities, including physical separation, extraction with solvents and adsorption.

13.3.1.1 Physical Separation

Since detoxification of mycotoxins by chemical applications is not an acceptable practise in some regions, physical separation of contaminated crops is a very important option for the producer (Kabak et al. 2006).

Cleaning

Cleaning grains removes kernels with extensive mold growth, broken kernels and fine materials, which helps to reduce mycotoxin concentration (Bullerman and Bianchini 2007). Cleaning of the maize can remove 26.6–69.4% of the fumonisins (Sydenham et al. 2004), while a 40–80% reduction in aflatoxin levels has been reported after physical cleaning and separation of mould-damaged kernels and seeds (Park 2002). However, cleaning was not effective in removing DON; only 6–19% reduction was achieved in wheat by cleaning (Abbas et al. 1985).

Sorting and Segregation

Physical separation may lower mycotoxin concentrations by removal of contaminated material in the peanut and other nutmeat industries throughout the world. The principle is based on the identification of damaged kernels in the seed lots, because of the variations in size, shape, colour, and (more often) visible mould growth on the affected kernels (Sinha 1998). Maize screenings or broken maize kernels usually contain fumonisin levels about tenfold higher than intact maize kernels and so separation of screenings, based on size, can significantly reduce fumonisin concentrations (Murphy et al. 1993).

Manual selection is the simplest procedure for the physical removal of contaminated peanuts, pistachios, Brazil nuts and almonds. A combined hand-picking and electronic sorting of contaminated peanuts has been used in the peanut industry in the United States to reduce aflatoxin levels (Sinha 1998). It has been shown that methods involving the segregation of contaminated or mouldy grains by hand-picking and density segregation achieved 70–90% reduction in aflatoxin and fumonisin levels in the grains (Vasanthi and Bhat 1998). Similarly, the removal of damaged grain by density segregation reduces DON and ZEA levels in maize and wheat (Jackson and Bullerman 1999). Fluorescence sorting can be used for screening maize, cottonseed and dried figs, where contamination can be observed by fluorescence following illumination with UV light, with a positive correlation being reported between the observation (bright greenish-yellow fluorescence under longwave (365 nm) UV light) and presence of aflatoxin in these commodities (Scott 1998).

Washing

Washing procedures, using distilled water, resulted in 65–69% reductions of DON and 2–61% reductions of ZEA in barley and maize, whereas using 1 M sodium carbonate solution for the first wash reduced DON by 72–74% and ZEA by 80–87% (Trenholm et al. 1992). This process might be a useful treatment before wet milling and brewing; otherwise, the cost of seed drying would be prohibitive. Such approaches are also cabaple of reducing patulin levels in the final juiced products (Acar et al. 1998).

Milling

Mycotoxin contamination may be redistributed and concentrated in certain mill fractions during the milling process, but there is no step that destroys mycotoxins (Bullerman and Bianchini 2007). It has been shown that bran and germ fraction contained the highest concentrations of fumonisins, while fractions including flaking grits and flour used for food production contained the lowest fumonisin levels (Brera et al. 2004). Wet milling is another basic process, widely used to prepare maize starch for human consumption. During the wet milling of maize, mycotoxins including aflatoxin, ZEA and FB_1 can be dissolved in the step water or distributed to the gluten fibre and germ, while the starch tends to be relatively free of these mycotoxins (Bullerman and Bianchini 2007).

13.3.1.2 Extraction with Solvents

Extraction with a variety of solvents including ethanol, aqueous isopropanol, methanol–water, and acetonitril–water has been shown to remove aflatoxins from contaminated commodities such as cottonseed and peanuts. On the other hand, high cost and problems related to disposal of the toxic extracts restrict its use for large-scale application (Rustom 1997).

13.3.1.3 Adsorption

Two of the most potent adsorbents for removal of mycotoxins are activated carbon (AC) and bentonite. When phosphate-buffered saline (PBS) and wine samples contaminated with 5 ng ml^{-1} OTA were treated with 1 mg ml^{-1} AC, 100% and 87% of the available toxin were adsorbed by the sorbent respectively (Var et al. 2008). In relation to other mycotoxins, AC has been shown to considerably decrease patulin levels in apple juice (Artık et al. 1995). Bentonite, which has a negative charged surface, for its part showed a very poor affinity for OTA (Var et al. 2008), DON and NIV (Avantaggiato et al. 2004), while Diaz et al. (2002) have observed that bentonite is effective in removing AFB_1 in the range 95–98.1%.

Yeasts have been focussed on the removal of mycotoxins in liquids in recent years. Cecchini et al. (2006) demonstrated that the percentage of OTA removal during fermentation was between 46.83% and 52.16% in white wine and between 53.21% and 70.13% in red wine, depending on the yeast strain used. Similarly, Caridi et al. (2006) have reported that the removal of OTA in wines by 20 different *Saccharomyces sensu stricto* strains, using a naturally and spiked OTA-containing grape must (1.58 and 7.63 ng ml^{-1} respectively), after 90 days of fermentation was between 39.9% and 92.1% and between 67.9% and 83.4% respectively.

13.4 Inactivation of Mycotoxins in Contaminated Commodities

13.4.1 Physical Methods

Physical strategies including thermal processing (cooking, boiling, baking, frying, roasting, microwave heating, extrusion) and irradiation have been applied for inactivation of the toxin or to reduce its content in foods and feeds.

13.4.1.1 Thermal Treatment

Most mycotoxins are heat-resistant within the range of conventional food processing temperatures (80–121°C), so little or no reduction in overall toxin levels occurs as a result of normal cooking conditions such as boiling and frying, or even following pasteurization (Scott 1984). The initial level of contamination, type of mycotoxin and its concentration, heating temperature and time, and the degree of heat penetration, as well as the moisture content, pH and ionic strength of food, among other factors, play a significant role in the achievement of toxin degradation (Samarajeewa et al. 1990; Rustom 1997). Effects of various thermal treatments on mycotoxin degradation are summarised in Table 13.1.

13.4.1.2 Irradiation

Radiation involving X-rays, γ-rays, electron-beam, UV radiation and solar energy has been shown to destroy mycotoxins. An earlier study demonstrated that when artificially and naturally contaminated groundnuts were subjected to irradiation at 2 kGy, the levels of aflatoxin decreased from 14.4 and 6.32 $\mu g\ g^{-1}$ to 2.82 and 1.67 $\mu g\ g^{-1}$ respectively (Patel et al. 1989). Additionally, treatment of peanut seeds with γ-irradiation (15–30 kGy) destroyed 55–74% of AFB_1 (Prado et al. 2003). In another study by Refai et al. (2003), at an irradiation dose level of 5 kGy all basterma samples and its components were free from aflatoxins.

13.4.2 Chemical Methods

A variety of chemicals, including acids, bases, oxidizing reagents, reducing agents, chlorinating agents, and miscellaneous reagents have been tested to detoxify mycotoxins. The success of detoxification process by chemical treatments highly depends on the type of food and/or feed. The use of chemicals in combination with physical treatments such as thermal processing for the detoxification of food products contaminated with mycotoxins could be increase the efficacy of mycotoxin degradation.

Table 13.1 Effects of different thermal processing on the mycotoxin content of food

Method	Food	Mycotoxin	Initial level ($\mu g\ g^{-1}$)	Reduction (%)	Reference
Heating: 190°C, 60 min	Maize meal	F ($B_1 + B_2$)	–	60–80	Scott and Lawrance (1994)
Heating: 220°C, 25 min	Maize meal	F ($B_1 + B_2$)	–	100	Scott and Lawrance (1994)
Heating: 90–100°C, 20 min	Apple juice	Patulin	0.22	19–26	Kadakal and Nas (2003)
Heating: 60°C, 120 min	Milk	CPA	1	9–17	Prasongsidh et al. (1998)
Heating: 80°C, 120 min	Milk	CPA	1	20–34	Prasongsidh et al. (1998)
Heating: 100°C, 120 min	Milk	CPA	1	49–50	Prasongsidh et al. (1998)
Pasteurization: 62°C, 30 min	Milk	F ($B_1 + B_2$)	0.05	No reductions	Maragos and Richard (1994)
Baking: 175–200°C, 20 min	Maize muffins	FB_1	5	16–28	Jackson et al. (1997)
Baking: 232°C, 20 min	Maize muffins	FB_1	5	48	Castelo et al. (1998b)
Baking: 204°C, 20 min	Maize muffins	FB_1	5	No reductions	Castelo et al. (1998b)
Baking: 204°C, 20 min	Maize muffins	MON	5	42	Pineda-Valdes et al. (2003)
Frying: 190°C, 15 min	Tortilla chips	FB_1	–	67	Jackson et al. (1997)
Frying: 169°C, 15 min	Flour	DON	0.26	>66	Samar et al. (2007)
Frying: 243°C, 1 min	Flour	DON	0.26	38	Samar et al. (2007)
Frying: 169°C, 15 min	Flour	DON	1.2	28	Samar et al. (2007)
Frying: 205°C, 2.5 min	Flour	DON	1.2	21	Samar et al. (2007)
Roasting: 218°C, 15 min	Maize meal	FB_1	5	100	Castelo et al. (1998b)
Roasting: 218°C, 15 min	Maize meal	MON	5	45	Pineda-Valdes et al. (2003)
Roasting: 150°C, 30 min	Pistachio kernels	AFB_1	0.044	66	Yazdanpanah et al. (2005)
Roasting: 150°C, 30 min	Pistachio kernels	AFB_2	0.006	63	Yazdanpanah et al. (2005)
Roasting: 150°C, 30 min	Pistachio kernels	AFB_1	0.213	24	Yazdanpanah et al. (2005)
Roasting: 150°C, 30 min	Pistachio kernels	AFB_2	0.012	33	Yazdanpanah et al. (2005)
Roasting: 260°C, 5 min	Green coffee	OTA	0.002	72	Pérez de Obanos et al. (2005)
Roasting: 260°C, 5 min	Green coffee	OTA	0.001	43	Pérez de Obanos et al. (2005)
Roasting: 200°C, 5.3 min	Coffee	OTA	0.05	83	Suárez-Quiroz et al. (2005)
Roasting: 250°C, 3.3 min	Coffee	OTA	0.05	80	Suárez-Quiroz et al. (2005)
Autoclaving: 121°C, 65 min	Creamed maize	MON	5	10	Pineda-Valdes et al. (2003)
Evaporation: 70°C, 20 min	Apple juice	Patulin	0.22	10	Kadakal and Nas (2003)
Evaporation: 80°C, 20 min	Apple juice	Patulin	0.22	14	Kadakal and Nas (2003)
Microwave roasting: 0.7 kw, 8.5 min	Peanuts	AFB_1	1	50–60	Pluyer et al. (1987)

Microwave roasting: 0.7 kw, 8.5 min	Peanuts	AFG_1	1	32–40	Pluyer et al. (1987)
Microwave roasting: 0.9 kw, 1.5 min	Peanuts	AFB_1	1	31–50	Ozkarslı (2003)
Extrusion: 120–160°C, 2 min	Maize grits	FB_1	5	31–68	Castelo et al. (1998a)
Extrusion: 100°C, 5 min	Maize grits	DON	4	53	Wolf-Hall et al. (1999)
Extrusion: 120–160°C, 2 min	Maize grits	ZEA	4.4	66–83	Ryu et al. (1999)
Extrusion: 180°C, 15% moisture	Maize flour	DON	5	>95	Cazzaniga et al. (2001)
Extrusion: 180°C, 15% moisture	Maize flour	AFB_1	0.05	10–25	Cazzaniga et al. (2001)
Extrusion: 200°C, 17% moisture	Wheat grain	OTA	0.05	42	Scudamore et al. (2004)

13.4.2.1 Acid Treatment

It is clear from the accumulated evidence that treatment of aflatoxins with strong acids destroys the biological activity of AFB_1 and AFG_1 by converting them to the hemiacetal forms AFB_{2a} and AFG_{2a} respectively, due to acid-catalysed addition of water across the double bond in the furan ring (Heathcote and Hibbert 1978). Treatment with HCl (pH 2) has been shown to reduce AFB_1 levels by 19.3% within 24 h (Doyle et al. 1982).

13.4.2.2 Treatment with Bases

Among bases and other chemicals, ammoniation has proved to be an effective method for detoxifying aflatoxin-contaminated agricultural products and animal feeds. The ammoniation process, using either ammonium hydrochloride or gaseous ammonia (NH_3), is equally effective in the detoxification of aflatoxins in maize, and has been shown in some cases to decrease aflatoxin levels by more than 75% (Burgos-Hernández et al. 2002). Ammoniation caused a 79% reduction of FB_1 in contaminated maize (Park et al. 1992). It has also been reported that ammoniation almost completely decomposed OTA in maize, wheat and barley (Scott 1996).

The ammoniation process does not leave toxic metabolites of mycotoxins in feed (Scott 1998), but the relatively long period of aeration and its cost, which can increase the price of the product by 5–20%, could restrict its use in animal feeds (Peraica et al. 2002). In addition, some undesirable effects in the sensory and nutritional quality of the feed, such as brown colour of the treated feed, a decrease in lysine and sulphur-containing amino acids, cannot be overlooked (Piva et al. 1995; Scott 1998).

Some authors have also tested the efficacy of nixtamalization, a traditional alkaline treatment of maize widely used to manufacture tortillas consisting of cooking the maize in boiling water supplemented with calcium hydroxide (Lopez-Garcia et al. 1999). Combination of heat treatment with $NaHCO_3$ and H_2O_2 alone or with $Ca(OH)_2$ reduced fumonisin levels by up to 100% in contaminated maize with 100 mg kg^{-1} FB_1 (Park et al. 1996). Additionally, treatment of naturally contaminated maize (127 μg of aflatoxin per kilogram) with $Ca(OH)_2$ resulted in up to 46% decrease of aflatoxin levels; however, the residual molecules can either be reformed by acidification (Price and Jorgensen 1985).

13.4.2.3 Oxidizing Agents

It is well-known that aflatoxins such as AFB_1, AFG_1 and AFM_1 which possess a terminal double bond in the dihydrofuran ring are more susceptible to attack by ozone (O_3) and other oxidizing agents than AFB_2, AFG_2 and AFM_2, which lack this double bond (McKenzie et al. 1997). Ozone is reported to reduce AFB_1 and AFG_1 levels by 77% and 80% respectively in peanuts after treatment at 75°C for 10 min, while the maximum degradation was 51%, occurring for AFB_2 and AFG_2 in peanuts, regardless of the exposure times (Proctor et al. 2004). In another study, the reductions of AFB_1 in paprika were 80% and 93% after exposures to 33 mg l^{-1} O_3 and 66 mg l^{-1} O_3 for 60 min respectively (Inan et al. 2007). However, limited experiments with other mycotoxins have shown that patulin, CPA, OA, FB_1 and ZEA were effectively degraded after treatment with O_3 at 10% for 15 s (McKenzie et al. 1997).

H_2O_2, one of the oxidizing agents, has been used on a commercial scale to detoxify aflatoxin. Treatment of figs with H_2O_2 at 0.2% caused a 65.5% reduction in AFB_1 levels following 72 h storage (Altuğ et al. 1990). Additionally, citrinin can be completely detoxified by H_2O_2 at 0.05% for 30 min at room temperature, whereas OTA was not detoxified by treatment with 0.05–0.1% H_2O_2 (Fouler et al. 1994). In another study, Abd Alla (1997) revealed that ZEA was degraded by 83.9% when using 10% H_2O_2 at 80°C for 16 h.

13.4.2.4 Reducing Agents

Sodium bisulfite ($NaHSO_3$) has been shown to destroy mycotoxins, primarily AFB_1 in maize (Doyle et al. 1982) and dried figs (Altuğ et al. 1990). Additionally, $NaHSO_3$ solutions were able to reduce DON level (85%) in contaminated maize (4.4 mg kg^{-1}) and form a DON-sulfonate conjugate when the treatment was performed at 80°C for 18 h (Young et al. 1987). Also, sodium metabisulfite at 10 g kg^{-1} is reported to be an effective tool for overcoming the depressing effects of DON on feed-intake in piglets (Dänicke et al. 2005). Alternatively, the reaction of FB_1 with reducing sugars such as D-glucose, D-fructose at 65°C for 48 h can block the primary amino group of FB_1, and seems to prevent FB_1-induced toxicity on cell tissue cultures on rats and swine (Fernández-Surumay et al. 2005).

13.4.2.5 Chlorinating Agents

Chlorine at a gas concentration of 11 mg g^{-1} achieved more than 75% degradation of AFB_1 in spiked copra meal (Samarajeewa et al. 1991), while about 90% destruction of 100 μg AFB_1 was observed after chlorinating, within 10 min (Sen et al. 1988). In another study, the trichothecene mycotoxins verrucarin A and roridin A (1 or 2 mg l^{-1}) were completely inactivated after treatment with 1,000 mg l^{-1} chlorine dioxide for 2 h (Wilson et al. 2005).

13.4.2.6 Other Chemicals

Other chemicals, including formaldehyde, aqueous ethanol, sodium chloride and various sulphur-containing compounds such as sulphur dioxide, have been reported to be effective in destroying several mycotoxins. On the other hand, yet no adequate research has been conducted to examine the efficacy and applicability of these treatments.

Even though chemicals have been shown to be effective in destroying various mycotoxins, they do not fulfil FAO requirements, especially those concerning the safety of reaction products and safeguarding of the nutritional and organoleptic properties of the treated foods and feeds, so their use is restricted (Galvano et al. 2001). It should be also noted that chemical treatment is not allowed within the EC for commodities destined for human consumption. This has led to the search for alternative strategies such as biological agents.

13.4.3 Biological Methods

An alternative approach to removing the toxic and carcinogenic potential of mycotoxins is the biological detoxification, intended as enzymatic degradation or modifying of toxins that leads to less toxic products. Studies in this area have been dramatically increased with the recent advances in molecular biology, genetic engineering and microbial genomics, coupled with the discovery of the catabolic capabilities of microbial populations. Detoxification of mycotoxins by microorganisms is reviewed extensively by Bata and Lásztity (1999) and Karlovsky (1999). Many species of bacteria have been reported to degrade mycotoxins. Earlier work by Ciegler et al. (1966) identified *Flavobacterium aurantiacum* NRRL B-184, which could irreversibly remove AFB_1 from a variety of food products including milk, oil, peanut butter, peanuts and maize without leaving toxic by-products. On the other hand, the bright orange pigmentation associated with *F. aurantiacum* restricts its use in food and feed fermentations (Line et al. 1994). Apart from *F. aurantiacum*, a variety of lactic acid bacteria originating from fermented products have been reported to inhibit mutagenic activity of AFB_1 (Park and Rhee 2001).

There have been many reports on the fate of mycotoxins during the fermentation process. Earlier work demonstrated that more than 99% of patulin (50 $\mu g\ l^{-1}$) can be removed during alcoholic fermentation of apple juice, while only 10% decrease was observed in the control sample (Stinson et al. 1978). Later, three commercial cider strains of *S. cerevisiae* degraded patulin during active fermentative growth (Moss and Long 2002). With respect to other mycotoxins, fermentation by *S. cerevisiae* of wort containing ZEA resulted in conversion of 69% of the toxin to β-zearalenol and 8.1% to α-zearalenol (Scott et al. 1992). Similarly, cultures of *Candida tropicalis*, *Torulaspora delbrucki*, *Zygosaccharomyces rouxii*, and seven *Saccharomyces* strains were able to convert ZEA to α- and β-zearalenol (Böswald et al. 1995). In another study, OTA, FB_1 and FB_2 at the levels of 0.19, 0.95 and 0.95 $\mu g\ ml^{-1}$ respectively were degraded in the range of 87–91% by three strains of *S. cerevisiae* during fermentation of wort at 25°C for 8 days (Scott et al. 1995). Additionally, some losses (<40%) of OTA occurred during fermentation (Baxter et al. 2001), while alcoholic fermentation of malt by *S. cerevisiae* resulted in an average of 53% decrease in the initial contamination level of DON and T-2 toxin (Garda et al. 2005).

13.5 Inhibition of Mycotoxin Absorption in the Gastrointestinal Tract

The most recent approach to reducing the exposure to mycotoxins has been the use of binding agents added to the diet that sequester the mycotoxin in the gastrointestinal tract, thus reducing their bioavailability. Previously, various adsorbents have been tested, such as AC, hydrated sodium calcium aluminosilicate (HSCAS), zeolite, bentonite and certain clays, and showed good potential for use in animal feed to help overcome mycotoxicosis (Galvano et al. 2001). Results of the *in vivo* screening test to determine the protective effect of adsorbents against mycotoxicosis in various test animals are listed in Table 13.2. The molecular size and the physicochemical properties of mycotoxins (Castellari et al. 2001), as well as the physical structure of the adsorbent including the total charge and charge distribution, the size of the pores, together with the surface area, among other factors; play a significant role in the achievement of mycotoxin-binding by adsorbent materials (Huwig et al. 2001).

13.5.1 Activated Carbon

Activated carbon, which is a nonsoluble powder formed by pyrolysis of different kinds of organic materials, shows different adsorbing properties depending on its origin (Ramos et al. 1996). Concerning the efficacy of AC against mycotoxins in the simulated gastrointestinal model, Avantaggiato et al. (2003) demonstrated that 2% AC was able to reduce zearalenone intestinal absorption from 32% of zearalenone (247 μg) intake to 5%. AC has also shown to be effective in reducing AFB_1 carry-over from feed to milk as AFM_1 in lactating animals but lower than that effect in vitro. Galvano et al. (1996) revealed decreases in the AFM_1 excreted in milk ranging from 22% to 45% resulting from addition to feed containing 11.28 μg kg^{-1} AFB_1 of 2% of the AC. However, the high cost of AC and the tendency of ACs to blacken the environment, the animals and the feed restrict its use to feeds. Also, it is not known whether its long-term use might lead to essential nutrient deficiency (i.e., vitamins and minerals) in domestic animals (Galvano et al. 2001).

13.5.2 Hydrated Sodium Calcium Aluminosilicate

Among the aluminosilicates, HSCAS, a phyllosilicate derived from natural zeolite, is the most extensively studied adsorbent due to its high aflatoxin-binding

Table 13.2 Effect of adsorbent materials against mycotoxicosis in several animals

Adsorbent (%)	Mycotoxin (mg kg^{-1})	Results observed	Reference
AC (0.5)	AFB_1 (0.1)	Elimination of histopathologic lesions in the liver of mink	Bonna et al. (1991)
AC (0.5)	AFB_1 (0.75)	Reduced excretion of AFM_1 in turkey poults, no protective effects against aflatoxicosis	Edrington et al. (1996)
AC (2)	AFB_1 (0.01)	Carry-over reduction of AFM_1 in milk of lactating cows diminished by 50%	Galvano et al. (1996)
AC (1)	AFB_1 (0.6)	76% reduction in AFM_1 concentration in milk of lactating goats	Nageswara Rao and Chopra (2001)
AC (0.25)	AFB_1 (1.5)	No effect on AFM_1 transmission to milk	Diaz et al. (2004)
HSCAS (0.5)	AFB_1 (7.5)	Decrease of growth inhibitory effects in broiler chicks	Phillips et al. (1988)
HSCAS (0.5)	AFB_1 (0.1)	Elimination of histopathologic lesions in the liver of mink	Bonna et al. (1991)
HSCAS (0.5)	AFB_1 (3.5)	Growth inhibitory effects on broiler chicks diminished by 57%	Huff et al. (1992)
HSCAS (0.5)	AFB_1 (0.25–1)	Significant reduction of urinary excretion of AFM_1 in rats	Sarr et al. (1995)
HSCAS (0.5)	AFB_1 (0.75)	Significant decrease of urinary excretion of AFM_1 in turkey poultry	Edrington et al. (1996)
HSCAS (2)	AFB_1 (0.01)	Carry-over reduction of AFM_1 in milk of lactating cows diminished by 36%	Galvano et al. (1996)
HSCAS (0.5)	AFB_1 (2)	Reduced excretion of AFM_1 in urine of rats	Abdel-Wahhab and Nada (1998)
HSCAS (0.5)	AFB_1 (2)	Significant decrease of urinary excretion of AFM_1 in rats	Mayura et al. (1998)
HSCAS (0.5)	AF (2.5)	Decrease of bioavailability of AF in the gastrointestinal tract of rats; protection against aflatoxicosis in animals	Abdel-Wahhab et al. (2002)
HSCAS (0.5)	T-2 toxin (3.5)	No significant effect (broilers)	Kubena et al. (1990)
HSCAS (0.5)	OTA (2)	No significant effect (broilers)	Huff et al. (1992)
HSCAS (1)	CPA (45)	No effect on CPA toxicity in the broilers	Dwyer et al. (1997)
Zeolite (0.5)	AFB_1 (2)	No significant effect (rats)	Mayura et al. (1998)
Clinoptilolite (1.5–2.5)	AF (2.5)	Protection against aflatoxicosis (broilers)	Oğuz et al. (2000)
Clinoptilolite (1)	CPA (45)	No effect on CPA toxicity in the broilers	Dwyer et al. (1997)
EG (0.05)	AFB_1 (1.5)	59% reduction in AFM_1 transmission to milk	Diaz et al. (2004)
Bentonite (2)	AFB_1 (0.02)	80% reduction of AFB_1 level in liver and kidney of fish	Ellis et al. (2000)

Bentonite (1)	AFB_1 (0.6)	66% reduction in AFM_1 level in milk of lactating goats	Nageswara Rao and Chopra (2001)
Ca-bentonite (1.2)	AFB_1 (1.5)	31% reduction in AFM_1 transmission to milk	Diaz et al. (2004)
Na-bentonite (1.2)	AFB_1 (1.5)	50–61% reduction in AFM_1 transmission to milk	Diaz et al. (2004)
Bentonite (5–10)	T-2 toxin (3)	Prevention of T-2 toxicosis in rats	Carson and Smith (1983)
Montmorillonite (0.5)	AF (2.5)	Reduction of bioavailability of AF in the gastrointestinal tract of rats; protection against aflatoxicosis in rats	Abdel-Wahhab et al. (2002)
EG (0.1)	AF (2)	Diminished the adverse effect of AF in broilers, improved body weight gains (59%)	Basmacoglu et al. (2005)

capacities (Ramos et al. 1996). An extensive review on the prevention of aflatoxicosis in farm animals including chickens, cows, goats, lambs, minks, pigs and turkey poultry by HSCAS was presented by Ramos and Hernández (1997). While HSCAS is effective in reducing the toxicity of aflatoxin, its efficacy against other mycotoxins including OTA (Santin et al. 2002), T-2 toxin (Kubena et al. 1990) and ZEA (Bursian et al. 1992) is very low. However, Abbés et al. (2006) showed a clear protective effect of the HSCAS against haematological, biochemical and pathological (hepatic and renal tissues) changes induced by ZEA.

13.5.3 Zeolites

There are conflicting reports in the literature concerning the adsorption ability of zeolites. The use of a zeolite, clinoptilolite (1.5%), was shown to reduce adverse effects of aflatoxins (2.5 mg kg^{-1}) on performance of broiler chickens (Oğuz and Kurtoğlu 2000), whereas the toxicity of high dietary concentrations of aflatoxin (3.5 mg kg^{-1}) was not significantly diminished by the utilization of clinoptilolite (0.5%) (Harvey et al. 1993). As regards mycotoxins other than aflatoxins, synthetic anion exchange zeolite was found to be effective in protecting rats against ZEA (0.25 μg kg^{-1}) when fed at 5% of the diet (Smith 1980).

13.5.4 Bentonite

Bentonite is reported to be effective in reducing toxic effects of aflatoxin. Lindeman et al. (1993) showed that the addition of sodium bentonite (0.5%) to the diet of pigs containing 0.84 mg kg^{-1} AFB_1 in a 49-day trial resulted in an improvement of average daily gain and feed intake. In another work, Santurio et al. (1999) concluded that the absorption of aflatoxins (3 mg kg^{-1}) from the intestinal lumen of the chicken was reduced by 0.5% of sodium bentonite, thus preventing its deleterious effect on broiler productivity. It has been also demonstrated that the addition of bentonite to the feed of growing fish contaminated with 20 μg kg^{-1} in a 7-day test period yielded in an 80% reduction of AFB_1 in liver and kidney tissues, compared to control fish not fed bentonite (Ellis et al. 2000). Regarding mycotoxins other than aflatoxins, Carson and Smith (1983) concluded that bentonite feeding (5–10%) prevents T-2 toxicosis by reducing intestinal absorption and increasing fecal excretion of the toxin, but its concentration of incorporation in the diet must be relatively high (about 10 times more) than the efficient level of aflatoxins.

13.5.5 Cholestyramine

Cholestyramine, a synthetic anion exchange resin, was reported to have high affinity for OTA (Kerkadi et al. 1998), ZEA (Ramos et al. 1996) and fumonisins (Solfrizzo et al. 2001), but not for DON and NIV (Avantaggiato et al. 2004). However, the high cost of cholestyramine would make its commercial use economically prohibitive (Galvano et al. 2001).

13.5.6 Yeast and Yeast-Derived Products

Live yeast (*S. cerevisiae*) has been used as a general performance promoter in the poultry industry from the early 1990s, and has recently been found to have beneficial effects on weight gain and immunoresponse in broilers exposed to aflatoxin (Stanley et al. 1993). A recent study by Madrigal-Santillán et al. (2006) found that *S. cerevisiae* improved weight gain and reduced genotoxicity produced by AFB_1 in mouse during a 6-week assay, while dietary *Trichosporon mycotoxinivorans* completely blocked OTA-induced immuno suppression in broiler chicks (Politis et al. 2005). The adsorption of mycotoxins could be enhanced with the use of yeast cell wall components instead of whole cells (Huwig et al. 2001). The addition of a yeast component, esterified glucomannan (EG) (0.1%) to an aflatoxin-containing diet (2 mg kg^{-1}) significantly improved the adverse effects of aflatoxins on haematological parameters, total protein, albumin values and aspartate-aminotransferase activity in broiler chickens. EG also partially improved body weight gains (59%) and the other biochemical parameters (Basmacoglu et al. 2005), and decreased the number and severity of pathological lesions influenced by aflatoxin treatment (Karaman et al. 2005). EG was also shown to have a protective effect against OTA, ZEA and T-2 toxin (Raju and Devegowda 2000; Aravind et al. 2003), but not against DON (Dänicke et al. 2007).

13.5.7 Probiotic Bacteria

Many studies are available describing the removal of various mycotoxins from contaminated PBS by probiotic bacteria (Kabak and Var 2004), but so far reduction under gastrointestinal conditions of the bioavailability of mycotoxins by probiotic bacteria has not been fully investigated. Limited *in vivo* experiments with lactic acid bacteria have shown that addition of some dairy strains of lactic acid bacteria to the diet resulted in reduced mycotoxin toxicities, indicating possible stability of the bacterial complex through the gastrointestinal tract. El-Nezami et al. (2000) demonstrated a 74% reduction in the uptake of AFB_1 by the intestinal tissue, in the presence of *Lb. rhamnosus* strain GG, taking place within 60 min.

13.6 Conclusions

The preferred strategy for reducing mycotoxin levels in foods and feeds is the prevention of mycotoxin contamination during the pre-harvest or post-harvest period. It is clear that a combination of the development of crop species resistance to toxigenic fungi and biocontrols using nontoxigenic biocompetitive agents may yield one of the most effective strategies for prevention of mycotoxin formation. The importance of drying and moisture control during storage is generally well-understood by the industry, in terms of the importance of prevention of fungal contamination.

Failure to prevent fungal attack and toxin production in the field or in storage will inevitably lead to a health risk to the consumer and to economic loss. It is therefore important to prevent and/or reduce human and animal exposure by developing practical, safe and effective methods for detoxification of mycotoxin-containing foods and feeds.

Even though certain treatments have been shown to reduce the levels of specific mycotoxins, it is clear that no single currently available method exists which would destroy all mycotoxins in food and animal feeds without leaving their metabolites and without changing the nutritional value of food and feed. A range of chemical compounds can destroy mycotoxins, but most are impractical or potentially unsafe to use because of the formation of toxic metabolites and/or the effect on nutritional quality of the product. It is also important to emphasise that chemical treatment is not allowed within the EC for commodities destined for human consumption. In addition to physical and chemical methods, some bacteria, yeast and fungi can remove the toxic and carcinogenic potential of mycotoxins.

Among the strategies, the use of mycotoxin-binding adsorbents including AC, HSCAS, zeolite, bentonite, cholestyramine, other clays, yeast and yeast-derived products is the most-applied physical method of protecting animals against the harmful effects of mycotoxin-contaminated feed. Additionally, probiotic bacteria have been shown to bind aflatoxins and other mycotoxins in aqueous solution. On the other hand, further *in vivo* studies are needed to determine the power of more probiotic bacteria to bind mycotoxin and thus reduce exposure of humans and/or animals to these highly toxic contaminant.

References

Abbas HK, Mirocha CJ, Pawlosky RJ, Pusch DJ (1985) Effect of cleaning, milling, and baking on deoxynivalenol in wheat. Appl Environ Microbiol 50:482–486

Abbés S, Ouanes Z, Salah-Abbés J, Houas Z, Oueslati R, Bacha H, Othman O (2006) The protective effect of hydrated sodium calcium aluminosilicate against haematological, biochemical and pathological changes induced by zearalenone in mice. Toxicon 47:567–574

Abd Alla ES (1997) Zearalenone: incidence, toxigenic fungi and chemical decontamination in Egyptian cereals. Nahrung 41:362–365

Abdel-Wahhab MA, Nada SA (1998) Effect of phyllosilicate clay and zeolite on the metabolic profile of aflatoxin B_1 in Sprague-Dawley rats. Toxicol Lett 95:153–154

Abdel-Wahhab MA, Nada SA, Khalil FA (2002) Physiological and toxicological responses in rats fed aflatoxin-contaminated diet with or without sorbent materials. Anim Feed Sci Technol 97:209–219

Abramson D (1998) Mycotoxin formation and environmental factors. In: Sinha KK, Bhatnagar D (eds) Mycotoxins in agriculture and food safety. Marcel Dekker, New York, pp 255–277

Acar J, Gökmen V, Taydaş EE (1998) The effects of processing technology on the patulin content of juice during commercial apple juice concentrate production. Z Lebensm Unters Forsch A 207:328–331

Aldred D, Magan N (2004) Prevention strategies for trichothecenes. Toxicol Lett 153:165–171

Altuğ T, Yousef AE, Marth EH (1990) Degradation of aflatoxin B_1 in dried figs by sodium bisulfite with or without heat, ultraviolet energy or hydrogen peroxide. J Food Prot 53:581–582

Aravind KL, Patil VS, Devegowda G, Umakantha B, Ganpule SP (2003) Efficacy of esterified glucomannan to counteract mycotoxicosis in naturally contaminated feed on performance and serum biochemical, haematological parameters in boilers. Poultry Sci 82:571–576

Artk N, Cemeroğlu B, Aydar G, Sağlam N (1995) Elma suyu konsantresinde aktif kömür kullanm üzerinde araştrmalar. Tr J Agric Forestry 19:259–265

Avantaggiato G, Havenaar R, Visconti A (2003) Assessing the zearalenone-binding activity of adsorbent materials during passage through a dynamic in vitro gastrointestinal model. Food Chem Toxicol 41:1283–1290

Avantaggiato G, Havenaar R, Visconti A (2004) Evaluation of the intestinal absorption of deoxynivalenol and nivalenol by an in vitro gastrointestinal model, and the binding efficacy of activated carbon and other adsorbent materials. Food Chem Toxicol 42:817–824

Aziz NH, El-Zeany SA, Moussa LAA (2002) Influence of g-irradiation and maize lipids on the production of aflatoxin B_1 by *Aspergillus flavus*. Nahrung 5:327–331

Basmacıoglu H, Oguz H, Ergul M, Col R, Birdane YO (2005) Effect of dietary esterified glucomannan on performance, serum biochemistry and haematology in broilers exposed to aflatoxin. Czech J Anim Sci 50:31–39

Bata A, Lásztity R (1999) Detoxification of mycotoxin-contaminated food and feed by microorganisms. Trends Food Sci Technol 10:223–228

Baxter ED, Slaiding IR, Kelly B (2001) Behavior of ochratoxin A in brewing. J Am Soc Brew Chem 59:98–100

Blandino M, Reyneri A, Vanara F (2008) Influence of nitrogen fertilization on mycotoxin contamination of maize kernels. Crop Prot 27:222–230

Bonna RJ, Aulerich RJ, Bursian SJ, Poppenga RH, Braselton WE, Watson GE (1991) Efficacy of hydrated sodium aluminosilicate and activated charcoal in reducing the toxicity of dietary aflatoxin to milk. Arch Environ Contam Toxicol 20:441–447

Boutrif E (1998) Prevention of aflatoxin in pistachios. Food Nutr Agric 21:32–39

Böswald C, Engelhardt G, Vogel H, Walnöfer PR (1995) Metabolism of the *Fusarium* mycotoxins zearalenone and deoxynivalenol by yeast strains of technological relevance. Nat Toxins 3:138–144

Brera C, Debegnach F, Grossi S, Miraglia M (2004) Effect of industrial processing on the distribution of fumonisin B_1 in dry milling corn fractions. J Food Prot 67:1261–1266

Bullerman LB, Bianchini A (2007) Stability of mycotoxins during food processing. Int J Food Microbiol 119:140–146

Burgos-Hernández A, Price RL, Jorgensen-Kornman K, López-García R, Njapau H, Park DL (2002) Decontamination of aflatoxin B_1-contaminated corn by ammonium persulphate during fermentation. J Sci Food Agric 82:546–552

Bursian SJ, Aulerich RJ, Cameron JK, Ames NK, Steficek BA (1992) Efficacy of hydrated sodium calcium aluminosilicate in reducing the toxicity of dietary zearalenone to mink. J Appl Toxicol 12:85–90

Cairns-Fuller V, Aldred D, Magan N (2005) Water, temperature and gas composition interactions affect growth and ochratoxin A production by isolates of *Penicillium verrucosum* on wheat grain. J Appl Microbiol 99:1215–1221

Caridi A, Galvano F, Tafuri A, Ritieni A (2006) Ochratoxin A removal during winemaking. Enzyme Microb Technol 40:122–126

Carson MS, Smith TK (1983) Role of bentonite in prevention of T-2 toxicosis in rats. J Anim Sci 57:1498–1506

Castellari M, Versari A, Fabiani A, Parpinello GP, Galassi S (2001) Removal of ochratoxin A in red wines by means of adsorption treatments with commercial fining agents. J Agric Food Chem 49:3917–3921

Castelo MM, Katta SK, Sumner SS, Hanna MA, Bullerman LB (1998a) Extrusion cooking reduces recoverability of fumonisin B_1 from extruded corn grits. J Food Sci 63:696–698

Castelo MM, Sumner SS, Bullerman LB (1998b) Stability of fumonisins in thermally processed corn products. J Food Prot 61:1030–1033

Cazzaniga D, Basílico JC, González Torres RL, De Greff DM (2001) Mycotoxins inactivation by extrusion cooking of corn flour. Lett Appl Microbiol 33:144–147

Cecchini F, Morassut M, Moruno EG, Di Stefano R (2006) Influence of yeast strain on ochratoxin A content during fermentation of white and red must. Food Microbiol 23:411–417

Ciegler A, Lillehoj EB, Peterson RE, Hall HH (1966) Microbial detoxification of aflatoxin. Appl Microbiol 14:934–939

Cleveland T, Dowd PF, Desjardins AE, Bhatnagar D, Cotty PJ (2003) United States department of agriculture. Agricultural research service research on pre-harvest prevention of mycotoxins and mycotoxigenic fungi in US crops. Pest Manag Sci 59:629–642

Cotty PJ (1994) Influence of field application of an atoxigenic strain of *Aspergillus flavus* on the populations of *Aspergillus flavus* infecting cotton bolls and on the aflatoxin content of cottonseed. Phytopathology 84:1270–1277

Dahleen LS, Okubara PA, Blechl AE (2001) Transgenic approaches to combat *Fusarium* head blight in wheat and barley. Crop Sci 41:628–637

Dänicke S, Goyarts T, Valenta H (2007) On the specific and unspecific effects of a polymeric glucomannan mycotoxin adsorbent on piglets when fed with uncontaminated or with *Fusarium* toxins contaminated diets. Arch Anim Nutr 61:266–275

Dänicke S, Valenta H, Gareis M, Lucht HW, Reichenbach H (2005) On the effects of a hydrothermal treatment of deoxynivalenol (DON)-contaminated wheat in the presence of sodium metabisulfite ($Na_2S_2O_5$) on DON reduction and on piglet performance. Anim Feed Sci Technol 118:93–108

Diaz DE, Hagler WM Jr, Blackwelder JT, Eve JA, Hopkins BA, Anderson KL, Jones FT, Whitlow LW (2004) Aflatoxin binders II: reduction of aflatoxin M_1 in milk by sequestering agents of cows consuming aflatoxin in feed. Mycopathologia 157(2):233–241

Diaz DE, Hagler WM Jr, Hopkins BA, Whitlow LW (2002) Aflatoxin binders I: in vitro binding assay for aflatoxin B_1 by several potential sequestering agents. Mycopathologia 156:223–226

D'Mello JPF, Macdonald AMC, Postel D, Dijksma WTP, Dujardin A, Placinta CM (1998) Pesticide use and mycotoxin production in *Fusarium* and *Aspergillus* phytopathogens. Eur J Plant Pathol 104:741–751

Dorner JW, Cole RJ, Connick WJ, Daigle DJ, Mcguire MR, Shasha BS (2003) Evaluation of biological control formulations to reduce aflatoxin contamination in peanuts. Biol Control 26:318–324

Dorner JW, Cole RJ, Wicklow DT (1999) Aflatoxin reduction in corn through field application of competitive fungi. J Food Prot 62:650–656

Dowd PF (2003) Insect management to facilitate preharvest mycotoxin management. J Toxicol 22:327–350

Doyle MP, Applebaum RS, Brackett RE, Marth EM (1982) Physical, chemical and biological degradation mycotoxins in foods and agricultural commodities. J Food Prot 45:964–971

Dwyer MR, Kubena LF, Harvey RB, Mayura K, Sarr AB, Buckley S, Bailey RH, Phillips TD (1997) Effects of inorganic adsorbents and cyclopiazonic acid in broiler chickens. Poultry Sci 76:1141–1149

Edrington TS, Sarr AB, Kubena LF, Harvey RB, Phillips TD (1996) Hydrated sodium calcium aluminosilicate (HSCAS), acidic HSCAS, and activated charcoal reduce urinary excretion of AFM_1 in turkey poults. Lack of effect by activated charcoal on aflatoxicosis. Toxicol Lett 89:115–122

Edwards SG (2004) Influence of agricultural practices on fusarium infection of cereals and subsequent contamination of grain by trichothecene mycotoxins. Toxicol Lett 153:29–35

Ellis RW, Clements M, Tibbetts A, Winfree R (2000) Reduction of the bioavailability of 20 μg/kg aflatoxin in trout feed containing clay. Aquaculture 183:179–188

El-Nezami H, Mykkänen H, Kankaanpää P, Salminen S, Ahokas J (2000) Ability of *Lactobacillus* and *Propionibacterium* strains to remove aflatoxin B_1 from the chicken duodenum. J Food Prot 63:549–552

European Commission (EC) report (1999) Opinion on the relationship between the use of plant protection products on food plants and the occurrence of mycotoxins in foods. European Commission SCP/RESI/063, Belgium

Fernández-Surumay G, Osweiler GD, Yaeger MJ, Rottinghaus GE, Hendrich S, Buckley LK, Murphy PA (2005) Fumonisin B-glucose reaction products are less toxic when fed to swine. J Agric Food Chem 53:4264–4271

Food and Agriculture Organization (1996) Basic facts on the world cereal situation. Food outlook, 5/6, Rome

Fouler SG, Triverdi AB, Kitabatake N (1994) Detoxification of citrinin and ochratoxin A by hydrogen peroxide. J AOAC Int 77:631–636

Galvano F, Pietri A, Bertuzzi T, Fusconi G, Galvano M, Piva A, Piva G (1996) Reduction of carryover of aflatoxin from cow feed to milk by addition of activated carbons. J Food Prot 59:551–554

Galvano F, Piva A, Ritieni A, Galvano G (2001) Dietary strategies to counteract the effects of mycotoxins: a review. J Food Prot 64:120–131

Garda J, Macedo RM, Faria R, Bernd L, Dors GC, Badiale-Furlong E (2005) Alcoholic fermentation effects on malt spiked with trichothecenes. Food Control 16:423–428

Harvey RB, Kubena LF, Elissalde MH, Phillips TD (1993) Efficacy of zeolitic ore compounds on the toxicity of aflatoxin growing broiler chickens. Avian Dis 37:67–73

Heathcote JG, Hibbert JR (1978) Aflatoxin chemical and biological aspects. Elsevier Scientific, Amsterdam

Huff WE, Kubena LF, Harvey RB, Phillips TD (1992) Efficacy of hydrated sodium calcium aluminosilicate to reduce the individual and combined toxicity of aflatoxin and ochratoxin A. Poultry Sci 71:64–69

Huwig A, Freimund S, Kappeli O, Dutler H (2001) Mycotoxin detoxification of animal feed by different adsorbents. Toxicol Lett 122:179–188

Inan F, Pala M, Doymaz I (2007) Use of ozone in detoxification of aflatoxin B_1 in red pepper. J Stored Prod Res 43:425–429

Ioos R, Belhadj A, Menez M, Faure A (2005) The effects of fungicides on *Fusarium* spp. and *Microdochium nivale* and their associated trichothecene mycotoxins in French naturally-infected cereal grains. Crop Prot 24:894–902

Jackson LS, Bullerman LB (1999) Effect of processing on *Fusarium* mycotoxins. Adv Exp Med Biol 459:243–261

Jackson LS, Katta SK, Fingerhut DD, De Vries JW, Bullerman LB (1997) Effects of baking and frying on the fumonisin B_1 content of corn-based foods. J Agric Food Chem 45:4800–4805

Jemmali M (1979) Decontamination and detoxification of mycotoxins. Pure Appl Chem 52: 175–181

Jouany JP (2007) Methods for preventing, decontaminating and minimizing the toxicity of mycotoxins in feeds. Anim Feed Sci Technol 137:342–362

Kabak B, Dobson ADW, Var I (2006) Strategies to prevent mycotoxin contamination of food and animal feed: a review. Crit Rev Food Sci Nutr 46:593–619
Kabak B, Var I (2004) Binding of aflatoxin M_1 by *Lactobacillus* and *Bifidobacterium* strains. Milchwissenschaft 59:301–303
Kadakal C, Nas S (2003) Effect of heat treatment and evaporation on patulin and some other properties of apple juice. J Sci Food Agric 83:987–990
Karaman M, Basmacoglu H, Ortatatl M, Oguz H (2005) Evaluation of the detoxifying effect of yeast glucomannan on aflatoxicosis in broilers as assessed by gross examination and histopathology. Brit Poultry Sci 46:394–400
Karlovsky P (1999) Biological detoxification of fungal toxins and its use in plant breeding, feet, and food production. Nat Toxins 7:1–23
Kerkadi A, Barriault C, Tuchweber B, Frohlich AA, Marquardt RR, Bouchard G, Yousef IM (1998) Dietary cholestyramine reduces ochratoxin A-induced nephrotoxicty in the rat by decreasing plasma levels and enhancing faecal excretion of the toxin. J Toxicol Environ Health 53:231–250
Kubena LF, Harvey RB, Huff WE, Corrier DE, Phillips TD, Rottinghaus GE (1990) Efficacy of a hydrated sodium calcium aluminosilicate to reduce the toxicity of aflatoxin and T-2 toxin. Poultry Sci 69:1078–1086
Lacey J, Bateman GL, Mirocha CL (1999) Effects of infection time and moisture on the development of ear blight and deoxynivalenol production by *Fusarium* spp. in wheat. Ann Appl Biol 134:277–283
Langesth W, Rundberget T (1999) The occurrence of HT-2 toxin and other trichothecenes in Norwegian cereals. Mycopathologia 147:157–165
Lindeman MD, Blodgett DJ, Kornegay ET, Schurig GG (1993) Potential ameliorators of aflatoxicosis in weanling/growing swine. J Anim Sci 71:171–178
Line JE, Brackett RE, Wilkinson RE (1994) Evidence for degradation of aflatoxin B_1 by *Flavobacterium aurantiacum*. J Food Prot 57:788–791
Lopez-Garcia R, Park DL, Phillips TD (1999) Integrated mycotoxin management systems. Food Nutr Agric 23:38–47
Madrigal-Santillán E, Madrigal-Bujaidar E, Márquez-Márquez R, Reyes A (2006) Antigenotoxic effect of *Saccharomyces cerevisiae* on the damage produced in mice fed with aflatoxin B_1 contaminated corn. Food Chem Toxicol 44:2058–2063
Maragos CM, Richard JL (1994) Quantitation and stability of fumonisins B_1 and B_2 in milk. J AOAC Int 77:1162–1167
Martin RA, MacLeod JA, Caldwell C (1991) Influences of production inputs on incidence of infection by *Fusarium* species on cereal seed. Plant Dis 75:784–788
Mayura K, Abdel-Wahhab MA, McKenzie KS, Sarr AB, Edwards JF, Naguib K, Phillips TD (1998) Prevention of maternal and developmental toxicity in rats via dietary inclusion of common aflatoxin sorbents: potential for hidden risks. Toxicol Sci 41:175–182
McKenzie KS, Sarr AB, Mayura K, Bailey RH, Miller DR, Rogers TD, Norred WP, Voss KA, Plattner RD, Kubena LF, Phillips TD (1997) Oxidative degradation and detoxification using a novel source of ozone. Food Chem Toxicol 35:807–820
Meidaner T, Reinbrecht C, Lauber U, Schollonberger M, Geiger H (2001) Effects of genotype and genotype-environment interaction on deoxynivalenol accumulation and resistance to *Fusarium* head blight in rye, triticale, and wheat. Plant Breeding 120:97–105
Moss MO (1992) Secondary metabolism and food intoxication moulds. J Appl Bacteriol 73:80–88
Moss MO, Long MT (2002) Fate of patulin in the presence of yeast *Saccharomyces cerevisiae*. Food Addit Contam 19:387–399
Murphy PA, Rice LG, Ross PF (1993) Fumonisin B_1, B_2, and B_3 content of Iowa, Wisconsin, and Illinois corn and corn screenings. J Agric Food Chem 41:263–266
Nageswara Rao SB, Chopra RC (2001) Influence of sodium bentonite and activated charcoal on aflatoxin M_1 excretion in milk of goats. Small Ruminant Res 41:203–213
Nesci A, Gsponer N, Etcheverry M (2007) Natural maize phenolic acids for control of aflatoxigenic fungi on maize. J Food Sci 72:180–185

Oğuz H, Keçeci T, Birdane YO, Onder F, Kurtoğlu V (2000) Effect of clinoptilolite on serum biochemical and haematoligical characters of broiler chickens during aflatoxicosis. Res Vet Sci 69:89–93
Oğuz H, Kurtoğlu V (2000) Effect of clinoptilolite on performance of broiler chickens during experimental aflatoxicosis. Brit Poultry Sci 41:512–517
Ozkarslı M (2003) Yer fıstıklarında aflatoksin B_1 üzerine geleneksel ve mikrodalgada kavurmanın etkisi. Cukurova Universitesi Fen Bilimleri Enstitüsü Yüksek Lisans Tezi, Adana, Turkey
Palazzini JM, Ramirez ML, Torres AM, Chulze SN (2007) Potential biocontrol agents for *Fusarium* head blight and deoxynivalenol production in wheat. Crop Prot 26:1702–1710
Park DL (2002) Effect of processing on aflatoxin. Adv Exp Med Biol 504:173–179
Park DL, Lopez-Garcia R, Trujillo-Preciado S, Price RL (1996) Reduction of risks associated with fumonisin contamination in corn. In: Jackson LS, Dewries JW, Bullerman LB (eds) Fumonisins in food. Plenum, New York, pp 335–344
Park DL, Njapau H, Boutrif E (1999) Minimizing risks posed by mycotoxins utulizing the HACCP concept. Food Nutr Agric 23:49–56
Park DL, Rua SM, Mirocha CJ Jr, Abd-Alla E-SAM, Weng CJ (1992) Mutagenic potentials of fumonisin contaminated corn following ammonia decontamination procedure. Mycopathologia 117:105–108
Park H-D, Rhee C-H (2001) Antimutagenic activity of *Lactobacillus plantarum* KLAB21 isolated from *kimchi* Korean fermented vegetables. Biotechnol Lett 23:1583–1589
Patel UD, Govindarajan P, Dave PJ (1989) Inactivation of aflatoxin B_1 by using the synergistic effect of hydrogen peroxide and gamma radiation. Appl Environ Microbiol 55:465–467
Peraica M, Domijan A-M, Jurjević Ž, Cvjetković B (2002) Prevention of exposure to mycotoxins from food and feed. Arh Hig Rada Toxikol 53:229–237
Pérez de Obanos A, González Peñas E, López de Cerain A (2005) Influence of roasting and brew preparation on the ochratoxin A content in coffee infusion. Food Addit Contam 22:463–471
Phillips TD, Kubena LF, Harvey RB, Taylor DR, Heidelbaugh ND (1988) Hydrated sodium calcium aluminosilicate. A high affinity sorbent for aflatoxin. Poultry Sci 67:243–247
Pineda-Valdes G, Ryu D, Hanna MA, Bullerman LB (2003) Reduction of moniliformin in corn by heat processing. J Food Sci 68:1031–1035
Pitt JI, Dyer SK, McCamon S (1991) Systematic invasion of developing peanut plants by *Aspergillus flavus*. Lett Appl Microbiol 13:16–20
Piva G, Galvano F, Pietri A, Piva A (1995) Detoxification methods of aflatoxins. A review. Nutr Res 15:767–776
Pluyer HR, Ahmed EM, Wei CI (1987) Destructions of aflatoxins on peanuts by oven- and microwave-roasting. J Food Prot 50:504–508
Politis I, Fegeros K, Nitsch S, Schatzmayr G, Kantas D (2005) Use of *Trichosporon mycotoxinivorans* to suppress the effects of ochratoxicosis on the immuno system of broiler chicks. Brit Poultry Sci 46:58–65
Prado G, Pinheiro de Carvalho E, Oliveira MS (2003) Effect of gamma irradiation on the inactivation of aflatoxin B_1 and fungal flora in peanuts. Braz J Microbiol 34:138–140
Prasongsidh BC, Kailasapathy K, Skurray GR, Breyden WL (1998) Kinetic study of cyclopiazonic acid during the heat-processing of milk. Food Chem 62:467–472
Price RL, Jorgensen KV (1985) Effects of processing on aflatoxin levels and on mutagenic potential of tortillas made from naturally contaminated corn. J Food Sci 50:347–349
Proctor AD, Ahmedna M, Kumar JV, Goktepe I (2004) Degradation of aflatoxins in peanut kernels/flour by gaseous ozonation and mild heat treatment. Food Addit Contam 21:786–793
Raju MVLN, Devegowda G (2000) Influence of esterified-glucomannan on performance and organ morphology, serum biochemistry and haemotology in broilers exposed to individual and combined mycotoxicosis (aflatoxin, ochratoxin and T-2 toxin). Brit Poultry Sci 41:640–650
Ramos AJ, Fink-Gremmels J, Hernández E (1996) Prevention of toxic effects of mycotoxins by means of nonnutritive adsorbent compounds. J Food Prot 59:631–641

Ramos AJ, Hernández E (1997) Prevention of aflatoxicosis in farm animals by means of hydrated sodium calcium aluminosilicate addition to feedstuffs: a review. Anim Feed Sci Technol 65:197–206

Refai MK, Niazi ZM, Aziz NH, Khafaga NEM (2003) Incidence of aflatoxin B_1 in the Egyptian cured meat basterma and control by γ-irradiation. Nahrung 6:377–382

Rustom IYS (1997) Aflatoxin in food and feed: occurrence, legislation and inactivation by physical methods. Food Chem 59:57–67

Ryu D, Hanna MA, Bullerman LB (1999) Stability of zearalenone during extrusion of corn grits. J Food Prot 62:1482–1484

Samapundo S, De Meulenaer B, Osei-Nimoh D, Lamboni Y, Debevere J, Devlieghere F (2007) Can phenolic compounds be used for the protection of corn from fungal invasion and mycotoxin contamination during storage? Food Microbiol 24:465–473

Samarajeewa U, Sen AC, Cohen MD, Wei CI (1990) Detoxification of aflatoxins in foods and feeds by physical and chemical methods. J Food Prot 53:489–501

Samarajeewa U, Sen AC, Fernando SY, Ahmed EM, Wei CI (1991) Inactivation of aflatoxin B_1 in corn meal, copra meal and peanuts by chlorine gas treatment. Food Chem Toxicol 29:41–47

Samar M, Resnik SL, González HHL, Pacin AM, Castillo MD (2007) Deoxynivalenol reduction during the frying process of turnover pie covers. Food Control 18:1295–1299

Santin E, Paulillo AC, Maiorka PC, Alessi AC, Krabbe EL, Maiorka A (2002) The effects of ochratoxin/aluminosilicate interaction on the tissues and humonal immune response of broilers. Avian Pathol 31:73–79

Santurio JM, Mallmann CA, Rosa AP, Appel G, Heer A, Dageförde S, Böttcher M (1999) Effect of sodium bentonite on the performance and blood variables of broiler chickens intoxicated with aflatoxins. Brit Poultry Sci 40:115–119

Sarr AB, Mayura K, Kubena LF, Harvey RB, Phillips TD (1995) Effects of phyllosilicate clay on the metabolic profile of aflatoxin B_1 in Fischer-344 rats. Toxicol Lett 75:145–151

Schaafsma AW, Tamburic-Ilinic L, Miller JD, Hooker DC (2001) Agronomic considerations for reducing deoxynivalenol in wheat grain. Can J Plant Pathol 23:279–285

Schmidt W, Nitzsche O (2004) Reducing *Fusarium* risk in maize rotations: rotating tillage and cultivar choice. Mais 32:8–11

Scott PM (1984) Effects of food processing on mycotoxins. J Food Prot 47:489–499

Scott PM (1996) Effects of processing and detoxification treatments on ochratoxin A: introduction. Food Addit Contam 13:19–21

Scott PM (1998) Industrial and farm detoxification processes for mycotoxins. In: Le Bars J, Galtier P (eds) Mycotox'98 international symposium, 2–4 July, Toulouse, France, pp 543–548

Scott PM, Kanhere SR, Dailey EF, Farber JM (1992) Fermentation of wort containing deoxynivalenol and zearalenone. Mycotoxin Res 8:58–60

Scott PM, Kanhere SR, Lawrance GA, Daley EF, Farber JM (1995) Fermentation of wort containing added ochratoxin A and fumonisins B_1 and B_2. Food Addit Contam 12:31–40

Scott PM, Lawrance GA (1994) Stability and problems in recovery of fumonisins added to corn-based foods. J AOAC Int 77:541–545

Scudamore KA, Banks JN, Guy RCE (2004) Fate of ochratoxin A in the processing of whole wheat grain during extrusion. Food Addit Contam 21:488–497

Sen AC, Wei CI, Fernando SY, Toth J, Ahmed EM (1988) Reduction of mutagenicity and toxicity of aflatoxin B_1 by chlorine gas treatment. Food Chem Toxicol 26:745–752

Sinha KK (1998) Detoxification of mycotoxins and food safety. In: Sinha KK, Bhatnagar D (eds) Mycotoxins in agriculture and food safety. Marcel Dekker, New York, pp 381–406

Smith TK (1980) Influence of dietary fiber, protein and zeolite on zearalenone toxicosis in rats and swine. J Anim Sci 50:278–285

Snijders CHA (2004) Resistance in wheat to *Fusarium* infection and trichothecene formation. Toxicol Lett 153:37–46

Solfrizzo M, Visconti A, Avantaggiato G, Torres A, Chulze S (2001) *In vitro* and *in vivo* studies to assess the effectiveness of cholestyramine as a binding agent for fumonisins. Mycopathologia 151:147–153

Stanley VG, Ojo R, Woldesenbet S, Hutchinson DH, Kubena LF (1993) The use of *Saccharomyces cerevisiae*, to suppress the effects of aflatoxicosis in broiler chicks. Poultry Sci 72: 1867–1872

Stinson EE, Osman SF, Huhtanen CN, Bills DD (1978) Disappearance of patulin during alcoholic fermentation of apple juice. Appl Environ Microbiol 36:620–622

Suárez-Quiroz M, De Louise B, Gonzalez-Rios O, Barel M, Guyot B, Schorr-Galindo S, Guiraud J-P (2005) The impact of roasting on the ochratoxin A content of coffee. Int J Food Sci Technol 40:605–611

Sydenham EW, Van Der Westhuizen L, Stockenström S, Shephard GS, Thiel PG (2004) Fumonisin-contaminated maize: physical treatment for the partial decontamination of bulk shipments. Food Addit Contam 11:25–32

Tekauz A (2002) *Fusarium* head blight of oat in Western Canada — preliminary studies on *Fusarium* species involved and level of mycotoxin(s) in grain. J Appl Genet 43:197–206

Trenholm HL, Charmley LL, Prelusky DB, Warner RM (1992) Washing procedures using water or sodium carbonate solutions for the decontamination of three cereals contaminated with deoxynivalenol and zearalenone. J Agric Food Chem 40:2147–2151

Var I, Kabak B, Erginkaya Z (2008) Reduction in ochratoxin A levels in white wine, following treatment with activated carbon and sodium bentonite. Food Control 19:592–598

Vasanthi S, Bhat RV (1998) Mycotoxins in foods: occurrence, health, and economic significance and food control measures. Indian J Med Res 108:212–224

Wilson SC, Brasel TL, Martin JM, Wu C, Andriychuk L, Douglas DR, Cobos L, Straus DC (2005) Efficacy of chlorine dioxide as a gas and in solution in the inactivation of trichothecene mycotoxins. Int J Toxicol 24:181–186

Wolf-Hall CE, Hana MA, Bullerman LB (1999) Stability of deoxynivalenol in heat-treated foods. J Food Prot 62:962–964

Yazdanpanah H, Mohammadi T, Abouhossain G, Majid Cheragjali A (2005) Effect of roasting on degradation of aflatoxins in contaminated pistachio nuts. Food Chem Toxicol 43:1135–1139

Young JC, Trenholm HL, Friend DW, Prelusky DB (1987) Detoxification of deoxynivalenol with sodium bisulfite and evaluation of the effects when pure mycotoxin or contaminated corn was treated and given to pigs. J Agric Food Chem 35:259–261

Chapter 14
Mycotoxigenic Fungi and Mycotoxins in Animal Feed in South American Countries

Carina Elizabeth Magnoli, Lilia Renée Cavaglieri, Carlos Alberto da Rocha Rosa, and Ana María Dalcero

14.1 Introduction

The animal feed industry is an integral and growing segment of the food supply chain in Latin-American countries. It supplies the necessary feed ingredients to produce healthy animals, which provide food protein and energy that are essential to human beings. Pre- and post-harvest contamination of food and feed crops by mycotoxigenic fungi is a common problem. Although there are geographic and climatic differences in the production and occurrence of mycotoxins, exposure to these substances is worldwide (Kuiper-Goodman 2004).

The development of modern agricultural methods together with large-scale processing has increased the problem, resulting in both acute diseases called mycotoxicoses, and chronic conditions, often recognized as situations involving mycotoxins impair the health of the animal (Binder et al. 2007).

Mycotoxins constitute the most significant risk of mycotoxicosis in different animals (chickens, hens, ducks, turkeys, pigs, cows, pets and rabbits) and humans, and are produced by species of *Aspergillus*, *Fusarium* and *Penicillium* genus. The most significant from the first genera are aflatoxins (AFs) and ochratoxin A (OTA); from *Fusarium* there are zearalenone (ZEA), vomitoxin or deoxynivalenol (DON), diacetoxyscirpenol (DAS), fumonisins (FBs), T-2 toxin, monoacetoxyscirpenol, triacetoxyscirpenol, and escirpentriol; and from *Penicillium* genus there are citrinin, patulin and OTA. Patulin must be taken into account due to the potential risk for humans. However, there is not enough evidence in order to classify this mycotoxin as of high risk for animals (CAST 2003; Morgavi and Riley 2007). Fungal species such as *Aspergillus fumigatus*, *Byssochlamys nivea*, *Monascus* spp., *Penicillium roqueforti* and *Trichoderma* spp. are the most frequent contaminants on

C.E. Magnoli (✉)
Departamento de Microbiología e Inmunología, Facultad de Ciencias Exactas Físico Químicas y Naturales Universidad Nacional de Río Cuarto, Río Cuarto, Córdoba, Argentina
e-mail: cmagnoli@exa.unrc.edu.ar

M. Rai and A. Varma (eds.), *Mycotoxins in Food, Feed and Bioweapons*,
DOI: 10.1007/978-3-642-00725-5_14, © Springer-Verlag Berlin Heidelberg 2010

ensilaged feed (Boysen et al. 2000; El-Shanawany et al. 2005; Garon et al. 2006; González Pereyra et al. 2008a). Mycotoxins produced by *A. fumigatus* include tremorgens and clavine alkaloids that seem to be responsible for abortions and gliotoxin that affect immune systems (Yamada et al. 2000). *Aspergillus fumigatus* with toxin-producing abilities are the predominant *Aspergillus* spp. in silage from Argentina (Cavaglieri et al. 2005; Pereyra et al. 2007b; 2008).

The purpose of this chapter is to inform about the mycoflora, toxigenic fungi and mycotoxins in raw materials and ready animal feed in Argentina and South American countries.

14.2 Occurrence of Toxigenic Fungal Species and Mycotoxins in Raw Materials and Feedstuffs

14.2.1 Silage and Ready Dairy Cattle Feed

Silages made from a variety of forage crops constitute important components of ruminant diets in many areas of the world. Maize silage represents the major proportion, followed by grass and legume silage and whole-crop cereal silage (Driehius and Oude Elferink 2000; Amigot et al. 2006; González-Pereyra et al. 2008a). This practice allows forage to be preserved through spontaneous lactic fermentation under anaerobic conditions. Poor management during silage processing can result in excessive moisture or dryness, condensation, heating, leakage of rainwater and insect infestation of the silo, leading to undesirable growth of microaerobic acid-tolerant fungi (dos Santos et al. 2003). Hay harvested in good conditions has a limited and balanced associated microflora of field fungi (pre-harvest period), intermediate fungi (during harvest), and storage fungi (storage after harvest). Intermediate and storage fungi are more numerous and diverse in hay harvested and stored in humid conditions. Some of these molds that contaminate silage can lead to the loss of nutritive substances by saprophytic action and may also result in biosynthesis of mycotoxins. Because the use of silage is increasing, risk assessment for mycotoxins becomes important as it concerns human and animal safety as well as animal performance.

Silage removal for animal feeding by shoveling breaks compaction of the silo on the extraction area and favors aeration changing the redox potential. Generally the fungal growth and mycotoxin production occur when the vegetable in the silo is not well packed, the oxygen tension is high and pH values range from 6 to 7 (Gotlieb 1997). Other researchers reported that conditions under mycotoxin production in silage remain uncertain (Oude Elferink et al. 2000).

In hay harvested and stored in humid conditions, hygrophilic and heat-tolerant species are predominant. *Aspergillus fumigatus*, which causes respiratory troubles and could produce gliotoxin, is the most important, followed by *Stachybotrys atra*, producing G and H satratoxins which cause stachybotryotoxicosis. In addition, numerous highly toxicogenic species of *Aspergillus*, *Penicillium* and *Fusarium* zearalenone producers have been detected in damp hay and straw (Yiannikouris and Jouany 2002). *Penicillium roqueforti* strains are considered micro-aerophilic or indifferent to oxygen presence. In a series of trials with grass and whole-crop maize, under all the oxygen status conditions, *P. roqueforti* was the only filamentous fungus found after 60 days of storage in Europe (Auerbach et al. 1998). Apart from aflatoxins and sterigmatocystin that can be found in insufficiently dried hay and straw, another toxigenic fungi, patulin producer *Byssochlamys nivea*, appears in silos after about 6 months of storage. Other fungi adapted to ensiling condition and frequently isolated on ensilaged feed are *Monascus* spp. and *Trichoderma* spp. (El-Shanawany et al. 2005; Garon et al. 2006; González Pereyra et al. 2008a).

Another factor that may contribute to the succession of silage mycoflora during fermentation is pH changes caused by the natural production of organic acids, such as lactic, acetic, propionic and butyric acids. The resistance of fungal conidia towards organic acid has been shown to differ among genera and species. Lactic acid has no detrimental effect, whereas propionic acid and butyric acid are potent mold inhibitors. Conidia of *P. roqueforti* proved to be less impaired by propionic acid than those of other Penicillia or Aspergilli (El-Gazzar et al. 1987).

Total fungal mold count over 10^9 CFU g^{-1} forage may be the reason for several health problems. Therefore, fungal propagule counts are another indicator of forage quality (Di Costanzo et al. 1995).

In beef cattle, mycotoxin consumption is associated with a decrease in feed intake, weight loss, reduced milk production, lack of response to diet change and therapies, and the possibility of mycotoxins (aflatoxins) transferred to milk or beef (Yiannikouris and Jouany 2002; Seglar 2003). The evaluation of *A. fumigatus* presence is very important; it is often associated with spoilage and heating forage (Scudamore and Livesey 1998). *Aspergillus fumigatus* has been proposed as the pathogenic agent associated with mycotic hemorrhagic bowel syndrome (HBS) in dairy cattle, mainly in immunosuppressed animals (Puntenney et al. 2003). Mycotoxins produced by *A. fumigatus* include tremorgens (Land et al. 1993), clavine alkaloids that appear to be responsible for abortions (Moreau 1979) and gliotoxin that affect immune systems (Belkacemi et al. 1999; Yamada et al. 2000). Numerous syndromes in ruminants are likely to be produced by fungal or toxin ingestion present in spoiled silage (Seglar 1999).

14.2.1.1 Mycoflora and Mycotoxin Incidence

González Pereyra et al. (2008a) studied the mycological quality of pre- and post-fermented corn silage samples in Central Argentina. This region is the most important area in terms of agricultural activities where silage practice is developed, mainly open-field farming, dairy and beef cattle breeding, and sowing. The mycological analysis showed that nine genera of filamentous fungi and yeasts were isolated from corn post-fermented silage samples. *Aspergillus* spp. was the most frequent (78%), followed by yeasts (65%), *Fusarium* spp. (62%), Mucorales spp.

(59%), *Penicillium* spp. and *Eurotium* spp. (34%). Predominant *Aspergillus* spp. were *A. flavus* (53%) and *A. fumigatus* (30%) followed by *A. niger* aggregate (Table 14.1). Eight *Fusarium* species were identified, and *F. verticillioides* was predominant (77%).

All tested samples from pre-fermented silo were negative for AFB_1, but this mycotoxin was detected in variable levels from post-fermented silage samples. It was noticed that AFB_1 positive samples from the upper section (visibly heavily infested) and lower section showed low AFB_1 contamination levels ($<10\ \mu g\ kg^{-1}$). Only six samples from the middle section of post-fermented silo (17%) were positive for AFB_1 contamination, showing toxin levels varying from 1.43 to 155.78 $\mu g\ kg^{-1}$. The corn plants arrive in the silo contaminated with *Aspergillus* section *Flavi* strains and could cause the initial aflatoxin levels present in the silage. In this study, although the corn plant did not show aflatoxin contamination, AFB_1 levels were present in some post-fermented silage samples. This result suggests that the prevailing environmental conditions could allow aflatoxin production by the potential aflatoxin producers initially present.

Mycotoxin levels between the two types of samples showed significant differences ($P < 0.05$) between FB_1 and DON. Fumonisin levels varied from 120 to 1,840 $\mu g\ kg^{-1}$ in pre-fermented silage samples and from 340 to 2,490 $\mu g\ kg^{-1}$ in post-fermented silage samples. Deoxynivalenol contamination followed the same tendency as FB_1. Levels varied from 100 to 230 $\mu g\ kg^{-1}$ in pre-fermented silage, and from 30 to 870 $\mu g\ kg^{-1}$ in post-fermented silage samples. Zearalenone levels did not show any significant difference between both kinds of samples tested. Zearalenone, DON and FB_1 co-occurred in 83.5% samples, whereas 16.5% samples showed only ZEA and FB_1 co-contamination (Table 14.2) (González Pereyra et al. 2008a).

Amigot et al. (2006) analyzed the mycoflora and mycotoxins from 147 samples of fermented forages for dairy feeding cattle from the central region of the province of Santa Fe and northeast of the province of Córdoba (Argentina dairy region) during the 1998/1999 period. The highest counts (higher than 10^9 CFU g^{-1}) were registered in 25.6, 18.4 and 16.4% of maize, lucerne and sorghum samples respectively. All the analyzed forages showed a prevalence of the potentially toxigenic genera as follows: *Aspergillus* spp. (around 17% of all the isolates), *Penicillium* spp. (9.7%) in maize, *Fusarium* spp. (12.3%) in lucerne, and *Byssochlamys* spp. (8.5%), *Fusarium* spp. and *Geotrichum* spp. (6.1% each) in sorghum. Forty percent of the isolated species were potentially toxigenics, with an important register of *Aspergillus* section *Flavi*. *Aspergillus fumigatus* had not been isolated from haylages. Sorghum showed the largest percentage of *A. fumigatus* isolates (9.6% from the overall isolated molds), mainly from wet ensilaged grains.

The lucerne and maize feedstuffs presented a high mycotoxin incidence and the simultaneous presence of AFs and DON was detected in the majority of the samples; the mean concentrations were not too high (AFs: 3.75 and 2.56 $\mu g\ kg^{-1}$, DON: 187 and 167.71 $\mu g\ kg^{-1}$ in lucerne and maize, respectively). The sorghum silages, in accordance with the low percentage observed in samples with high fungal counts, presented a smaller number of samples with risk mycotoxin values

Table 14.1 Argentinean and Brazilian feedstuffs contaminated with *Aspergillus* species

Fungal species	Feedstuffs samples percentage (%)								References
	Poultry		Pig	Rabbit	Equine		Silage		
	Argentina	Brazil	Argentina	Argentina	Argentina	Brazil	Argentina	Brazil	
A. flavus	8–36	25–50	28–35	15–30	40	35	53	36	Dalcero et al. 1997, 1998; Magnoli et al. 2002, 2005; Rosa et al. 2006; Oliveira et al. 2006; González-Pereyra et al. 2008a, b, c; Keller et al. 2007; Fraga et al. 2007
A. parasiticus	10–33	2.5	5–10	12–18	–	–	5		Dalcero et al. 1997, 1998; Magnoli et al. 2005; González-Pereyra et al. 2008a, b; Fraga et al. 2007
A. niger aggregate	5–30	4–15	5–15	8–12	–	15	10	16	Dalcero et al. 1997, 1998; Magnoli et al. 2002, 2005; Rosa et al. 2006; Oliveira et al. 2006; González-Pereyra et al. 2008a, b; Keller et al. 2007,
A. fumigatus	10–20	3–8	8–32	8	–	–	30	4	Dalcero et al. 1997, 1998; Magnoli et al. 2002, 2005; Rosa et al. 2006; Oliveira et al. 2006; González-Pereyra et al. 2008a; Keller et al. 2007; Fraga et al. 2007
A. candidus	15–42	4–8	32–35	5	–	15	–	16	Dalcero et al. 1997, 1998; Magnoli et al. 2002, 2005; Keller et al. 2007; Fraga et al. 2007

Table 14.2 Incidence of mycotoxins on feedstuff: Argentinean and Brazilian samples

Country and Feed	AFB$_1$		ZEA		DON		FBs		OTA		References
	Positive samples[a]	Levels (ng/g)	Positive samples[a]	Levels (ng/g)	Positive samples [a]	Levels (ng/g)	Positive samples[a]	Levels (ng/g)	Positive samples[a]	Levels (ng/g)	
Argentina Poultry feed	14/300	17–197[£] (89 ± 72)[b]	1/300	3–280[£] (143 ± 126)[b]	8/300	240–400[£] (309 ± 77)[b]	FB$_1$ 97/120	136–4,270[£] (1,839 ± 1,449)[b]	38/120	25–30[£] (27 ± 2.64)[b]	Dalcero et al. 1997, 2002
	48/130	10–123[£] (27 ± 33)[b]	15/130	327–5,850[£] (2,544 ± 2,346)[b]	–	–	FB$_2$ 97/120	39–1,710[£] (676.5 ± 582)[b]	–	–	Dalcero et al. 1998
	97/120	14–174[£] (76.5 ± 56)[b]	17.7/120	1550–4507[£] (2,581 ± 1,668)[b]	6.15/120	124[£] (124 ± 0.0)[b]	FB$_3$ 97/120	63–955[£] (283.6 ± 269)[b]	–	–	Magnoli et al 2002
Pig feed	44/120	(70 ± 60)[b]	–	–	–	–	FB$_1$ 100/120	<100–1,500	25/120	34[£] (34 ± 0.0)[b]	Dalcero et al. 2002
	–	–	–	–	–	–	FB$_1$ 100/240	<100–1,000	–	–	González-Pereyra et al. 2008b
Rabbit feed	–	–	–	–	–	–	–	–	13/120	18.5–25[£] (21.7 ± 4.6)[b]	Dalcero et al. 2002
Silage	ND	ND	83.5/30	10–25[£] (18 ± 7)[b]	83.5/30	100–230[£] (150 ± 60) [b]	FB$_1$ 83.5/30	120–1,840[£] (600 ± 440) [b]	–	–	González-Pereyra et al. 2008a
Argentina Silage[§]	17/30	1.43–155.78[£]	83.5/30	ND–350[£] (50 ± 60) [b]	83.5/30	30–870[£] (276 ± 130)[b]	FB$_1$ 83.5/30	340–2,490[£] (1,110 ± 500)[b]	–	–	González-Pereyra et al. 2008a
Equine feed	35/50	5.36–64.40[£]	29/50	DNQ	–	–	ND	ND	–	–	González-Pereyra et al. 2008c.
Ready pet food	21/24	0.066–339	–	–	–	–	86/21	9.75–166.9	–	–	Fernández Juri (Pers. Comm.)
Brazil Poultry feed	66.7/480	1.2–17.5[£]	77/480	100–7,000[£]	–	–	FB$_1$ 97.8/480	1,500–5,500[£]	–	–	Oliveira et al. 2006.
Brewer's grain[c]	33.75/80	1–3[£]	–	–	–	–	–	–	ND	ND	Simas et al 2006
Brewer's grain[c]	–	–	–	–	–	–	FB$_1$ + FB$_2$ 72.5/82	50.30–908.47[£]	–	–	Batatinha et al. 2007
Brewer's grain[c]	–	–	–	–	–	–	–	–	45/22	27–439	Rosa et al. 2009

Corn	–	–	–	–	–	–	–	–	31/26	4.9–132	Rosa et al. 2006
Barley rootlets	–	–	–	–	–	–	–	–	22/37	20–637	Ribeiro et al. 2006
Cow's feed	–	–	–	–	–	–	–	–	25/24	12–324	Rosa et al. 2009
Equine feed	–	0.01–99.4£	–	–	–	–	FB_1 –	10–749£	–	–	Keller et al. 2007
Brazil Lucerne[c]	100/49	1.4–11.80£	–	–	100/49	100–2,000£	–	–	–	–	Amigot et al. 2006.
Sorghum[c]	16.4/55	1–80	–	–	9.1/55	100–500	–	–	–	–	Amigot et al. 2006.
Maize[c]	100/43	0.5–13.8£	–	–	100/43	100–2,000£	–	–	–	–	Amigot et al. 2006.
Raw material pet food¢	70/168	0.1–24.5£	–	–	–	–	–	–	–	–	Campos et al. 2008
Cat, dog and birds food	12/100	15–374.5£	–	–	–	–	–	–	–	–	Maia et al. 2002
Standard dog food	94.4/60	>0.3–9.43£	–	–	–	–	–	–	–	–	Campos et al. 2009
Premium dog food	57.1/60	>0.3–8.11£	–	–	–	–	–	–	–	–	Campos et al. 2009
Super premium dog food	60.0/60	>0.3–6.38£	–	–	–	–	–	–	–	–	Campos et al. 2009

[a]Percentage of positive samples and number of total samples assayed. £range of mycotoxin concentration based only on positive samples.
[b]Means of mycotoxin concentration ± SD (standard deviation). pre-fermented silage samples. §post-fermented silage samples.
[c]Used as dairy cattle feed. ¢corn grains, corn and sorghum meal, corn meal and gluten (21%).
ND: not detected. DNQ: detectable, non-quantifiable, quantification limit 5 ppb. NQ: non-quantitative levels. (-) percentage/number of samples or mycotoxin levels has not been informed

(AFs: 16.4% and DON: 9.1%). However, the maximum AFs concentration was higher for the sorghum (80 μg kg^{-1}).

Contamination with aflatoxins in cattle feed has been mainly registered in seeds stored in warm climates. In a study conducted by da Silva et al. (2000) from Brazilian stored sorghum samples, similar levels of aflatoxin incidence were reported. From 104 samples, AFB_1 contaminated 12.8% with values that ranged from 7 to 33 μg kg^{-1}.

In recent works, González-Pereyra et al. (2008a,b) and Alonso et al. (2007), evaluated toxigenic fungi from raw materials and ready dairy cattle feed from Argentina. Samples included raw materials (corn grains, silage, cotton seeds and malt) and ready feed. The determination of physical properties of the samples revealed strait ranges of a_W and temperature among samples obtained at different sampling places. Water activity levels ranged between 0.929 ± 0.090 and 0.976 ± 0.010, and ready feed had 0.973a_W. Temperature levels did not show great variations and had an average of 25.6°C. The pH values showed the most important difference; the middle section had 3.88 whereas upper and lower sections obtained 5.8 and 5.5 values, respectively.

Total counts of all samples obtained from dichloran rose bengal chloramphenicol medium (DRBC) and counts of xerophilic fungi obtained from dichloran chloramphenicol 18% glicerol medium (DG18) were over 1×10^6 CFU g^{-1}, showing a high contamination degree. The statistical analysis (LSD test) revealed no significant differences among counts from all ready dairy cattle feed samples and corn silage at different sampling places from DRBC and DG18 culture media. *Fusarium* spp. counts obtained from Nash Snyder medium were from 1×10^2 to 1×10^3 CFU g^{-1}. *Aspergillus* spp. was the most prevalent genus; *A. fumigatus* relative density was high in both the middle section of the silo and the ready dairy cattle feed. This species was not present in the lowest silo section. *Aspergillus* section *Flavi* showed the highest relative density (83%) in the upper section. Cottonseeds had 100% of *Aspergillus* spp. followed by ready cattle feed (74%) and corn grains (40%). Malt samples did not show *Aspergillus* spp. contamination. Relative densities of *Aspergillus* section *Flavi* ranged from 43 to 50% in raw materials.

Pereyra et al. (2007a,b; 2008) evaluated the toxigenic capacity of *A. fumigatus* strains isolated from ready cattle feed and silage samples. They showed that a high percentage of these strains was able to produce gliotoxin and a small percentage produced fumitremorgen B. The same percentage of *A. fumigatus* strains able to produce gliotoxin, fumitremorgen B and C, and fumigaclavine B was obtained from corn silage. Only a low percentage of strains with capacity to produce fumigaclavine C was obtained from corn silage. Gliotoxin levels produced by *A. fumigatus* isolated from corn silage varied between 160 and 1000 μg l^{-1}, whereas *A. fumigatus* strains isolated from finished dairy cattle feed samples varied from 160 to 1600 μg l^{-1}. Fumigaclavine A and C, fumitremorgen B and C was detected at non-quantitative levels. Forty-five percent and 12% of *A. fumigatus* strains isolated from corn silage and ready cattle feed respectively were able to produce more than one mycotoxin.

Natural incidence of gliotoxin (μg g^{-1}) from corn silage and ready cattle feed was determined by Pereyra et al. (2008). Gliotoxin levels found in ready cattle feed were three times greater than in corn silage samples. Although mean levels were near 14 μg g^{-1}, some samples obtained values over than 40 μg g^{-1}.

The mycoflora and OTA production on raw materials and ready dairy cattle feed were evaluated by Rosa et al. (2009). Raw materials (109 samples) including corn, brewer's grain and barley rootlets used as ingredients and finished cow feed samples were analyzed. These samples were collected in Rio de Janeiro State, Brazil. The mycological survey of corn samples showed that *Aspergillus* spp. and *Fusarium* spp. were isolated from 60% of the samples, and *Penicillium* spp. in lower frequency (40%). These genera were prevalent in other cereals, the frequency of isolation varied according to the substrate. *Aspergillus flavus* was isolated from corn, brewer's grain and finished cow feed in 34, 14 and 32% of the samples respectively. *Aspergillus carbonarius* species was only isolated from brewer's grain in a low percentage of samples. *Aspergillus niger* and *A. ochraceus* were found in all substrates in 20–40% and 10–25% of the samples respectively. *Penicillium verrucosum* strains were isolated from corn, brewer's grain and barley rootlets in 8, 15 and 17% of the samples respectively.

The ability to produce OTA on yeast extract sucrose medium (YES) by ochratoxigenic strains isolated from raw materials including corn, brewer's grain and barley rootlets was evaluated. Seventy-one percent of *A. carbonarius* strains were OTA producers, 38% of *A. niger*, 33% of *A. ochraceus* and 13% of *P. verrucosum*. OTA levels ranged from 16 to 116 μg kg^{-1}. *Aspergillus ochraceus* and *P. verrucosum* strains produced the highest OTA levels followed by *A. carbonarius* and *A. niger*. High percentages of potential OTA-producer strains were found. They represent 40.6, 69, 46 and 38% in corn, brewer's grain, barley rootlets and finished cow feed respectively.

Ochratoxin A contamination of samples was analyzed by HPLC using immunoaffinity columns. This study showed that brewer's grain samples were the most contaminated (45%), showing OTA levels between 27 and 439 μg kg^{-1}. Thirty-one percent of corn and 22% of barley rootlets were contaminated with OTA at levels ranging from 4.9 to 132 μg kg^{-1} and 20 to 637 μg kg^{-1} respectively. Twenty-five percent of samples of finished cow's feed samples were positive for OTA with levels ranging from 12 to 324 μg kg^{-1} (Table 14.2) (Rosa et al. 2009).

From the same country, Simas et al. (2006), evaluated the mycoflora and mycotoxins from brewer's grains used to feed dairy cattle in the State of Bahía. *Aspergillus* spp was the most frequent isolated genus (42.5%), followed by *Mucor* spp. and *Rhizopus* spp. (32.5%), *Penicillium* spp (7.5%), and *Fusarium* spp (2.5%). Mycotoxin analysis did not show the presence of ochratoxins from Brazilian brewer's grains used as dairy cattle feed, but the presence of total aflatoxins was observed in 33.7% of the samples, with contamination levels ranging from 1 to 3 ng g^{-1}. Recently, from the same substrate from Paraná state (Brazil), Batatinha et al. (2007) reported on the presence of FBs (FB_1 and FB_2) in 72.5% of samples, with contamination levels ranging from 50.30 to 908.47 ng g^{-1} (Table 14.2).

In Argentina, Tapia et al. (2005), investigated the presence of patulin-producing *Penicillium* spp. in fermented corn silage and high moisture corn samples. The results showed that this genus was isolated from 6.6% of samples; of these strains, 1.1% were patulin producers on yeast extract sucrose and potato dextrose agar. The patulin-producing strains belonged to the *P. viridicatum* group. Other isolated genera isolated were *Mucor* spp. (2.8%), *Aspergillus* spp. (2.6%) and *Fusarium* spp. (1.8%) among others. Total mold counts higher than 10^5 CFU g^{-1} were obtained from 4.1% of feed samples, and 3.3% of them had more than 10^4 CFU g^{-1} of *Penicillium* spp.

In central Argentina, Pereyra et al. (2007b; 2008) showed that *Aspergillus* spp., *Fusarium* spp. and *Penicillium* spp., were the prevalent genera isolated from silage. *Aspergillus flavus* showed the highest relative density among *Aspergillus* species followed by *A. fumigatus* and *A. niger* aggregate species. The presence of *A. fumigatus* means a dual hazard from the ingestion of pathogenic spores and from potential mycotoxin production such as gliotoxina, fumigaclavine A, fumigaclavine C and several fumitremorgens. Recovered yeasts yielded higher percentages than those found by other researchers (El-Shanawany et al. 2005). Mor and Singh (2000), dos Santos et al. (2003), and Garon et al. (2006) have reported the isolation of *A. fumigatus* from cereal grains and other animal feed silage. Other researchers (Upperman et al. 2003; Watanabe et al. 2003; Morgavi et al. 2004; Boudra and Morgavi 2005) have stressed the importance of studying the capacity of *A. fumigatus* strains to produce mycotoxins, such as gliotoxin, the most toxic metabolite produced by *A. fumigatus*. The amount produced by some strains is enough to pose a potential animal health problem if found in feed.

Some *A. fumigatus* strains isolated from Argentinian silage were able to produce more than one mycotoxin (Pereyra et al. 2007a). The strains isolated from ready cattle feed produced higher gliotoxin levels than those isolated from corn silage. Boudra and Morgavi (2005) evaluated the capacity of 14 *A. fumigatus* strains to produce gliotoxin in Europe. After 3 days of incubation this toxin was found in ten isolates at concentrations ranging from 0.2 to 23.2 μg ml^{-1}. These levels were lower than those reported in Argentina. Dos Santos et al. (2003) reported on the toxin-producing capacity of *A. fumigatus* strains isolated from corn silage in Portugal. In contrast with Argentinian results, they found that fumitremorgen B was the most commonly produced, usually in combination with other toxins. Fumitremorgen B and C, and fumigaclavine C were found at similar levels, whereas gliotoxin was found only in 11% of strains. Other studies did not report mycotoxin presence.

Pereyra et al. (2007a,b 2008) are the only authors that have mentioned the gliotoxin incidence in raw materials and ready feed destined for dairy cattle. They found 100% of samples contaminated with gliotoxin and any samples of corn silage containing this mycotoxin.

Upperman et al. (2003) and Watanabe et al. (2003) observed that concentrations as low as 0.01 μg ml^{-1} of gliotoxin could induce immunosuppressive and apoptotic effects in vitro. The concentration of gliotoxin found in Argentinian silage and ready cattle feed was largely high. In conclusion, *A. fumigatus* was found at high density from feedstuffs and demonstrated to be highly toxicogenic. Most tested strains produced two or more of the five monitored mycotoxins. Gliotoxin was found at important concentrations in silage and ready cattle feeds and its presence could affect productivity and presents a health risk for animals.

14.2.2 Poultry, Pig, Rabbit and Equine Feed

14.2.2.1 Mycoflora Incidence

Mycological surveys of Argentinian poultry feed samples collected over a period of 2 years from factories showed the presence of three toxigenic fungal genera, *Aspergillus* spp (85–52% of the samples), *Penicillium* spp. (98–52%) and *Fusarium* spp. (87–70%) (Dalcero et al. 1997, 1998; Magnoli et al. 2002). These results were similar to those obtained by other researchers in Brazilian poultry feeds using the same culture media (Oliveira et al. 2006; Rosa et al. 2006). In Argentina, pelleted poultry and pig feeds sampling contain approximately 60% corn. Corn is replaced by alfalfa, oat and barley in rabbit feed, and depending on market availability, the protein value is achieved by adding sunflower pellets or soy.

Several authors have isolated *A. flavus*, *A. parasiticus*, *A. niger* aggregate, *A. fumigatus*, *A. carbonarius*, *A. ochraceus* and *A. versicolor* toxigenic species from different animal feeds (Dalcero et al. 1997; 1998; Magnoli et al. 2002; 2005; Rosa et al. 2006; Oliveira et al. 2006) (Table 14.1).

In a recent study, González-Pereyra et al. (2008b), determined the mycological quality of initial, growing, final feed and corn grains intended for pig feeding in central Argentina. Total mold counts obtained from DRBC medium and counts of xerophilic fungi from DG18 medium were higher than 1×10^5 CFU g^{-1}, showing a high contamination degree. *Fusarium* spp. counts were from $<1 \times 10^2$ to 1×10^5 CFU g^{-1} whereas *Aspergillus* spp. counts ranged between 2×10^3 to 4.3×10^5 CFU g^{-1}.

The mycological surveys showed that *Fusarium* spp. was the most frequent genus (100%) isolated in all pig feed and corn samples on DRBC as well as on DG18 media, followed by *Aspergillus* and *Penicillium* genera. Three *Fusarium* species were identified in pig feed and corn grains. *F. verticillioides* was the predominant species. It was found in all feed samples tested (100%). *F. subglutinans* was found in growing feed (2.5%) and in corn grain (2.5%) samples whereas *F. oxysporum* was only isolated from growing feed (2.5%). Three *Aspergillus* species were identified. *A. flavus* was the prevalent species followed by *A. niger* and *A. parasiticus*. *A. flavus* was found in all samples tested (Table 14.1). A great variety of *Penicillium* species from poultry, pig and rabbit feed has been reported in Argentina and Brazil. Species of biverticillate and terverticillate penicilli belonging to *Furcatum*, *Biverticillium* and *Penicillium* subgenera are the most prevalent in these substrates. The species *P. minioluteum*, *P. purpurogenum*, *P. implicatum*, *P. citrinum*, *P. solitum*, *P. crustosum* and *P. citrionigrum* have been isolated from the majority of animal feeds.

Recently, in Rio de Janeiro State, Brazil a survey was carried out to evaluate the mycoflora of raw materials (maize, initial poultry feeds, pelleted poultry feed) and trough feed samples (Fraga et al. 2007). Total maize mycobiota counts ranged between 1×10^4 and 7×10^4 CFU g^{-1} and total initial poultry feed mycobiota counts varied between 4×10^3 and 3×10^4 CFU g^{-1}. When the pelleting process was done these counts decreased significantly; nevertheless they were below the limit of detection (100 g^{-1}). In maize and initial poultry feed samples *A. flavus* and *E. chevalieri* were the more prevalent species isolated (28%), followed by *E. amstelodami*, *A. candidus* and *A. niger*. In trough feed samples, *A. flavus*, *E. amstelodami* and *E. chevalieri* were the more frequent species, isolated from 22% of the samples, followed by *A. niger*, *A. sydowii* and *A. versicolor*.

Keller et al. (2007) carried out the mycological examination of 30 samples of Brazilian equine feeds from five hypocenters. They showed seven fungal genera of which the most frequent was *Aspergillus* spp. and its teleomorphs (40.54%) followed by *Penicillium* spp. (18.38%) and *Fusarium* spp. (16.22%). From *Aspergillus* genus, the most prevalent was *A. flavus* (36%). In decreasing order *A. niger* (16%), *A. candidus* (16%), *A. ochraceus* (8%), *A. pumilus* (8%) and others were present (Table 14.1). Equine feed fungal colony counts (CFU g^{-1}) showed DRBC counts that ranged between less than 1×10^2 to 2×10^5 CFU g^{-1}. The highest values were obtained from non-pelletized feeds and oats, which ranged between 1.3×10^4 to 2×10^5 CFU g^{-1} and 6×10^3 to 4.8×10^5 CFU g^{-1} respectively. Counts of xerophilic fungi (DG18) were obtained from equine feeds at ranges between less than 1×10^2 to 1×10^5 CFU g^{-1}. Non-pelletized feed counts were higher than pelletized feeds. Oat counts from DG18 ranged between 1.8×10^4 to 3.8×10^5 CFU g^{-1}.

González Pereyra et al. (2008c) evaluated the mycological quality in 50 Argentinean equine feed samples (maize, oat and ensiled alfalfa) collected at random during 1 year from hippo centers (pure race, pole, sport mares and racially mixed horses) located in Córdoba province, central Argentine. The most frequent genus isolated was *Aspergillus* spp. and Mucorales (100%) followed by *Fusarium* spp. (60%). Yeasts were isolated at a frequency of 100%. From *Aspergillus* genus, the most prevalent was *A. flavus* (43%). From *Fusarium* spp., *F. verticillioides* was the predominant (Table 14.1). Total counts (DRBC) ranged between 2.2×10^4 to 7.3×10^5 CFU g^{-1}. Counts of xerophilic fungi (DG18) ranged between 1.3×10^4 to 1×10^6 CFU g^{-1}.

Pacin et al. (2002) reported the natural mycoflora in corn samples from farms at harvest (coastal and mountain regions) and samples of corn-based pelletized feed from Ecuador. The fungi associated with corn-based pellets destined for poultry, pig and cow feeds showed that *A. flavus* was the most prevalent fungus present followed by *F. graminearum*, *F. verticillioides* and *A. parasiticus*. In this substrate, *A. niger* was not isolated. The predominant *Aspergillus* spp. isolated from corn was *A. flavus* followed by *A. niger* and *A. parasiticus* isolated in 60, 20 and 10% respectively of the samples from the coastal region, whereas in the mountain region, these species were isolated in 25, 12.5 and 4.2% respectively of samples. Significant differences in the isolation relative density of *A. flavus* and *A. parasiticus* were observed between the two regions, but there was no significant difference for *A. niger*.

Céspedes and Díaz (1997) evaluated the natural occurrence of mycotoxins from mixed feeds (grain sorghum, maize, processed soybean, rice meal, cottonseed meal) and poultry and pig feeds taken from feed-manufacturing Colombian plants. Aflatoxins were detected in 24, 12, 36, 88, 40 and 43.7% from sorghum, maize, rice meal, cottonseed meal, poultry and pig feed samples.

14.2.2.2 Toxigenic Ability of Potential Mycotoxin Producer

The ability to produce aflatoxins was evaluated in 45 *Aspergillus* section *Flavi* strains isolated from Argentinian poultry feeds on natural substrate (rice kernels) and determined by TLC technique. Forty seven *A. flavus* strains produced AFs; among these, 24% of the strains produced AFB_1 and AFB_2 with levels of 181–14,545 and 6–3,640 ng g^{-1}, and 47% produced only AFB_1 with levels ranging from 10 to 920 ng g^{-1}. Regarding *A. parasiticus*, six aflatoxigenic profiles were identified and 50% of the assayed strains were AFs producers. The levels ranged from 30 to 24,545 ng g^{-1} for AFB_1, 45–982 ng g^{-1} for AFB_2, 11–10,300 ng g^{-1} for AFG_1 and 20–1,227 ng g^{-1} for AFG_2 (Magnoli et al. 1998).

Strains of *F. verticillioides*, *F. proliferatum* and *F. nygamai* isolated from Argentinian poultry feeds were evaluated for fumonisin-production ability on corn kernels and determined by HPLC technique. All strains produced FBs and there was a high degree of variability in the quantities of FB_1, FB_2 and FB_3 produced. Fumonisins B_1 was the mycotoxin produced at highest levels by all isolates. The total levels of FBs ranged from 7.41 to 5,922 ppm. The range of concentration varied from 5.4 to 3,991, 1.01 to 189 and 0.4 to 765 ppm per gram of corn for FB_1, FB_2 and FB_3 respectively (Magnoli et al. 1999).

Recently, Fraga et al. (2007) evaluated the potential for aflatoxins and ochratoxin A production by *Aspergillus* (141 strains) and *Eurotium* (88 strains) species isolated at different stages of poultry feed processing. Twenty-one strains of *Aspergillus* spp. and 2.2% of *Eurotium* spp. produced one or more AFs. Among them, 22% of *A. candidus*, 29% *A. flavus*, 100% *A. parasiticus* and 3.3% *E. chevalieri* strains were able to produce AFs. Regarding aflatoxin producers, 63% of them produced AFB_1, AFB_2, AFG_1 and AFG_2 levels, whereas 27%, belonging to *A. flavus* and *A. candidus* strains, produced AFB_1, AFB_2 and AFG_1 levels. *Aspergillus parasiticus* produced higher AFB_1 and AFG_1 levels than the strains tested previously.

The potential for OTA production in YES medium was evaluated by Magnoli et al. (2002) from different feeds (poultry pig and rabbit). Eighty samples were taken at random from factories located in Córdoba Province, Argentina, over a period of 8 months. The highest percentage of ochratoxigenic strains was isolated from rabbit feeds; 100% of *Aspergillus niger* aggregate strains were OTA producers, with levels ranging from 13 to 16.5 ng ml^{-1}. From pig feeds, 61% of *A. niger* aggregate strains were OTA producers with levels ranging from 13 to 23.6 ng ml^{-1}. In poultry feeds, the lowest percentage of OTA producer strains was detected.

Rosa et al. (2006) analyzed the potential OTA production on Czapek yeast extract agar (CYA) from different *Aspergillus* species and *P. verrucosum* isolated from Brazilian poultry feed samples. All species, *A. niger* (24%), *A. carbonarius* (71.4%), *A. ochraceus* (25.6%), *A. melleus* (39%) and *P. verrucosum* (13%), were OTA producers. The maximum produced levels ranged between 25 and 120 μg kg^{-1} OTA. Minimum levels varied from 8 to 50 μg kg^{-1} OTA. *Aspergillus ochraceus* and *P. verrucosum* strains produced the highest OTA levels on CYA medium, followed by *A. melleus A. carbonarius* and *A. niger* OTA levels.

González Pereyra et al. (2008b) analyzed the toxigenic ability of *A. flavus* strains isolated from oat samples. All strains were able to produce AFB_1 at levels between 20 and 35 ppm.

Aspergillus flavus was the most frequent species isolated from various feeds. The frequency of potential aflatoxin producers shows the potential risk of aflatoxigenic production in feeds when environmental conditions are adequate.

14.2.2.3 Mycotoxin Incidence

Dalcero et al. (1997) informed the mycotoxins of Argentinian poultry feed samples obtained from two factories in the period 1996/1997. Aflatoxin B_1 was the predominant toxin detected and 14% of the samples were contaminated with levels ranging from 17 to 197 ng g^{-1}. Ten percent of the samples contained aflatoxin G_1 (AFG_1) and aflatoxin G_2 (AFG_2). *Fusarium* mycotoxins, DON and ZEA, were found in 8% and 1% of the samples at levels ranging from 240 to 410 ng g^{-1} and 82 to 99 ng g^{-1} respectively. Sampling in the 1997/1998 period showed that AFB_1 was the most frequent mycotoxins (48%), followed by ZEA in 15% of the samples (Dalcero et al. 1998) (Table 14.2).

The mycotoxin analysis done by Magnoli et al. (2002) in recent years showed that FBs had the highest incidence in the period 1998/1999, and were found in 97% of the analyzed samples, followed by AFB_1 (46% of the samples). Zearalenone and DON were detected in 17.7% and 6% of the samples, in levels from 1,550 to 4,507 ng g^{-1} and 124 ng g^{-1} respectively. With regard to the co-occurrence of mycotoxins, all of the FBs-contaminated feed samples were co-contaminated with AFB_1 (Table 14.2).

The natural incidence of OTA in corn based feeds (poultry, pig and rabbit feed) from Argentina was analyzed by Dalcero et al. (2002). This toxin was found in 38% of the poultry feed samples with levels ranging from 25 to 30 ng g^{-1}. From rabbit feed samples, 25% contained OTA, and the levels ranged from 18.5 to 25.5 ng g^{-1}. Only 13% of pig feed samples were contaminated with similar levels of toxins. In a later study in the same substrates OTA was detected in 10%, 15% and 12% of poultry, pig and rabbit feed samples, respectively (Table 14.2) (Magnoli et al. 2005).

In a recent work, González-Pereyra et al. (2008b), determined mycotoxin presence in initial, growing and final feed and corn grains intended for pig feeding in Argentina. Home-corn grains used in farms for feed elaboration showed field contamination and provided initial FBs levels present in the pig feed ($<10\ \mu g\ g^{-1}$). The same corn did not show AFs contamination. However, it was present in pig feed at levels from 30 to 70 ng g^{-1}. This study showed the simultaneous occurrence of two carcinogenic mycotoxins, AFB_1 and FB_1 in feed intended for pig consumption and suggested that the prevailing environmental conditions should be influencing AFs production by the potential aflatoxin producers present. The frequency of *F. verticillioides* and FB_1 contamination showed a positive correlation whereas *A. flavus* frequency and AFs contamination showed a negative correlation. On the other hand, high levels of AFB_1 were found in the contaminated samples (Table 14.2). Moreover, a negative correlation between FB_1 and AFB_1 contents was found, in agreement with Yoshizawa et al. (1996) and Picco et al. (1999).

The GMP regulations on standard products in the animal feed sector established that the current maximum permitted level for AFB_1 for pigs is 0.02 $\mu g\ g^{-1}$ (GMP 2005). Aflatoxin B_1 levels were higher than the recommended limits for complementary feeding stuffs for pigs. In our study, many samples contained FB_1 and FB_2 levels higher than the permissible limits.

In Río de Janeiro State, Brazil, Fraga et al. (2007) evaluated the natural occurrence of OTA from 144 samples of maize, initial poultry feeds, pelletized poultry feed and trough feed samples. Final poultry feed samples showed that 100% were contaminated with OTA at levels from 17 to 197 $\mu g\ kg^{-1}$. These authors concluded that although the pelleting process produced an important fungal reduction, this did not occur with mycotoxins. These results agree with Rosa et al. (2006), from the same investigation centre, who obtained the same percentage of OTA-contaminated samples at lower levels, which varied from 1.3 to 80 $\mu g\ kg^{-1}$.

The mycotoxin analyses of Brazilian equine samples showed that AFB_1 values ranged between 0.01 and 99.4 $\mu g\ kg^{-1}$. Fumonisin B_1 levels ranged between 0.01 and 7.49 $\mu g\ kg^{-1}$. (Table 14.2) (Keller et al. 2007).

From 50 Argentinian equine feed samples (maize, oat, ensiled alfalfa), González Pereyra et al. (2008c) found that 35% of the samples were contaminated with AFB_1 at levels that varied between 5 and 64 ppb. Twenty-eight percent of the samples were positive to ZEA contamination at detectable non-quantitative levels. Fumonisin B_1 was not detected. Oat samples were contaminated with AFB_1 and 20% of them showed levels over the recommended limit (20 ppb). All samples were contaminated with ZEA: maize (100%), oat (40%) and ensiled alfalfa (10%). The co-occurrence of AFB_1 and ZEA was observed in 40% of the analyzed samples (Table 14.2).

For animal feeds, the FDA advisory guidelines (Whitlow and Hagler, 2002) recommend that contaminated corn or corn by-products be limited to no more than 20% of the diet for equines. Furthermore, the recommended maximum concentration of total fumonisins in corn and corn by-products is 5 ppm for equines. Moreover, FDA has established nonbonding action levels as informal guidelines for its enforcements of AFs in feedstuffs. In general, all feedstuffs have an action level of 20 ppb. In the presented data, some sample feeds had higher AFB_1 levels than the established limit (20 ppb). In contrast, FB_1 levels in equine feeds did not exceed the limit levels. Even if the amount of detected mycotoxins is not enough to cause adverse effects in animals, the feed quality will be nutritionally lower.

14.2.3 Ready Pet Food

14.2.3.1 Mycoflora Incidence

Campos et al. (2008) evaluated the mycoflora from pet food ingredients [corn grains, corn and sorghum and corn meal and gluten (21%)] and ready pet foods. The total mold counts obtained from corn grains, corn and sorghum and corn meal and gluten (21%) were the highest in both media assayed. The isolated strains number ranged from 19 to 617. Sorghum meal and corn meal and gluten (21%) obtained the highest values whereas ready pet food had the lowest. All samples obtained at least one strain belonging to the main toxicogenic genera *Aspergillus* spp., *Penicillium* spp. and/or *Fusarium* spp. *Aspergillus* spp. were the prevalent genera in all substrates (65–89%). *Aspergillus flavus* was the most prevalent, followed by *A. sydowii*, *A. fumigatus* and *A. versicolor*. *Aspergillus flavus* frequencies ranged from 58% to 86% except from sorghum meal (1%). *A. fumigatus* was isolated from all samples (except sorghum meal) in frequencies lower than 25%. Corn meal and corn meal and gluten samples had 100% *F. verticillioides*. *F. verticillioides* (98%), *F. sporotrichioides* (1%) and *F. graminearum* (1%) were isolated from sorghum meal.

Another study of Brazilian commercial ready pet food (Campos et al. 2009) showed that fungal counts in DRBC and DG18 media ranged from not-detected (Nd) to 10^3 CFU g^{-1}. The main genus isolated was *Aspergillus* spp. (60%) followed by *Penicillium* spp. (20.5%) and *Fusarium* spp. (13.7%). Seven *Aspergillus* spp. were isolated. *A. flavus* and *A. parasiticus* were the most prevalent of all kinds of samples tested. They were followed by *A. niger and A. ochraceus* (lower than 15%), *A. fumigatus, A. candidus* and *A. flavipes* (lower than 8%). Similar results were informed in Argentina by Fernández-Juri et al. (2006), who analyzed raw materials and ready pet food obtained from a factory and commercial ready pet food (standard, premium and super premium).

Bueno et al. (2001) reported the mycoflora of 21 dry pet foods in Argentina (dogs and cats). The predominant genera were *Aspergillus* spp. (62%), *Rhizopus* spp. (48%), and *Mucor* (38%). The most prevalent species among *Aspergillus* spp. were *A. flavus* followed by *A. niger* and *A. terreus*.

14.2.3.2 Toxigenic Ability

On the other hand Campos et al. (2008) evaluated the toxigenic ability of *A. flavus* strains to produce AFs from pet food ingredients (corn grains, corn and sorghum and corn meal and gluten - 21%) and ready pet foods in Brazil. Seventy five percent of *A. flavus* strains isolated from raw materials were able to produce AFs, whereas toxigenic strains isolated from pet food were 57%. Aflatoxins toxigenic ability of *A. flavus* and *A. parasiticus* strains isolated from standard, premium and super-premium pet food samples was evaluated. One hundred percent of strains isolated from super premium samples were able to produce AFB_1, whereas toxigenic strains isolated from standard and premium samples varied from 80 to 100%. The results agree with Campos et al. (2009). The toxigenic capacity of *A. flavus* strains from Argentinian ingredients and commercial ready pet foods was reported. Fifty percent of strains were aflatoxin producers with levels ranging from 0.2 to 66.5 $\mu g\ g^{-1}$ (Fernández-Juri et al. 2006; 2007).

14.2.3.3 Mycotoxin Incidence

Aflatoxin analysis of 120 samples of raw materials and ready pet food from Brazil showed that AFs were detected in corn and sorghum meal, corn grain, corn meal and gluten at levels ranging from 0.1 to 24.5 ng g^{-1}. At least 90% of corn meal and sorghum meal samples were contaminated with AFs levels over 20 ng g^{-1}. Ready pet food samples did not show AFs contamination (Campos et al. 2008). Another study (Campos et al., 2009) showed that standard samples were the most contaminated with AFB_1. All kinds of samples tested had levels lower than 9.43 ng g^{-1}, and 90% of them did not achieve 4.7 ng g^{-1}. Fernández-Juri et al. (personal communication) showed 21% and 86% ready pet food samples contaminated with AFB_1 and FB_1 respectively (Table 14.2).

Maia and Pereira Bastos de Siqueira (2002) analyzed domestic pet foods (cat, dog and bird) to determine the occurrence of AFs as well as their risk to animal health in Paraná State, Brazil. The samples were collected from pet shops from southeast Brazil. Total aflatoxins were detected in 12% of the samples at levels ranging from 15 to 374 ng g^{-1}. Levels above the maximum limit established in Brazil (50 ng g^{-1}) for animal food were detected from 41.7% of the samples.

The high percentage of *A. flavus* aflatoxin-producing strains isolated from ready pet food indicates aflatoxicogenic resistance to manufacturing practices such as the extrusion process for obtaining finished pet food. These results are in agreement with those from conducted by Gunsen and Yaroglu (2002) and Maia and Pereira Bastos de Siqueira (2002), who found negative samples of aflatoxin contamination; the few positive samples contained <20 μg of aflatoxin B_1 per kilogram in commercial dog and cat foods. In contrast, Sharma and Marquez (2001) from Mexico found 88% contaminated pet food samples with high AFs levels.

14.3 Conclusions and Future Perspectives

In this chapter, have been reviewed mycotoxigenic fungi and mycotoxins in animal feed in South American countries.

Total mold counts of poultry, pig, rabbit, pet and equine feed samples showed that some samples analyzed exceeded the levels proposed as feed quality limits (1×10^{-4} CFU g^{-1}). The fungal counts for samples of various pet foods and ingredients were not over the value proposed for the three kind of feed (GMP 2005). The results showed that the *Aspergillus* section of *Flavi* species was isolated from various feeds at high frequency.

In general, from *Penicillium* genera, a great variety of species have been isolated from different feeds. *P. citrinum*, *P. rugulosum*, *P. crustosum*, *P. variabile*, *P. aurantiogriseum*, *P. verrucosum* and *P. purpurogenum* are potential toxic secondary metabolite producers, for example citrinin, OTA, penitren A, naftoquinones, xantomegnin, viomellein, vioxantin, rugulosins, rubratoxins and secalonic acid (Mills et al. 1995; Pitt and Hocking 1997). The toxicity of several of these compounds in experimental animals has been demonstrated. Therefore, *Penicillium* toxin presence involves potential animal health risk. Cereal grains naturally contaminated with OTA and citrinin could contain other minor secondary metabolites of these *Penicillium* species. There are few data on incidence of *Fusarium* species in animal feeds. Some of them come from poultry feed, and the predominant species are *F. verticillioides*, *F. proliferatum*, *F. nygamai* and *F. graminearum*.

Aspergillus toxigenic species are in co-occurrence with *Fusarium* spp. The simultaneous presence of mycotoxins has also become a significant issue with complex, and indeterminate implications for animal health and welfare.

The presence of many mycotoxins of toxicological importance in feedstuffs, such as aflatoxins and others, strongly affects animal production and health. In some substrates such as pet foods, AFB_1-producing strains (*A. flavus* and *A. parasiticus*) were isolated at a higher percentage in Latin America. Despite the fact that *A. parasiticus* was found in high frequency and there are data that show that this species is more stable regarding toxin production (Pitt and Hocking 1997).

The natural co-occurrence of toxigenic fungi and mycotoxins in different animal feeds has been demonstrated in South American countries. In animal production, the simultaneous occurrence of mycotoxins causes not only bad health in animals, but also low production.

Based on continuous testing of raw materials as well as ready product, feed should be continuously monitored. The known contamination of fungal species allows the prediction of the toxicological potential present in the feed. On the other hand, the presence of potential toxigenic species could indicate the presence of various mycotoxins. This fact represents potential risk to animals. Where action is necessary to detoxify contaminated materials, the choice depends on the consideration of the mycotoxins and species involved, in order to secure the safety and performance of farm animals in general and the whole food chain in particular. Blending with non-contaminated feed ingredients, re-routing contaminated grain to less susceptible animal species, and the addition of feed additives based on adsorptive capacity are widely used strategies to reduce mycotoxin-induced performance impairment.

References

Alonso VA, González Pereyra ML, Pereyra CM, Peirano S, Rosa CAR, Dalcero AM, Cavaglieri LR, Chiacchiera SM (2007) Mycoflora and aflatoxin/zearalenone contamination of animal feed and silage intended to dairy cattle feed. X International IUPAC Symposium. Mycotoxins and Phycotoxins. Istambul, Turkey

Amigot SL, Fulgueira CL, Bottai H, Basilico JC (2006) New parameters to evaluate forage quality. Postharvest Biol Technol 41:215–224

Auerbach H, Oldenburg E, Weissbach F (1998) Incidence of *Penicillium roqueforti* and roquefortine C in silages. J Sci Food Agric 76:565–572

Batatinha MJM, Simas MMS, Botura MB, Bitencourt TC, Reis TA, Correa B (2007) Fumonisins in brewers' grain (barley) used as dairy cattle feed in the State of Bahia, Brazil Food Control 18:608–612

Belkacemi L, Barton RC, Hopwood V, Evans EGV (1999) Determination of optimum growth conditions for gliotoxin production by *Aspergillus fumigatus* and development of a novel method for gliotoxin detection. Med Mycol 37:227–233

Binder EM, Tan LM, Chin LJ, Handl J, Richard J (2007) Worldwide occurrence of mycotoxins in commodities, animal feed and feed ingredients. Animal Feed Sci Technol 137:265–282

Boudra H, Morgavi DP (2005) Mycotoxin risk evaluation in feeds contaminated by *Aspergillus fumigatus*. Animal Feed Sci Technol 120:113–123

Boysen ME, Jacobsson KG, Schnürer J (2000) Molecular identification of species from the *Penicillium roqueforti* group associated with spoiled animal feed. Appl Environ Microbiol 66:1523–1526

Bueno DJ, Silva JO, Oliver GJ (2001) Mycoflora in commercial pet foods. Food Prot 64:741–743

Campos GS, Cavaglieri LR, Fernández-Juri MG, Krüger C, Dalcero AM, Magnoli CE, Rosa CAR (2008) Mycobiota and aflatoxins in raw materials and pet food in Brazil. J Animal Physiol Animal Nutr 92:377–383

Campos GS, Cavaglieri LR, Krüger C, Fernández-Juri MG, Dalcero AM, Magnoli CE, Rosa CAR (2009) Aflatoxigenic fungi and aflatoxins in commercial ready pet food in Rio de Janeiro State, Brazil. World Mycotox. J. 2:85–90

CAST, Council for Agricultural Science and Technology (2003) Mycotoxins: Risks in Plant Animal and Human Systems. Task Force Report No. 139. Ames, IA, USA

Cavaglieri LR, González Pereyra ML, Pereira CM, Magnoli CE, Chulze SN, Dalcero AM (2005) Fungal and mycotoxin contamination of cow feedingstuffs in Argentina. Reducing Impact of Mycotoxins in Tropical Agriculture Congress. Ghana, Africa

Céspedes AE, Diaz GJ (1997) Analysis of aflatoxins in poultry and pig feeds and feedstuffs used in Colombia. J AOAC Int 80:1215–1219

da Silva JB, Pozzi CR, Mallozzi MAB, Orteg EM, Corrêa B (2000) Mycoflora and occurrence of aflatoxin B_1 and fumonisin B_1 during storage of Brazilian sorghum. J Agric Food Chem 48:4352–4356

Dalcero A, Magnoli C, Chiacchiera S, Palacio G, Reynoso M (1997) Mycoflora and incidence of aflatoxin B_1, zearalenone and deoxynivalenol in poultry feeds in Argentina. Mycopathologia 137:179–184

Dalcero A, Magnoli C, Luna M, Ancasi G, Reynoso M, Chiacchiera SM, Miazo R, Palacio G (1998) Mycoflora and naturally ocurring mycotoxins in poultry feeds in Argentina. Mycopathologia 141:37–43

Dalcero A, Magnoli C, Hallak C, Chiacchiera SM, Palacio G, Da Rocha Rosa CA (2002) Detection of ochratoxin A in animal feeds and capacity to produce this mycotoxin by *Aspergillus* section *Nigri* in Argentina. Food Addit Contam 19:1065–1072

Di Costanzo A, Johnston L, Windels H, Murphy M (1995) A review of the effects of molds and mycotoxins in ruminants. Profess Anim Scient 12:138–150

Driehius F, Oude Elferink SJWH (2000) The impact of the quality of silage on animal health and food safety: a review. Vet Quart 22:212–217
dos Santos VM, Dorner JW, Carreira F (2003) Isolation and toxigenicity of *Aspergillus fumigatus* from moldy silage. Mycopathologia 156:133–138
El-Gazzar FE, Rusul G, Marth EH (1987) Growth and aflatoxin production by *Aspergillus parasiticus* NRRL 2999 in the presence of lactic acid and at initial different pH values. J Food Prot 50:940–944
El-Shanawany AA, Eman Mostafa M, Barakat A (2005) Fungal populations and mycotoxins in silage in Assiut and Sohag governorates in Egypt, with a special reference to characteristic Aspergilli toxins. Mycopathologia 159:281–289
Fernandez-Juri MG, Astoreca A, Barberis C, Asili R, Rosa CAR, Dalcero AM, Magnoli CE (2006) Mycoflora and aflatoxigenic strains in commercial pet food in Argentina. In: Tenth Congress of the European Society of Veterinary and Comparative Nutrition: Congress Proceedings. Nantes, APERDPA France, vol. 1. p 121
Fernández-Juri MG, Bressan F, Astoreca A, Barberis C, Asili R, Dalcero AM, Magnoli CE (2007) Evaluación de la micoflora y cepas aflatoxicogénicas en alimentos balanceados para perros en Argentina. XI Congreso Argentino de Ciencia y Tecnología de Alimentos (CYTAL), Buenos Aires, Argentina
Fraga ME, Curvello FA, Gatti MJ, Cavaglieri LR, Dalcero AM, Rosa CAR (2007) Potential aflatoxin and ochratoxin A production by *Aspergillus* species in poultry feed processing, Vet Res Comm 31:343–353
Garon D, Richard E, Sage L, Bouchart V, Pottier D, Lebailly P (2006) Mycoflora and multi-mycotoxin detection in corn silage: experimental study. J Agric Food Chem 54:3479–3484
Good Manufacturing Practices, GMP (2005) Regulations on product standards in the animal feed sector. GMP 14:17–08–05
González Pereyra ML, Alonso VA, Sager R, Morlaco MB, Magnoli CE, Astoreca AL, Rosa CAR, Chiacchiera SM, Dalcero AM, Cavaglieri LR (2008a) Fungi and selected mycotoxins from pre- and post-fermented corn silage. J Appl Microbiol 104:1034–1041
González Pereyra ML, Pereyra CM, Ramírez ML, Rosa CAR, Dalcero AM, Cavaglieri LR (2008b) Determination of mycobiota and mycotoxins in pig feed in central Argentine. Lett Appl Microbiol 46:555–561
González Pereyra ML, Tissera J, Manteña C, Cavaglieri LR, Keller KM, Rosa CAR, Magnoli CE, Dalcero AM (2008c). Mycotoxicological quality of equine feed in Argentina. In: Tenth Congress of the World Equine Veterinary Association, January 28–February 1, Moscow, Russia
Gotlieb A (1997) Causes of mycotoxins in silages. In: Proceedings from the 'Silage: field to feedbunk' North American Conference, Hershey, Pennsylvania February 11–13, pp 213–232
Gunsen U, Yaroglu T (2002) Aflatoxin in dog and horse feeds in turkey. Vet Human Toxicol 44:113–114
Keller KM, Queiroz BD, Keller LAM, Ribeiro JMM, Cavaglieri LR, González Pereyra ML, Pereyra CM, Dalcero AM, Rosa CAR (2007) The mycobiota and toxicity of equine feeds. Vet Res Comm 31:1037–1045
Kuiper-Goodman T (2004) Risk assessment and risk management of mycotoxins in food. In: Magan N, Olsen M (eds) Mycotoxins in food. detection and control. Woodhead Publishing Limited, England, pp 3–31
Land CJ, Lundstrom H, Wemer S (1993) Production of tremorgenic mycotoxins by isolated of *Aspergillus fumigatus* from sawmills in Sweden. Mycopathologia 124:87–93
Magnoli CE, Saenz MA, Ferrero S, Chiacchiera SM, Dalcero AM (1999) Natural occurrence of *Fusarium* species and fumonisins-production by toxigenic strains isolated from poultry feeds in Argentina. Mycopathologia 145:35–41
Magnoli C, Chiacchiera S, Miazzo R, Palacio G, Angeletti A, Hallak C, Dalcero AM (2002) The mycoflora and toxicity of feedstuffs from a production plant in Córdoba, Argentina. Mycotoxin Res 18:7–22

Magnoli C, Hallak C, Astoreca A, Ponsone L, Chiacchiera SM, Palacio G, Dalcero AM (2005) Surveillance of toxigenic fungi and ochratoxin A in feedstuffs from Córdoba Province, Argentina. Vet Res Comm 29:431–445

Maia PP, Pereira Bastos de Siqueira ME (2002) Occurrence of aflatoxins B_1, B_2, G_1 and G_2 in some Brazilian pet foods. Food Addit Contam 19:1180–1183

Mills JT, Seifert KA, Frisvad JC, Abramson D (1995) Nephrotoxigenic *Penicillium* species occurring on farm-stored cereal grains in western Canada. Mycopathologia 130:23–28

Mor S, Singh K (2000) Incidences of toxigenic molds in dairy foods and animal feeds. Indian J Anim Sci 70:766–768

Moreau C (1979) Moulds, toxins and food. Wiley, Chichester

Morgavi DP, Boudra H, Jouany JP, MichaletDoreau B (2004) Effect and stability of gliotoxin, an *Aspergillus fumigatus* toxin, on in vitro rumen fermentation. Food Addit Contam 21:871–878

Morgavi D, Riley RT (2007) An historical overview of field disease outbreaks known or suspected to be caused by consumption of feeds contaminated with *Fusarium* toxins. Review. Anim Feed Sci Technol 137:201–212

Oliveira GR, Ribeiro JMM, Fraga ME, Cavaglieri LR, Direito GM, Keller KM, Dalcero AM, Rosa CAR (2006) Mycobiota in poultry feeds and natural occurrence of aflatoxins, fumonisins and zearalenone in the Rio de Janeiro State, Brazil. Mycopahologia 162:355–362

Oude Elferink SJWH, Driehuis F, Gottschal JC (2000) Silage fermentation processes and their manipulation. In: FAO Electronic Conference on Tropical Silage, Rome. Silage making in the tropics with emphasis on smallholders. Proceedings. FAO, Rome, pp 17–30

Pacín AM, González HHL, Etcheverry M, Resnik SL, Vivas L, Espin S (2002) Fungi associated with food and feed commodities from Ecuador. Mycophatologia 156:87–92

Pereyra CM, Alonso VA, Fernández-Juri MG, González Pereyra ML, Chiacchiera SM, Rosa CAR, Dalcero AM, Cavaglieri LR (2007a) Survey of *Aspergillus fumigatus* and toxic metabolites from feed intended for animal production and pets. III Latin American Congress of Food Hygienists, Porto Seguro, Brazil

Pereyra CM, Alonso VA, Chiacchiera SM, Dalcero AM, Cavaglieri LR (2007b). Toxicidad de *Aspergillus fumigatus* aislados de ensilaje de maíz y alimento terminado. Incidencia natural de gliotoxina. XI Congreso Argentino de Microbiología, Córdoba, Argentina

Pereyra CM, Alonso VA, Rosa CAR, Chiacchiera SM, Dalcero AM, Cavaglieri SM (2008) Gliotoxin natural incidence and toxigenicity of *Aspergillus fumigatus* isolated from corn silage and ready dairy cattle feed. World Mycotox J 1:457–462

Picco M, Nesci A, Barros G, Cavaglieri LR, Etcheverry M (1999) Aflatoxin B_1 and fumonisin B_1 in mixed cultures of *Aspergillus flavus* and *Fusarium proliferatum* on maize. Nat Toxins 7:331–336

Pitt JI, Hocking AD (1997) Fungi and food spoilage. CSIRO Division of Food Science and Technology, Academic Press, Sydney, Australia

Puntenney SB, Wang Y, Forsberg NE (2003) Mycotic infections in livestock: recent insights and studies on etiology, diagnostics and prevention of Hemorrhagic Bowel Syndrome. In: Southwest Nutrition & Management Conference, Phoenix. University of Arizona, Department of Animal Science, Tuscon, pp 49–63

Ribeiro JMM, Cavaglieri LR, Direito GM, Fraga ME, Dalcero AM, Rosa CAR (2006) Influence of water activity, temperature and time on mycotoxins production on barley rootlets. Letts Appl Microbiol 42:179–184

Rosa CA, Ribeiro JMM, Fraga ME, Gatti MJ, Cavaglieri LR, Magnoli CE, Dalcero AM, Lopes CWG (2006) Mycoflora and ochratoxin-producing ability *Aspergillus* and *Penicillium* species isolated from poultry feed in Brazil. Vet Res Comm 113:89–96

Rosa CA, Keller KM, Keller LA, González Pereyra ML, Pereyra CM, Dalcero AM, Cavaglieri LR, López CW (2009) Mycological survey and ochratoxin A natural contamination of swine feedstuffs in Rio de Janeiro State, Brazil. Toxicon 53:283–288

Scudamore KA, Livesey CT (1998) Occurrence and significance of mycotoxins in forage crops and silage: a review. J Sci Food Agric 77:1–17

Seglar B (1999) Ruminant disorders associated with pathogens found within ensiled forages. The Bovine Proceedings 32:54–60

Seglar B (2003) Mycotoxin effects on dairy cattle. http://www.uwex. edu/ces/forage/wfc/proceedings2001/dairy mycotoxin.htm. Accessed 26 June 2008

Sharma M, Marquez C (2001) Determination of aflatoxins in domestic pet foods (dog and cat) using immunoaffinity column and HPLC. Anim Feed Sci Technol 93:109–114

Simas MMS, Botura MB, Correa B, Sabino M, Mallmann CA, Bitencourt TCBSC, Batatinha MJM (2006) Determination of fungal microbiota and mycotoxins in brewers grain used in dairy cattle feeding in the state of Bahia, Brazil. Food Cont 18:404–408

Tapia MO, Stern MD, Soraci AL, Meronuck R, Olson W, Gold S, Koski-Hulbert RL, Murphy MJ (2005) Patulin-producing molds in corn silage and high moisture corn and effects of patulin on fermentation by ruminal microbes in continuous culture. Anim Sci Technol 119:247–258

Upperman JS, Potoka DA, Zhang XR, Wong K, Zamora R, Ford HR (2003) Mechanism of intestinal-derived fungal sepsis by gliotoxin, a fungal metabolite. J Pediatric Surg 38:966–970

Watanabe A, Kamei K, Sekine T, Waku M, Nishimura K, Miyaji M, Kuriyama T (2003) Immunosuppressive substances in *Aspergillus fumigatus* culture filtrate J Infect Chemother 9:114–121

Whitlow LW, Hagler WM (2002) Mycotoxins in feeds. Feedstuffs 74:1–10

Yamada A, Kataoka T, Nagai K (2000) The fungal metabolite gliotoxin: immunosuppressive activity on CTL-mediated cytotoxicity. Immunol Lett 71:27–32

Yiannikouris A, Jouany JP (2002). Les mycotoxines dans les aliments des ruminants, leur devenir et leurs effets chez l'animal. INRA Prod Anim 15:3–16

Yoshizawa T, Yamashita A, Chokethaworn N (1996) Occurrence of fumonisins and aflatoxins in corn from Thailand. Food Addit Contam 13:163–168

Part III
Toxicology and Bioweapons

Chapter 15
Nivalenol: The Mycology, Occurrence, Toxicology, Analysis and Regulation

Yoshiko Sugita-Konishi and Takashi Nakajima

15.1 Introduction

Nivalenol (NIV) is a mycotoxin contaminated in wheat and barley affected by Fusarium head blight (FHB, Fig. 15.1). FHB is caused by the infection of wheat, barley and maize by *Fusarium*. *Fusarium* mycotoxins are produced in the infected plants and accumulate in agricultural products. The majority of *Fusarium* mycotoxins giving rise to food safety concerns are trichothecene mycotoxins possessing the 12,13-epoxy-trichothecene skeleton. These mycotoxins can be classified into two types: Type A including T2 toxin and HT2 toxin, diacetoxyscirpenol and neosolaniol, and Type B including deoxynivalenol (DON), NIV and fusarenon X (FX). The major pathogens causing FHB are *Fusarium graminearum* and *F. culmorum*, which are highly phytopathogenic fungi distributed worldwide, and which mainly produce DON and NIV. In Japan, the occurrence of these fungi has long been reported, and both DON and NIV were discovered by Japanese researchers. Taking advantage of the head blight epidemic across Western Japan in 1963, Tatsuno et al. (1968) successfully isolated the toxigenic fungus, *Fusarium nivale Fn-2B* (renamed *F. kyusyuense* in 1998, Aoki and O'Donnell 1998). In 1972, Yoshizawa and Morooka (1973) discovered DON in wheat and barley affected by FHB. The discovery of DON revealed the co-contamination with DON and NIV in domestic wheat and barley, and triggered co-contamination surveys around the world.

To date, there have been many reports of trichothecene co-contamination of wheat and barley. These reports have revealed that the regions suffering from mycotoxin contamination can be roughly divided into two types, those where DON is the main contaminant and those where co-contamination by DON and NIV is most common. Due to the global predominance of DON contamination,

Y. Sugita-Konishi (✉)
Division of Microbiology, National Institute of Health Sciences, 1–18–1, Kamiyoga, Setagaya-ku, Tokyo, 158–8501, Japan
e-mail: ykonishi@nihs.go.jp

M. Rai and A. Varma (eds.), *Mycotoxins in Food, Feed and Bioweapons*,
DOI: 10.1007/978-3-642-00725-5_15, © Springer-Verlag Berlin Heidelberg 2010

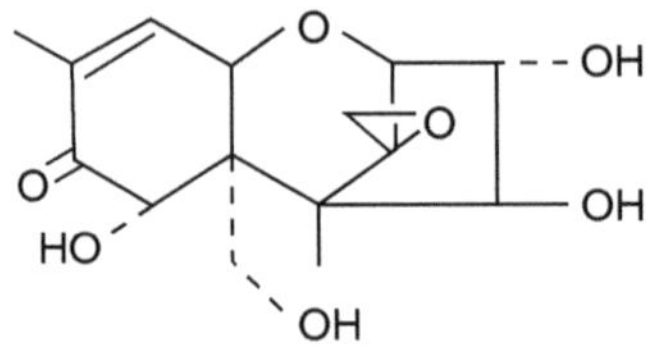

Fig. 15.1 Structure of NIV

DON has been extensively researched, and the risks it poses have already been evaluated by the Joint WHO/FAO Expert Committee on Food Additives (JECFA) (WHO 2001). However, the occurrence of NIV contamination is limited to some parts of Europe and Asia. Consequently, NIV has been poorly studied, and the risks have not been evaluated by JECFA. Between 1946 and 1958, Japan suffered frequent occurrences of food poisoning that were suspected to be caused by the presence of trichothecene mycotoxins, including NIV in *udon* (wheat noodles), *suiton* (wheat flour balls) and bread. In countries where NIV contamination of grains occurs, NIV-induced health damage is regarded as a serious problem and NIV is considered to be one of the mycotoxins that need to be regulated. In Europe, field surveys and risk assessment of trichothecene mycotoxins, including NIV, conducted from 2000 to 2002, revealed the existence of NIV contamination (European Commission Scientific Committee on Food 2003). The European Commission Scientific Committee on Food conducted a risk assessment of NIV and determined a provisional daily tolerable intake (European Commission. Scientific Committee on Food 2000). Nevertheless, the accumulation of further information will be of great importance in evaluating toxicity for NIV.

This chapter focuses on our studies and introduces the latest findings on the mycology, occurrences, toxicology, analytical methods, and control and regulation of NIV.

15.2 Mycology

FHB is a widespread and destructive disease that affects wheat, barley, and other cereals. This disease reduces grain yield and quality, and also results in the contamination of the grain with thrichothecene mycotoxins such as DON and NIV and other mycotoxins, all of which are harmful to human beings and animals. Several *Fusarium* species cause FHB; the pathogens can be divided into two chemotaxonomic groups, the DON chemotype and the NIV chemotype, based on the production of different trichothecenes (Ichinoe et al. 1983; Seo et al. 1996; Lee et al. 2002). A single gene (*Tri13*) that is responsible for the differential ability to produce DON or NIV has been identified (Lee et al. 2002). DON is the most prevalent mycotoxin in cereals (Placinta et al. 1999), and DON chemotypes of *Fusarium* are found worldwide. On the other hand, NIV chemotypes are found in more restricted regions; this chemotype has been found in Asia, Africa, and Europe, but not in North America (Mirocha et al. 1989; Kim et al. 2003). In this chapter, we will focus on NIV-producing *Fusarium* species, their mycological and phytopathological characteristics.

NIV was first isolated from *F. nivale* strain Fn-2B recovered from wheat infected with FHB in Kumamoto prefecture, Japan (Tatsuno et al. 1968). FX (Ueno et al. 1969) and diacetylnivalenol (Tatsuno et al. 1970) were isolated and characterized from the same strain. The toxin was named NIV after the fungal species "*nivale.*" However, Marasas et al. (1984, 1985) reexamined the toxigenic strains of *F. nivale* in the International Toxic Fusarium Reference Collection and reported that Fn-2B should be identified as *F. sporotrichioides* Sherb (W&R, G, B, J). Subsequently, Fn-2B was reclassified as a new species, *F. kyushuense* by Aoki and O'Donnell (1998) based on molecular phylogenetic analyses. *F. kyushuense* produces floccose mycelium with reddish pigmentation on potato dextrose agar (PDA), aseptate or septate conidia produced singly and holoblastically on the tips of sympodially proliferating aerial conidiophores, falcate to fusiform septate sporodochial conidia, and is characterized by the absence of chlamydospores and sclerotia (Aoki and O'Donnell 1998). The species *F. nivale* has been transferred from the genus *Fusarium* to the genus *Microdochium,* and is now considered to be *Microdochium nivale* (Ces. ex Berl. & Vogl.) Samuels & Hallett and its teleomorph is *Monographella nivalis* (Schaffnit) E. Müll.(Gams and Müller 1980; Gams 1989). *M. nivale* is a cryophilic fungus and a causal pathogen of pink snow mold and FHB in wheat, barley, and grasses; however, *M. nivale* does not produce any mycotoxins (Chelkowski et al. 1991; Nakajima and Naito 1995). We must now recognize that the trichothecene NIV is not associated with the fungal species name "*nivale.*"

The *F. graminearum* species complex (Ward et al. 2002), once considered a single species [*F. graminearum* Schwabe; teleomorph: *Gibberella zeae* (Schwein) Petch], is the most important causal organism of FHB (Parry et al. 1995). On PDA, the *F. graminearum* species complex produces abundant white mycelium that becomes yellow to brownish or rose as the cultures age; the color of the bottom surface of the colony is usually deep red. Macroconidia are usually long, slender, and slightly curved to straight, with five to six septa and a well-developed foot cell. Microconidia are absent; chlamydospores are rare but may form in macroconidia (Desjardins 2006). Molecular phylogenetic analyses using worldwide collections revealed that *F. graminearum* is a species complex consisting of at least nine biogeographically structured lineages (O'Donnell et al. 2000, 2004). Recently, the lineages attained species status (lineage 1: *F. austroamericanum*, lineage 2: *F. meridionale*, lineage 3: *F. boothii*, lineage 4: *F. mesoamericanum*, lineage 5: *F. acaciae-mearnsii*, lineage 6: *F. asiaticum*, lineage 7: *F. graminearum* s. str., lineage 8: *F. cortaderiae*, lineage 9: *F. brasilicum*; O'Donnell et al. 2004). The trichothecene metabolite profiles (chemotypes) of the *F. graminearum* species complex are strain-specific; some strains mainly produce DON and 3-acetyl deoxynivalenol (3ADON), some others produce DON and 15-acetyl deoxynivalenol (15ADON) and some produce NIV (Ichinoe et al. 1983; Seo et al. 1996; Lee et al. 2002). Phylogenetic analyses of the *F. graminearum* species complex have indicated that the trichothecene chemotype is not well correlated with the phylogeny inferred from sequences of six nuclear genes (O'Donnell et al. 2000; Ward et al. 2002) and that the three trichothecene chemotypes persisted through multiple speciation events (Ward et al. 2002). All three trichothecene chemotypes have been found in *F. graminearum* s. str. (Ward et al. 2002). However, regional differences in trichothecene chemotype compositions may be present in some species. The 3ADON type was not detected from 15 strains of *F. graminearum* s. str. and all 13 strains of *F. cortaderiae* were NIV type in New Zealand (Monds et al. 2005). Recently, *F. asiaticum* NIV producers have also been identified, and these may represent about 25% of the population in Louisiana, USA (Gale et al. 2005).

In Japan, *F. graminearum* s. str. is predominant in the northernmost island (Hokkaido district), while *F. asiaticum* is predominant in southern areas. In the northern part of the main island (Tohoku district), distinct co-localization of these species was observed. All 50 strains of *F. graminearum* s. str. were 15- or 3ADON types, while 173 (71%) out of 246 strains of *F. asiaticum* were NIV types. Suga et al. (2008) and Yoshida et al. (2004) hypothesized that the prevalence of the NIV chemotype of *F. asiaticum* on wheat and barley in the western part of Japan could be attributed to the double-cropping system for rice and wheat or barley in the area, because the NIV chemotype was associated with rice based on the mutual inoculation test. As for pathogenicity or aggressiveness, among the progeny of a cross between two strains of *F. graminearum*, DON producers were about twice as aggressive as NIV producers (Cumagun et al. 2004). In a Nepalese population, DON producers were more virulent than NIV producers (Desjardins et al. 2004). However, *F. asiaticum* NIV producers isolated from the western part of Japan were significantly more virulent than the most virulent DON chemotype *F. graminearum* s. str. strains (Nakajima and Yoshida 2007). There was no significant difference in the aggressiveness to wheat between DON producers and NIV producers (Carter et al. 2002). The level of pathogenicity or aggressiveness might be associated with not only trichothecene chemotype but also the amount of toxin produced (Atanassov et al. 1994).

Fusarium culmorum (W.G. Smith) Sacc. is the second most important pathogen of wheat, causing seedling blight, foot rot, and FHB. The pathogen is dominant in cooler areas such as Northern, Central and Western Europe, North America, and eastern Australia. *F. culmorum* produces abundant mycelium that turns yellow to rose as the culture ages on PDA. The color of the bottom surface of the colony is usually deep carmine red but may be brown in some strains. Macroconidia are usually abundant, short, stout, and thick walled, with three to five distinct septa. Chlamydospores are variably produced, and macroconidia are absent (Desjardins 2006). Strains of both DON producers and NIV producers were isolated from the same fields in England and Wales (Jennings et al. 2004). NIV chemotypes predominated in the south and west of England and Wales, whereas a greater proportion of DON chemotypes were found in the north and east of England (Jennings et al. 2004). In contrast, only DON producers were found in 42 of 42 strains of *F. culmorum* isolated from Western Canadian wheat (Abramson et al. 2001).

Fusarium poae (Peck) Wollenw. is usually considered a minor component of FHB in small grain cereals in Europe, North America, and Japan. *F. poae* produces dense white mycelium that may become reddish brown as the culture ages. The color of the bottom surface of the colony may vary from white to yellow to deep carmine red. Macroconidia are slightly curved to straight, have three septa, are wider above the middle septum, and are usually rare. Microconidia are globose to lemon-shaped, and are produced in abundance in distinctive clusters on monophialides (Desjardins 2006). In the northernmost area of Japan (Hokkaido district), where DON-producing *F. graminearum* s. str. prevail, NIV-producing *F. poae* were responsible for the contamination of grains with NIV (Sugiura et al. 1993). Furthermore, several isolates of *F. poae* distributed in Hokkaido possessed the ability to produce both type A (diacetoxyscirpenol) and type B [NIV and 4-acetyl-nivalenol (4ANIV)] trichothecenes (Sugiura et al. 1993). As in Japan, NIV-producing *F. poae* was frequently found in Sweden (Pettersson et al. 1995) and has become a major concern as a fungus responsible for NIV contamination in Scandinavia (Pettersson et al. 1995).

Fusarium crookwellense L.W. Burgess, P.E. Nelson & Toussoun was isolated for the first time in Australia in 1971 and described as a new species (Burgess et al. 1982). It is usually considered a minor component of FHB in small grain cereals in Poland, New Zealand, China, Canada, Japan, and other countries (Miller et al. 1991). *F. crookwellense* produces white to buff or yellow mycelium that may become grayish rose to red at the edge of the colony. The color of the bottom surface of the colony is deep red. Macroconidia are usually abundant, thick-walled, sickle-shaped, and widest at the midpoint, with five septa. Chlamydospores are abundant, but macroconidia are absent (Desjardins 2006). *F. crookwellense* isolated from scabby wheat in the northernmost area of Japan (Hokkaido district) produced NIV, 4ANIV, and zearalenone (ZEN) on rice medium (Sugiura et al. 1993). *F. crookwellense* isolated from dry rotted potato tubers in central Poland produced six mycotoxins (ZEN, α-trans-zearalenol, β-trans-zearalenol, fusarin C, and the trichothecenes FX and NIV) on both rice and corn substrates (Golinski et al. 1988). In Poland, reduction in grain yield by 32% and an accumulation of NIV at an average level of 0.15 mg kg^{-1} were reported after the artificial inoculation of oats with *F. crookwellense* (Mielniczuk et al. 2004).

15.3 Occurrence

15.3.1 Distribution of NIV Contamination

In contrast to the worldwide distribution of DON, NIV contamination is found only in limited areas. Table 15.1 shows a list of countries in which NIV contamination has been reported (Weidenborner 2008). Contamination of cereals (wheat, oats, barley, maize, rice, rye) has been frequently found in Europe (Germany, Poland,

Table 15.1 Global distribution of NIV in grains

Grains	Name of country
Wheat	Argentina, Brazil, Bulgaria, Canada, China, France, Finland, Germany, Greece, Hungary, Iran, Italy, Japan, Korea, Lithuania, The Netherlands, Nepal, Norway, Poland, Saudi Arabia, UK, USA, Yemen
Barley	Argentina, Canada, China, France, Finland, Germany, Italy, Japan, Korea, Lithuania, Nepal, Norway, Poland, Saudi Arabia, Yemen
Oats	Finland, Germany, Japan, Netherlands, Nepal, Norway
Rye	Canada, Finland, Germany, Korea, Lithuania, Netherlands
Maize	Canada, China, Indonesia, Italy, Japan, Korea, Nepal, South Africa, The Philippines, Poland
Rice	Australia, Japan, Korea, Nepal
Sorghum	Japan, Yemen
Soybean	Australia

From Weidenborner (2008)

Table 15.2 Occurrences of DON and NIV in commodities in EU country

Mycotoxin	Commodities	No. of country	No. of samples	Positive (%)	Range ($\mu g\ kg^{-1}$)	mean1 (<LOD = 0)	mean2 (<LOD = LOD)
DON	Total	11	11,022	57			
	Wheat-flour	11	6,358	61	2–500,000	205	293
	Barley	5	781	47	1.7–619	37	107
	Oats	5	595	33	2–5,004	95	253
	Ryes	5	271	41	2–595	42	95
	Corn	3	520	89	7–8,850	594	660
NIV	Total	7	4,166	16			
	Wheat flour	7	2,166	14	2–440	24	98
	Barley	4	521	8	1.7–351	15	76
	Oats	4	545	21	2–1,860	56	–
	Ryes	4	185	5	2–48	–	–
	Corn	2	268	35	7–340	–	–

From European Commission (2000)

Norway, Netherlands), Far East Asia (China, Korea, Japan), Southeast Asia (the Philippines, Vietnam), Oceania (New Zealand, Australia) and in Eastern Europe (Lithuania). The contamination of soybeans (50 $\mu g\ kg^{-1}$ of NIV) in Australia has also been reported (Yuwai et al. 1994).

Twelve European countries conducted a large-scale field survey of the contamination of food and food raw materials by trichothecenes (DON, NIV, FX, T2 and HT2 toxin, T2 triol, diacetoxyscirpenol, neosolaniol and verrucarol) (European Commission 2003). Table 15.2 shows the survey results regarding DON and NIV contamination in food raw materials. NIV was mainly detected in cereals; 16% of the 4,166 specimens were NIV positive. In detail of the contamination, 14% of 2,166 wheat and flour, 8% of 521 barley, 21% of 545 oats, 5% of 185 rye and 35% of 268 corn specimens were positive for NIV. The highest NIV concentration was observed in oats, although the concentration was not so high throughout the commodities surveyed. Compared with DON and NIV contamination, contamination with DON was higher than that with NIV, in terms both of level and frequency.

15.3.2 Occurrence of NIV in Japan

Japan, inhabited by DON/NIV-producing fungi, is known for frequent occurrence of DON/NIV co-contamination (Yoshizawa and Jin 1995). Since 2002, the Ministry of Agriculture, Forestry and Fisheries of Japan has been monitoring domestic wheat and barley. Tables 15.3 and 15.4 show the annual changes of DON and NIV concentrations and their contamination frequencies in wheat and barley. In response to the results of the 2002 surveillance, which revealed a relatively high DON level in domestic wheat, the Ministry of Health, Labor and Welfare of Japan have set 1.1 mg kg^{-1} as the provisional standard for DON in unpolished wheat. Since the setting of the provisional standard, the DON level in domestic wheat has decreased markedly, but the NIV level is relatively unchanged. In Europe, the level and frequency of DON contamination are often higher than those of NIV. Thus, it is generally believed that DON is more predominant than NIV and that their contamination levels shift in a parallel manner, i.e. when the DON level decreases, the NIV level also decreases. However, this is not the case in Japan, which indicates that there are some areas where the fungi-producing NIV only affects in Japan. These results imply the presence of a regional difference in DON/NIV-producing fungi. Areas frequently attacked by NIV contamination need to consider the establishment of a standard for NIV.

Table 15.3 Occurrences of DON in wheat and barley in Japan

Mycotoxin	Commodities	Fiscal year	Number of samples	Number of >LOQ	Max (mg kg^{-1})	Average (mg kg^{-1})	Frequency (% of >LOQ)
DON	Wheat	2002	199	81	2.1	0.16	40.7
	Wheat	2003	213	77	0.58	0.07	36.1
	Wheat	2004	226	81	0.93	0.04	35.8
	Wheat	2005	200	72	0.23	0.02	36.0
	Wheat	2006	100	84	0.88	0.13	84.0
	Barley	2002	50	–[a]	4.8	0.26	–
	Barley	2003	56	33	1.8	0.24	58.9
	Barley	2004	53	20	3.7	0.29	37.7
	Barley	2005	50	27	0.46	0.60	54.0
	Barley	2006	10	10	2.5	0.55	100.0

LOQ: 0.01 mg kg^{-1} (2002, 2004–2006), 0.05 mg kg^{-1} (2003), LOD: 0.002 mg kg^{-1} (2004–2006)
[a]not published
From Web sites of Ministry of Agriculture Forestry and Fisheries
http://www.maff.go.jp/syohi_anzen/kabi/pdf/20030509press_2.pdf Accessed 20 Dec 2007
http://www.maff.go.jp/syohi_anzen/kabi/pdf/20040427press_1.pdf. Accessed 20 Dec 2007
http://www.maff.go.jp/syohi_anzen/kabi/pdf/20050317press_5.pdf. Accessed 20 Dec 2007
http://www.maff.go.jp/j/press/cont2/20060523press_9b.pdf. Accessed 20 Dec 2007
http://www.maff.go.jp/j/press/2007/20070615press_3b.pdf. Accessed 20 Dec 2007

Table 15.4 Occurrences of NIV in wheat and barley in Japan

Mycotoxin	Commodities	Fiscal year	Number of samples	Number of >LOQ	Max ($mg\ kg^{-1}$)	Average ($mg\ kg^{-1}$)	Frequency
NIV	Wheat	2002	199	–[a]	0.64	0.06	–
	Wheat	2003	213	69	0.55	0.04	32.3
	Wheat	2004	226	108	0.55	0.003	47.8
	Wheat	2005	200	89	0.20	0.01	44.5
	Wheat	2006	100	70	1.0	0.09	70.0
	Barley	2002	50	–[a]	1.2	0.16	–
	Barley	2003	56	33	0.95	0.13	58.9
	Barley	2004	53	20	1.2	0.20	37.7
	Barley	2005	50	27	0.38	0.04	54.0
	Barley	2006	10	9	3.0	0.58	90.0

LOQ: 0.006 mg kg^{-1} (2004–2006), LOD: 0.05 mg kg^{-1} (2002–2003), 0.002 mg kg^{-1} (2004–2006)

[a]not published

From Web sites of Ministry of Agriculture Forestry and Fisheries
http://www.maff.go.jp/syohi_anzen/kabi/pdf/20030509press_2.pdf. Accessed 20 Dec 2007
http://www.maff.go.jp/syohi_anzen/kabi/pdf/20040427press_1.pdf. Accessed 20 Dec 2007
http://www.maff.go.jp/syohi_anzen/kabi/pdf/20050317press_5.pdf. Accessed 20 Dec. 2007
http://www.maff.go.jp/j/press/cont2/20060523press_9b.pdf. Accessed 20 Dec 2007
http://www.maff.go.jp/j/press/2007/20070615press_3b.pdf. Accessed 20 Dec 2007

15.4 Toxicology

15.4.1 Acute Toxicity

Studies on the acute toxicity of NIV using ddY mice revealed that oral administration (LD_{50} of 38.9 mg kg^{-1}) of NIV is less toxic than its interperitoneal (ip., LD_{50} of 7.4 mg kg^{-1}), intravenous (iv, LD_{50} of 7.3 mg kg^{-1}) or subcutaneous (sc, LD_{50} of 7.2 mg kg^{-1}) administrations. The reported causes of death were congestion and hemorrhage of the gastrointestinal tract (Ryu et al. 1988). In a study using F344 rats, the oral LD50 was 19.5 mg kg^{-1} bw (Kawasaki et al. 1990) but was 0.9 mg kg^{-1} bw via sc administration (Ueno 1983). Onji et al. (1989) reported that 87% of the dose was eliminated in the feces, so the relatively low oral toxicity can be attributed to the poor absorption of NIV. In our toxicokinetics study using NIV and a NIV precursor (FX), labeled with radioisotopes (Poapolathep et al. 2004), the plasma radioactivity reached a peak between 30 and 60 min after the administration of ^{3}H-NIV. The plasma peak level was five times higher, and the area under curve was ten times higher, in ^{3}H-FX-administered than in ^{3}H-NIV-administered mice. FX is absorbed from the gastrointestinal tract more rapidly and more efficiently than NIV. This study demonstrated that the higher oral toxicity of FX compared to NIV that has been observed in mice and rats is due to the more efficient absorption from the gastrointestinal tract of FX than NIV, followed by its rapid conversion to NIV by the liver and kidney. These results confirmed the poor absorption of NIV from the gastrointestinal tract compared to FX, supporting the explanation for the lower toxicity of orally administered NIV compared to that administered through other routes. Recent reports supported the finding that NIV absorption involves its binding to the ABC transporters in the gastrointestinal tract (Tep et al. 2007), providing an explanation for the poor absorption of NIV.

15.4.2 Chronic Toxicity

There are reports on 28-day, 1-year and 2-year repeated dose studies aimed at determining the no-observed-adverse-effect level (NOAEL), low-observed-effect level (LOEL) or low-observed-adverse-effect level LOAEL) of NIV, (Ryu et al. 1987; Ohtsubo et al. 1989; Yamamura et al. 1989; Kawasaki et al. 1990; Yabe et al. 1993), but the 1-year and 2-year studies were conducted using a rice medium containing NIV instead of purified toxin. Since the influence of components other than NIV cannot be eliminated from such data, the adequacy of the data for determining the NOAEL or NOEL has been questioned. Therefore, we isolated pure NIV from a culture of *F. kyusyuense* and used it in a 90-day repeated dose study using F344 rats (Takahashi et al. 2008).

Rats, 5 weeks old, ten males and ten females, were given powdered diet containing 0 (control), 6.25, 100 ppm NIV (corresponding to doses of 0.4, 1.5, 6.9 mg kg^{-1} day^{-1} in males and 0.4, 1.6, 6.4 mg kg^{-1} day^{-1} in females) for the treatment period of 13 weeks, except for a one-night fasting prior to the scheduled sacrifice. The animals were given the diet ad libitum, and observed daily for clinical signs and mortality, and body weights were measured every week during the study period. At the end of day 90, all animals were anesthetized with ether and weighed, and blood samples were collected for hematology and serum biochemistry tests.

In both sexes, body weight was suppressed at 100 ppm from week 1 of the experiment. Body weight suppression was also evident at 25 ppm from week 4 in females and from week 6 in males. With regard to organ weights, decrease in the absolute liver weight and increase in the relative weights of the brain, lungs, heart, kidneys and testes were found in males from 25 ppm. However, an increase in the relative liver weight and a decrease in the absolute weights in the brain, lung, heart, spleen and kidneys were observed at 100 ppm. In the thymus, both absolute and relative weights decreased at 100 ppm. In females, the relative weights of the lungs, heart, spleen and kidney increased in a dose-dependent manner from 25 ppm, but the absolute weight of these organs were unchanged or decreased. Similarly, a decrease in the absolute weight and an increase in the relative weight were observed in the brain, liver, and kidneys at 100 ppm.

Regarding hematology, in males, a significant increase of mean corpuscular volume was observed from 25 ppm, and an increase of mean corpuscular hemoglobin and decreases of red blood cell, platelet count (Plt) and white blood count (WBC) were detected at 100 ppm. In females, a decrease of WBC was observed from 6.25 ppm in a dose-dependent manner, and slight decreases of hemoglobin concentration and Plt were found at 100 ppm (Table 15.5).

Table 15.5 Hematological data for F344 rats fed diets containing NIV for 90 days

NIV in diet (ppm)	0	6.25	25	100
Males				
RBC($\times 10^4$ ml^{-1})	–	→	→	↓
Hb (g dl^{-1})	–	→	→	→
HCT (%)	–	→	→	→
MCH (pg)	–	→	↓	↓↓
Plt ($\times 10^4$ ml^{-1})	–	→	→	↓↓
WBC ($\times 10^4$ ml^{-1})	–	→	→	↓↓
Female				
RBC($\times 10^4$ ml^{-1})	–	→	→	→
Hb (g dl^{-1})	–	→	→	↓↓
HCT (%)	–	→	→	→
MCH (pg)	–	→	→	→
Plt ($\times 10^4$ ml^{-1})	–	→	→	↓↓
WBC ($\times 10^4$ ml^{-1})	–	↓	↓	↓↓

→ no change compared to control (0 ppm)
↓ significant reduction ($p < 0.05$)
↓↓ significant reduction ($p < 0.01$)
RBC red blood cell count, *Hb* hemoglobin concentration, *Ht* hematocrit, *MCV* mean corpuscular volume, *MCH* mean corpuscular hemoglobin, *MCHC* mean corpuscular hemoglobin concentration, *Plt* platelet count, *WBC* white blood cell count

Treatment-related typical changes in histopathological data were recognized in immune-competent organs and reproductive organs (Fig. 15.2). In both sexes, atrophy of the thymus and reduction of hematopoietic cells in the bone marrow were apparent at 100 ppm. In the reproductive organs, an increase of ovarian atretic follicles and interstitial glands was apparent at 100 ppm, the lack of corpora lutea development being apparent in severely affected cases, although the number of secondary follicles remained unchanged in the control. Atrophy with diestrus endometrial mucosal change of the uterus was observed at 100 ppm, and the vagina of these animals demonstrated proestrus or diestrus of the estrus cycle (Fig. 15.3).

In summary, this study showed that NIV targets endocrine organs (especially the anterior pituitary in both sexes and the female reproductive organs) and hematopoietic and immune organs in rats after 90-day exposure through diet. Based on the hematological data, the NOAEL of NIV was determined to be less than 6.25 ppm (0.4 mg kg^{-1} day^{-1}).

15.4.3 Immunotoxicity

The immunotoxicity of NIV has been investigated, and Hinoshita et al. (1997) published an interesting report on the induction of IgA nephropathy by NIV in mice, while Hedman et al. (1997) reported that NIV decreased the numbers of CD4+, CD8+T cell and IgM+subpopulations in the spleen of swine in a 3-week dosing study. Choi et al. (2000) have reported the *in vitro* immunotoxicity of NIV.

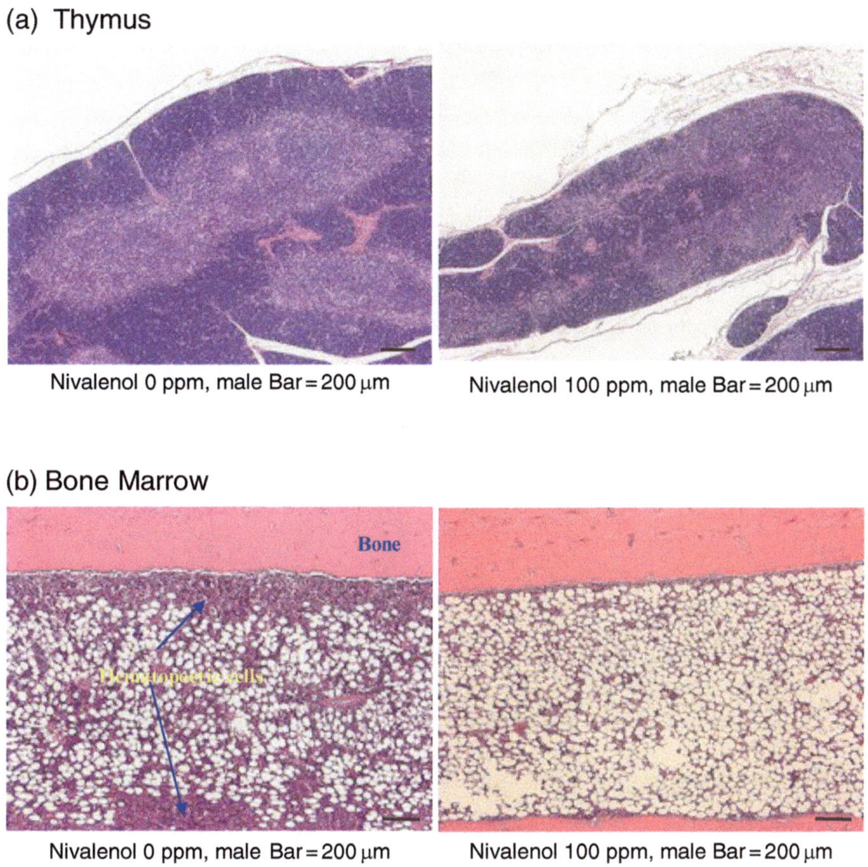

Fig. 15.2 The histopathological changes in the hematological organ. **a** Thymus. **b** Bone marrow

In our study (Kubosaki et al. 2008), we analyzed the influence of NIV on immunoglobulins, B and T cell populations and NK cells using the male rats used in the 90-day repeated dose study already described above (Takahashi et al. 2008).

IgG, IgM and IgA subclass antibodies in the serum samples were measured by enzyme-linked immunosorbent assay. A slight increase was observed in the IgM level at 100 ppm, but the mean levels of IgG and IgA were not affected by NIV treatment. Although dysregulation of IgA production has been observed in mice after NIV feeding for 8 weeks (Hinoshita et al. 1997), there were no changes in the production of IgA in the rat. Accordingly, it is suggested that there is a species difference for the development of IgA nephropathy.

For the spleen cell subset analysis, spleen cells were stained with a combination of the mAb mouse monoclonal cell-surface antibodies with fluorescein isothiocyanate or phycoerythrin. Table 15.6 summarizes the immunological effect of NIV. Flow cytometry analysis of the T and B lymphocyte populations in the spleen revealed that the T lymphocyte/B lymphocyte ratios ($CD3^+/B220^+$) were

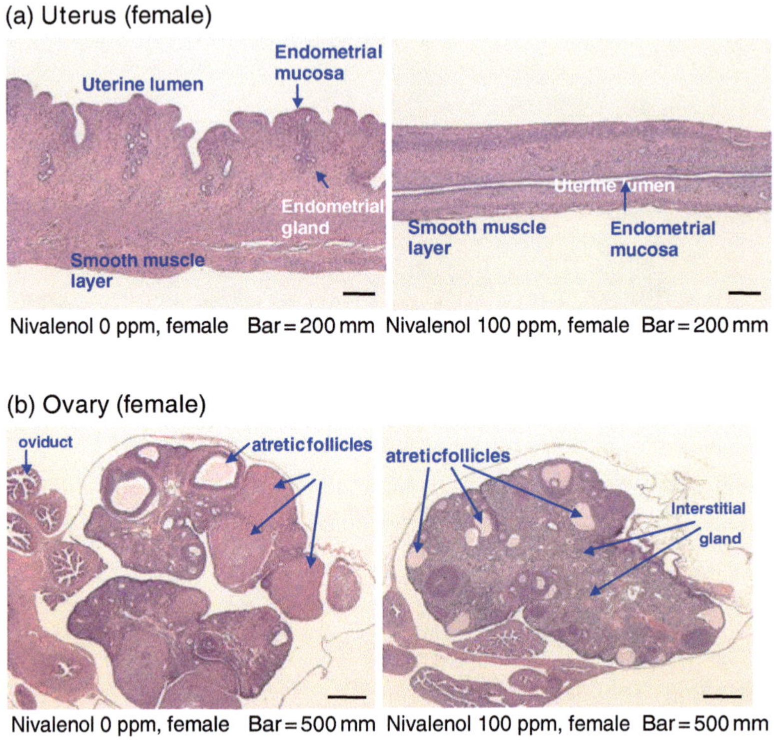

Fig. 15.3 The histopathological changes in the reproductive organ. **a** Uterus (female). **b** Ovary (female)

Table 15.6 The immunological effect of NIV in F344 rats after oral administration for 90 days

Nivaleno in diet	T cell/B cell ratio	CD4/CD8	NK activity ratio
0 ppm	–	–	–
6.25 ppm	→	→	↑
25 ppm	→	↑	↑
100 ppm	↓	↑	↑

These experiments were performed in male rats
→ No change compared to control (0 ppm)
↓ Significant reduction ($p < 0.05$)

decreased in a dose-dependent manner from 25 ppm, while the means of the $CD4^+$/$CD8^+$ratios were not changed at 6.25 and 25 ppm as compared with the untreated control value, but were increased significantly at 100 ppm.

NK activity in spleen cells was assayed according to the method of Arase et al. (1999) using YAC-1 target cells labeled with PKH2 green fluorescent dye. Following incubation, dead cells were stained with propidium iodide (PI), and determined by analyzing with a FACSCalibur. The population of NKR-P1A^+cells as an indicator of NK cells decreased at 100 ppm. The results obtained from this study are parallel to the results of the previous in vitro study using human peripheral blood mononuclear cells (Berek et al. 2001). On the other hand, splenic NK activity was unexpectedly enhanced by exposure to NIV from the lowest dose level in the present study. Since the activity of NK cells in athymic nude mice is enhanced compared to normal mice, enhanced NK activity in NIV-treated rats may reflect complementation to the reduction/depletion of T lymphocytes (Hasui et al. 1989). With regard to the immunotoxicity, an effect was apparent from 25 ppm judging from the increase in the splenic B-cell population, while an increase in NK activity was apparent from 6.25 ppm as a signature of beneficial host-defense responses.

15.4.4 Carcinogenesis

Although there have been reports of NIV being responsible for developmental toxicity (Ito et al.1988) and carcinogenicity (Ohtsubo et al. 1989) the International Agency for Research on Cancer (IARC 1993) concluded that NIV is not carcinogenic. There have since been several reports that NIV acts as a promoter (Ueno et al. 1991, 1992; Hsia et al. 2004). These reports support the hypothesis that NIV causes an enhancing effect on aflatoxin B_1-induced hepatocarcinogesis.

15.5 Analysis

15.5.1 Detection Methods for Trichothecenes Including NIV

The most effective way to minimize the consumption of mycotoxins is to set regulations for mycotoxins in foods and constantly monitor them. In order to consolidate such monitoring systems, accurate, reliable analytical methods are needed. Moreover, risk assessments conducted for setting standards require estimation of mycotoxin intake. To obtain accurate results from contamination surveys, analytical methods with higher sensitivity and reliability are required.

Since NIV occurs as a co-contaminant with other trichothecene mycotoxins, it is often analyzed simultaneously with the co-contaminants rather than alone. Analytical methods developed so far include: thin layer chromatography (TLC); capillary gas chromatography (GC) with electron-capture detection, flame ionization detection (FID) or mass spectrometric detection (GC/MS); high-performance liquid chromatography (HPLC) with ultra violet (UV), fluorescence or mass spectrometric detection; supercritical fluid chromatography; and time-of-flight mass spectrometry (LC/TOF-MS). The most recently developed methods (listed last) are expected to be continuously improved.

The GC method is typically used for simultaneous analysis for trichothecene mycotoxins and ZEN, but requires trimethylsilyl derivatization before the analysis.

However, a more recently developed analytical method using LC/MS can also simultaneously analyze for several trichothecene mycotoxins and ZEN, but requires no TMS derivatization pretreatment of the extract.

To date, two official methods for analyzing trichothecenes have been established by the AOAC International; one is method No. 986.17 based on TLC, and the other is the GC method validated as method No. 986.18.

So far, there are no methods validated as a common European standard method, but the GC-FID method produced good results in a validation in the European Union Standards, Measurements and Testing Programme (Schothorst and Jekel 2001). The European Committee for Standardization (CEN) Technical Committee 275, Working Group 5 "Biotoxins" has established minimum requirements for the performance characteristics that mycotoxin methods should meet (European Commission. Scientific Committee on Food 1999). According to these standards, analytical methods for DON and NIV contained at concentrations of 100 μg kg^{-1} or higher are required to have a recovery of 100% and $R.S.D_r$ and relative between laboratory standard deviation ($R.S.D_R$) of 20 and 40% or less, respectively.

In Japan, a method for measuring DON by HPLC-UV was validated when the provisional standard for DON was set in 2002 (Sugita-Konishi et al. 2006). Methods capable of measuring NIV have also been established. These include the method using LC/MS, which can measure six trichothecenes and ZEN in wheat, and a method using LC/TOF-MS, which can simultaneously analyze for 13 mycotoxins including aflatoxins (Tanaka et al. 2006).

15.5.1.1 LC-MS Analytical Method for the Determination of Trichothecenes

We compared three ionization methods; electrospray ionization (ESI), atmospheric pressure chemical ionization (APCI) and atmospheric pressure photo ionization (APPI) in LC-MS for determination of NIV. The ESI method ionizes the analytical samples through collision with thermions; the APCI method ionizes the analytical samples by corona discharge and is suitable for ionizing medium-polarity compounds; and the APPI method ionizes the samples with light and is suitable for ionizing low- to medium-polarity compounds.

Since trichothecene mycotoxins such as NIV and DON are low- to medium-polarity compounds, the APCI and APPI methods are expected to achieve higher analytical selectivity and sensitivity than the ESI method. To confirm this, we compared the three ionization methods coupled with LC-MS in the simultaneous analysis of six trichothecenes, and a frequent co-contaminant, ZEN, in wheat. Clean-up was performed by solid-phase extraction using a multifunctional column (Multisep #226 AflaZon+, Romer Labs Inc.).

HPLC was performed under the following conditions for all three ionization methods:
Column: Shimadzu Shim-Pack VP-ODS (150 mm × 2 mm)

Mobile phase: A: methanol, B: 5 mM ammonium acetate, 10% A/B – (5 min) – 10% A/B – (15 min) – 100% A/B – (10 min) – 100% A/B – (30 min) – 10% A/B
Flow rate: 0.1 ml min^{-1}, column oven temperature: 40°C, injection volume 10 μl.

In order to examine the extraction and purification efficiencies for seven mycotoxins (six trichothecenes and ZEN) from wheat, the wheat matrix was subjected to shaking extraction in acetonitrile-water (85:15, v/v) for 30 min, the obtained extract was centrifuged, and the supernatant was loaded onto the multifunctional solid-phase extraction column. Nine milliliter of the eluant from the column was collected in 1-ml fractions, and each fraction was concentrated to dryness under a nitrogen stream. After drying, each fraction was dissolved in 1 ml of 5 mM ammonium acetate and subjected to LC/MS measurement.

For the ESI method, NIV 371(−), DON 355(−), FX 413(−), 3ADON 397(−), HT2 483(−), T2 484(+) and ZEN 317(−) were selected as the monitoring ions. The standard curves for all seven mycotoxins showed linearity at concentrations above 5 ng ml^{-1} (correlation coefficients of 0.99 or higher), and the limits of quantification (S/N ratio of 10) and detection (S/N ratio of 3) were 0.79 and 0.24 ng ml^{-1} for NIV, 2.84 and 0.85 ng ml^{-1} for DON, 2.55 and 0.77 ng ml^{-1} for FX, 0.32 and 0.10 ng ml^{-1} for 3ADON, 2.16 and 1.35 ng ml^{-1} for HT2, 4.50 and 0.65 ng ml^{-1} for T2 and 0.56 and 0.17 ng ml^{-1} for ZEN, respectively.

For the APCI method, NIV 371(−), DON 355(−), FX 413(−), 3ADON 339(+), HT2 483(−), T2 484(+) and ZEN 317(−) were selected as the monitoring ions, based on the results of a full scan analysis.

The standard curves for all seven mycotoxins exhibited linearity at concentrations above 3 ng ml^{-1} (correlation coefficients of 0.99 or higher), and the limits of quantification and detection were 0.84 and 0.25 ng ml^{-1} for NIV, 1.50 and 0.45 ng ml^{-1} for DON, 2.11 and 0.63 ng ml^{-1} for FX, 2.78 and 0.83 ng ml^{-1} for 3ADON, 2.29 and 0.50 ng ml^{-1} for HT2, 1.67 and 0.68 ng ml^{-1} for T2 and 1.43 and 0.43 ng ml^{-1} for ZEN, respectively.

For the APPI method, NIV 371(−), DON 355(−), FX 413(−), 3ADON 397(−), HT2 483(−), T2 484(+) and ZEN 317(−) were selected as the monitoring ions. The standard curves for all seven mycotoxins exhibited linearity at concentrations above 3 ng ml^{-1} (correlation coefficients of 0.99 or higher), and the limits of quantification and detection were 0.63 and 0.18 ng ml^{-1} for NIV, 0.83 and 0.25 ng ml^{-1} for DON, 2.58 and 0.77 ng ml^{-1} for FX, 0.70 and 0.21 ng ml^{-1} for 3ADON, 9.71 and 2.91 ng ml^{-1} for HT2, 1.45 and 0.43 ng ml^{-1} for T2 and 0.46 and 0.13 ng ml^{-1} for ZEN, respectively.

For NIV, DON, T2 and ZEN, which are relatively frequently occurring contaminants, the APPI method exhibited a higher sensitivity compared to the ESI and APCI methods.

When wheat was used as the matrix the elution pattern from the clean-up columns (Multisep #226 AflaZon+, Romer Labs Inc.) was such that reasonable recovery rates (80% or higher) for the seven mycotoxins could be achieved in the fourth and later fractions.

The results of the present study demonstrated that the method using multifunctional solid-phase extraction columns (MultiSep #226) for sample clean-up and measurement by LCMS with APPI ionization can be used to analyze for the six tested trichothecenes including NIV, and for ZEN, a frequent co-contaminant with these mycotoxins.

15.6 Regulation

Establishment of regulation requires risk assessment. Among the trichothecene mycotoxins, only DON, T2 and HT2 toxin have been subjected to JECFA risk assessments (WHO 2001). The EC/SCF conducted a risk assessment on NIV, in addition to the above three trichothecene mycotoxins, at a meeting held in 2002. In response to this, 37 countries have imposed regulations on DON, and about ten European countries (other than EU member countries) have established regulation values for T2 toxin. Recently, the EU has begun to move toward establishing standards for T2 and HT2 toxin. Meanwhile, no country has established a standard for NIV.

As a result of the EC/SCF risk assessment on NIV, the daily tolerable intake (PMTDI) for NIV has been set at 0.7 $\mu g\ kg^{-1}$ bw day^{-1}, based on the results of the 1- and 2-year repeated dose studies reported by Ryu et al. (1988) and Ohtsubo et al. (1989) respectively in 1990. However, the PMTDI should be reviewed, based on recent reports on the sub-acute repeated dose toxicity studies using pure NIV in rats and mice (Takahashi et al. 2008; Gouze et al. 2007; Kubosaki et al. 2008) (Table 15.7).

15.7 Conclusion

- Several *Fusarium* species are responsible for FHB. The pathogens can be divided into two chemotaxonomic groups, the DON chemotype and the NIV chemotype, based on the production of different trichothecenes. A single gene (*Tril* 3) responsible for the differential ability to produce DON or NIV has been identified.
- In limited areas of the world, namely Far East Asia, Southeast Europe and Oceania, DON and NIV are co-contaminated in wheat.
- NIV possesses acute, chronic, immunological toxicities as well as DON. The effect of NIV on hematopoietic organ has been reported to be stronger than those of DON. However, as for carcinogenicity of trichothecenes including DON and NIV, the International Agency for Research on Cancer has concluded that trichothecens are not carcinogenic.
- The state of the art of analytical methods for determination of trichothecenes is simultaneous analysis using LC-MS, LS-MS/MS and LC-TOF /MS etc. Since it is generally accepted that *Fusarium* species produce some trichothecenes simultaneously, the methods would be effective for surveying trichothecenes occurrence.

Table 15.7 NOAEL or LOAEL data for NIV vs oral administration

Animal	Route	Sex	Critical effect	NOAEL	LOAEL	Reference
Rat (15–30 days)	Gavage (pure)	♀	Increased liver and spleen weight	0.4 mg kg^{1} bw	2 mg kg^{1} bw	Kawasaki et al. (1990)
Rat (2–4 weeks)	Diet	♂	Organ weights, drug metabolize enzyme		6 mg kg^{1}	Yabe et al. (1993)
Rat (8 weeks)	Diet	♂	Decreased body weight		50 mg $diet^{1}$	Onji et al. (1990)
Rat (13 weeks)	Diet (pure)	♀	Leukopenia		0.4 mg kg^{1} bw	Takahashi et al. (2008)
Rat (13 weeks)	Diet (pure)	♂	Increased NK cell activity		0.4 mg kg^{1} bw	Kubosaki et al. (2008)
Mouse (24 days)	Diet	♀	erythrocytopenia	10 mg kg^{1} bw	30 mg kg^{1} bw	Ryu et al. (1987)
Mouse (12 weeks)	Diet	♂♀	Increased alkaline phosphatase activity		6 mg kg^{1} bw	Yamamura et al. (1989)
Mouse (12 weeks)	Diet (3 times/ week, 4 weeks	♂	Increased glutathione transferase activity	1.774 mg kg^{1} bw	0.87 mg kg^{1} bw	Gouze et al. (2007)
Mouse (6 months– 1 year)	Diet	♀	Reduced body weight, leukopenia		0.68 – 0.76 mg kg^{1} bw	Ryu et al. (1988)
Mouse (2 years)	Diet	♀	Reduced body weight		0.66 mg kg^{1} bw	Ohtsubo et al. (1989)
Mouse (8 weeks)	Diet	♀	Elevated IgA levels		6 mg kg^{1} bw	Hinoshita et al. (1997)
Swine (3 weeks)	Diet	♂	Reduced spleen cell number		0.1 mg kg^{1} bw	Hedman et al. (1997b)
Swine (3 weeks)	Diet	♂	Decreased IgG, increased IgA		0.1 mg kg^{1} bw	Hedman et al. (1997a)
Chicken (20 days)	Diet	♂	Gizzard erosions, reduced body weight		3 mg kg^{1} bw	Hedman et al. (1995)

- There are now regulatory guidelines for DON; however, no country has regulated for NIV. Since NIV has a toxicity similar to or stronger than DON, and is a serious problem in some countries, NIV should be evaluated and regulated where necessary.

References

Abramson D, Clear RM, Gaba D, Smith DM, Patrick SK, Saydak D (2001) Trichothecene and moniliformin production by *Fusarium* species from Western Canadian Wheat. J Food Prot 64(8):1220–1225

Aoki T, O'Donnell K (1998) *Fusarium kyushuense* sp. nov. from Japan. *Mycoscience* 39(1):1–6

Arase K, Saijo K, Watanabe H, Konno A, Arase H, Saito T (1999) Ablation of a specific cell population by the replacement of a uniquely expressed gene with a toxin gene. Proc Natl Acad Sci USA 96(16):9264–9268

Atanassov Z, Nakamura C, Mori N, Kaneda C, Kato H, Jin Y-Z, Yoshizawa T, Murai K (1994) Mycotoxin production and pathogenicity of *Fusarium* species and wheat resistance to Fusarium head blight. Can J Bot 72(2):161–167

Berek L, Petri IB, Mesterhazy Á, Téren J, Molnár J (2001) Effects of mycotoxins on human immune functions in vitro. Toxicol In Vitro 15(1):25–30

Burgess LW, Nelson PE, Toussoun TA (1982) Characterization, geographic distribution and ecology of *Fusarium crookwellense* sp. Nov. Trans Br Mycol Soc 79(3):497–505

Carter JP, Rezanoor HN, Holden D, Desjardins AE, Plattner RD, Nicholson P (2002) Variation in pathogenicity associated with the genetic diversity of *Fusarium graminearum*. Eur J Plant Pathol 108(6):573–583

Chelkowski J, Golinski P, Perkowski J, Visconti A, Rakowska M, Wakulinski W (1991) Toxinogenicity of *Microdochium nivale* (*Fusarium nivale*) isolates from cereals in Poland. Mycotoxin Res 7:140–145

Choi C-Y, Nakajima-Adachi H, Kaminogawa S, Sugita-Konishi Y (2000) NIV inhibits total and antigen-specific IgE production in mice. Toxicol Appl Pharmacol 165(1):94–98

Cumagun CJR, Bowden RL, Jurgenson JE, Leslie JF, Miedaner T (2004) Genetic mapping of pathogenicity and aggressiveness of *Gibberella zeae* (*Fusarium graminearum*) toward wheat. Phytopathology 94(5):520–526

Desjardins AE, Jarosz AM, Plattner RD, Alexander NJ, Brown DW, Jurgenson JE (2004) Patterns of trichothecene production, genetic variability, and virulence to wheat of *Fusarium graminearum* from smallholder farms in Nepal. J Agric Food Chem 52(20):6341–6346

Desjardins AE (2006) Fusarium mycotoxins: chemistry, genetics, and biology. American Phytopathological Society Press, St. Paul, MN, USA

European Commission (2003) Collection of occurrence data of Fusarium toxins in food and assessment of dietary intake by the population of EU member states. Report on tasks for scientific cooperation (SCOOP) 3.2.10, European Commission, Brussels. http://ec.europa.eu/food/fs/scoop/task3210.pdf. Accessed 20 Dec 2007

European Commission Scientific Committee on Food (2000) Opinion of the scientific committee on food on Fusarium toxins. Part 4: Nivalenol (NIV) SCF/CS/CNTM/MYC/26 Final. http://Europe.eu.int/comm./food/fs/sc/scf/index_en.html. Accessed 20 Dec 2007

European Commission Scientific Committee on Food (1999) Opinion of *Fusarium* toxins-Part 1: Deoxynivalenol (DON) (expressed on 2 December 1999). http://europa.eu.int/comm/food/fs/sc/scf/out44_en.pdf

Gale LR, Ward TJ, O'Donnell K, Harrison SA, Kistler HC (2005) Fusarium head blight of wheat in Louisiana is caused largely by nivalenol producers of *Fusarium graminearum* and *Fusarium asiaticum*. Proceedings of the 2005 National Fusarium Head Blight Forum, p 159

Gams W (1989) Taxonomy and nomenclature of *Microdochium nivale* (*Fusarium nivale*). In: Chelkowski J (ed) Fusarium mycotoxins, taxonomy and pathogenicity. Elsevier, Amsterdam, pp 195–198

Gams W, Müller E (1980) Conidiogenesis of *Fusarium nivale* and *Rhynchosporium oryzae* and its taxonomic implications. Eur J Plant Pathol 86(1):45–53

Golinski P, Vesonder RF, Latus-Zietkiewicz D, Perkowski J (1988) Formation of fusarenone X, nivalenol, zearalenone, a-*trans*-zearalenol, b-*trans*-zearalenol, and fusarin C by *Fusarium crookwellense*. Appl Environ Microbiol 54(4):2147–2148

Gouze ME, Laffitte J, Printon P, Dedieux G, Galinier A, Thouvenot JP, Loiseau N, Oswald IP, Galtier P (2007) Effect of subacute oral doses of nivalenol on immune and metabolic defence systems in mice. Vet Res 38(4):635–646

Hasui M, Saikawa Y, Miura M, Takano N, Ueno Y, Yachie A, Miyawaki T, Taniguchi N (1989) Effector and precursor phenotypes of lymphokine-activated killer cells in mice with severe combined immunodeficiency (scid) and athymic (nude) mice. Cell Immunol 120(1):230–239

Hedman R, Pettersson H, Lidberg JE (1997) Absorption and metabolism of nivalenol in pigs. Archiv für Tierernährung 50(1):13–24

Hinoshita F, Suzuki Y, Yokohama K, Hara S, Yamada A, Ogura Y, Hashimoto H, Tomura S, Marumo F, Ueno Y (1997) Experimental IgA nephropathy induced by a low-dose environmental mycotoxin, nivalenol. *Nephron* 75(4):469–478

Hsia CC, Wu ZY, Li YS, Zhang F, Sun ZT (2004) Nivalenol, a main *Fusarium* toxin in dietary foods from high-risk areas of cancer of esophagus and gastric cardia in China, induced benign and malignant tumors in mice. Oncol Rep 12(2):449–456

IARC (1993) Monographs on the evaluation of carcinogenic risks to humans. Some naturally occurring stubstances: food items and constituents, heterocyclic aromatic amines and mycotoxins. vol 56. International Agency for Research on Cancer, World Health Organization, Lyon, pp 397–443

Ichinoe M, Kurata H, Sugiura Y, Ueno Y (1983) Chemotaxonomy of Gibberella zeae with special reference to production of trichothecenes and zearalenone. Appl Env Microbiol 46(6):1364–1369

Ito Y, Ueno Y, Tanaka T, Nakamura K, Ohtsubo K (1988) Embryotoxicity of oral nivalenol in mice. Mycotoxins (Tokyo) 27:33–36

Jennings P, Coates ME, Turner JA, Chandler EA, Nicholson P (2004) Determination of deoxynivalenol and nivalenol chemotypes of *Fusarium culmorum* isolates from England and Wales by PCR assay. Plant Pathol 53(2):182–190

Kawasaki Y, Uchida O, Sekita K, Matsumoto K, Ochiai T, Usui A, Nakaji Y, Furuya T, Kurokawa Y, Tobe M (1990) Single and repeated oral administration toxicity studies of nivalenol in F344 rats. J Food Hyg Soc Jpn 31(2):144–154

Kim HS, Lee T, Dawlatana M, Yun SH, Lee YW (2003) Polymorphism of trichothecene biosynthesis genes in deoxynivalenol- and nivalenol-producing *Fusarium graminearum* isolates. Mycol Res 107(02):190–197

Kubosaki A, Aihara M, Park BJ, Sugiura Y, Shibutani M, Hirose M, Suzuki Y, Takatori K, Sugita-Konishi Y (2008) Immunotoxicity of Nivalenol after subchronic dietary exposure to rats. Food Chem Toxicol 46(1):253–258

Lee T, Han YK, Kim KH, Yun SH, Lee YW (2002) *Tri 13* and *Tri 7* determine deoxynivalenol- and nivalenol- producing chemotypes of *Gibberella zeae*. Appl Environ Microbiol 68(5):2148–2154

Marasas WFO, Nelson PE, Toussoun TA (1984) Identify and mycotoxicology. In:Toxigenic *Fusarium* species. Pennsylvania State University Press, University Park, PA, USA, pp 273–310

Marasas WFO, Nelson PE, Toussoun TA (1985) Taxonomy of toxigenic fusaria. In: Lacey J (ed) Tricothecenes and other mycotoxins. Wiley, New York, pp 3–14

Mielniczuk E, Kiecana I, Perkowski J (2004) Susceptibility of oat genotypes to *Fusarium crookwellense* Burgess, Nelson and Toussoun infection and mycotoxin accumulation in kernels. Biologia 59(6):809–816

Miller JD, Greenhalgh R, Wang Y-Z, Lu M (1991) Trichothecene chemotypes of three *Fusarium* species. Mycologia 83(2):121–130

Mirocha CJ, Abbas HK, Windels CE, Xie W (1989) Variation in deoxynivalenol, 15-deoxynivalenol, 3-acetyldeoxynivalenol and zearalenone production by *Fusarium graminearum* isolates. Appl Environ Microbiol 55(5):1315–1316

Monds RD, Cromey MG, Lauren DR, Menna M, Marshall J (2005) *Fusarium graminearum*, *F. cortaderiae* and *F. pseudograminearum* in New Zealand: molecular phylogenetic analysis, mycotoxin chemotypes and co-existence of species. Mycol Res 109(04):410–420

Nakajima T, Yoshida M (2007) Mycotoxin productivity and virulence of *Fusarium graminearum* species complex causing Fusarium head blight on wheat and barley in the western part of Japan. Jpn J Phytopathol 73(2):106–111 (in Japanese with English abstract)

Nakajima T, Naito S (1995) Reassessment of mycotoxin productivity of *Microdochium nivale* in Japan. Ann Phytopathol Soc Jpn 61(44):357–361

O'Donnel K, Kistler HC, Tacke BK, Casper HH (2000) Gene genealogies reveal global phylogeographic structure and reproductive isolation among lineages of *Fusarium graminearum*, the fungus causing wheat scab. Proc Natl Acad Sci U S A 97(14):7905–7910

O'Donnel K, Ward TJ, Geiser DM, Kistler HC, Aoki T (2004) Geneological concordance between the mating type locus and seven other nuclear genes supports formal recognition of nine phylogenetically distinct species within the *Fusarium graminearum* clade. Fungal Genet Biol 41:600–623

Ohtsubo K, Ryu LC, Nakamura K, Izumiyama N, Tanaka T, Yamamura H, Kobayashi T, Ueno Y (1989) Chronic toxicity of of nivalenol in female mice: a 2-year feeding study with *Fusarium nivale* Fn 2b moulded rice. Food Chem Toxicol 27(9):591–598

Onji Y, Dohi Y, Aoki Y, Moriyama T, Nagami H, Uno M, Tanaka T, Yamazoe Y (1989) Deepoxynivalenol: a new metabolite of nivalenol found in the excreta of orally administered rats. J Agric Food Chem 37:478–481

Parry DW, Jenkinson P, McLeod L (1995) Fusarium ear blight (scab) in small grain cereals – a review. Plant Pathol 44(2):207–238

Pettersson H, Hedman R, Engstrom B, Elwinger K, Fossum O (1995) Nivalenol in Swedish cereals — occurrence, production and toxicity towards chickens. Food Addit Contam 12(3):373–376

Placinta CM, D'Mello JBF, Macdonald AMC (1999) A review of world contamination of cereal grains and animal feeds with *Fusarium* mycotoxins. Anim Feed Sci Technol 78(1):21–37

Poapolathep A, Sugita-Konishi Y, Phitsanu T, Doi K, Kumagai S (2004) Placental and milk transmission of trichothecene mycotoxins, nivalenol and fusarenon-X, in mice. Toxicon 44(1):111–113

Ryu J-C, Ohtsubo K, Izumiyama N, Mori M, Tanaka T, Ueno Y (1987) Effects of nivalenol on the bone marrow in mice. J Toxicol Sci 12(1):11–21

Ryu J-C, Ohtsubo K, Izumiyama N, Mori M, Tanaka T, Yamamura H, Ueno Y (1988) The acute and chronic toxicities of nivalenol in mice. Fundam Appl Toxicol 11(1):38–47

Schothorst RC, Jekel AA (2001) Determination of trichothecenes in wheat by capillary gas chromatography with flame ionisation detection. Food Chem 73(1):111–117

Seo JA, Kim JC, Lee DH, Lee YW (1996) Variation in 8-ketotrichothecenes and zearalenone production by *Fusarium graminearum* isolates from corn and barley in Korea. Mycopathologia 134(1):31–37

Suga H, Karugia GW, Ward T, Gale LR, Tomimura K, Nakajima T, Miyasaka A, Koizumi S, Kageyama K, Hyakumachi M (2008) Molecular characterization of the *Fusarium graminearum* species complex in Japan. Phytopathology 98(1):159–166

Sugita-Konsihi Y, Tanaka T, Tabata S, Nakajima M, Nouno M, Nakaie Y, Chonan T, Aoyagi M, Kibune N, Mizuno K, Ishikuro E, Kanamaru N, Minamisawa M, Aita N, Kushiro M, Tanaka K, Takatori K (2006) Validation of an HPLC analytical method coupled to a multifunctional clean-up column for the determination of deoxynivalenol. Mycopathologia 161(4):239–243

Sugiura Y, Fukasaku K, Tanaka T, Matsui Y, Ueno Y (1993) *Fusarium poae* and *Fusanium crookwellense*, fungi responsible for the natural occurrence of nivalenol in Hokkaido. Appl Environ Microbiol 59(10):3334–3338

Takahashi M, Shibutani M, Sugita-Konishi Y, Aihara M, Inoue K, Woo G-H, Fujimoto H, Hirose M (2008) A 90-day subchronic toxicity studyof nivalenol, a trichothecene mycotoxin, in F344 rats. Food Chem Toxicol 46(1):125–135

Tanaka H, Takino M, Sugita-Konishi Y, Tanaka T (2006) Development of a liquid chromatography/time of flight mass spectrometric method for the simultaneous determination of trichothecenes, zearalenone and aflatoxins in foodstuffs. Rapid Commun Mass Spectrom 20(9):1422–1428

Tatsuno T, Morita Y, Tsunoda H, Umeda M (1970) Rechereches toxicologiques des substances métaboliques du *Fusarium nivale* VII. La troisiéme substance metabolique de *F. nivale*, le diacetate de nivalenol. Chem Pharm Bull 18(7):1485–1487

Tatsuno T, Saito M, Enomoto M, Tsunoda H (1968) Nivalenol, a toxic principle of *Fusarium nivale*. Chem Pharm Bull 16(12):2519–2520

Tep J, Videmann B, Mazallon M, Balleydier S, Cavret S, Lecoeur S (2007) Transepithelial transport of fusariotoxin nivalenol: mediation of secretion by ABC transporters. Toxicol Lett 170(3):248–258

Ueno Y, Ueno I, Tatsuno T, Ohkubo K, Tsunoda H (1969) Fusarenon-X, a toxic principle of *Fusarium nivale* culture filtrate. Experientia 25(10):1062

Ueno Y (1983) General toxicology. In: Ueno Y (ed) Developments of food science. IV Trichothecenes-chemical, biological and toxicological aspects. Kodansya, Tokyo, pp 135–146

Ueno Y, Kobayashi T, Yamamura H, Kato T, Tashiro F, Nakamura K, Ohtsubo K (1991) Effect of long-term feeding of nevalenol on aflatoxin B_1-initiated hepatocarcinogenesis in mice. Relevance to human cancer of *N*-nitroso compounds,tobacco and mycotoxins. IARC Scientific Publications No.105. International Agency for Research on Cancer, World Health Organization, Lyon, pp 420–423

Ueno Y, Yabe T, Hashimoto H, Sekijima M, Masuda T, Kim DJ, Hasegawa R (1992) Enhancement of GST-P-positive liver cell foci development by nivalenol, a trichothecene mycotoxin. Carcinogenesis 13(5):787–791

Ward TJ, Bielawski JP, Kistler HC, Sullivan E, O'Donnell K (2002) Ancestral polymorphism and adaptive evolution in the trichothecene mycotoxin gene cluster of phytopathogenic *Fusarium*. Proc Natl Acad Sci U S A 99(14):9278–9283

Weidenborner M (2008) Mycotoxins in food stuffs. Springer Science+Business Media LLC, Cambridge, MA,USA

World Health Organization (2001) Safety evaluation of certain mycotoxins in food. WHO Food Additives series No. 47. Prepared by the Fifth-Sixth meeting of the joint FAO/WHO Expert committee on food additives (JECFA), WHO, Geneva, pp 419–556

Yabe T, Hashimoto H, Sekijima M, Degawa M, Hashimoto Y, Tashiro F, Ueno Y (1993) Effects of nivalenol on hepatic drug-metabolizing activity in rats. Food Chem Toxicol 31(8):573–581

Yamamura H, Kobayashi T, Ryu J-C, Ueno Y, Nakamura K, Izumiyama N, Ohtsubo K (1989) Subchoronic feeding studies with nivalenol in C57BL/6 mice. Food Chem Toxicol 27(9):585–590

Yoshida M, Nakajima T, Arai M, Tomimura K (2004) Symptom and myvotoxin accumulation in rice inoculated with *Fusarium graminearum* isolated from wheat and barley. Proceedings of 2nd International Symposium on Fusarium Head Blight p 539

Yoshizawa T, Jin YZ (1995) Natural occurrence of acetylated derivatives of deoxynivalenol and nivalenol in wheat and barley in Japan. Food Addit Contam 12(5):689–694

Yoshizawa T, Morooka N (1973) Deoxynivalenol and its monoacetate: new mycotoxins from *Fusarium roseum* and moldy barley. Agric Biol Chem 37(12):2933–2934

Yuwai KE, Rao KS, Singh K, Tanaka T, Ueno Y (1994) Occurrence of nivalenol, deoxynivalenol, and zearalenone in imported cereals in Papua, New Guinea. Nat Toxins 2(1):19–21

Chapter 16
Yeast Killer Toxins Technology Transfer

Walter Magliani, Stefania Conti, Laura Giovati, and Luciano Polonelli

16.1 Introduction

Among the different competitive mechanisms which microorganisms can use to multiply and survive in natural communities, the production of antimicrobial toxins represents a common, efficient and specific ecological way to eliminate competitor strains/species from the same habitat. By killing or severely reducing the fitness of sensitive strains, toxin-producing microorganisms which are self-immune can be selected and so dominate in specific environments. Similar examples of interference competition have been described in bacteria (bacteriocins) as well as in eukaryotic microorganisms. In particular, the production of killer toxins (KTs) by yeasts ("yeast killer phenomenon") has aroused great interest, because of its implications, technological transfer and intriguing perspectives.

Research into killer systems has provided important insights into basic and general aspects of eukaryotic cell, such as molecular structure and assembly of fungal cell wall, intracellular processing, maturation and secretion of proteins, and virus-host cell interaction. Killer strains secrete low molecular mass proteins or glycoproteins, which kill sensitive cells of the same or related genera, in a specific KT receptor (KTR)-mediated fashion without direct cell-to-cell contact. For yeasts in natural habitats, KT production can represent a sophisticated biological mechanism for restricting the growth of sensitive competitors for limited available nutrients. Soon after its discovery in *Saccharomyces cerevisiae*, the killer phenomenon was found to be widespread among yeasts. Different KTs can display different modes and spectra of antimicrobial activity, including strains of the same (narrow spectrum), or taxonomically related and unrelated species (broad spectrum). As these aspects have been previously comprehensively revised (Magliani et al. 1997b;

W. Magliani (✉)
Dipartimento di Patologia e Medicina di Laboratorio, Sezione di Microbiologia, Università degli Studi Parma, Viale A. Gramsci, 14–43100 Parma, Italy
e-mail: magliwal@unipr.it

M. Rai and A. Varma (eds.), *Mycotoxins in Food, Feed and Bioweapons*,
DOI: 10.1007/978-3-642-00725-5_16,

Schmitt and Breinig 2006), this chapter focuses on some aspects of technology transfer of yeast killer systems and their possible applications in biotyping, industrial processes, biotechnology and biomedicine.

16.2 Microbial Biotyping by the Yeast Killer System

Microbial typing is a useful tool not only for epidemiological purposes in clinical microbiology, but also in control laboratories for the discrimination of strains of industrial importance or those covered by patent. Besides the identification at the species level, the intraspecific differentiation into "types" or "biotypes" has become a pressing need for determining strain clonality. Several typing methods have been applied. They include traditional phenotypic approaches such as serotyping, phagetyping, antibiotype determination, as well as more sophisticated molecular methods such as plasmid profiling, DNA restriction endonuclease analysis, and others. While phenotypic methods are usually considered to be sometimes too variable, labor-intensive, and time-consuming to be of practical value, molecular approaches require the availability of specialized facilities and technologies uncommon in small laboratories. Thus, phenotypic methods are usually used as an appropriate first step and molecular ones as a second level of analysis.

In this context, since the 1980s microbial biotyping by yeast KTs has been proposed as an easy and effective non-molecular method based on the determination of killer sensitivity patterns towards panels of selected killer strains or KTs. After standardization of medium composition, inoculum size, and modalities of test and incubation, the simplest biotyping laboratory procedure involves streaking killer strains directly onto the surface of an agar plate, which has been previously inoculated with the strain of interest. Upon growth, the killer yeasts secrete KTs into the agar and, after proper incubation, sensitivity can be simply visualized by the presence of a halo of inhibition around each selected killer yeast (Fig. 16.1a and b) allowing the total spectrum of susceptibility to be established.

In its original version, a yeast killer system based on nine selected killer yeasts grouped into three triplets (Table 16.1a) was used to biotype *Candida albicans* clinical isolates. A conventional three-number code (Table 16.1b) was assigned to each strain (e.g., 111, 223, etc.), allowing the identification of different strain types. The most frequently encountered one (111, 52%) was sensitive to all nine killer yeasts (Polonelli et al. 1983). By increasing the number of triplets the sensitivity of the system was increased, making it possible to split the most common biotypes (Table 16.1c) (Caprilli et al. 1985) and to distinguish strain types among other pathogenic yeasts, such as *C. kefyr* (*C. pseudotropicalis*), *C. tropicalis*, *C. glabrata* and *Cryptococcus neoformans* (Morace et al. 1984).

Even though the procedure based on streaking killer strains is simple and easy to perform, it can lead to conflicting results due to the difficulty of standardizing the actual amount of secreted KT. The use of spots containing standardized suspensions of killer cells (Fig. 16.1c) or wells with more or less partially purified KTs

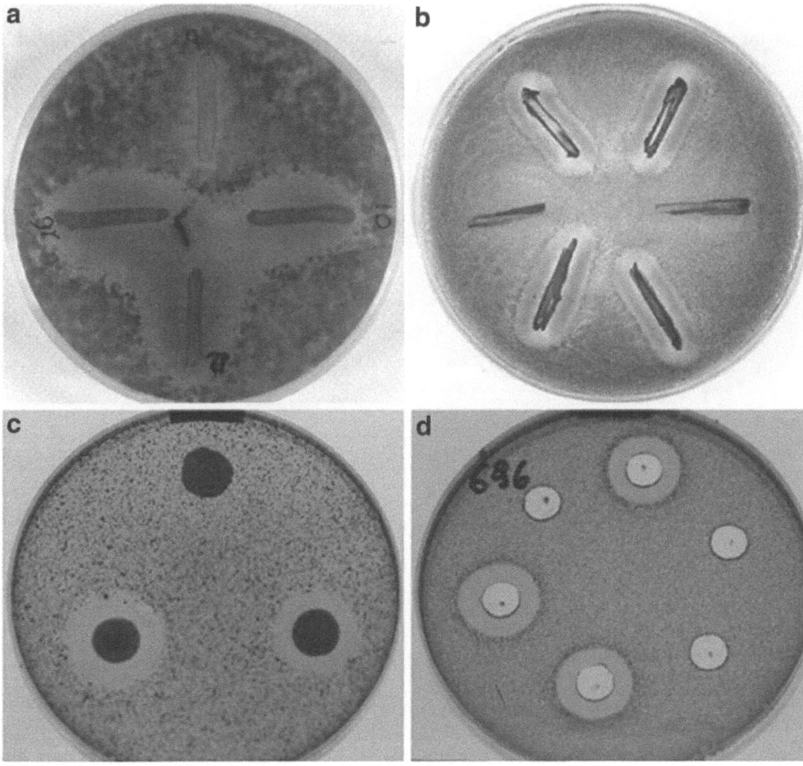

Fig. 16.1 Yeast killer activity on different pathogenic fungi, as evaluated by different methods. Killer strains are streaked directly (**a**,**b**) or placed as 50-μl drops of standardized suspensions (**c**) on the surface of agar plates previously inoculated with a mould (**a**) or yeast (**b**,**c**,**d**) strain of interest. **d** 10 mm diameter wells are punched in the inoculated agar and filled with 100 μl of concentrated KT. After proper incubation, the presence of a halo of growth inhibition demonstrates sensitivity to each killer strain

(Fig. 16.1d) made the method independent of the different growth conditions of concurring killer yeasts and KT-sensitive strains. The diameters of inhibition halos have been used to define the biotype of the investigated strains and the application of computer-aided programs has made it possible to further improve the reproducibility of the method (Polonelli et al. 1985).

Effective KT-biotyping has been more recently reported among *C. albicans* and non-*albicans* strains, in that sensitivity to KTs is apparently expressed more consistently than either germ tube formation or chlamydospore production, and among yeast strains of different origin and species (genera *Debaryomyces*, *Kluyveromyces*, *Saccharomyces* and *Zygosaccharomyces*) (Buzzini et al. 2007). In *C. neoformans* complex, the new species *C. gattii* (formerly *C. neoformans* var. *gattii*, serotypes B and C) and varieties *C. neoformans* var. *grubii* (formerly *C. neoformans* serotype A) and var. *neoformans* (formerly serotype D) have been successfully differentiated (Fuentefria et al. 2007).

Table 16.1 Biotyping by the killer system. **a** Original killer system: list of the killer yeasts grouped into triplets. **b** Numerical code of the activity of the selected triplets. **c** Splitting of *Candida albicans* biotype 111 isolates (most frequently encountered) by using an amplified killer system

a) Killer yeasts

First triplet	*Pichia* sp. Stumm 1034
	Pichia sp. Stumm 1035
	P. anomala UM3
Second triplet	*P. anomala* CBS 5739
	P. anomala Ahearn Um866
	P. californica Ahearn WC40
Third triplet	*P. canadensis* Ahearn WC41
	P. dimennae Ahearn WC44
	Williopsis mrakii Ahearn WC51

b) Killer activity

1st killer strain	2nd killer strain	3rd killer strain	Code
+	+	+	1
+	+	–	2
+	–	+	3
–	+	+	4
+	–	–	5
–	+	–	6
–	–	+	7
–	–	–	8

c) Amplified killer system

C. albicans isolate	Serotype	Specimen	Patient	Killer system biotype
UCSC 110	A	kidney	B.M.L.	111 688 888 887
UCSC 111	A	blood	M.M.C.	111 688 888 457
UCSC 112	A	cutaneous lesion	C.	111 688 788 857
UCSC 113	A	cutaneous lesion	B.E.	111 688 888 887
UCSC 114	A	urine	B.E.	111 688 888 457
UCSC 115	A	urine	R.M.	111 688 788 457
UCSC 116	A	kidney	M.M.C.	111 688 888 887
UCSC 117	A	pus	C.C.	111 688 888 887
UCSC 118	A	throat	G.M.[a]	111 688 888 457
UCSC 122	B	throat	C.L.[b]	111 688 888 457
UCSC 123	A	throat	O.I.[b]	111 688 888 457
UCSC 124	A	throat	T.E.[b]	111 688 888 687
UCSC 125	A	throat	N.C.[a]	111 688 888 687

UCSC, Università Cattolica Sacro Cuore, Istituto Microbiologia, Rome, Italy
[a]patients without candidiasis, [b]nurses

Soon after the first studies, the potential use of the killer phenomenon has been re-evaluated, since it became evident that the susceptibility to KTs, initially reputed to be limited to taxonomically related species, is widespread also in microorganisms belonging to different phyla (yeasts and molds) and kingdoms (bacteria) (Polonelli and Morace 1986). Once standardized to meet with the different growth requirements of the sensitive microorganisms to be investigated, the yeast killer system was successfully used as unique epidemiological marker for investigating cases of nosocomial infections, sexually transmitted diseases, post-transplantation infections, and epidemics. It has proved to be particularly effective for the biotyping of molds for which other phenotypic methods for intraspecific differentiation are not readily available, such as *Aspergillus fumigatus* and its related taxa, *Penicillium camemberti*, *Pseudoallescheria boydii*, *Sporothrix schenckii*, species belonging to the genus *Fusarium* (Fanti et al. 1989; Polonelli et al. 1989), and *Microsporum canis* (Maia et al. 2001).

The use of a modified killer technique, based on the co-incubation in a liquid medium of a KT with the test strain, permitted the killer activity to be evaluated by comparing the number of colony-forming units (CFUs) with the proper control without KT. The procedure, quantitatively very sensitive and statistically reproducible, could be adopted also for bacteria, other than yeasts and molds. Although the nature of the interaction(s) between killer yeasts and other microbial groups and the occurrence of specific KTRs in microorganisms different from fungi still need to be elucidated, biotyping based on KTs and CFU determination has been adapted to the growth characteristics of Gram negative and Gram positive bacteria, *Mycobacteria*, aerobic actinomycetes and anaerobic bacteria, for their intraspecific differentiation. Interestingly, killer yeasts were able to exert their activity under various culture conditions, offering great flexibility for use in epidemiological investigations. In particular cases killer biotyping has allowed the discrimination of strains belonging to the same serotype (Morace et al. 1989; Polonelli et al. 1992). In particular cases, as for *N. asteroides* complex (*N. asteroides*, *N. farcinica*, *N. nova*), KT biotyping allowed the assignation of isolates to the different species, otherwise difficult by other more classical methods (Provost et al. 1995).

Due to its wide flexibility, the killer system may be a unique and convenient biotyping method for microbiological laboratories with some limitations, such as previous practices in standardization and determination of the optimal number of KTs to be used. It only requires commercially available media and a set of suitably standardized killer yeasts. In addition, it can be used on a large number of isolates to monitor nosocomial infections, to establish the exogenous or endogenous nature of an infection, and to correlate isolates from various industrial sources and geographical areas.

16.3 Industrial and Biotechnological Applications of the Yeast Killer System

The growing understanding of the complex genetics, regulating mechanisms, and compatibility of the mechanisms of KT production and secretion have suggested several industrial and biotechnological applications. In recombinant DNA, killer plasmids from *S. cerevisiae*, *K. lactis* and others have been used as cloning vectors for the effective expression and secretion of foreign eukaryotic proteins, not only in *S. cerevisiae*, but also in other host cells such as *Schizosaccharomyces pombe* and *Pichia pastoris* (Heintel et al. 2001; Bennett et al. 2002; Giga-Hama et al. 2007). Since recombinant proteins are not always secreted from yeast, the addition of a suitable secretion-signal sequence onto a mature protein could be essential for the final yield of the active protein. Besides the original one of the desired protein, when available, other yeast-secretion and processing signal sequences have been used, which can be applied also to non-secretory proteins. Most KTs are synthesized as precursors (pre-protoxins, pptoxs), which are characterized by the presence of N-terminal leader peptides necessary for their post-translational translocation into the lumen of the endoplasmic reticulum and subsequent sorting into the golgi network. During passage through the yeast secretory pathway, pptoxs are processed and the signal peptides cleaved by specific peptidases, allowing the secretion of the mature and biologically active KT. Thus, different pptox-secretion and processing signals have been inserted into recombinant proteins to direct their entry into the eukaryotic secretion pathway and successful secretion (Eiden-Plach et al. 2004; Gibbs et al. 2004).

The competitive advantage that killer yeasts have over sensitive strains may be responsible for their abundance in natural habitats, and consequently, for their ability to frequently contaminate processes of food and beverage production, sometimes with unwanted organoleptic effects. In wine-making, the quality of the product is largely influenced by the properties of the strain which completes the fermentation. Killer wild wine yeasts could negatively affect the sensory quality of the wine by acting as causal agents of stuck and/or sluggish fermentation (Carrau et al. 1993; Pérez et al. 2001). The inoculation of selected killer yeasts as starter cultures in beer and wine fermentations has been suggested to control the growth of contaminant yeasts during the early stages of fermentation and to prevent spoilage. Natural or genetically modified killer strains with desirable oenological properties have been constructed or selected to improve the quality of beverages (Van Vuuren and Jacobs 1992; Yap et al. 2000).

Some KTs have been envisaged as biopreservative agents for wine and grapes, such as a *K. phaffii* KT for controlling apiculate wine yeasts (Ciani and Fatichenti 2001), *P. anomala* and *K. wickerhamii* KTs as new tools against spoilage yeasts during wine ageing and storage (Comitini et al. 2004), and *P. membranifaciens* KT for the biocontrol of *Botrytis cinerea* and grey mold disease of grapevine (Santos and Marquina 2004). The use of killer strains has been suggested in the control of ochratoxin-producing black aspergilli naturally occurring on grapes. Indirectly, this could have a great impact on the reduction of ochratoxin A, a potent nephrotoxic, carcinogenic, teratogenic and immunotoxic mycotoxin, frequently reported in wine and grape juice (Bleve et al. 2006).

Killer strains and KTs have also been identified or tested in other processes, such as the baking industry, maize and wheat silage, and yoghurt spoilage, and envisaged as bio-control agents in the preservation of foods, due to their ascertained killer activity against yeasts that cause trouble in the food industry (Lowes et al. 2000; Druvefors et al. 2002). KTs have been also evaluated as antifungal agents against disease-causing phytopathogens by inserting and expressing KT-encoding genes in transgenic plants. That has been the case with *Ustilago maydis* KTs, which are lethal to *Ustilaginales* strains, including pathogens responsible for relevant plant diseases, such as loose smut (*U. tritici*) and stinking smut (*Tilletia caries*) (Clausen et al. 2000; Schlaich et al. 2007). These observations might suggest novel strategies for engineering biological control of fungal pathogens and studying its potential in plant protection.

16.4 Yeast Killer System Applications in Biomedicine

In addition to the above-reported potential applications, it is perhaps in the field of biomedicine that the growing knowledge of the killer phenomenon has opened up the most intriguing new potential perspectives for anti-infective immunoprevention and therapy. With the increasing development of antimicrobial resistance among important pathogenic microorganisms, there is a pressing need for new classes of antimicrobial compounds with improved efficacy and safety and KTs could represent a potential source.

16.4.1 Killer Toxins as Potential Antimicrobial Agents

Due to their wide spectrum of activity, some KTs have attracted increasing attention and have been exploited as new potential therapeutic tools in the control of infectious diseases. In particular, chromosomally encoded KTs produced by killer strains belonging to species of *Pichia* and *Williopsis*, have demonstrated lethal activity against a significantly wide range of unrelated eukaryotic and prokaryotic microorganisms, including important human, animal and plant pathogens (Magliani et al. 1997b; Marquina et al. 2002; Passoth et al. 2006).

Panomycocin, a 49 kDa highly stable *P. anomala*-glycosylated KT with an exo-*β*-1,3-glucanase activity, demonstrated fungicidal activity against human isolates of different pathogenic *Candida* spp. and dermatophytes (*Microsporum* and *Trichophyton* spp.) (Izgü et al. 2007a, b). Wicaltin, a 35–37 kDa killer glycoprotein from *W. californica*, showed antifungal activity against 14 yeast genera, including human pathogens, such as *C. albicans*, *C.* non-*albicans* and *S. schenckii* (Theisen et al. 2000). HM-1 and HYI are strongly cytocidal small KTs from *W. saturnus* var. *mrakii* and var. *saturnus* respectively, which share the same mechanism of killing against sensitive yeasts by inhibiting β-glucan synthase activity, thus causing osmotic cell lysis (Komiyama et al. 1998; Miyamoto et al. 2006).

*Wm*KT, a 85-kDa glycoprotein produced by *W. saturnus* var. *mrakii*, shares structural similarities with yeast cell wall proteins suspected of exhibiting glucosidase activity. It induces rapid cell permeation and death soon after its binding to target cells by interaction with cell wall β-glucans, and appears to be antigenically related to a previously described KT produced by *P. anomala* ATCC 96603 (*Pa*KT), formerly designated *Hansenula anomala* UCSC 25F (Guyard et al. 2001, 2002). *Pa*KT, a glycoprotein with an apparent molecular weight of 105 kDa, demonstrated a wide spectrum of antimicrobial activity by direct interaction with specific cell wall KTRs, putatively constituted by β-glucans (Magliani et al. 1997b).

Even though KTs have been suggested as potential antimicrobial agents, most of them exhibit their cytotoxic activity only within narrow pH and temperature ranges, usually below the physiological ones, and are characterized by antigenicity and toxicity. Apart from a study which demonstrated a significant inhibition of *P. carinii*'s infectivity to severe combined immunodeficient (SCID) mice subsequent to in vitro pre-incubation of mouse-derived parasites with *Pa*KT (Séguy et al. 1996), a unique study evaluating a direct therapeutic use of a KT against fungal infections has been reported (Polonelli et al. 1986). The topical application of concentrated crude *Pa*KT demonstrated therapeutic effectiveness in different experimental infections by *Malassezia furfur* and *M. pachydermatis*. However, systemic use of *Pa*KT was proved to be unsuitable, in that, apart from its lability in physiological conditions, it demonstrated significant toxic effects on intestinal fluid homeostasis and electrolyte balance in rat (Pettoello-Mantovani et al. 1995).

Thus, the killer phenomenon has been envisaged as a potential, attractive source of novel tools for combatting clinically relevant microbial infections, while excluding a direct use of KTs as therapeutic agents. In the attempt to overcome these limitations, an innovative approach on the basis of the idiotypic network hypothesis has been developed by which immunological derivatives mimicking the action of *Pa*KT have been generated.

16.4.2 Yeast Killer Toxin-Neutralizing Antibodies and Idiotypic Vaccination

Taking advantage of the *Pa*KT immunogenicity, anti-*Pa*KT monoclonal antibodies (mAbs) has been obtained by conventional fusion of splenocytes from mice primed with a crude extract of KT. One of them (designated KT4) was able to neutralize *Pa*KT killer activity and, on the basis of the theory of the idiotypic (Id) network, has been used to immunize animals (rabbit, mouse, rat). Anti-Id Abs have been raised which were able to compete with *Pa*KT for the binding site of mAb KT4 and, more importantly, to inhibit the growth of a susceptible *C. albicans* reference strain, chosen as a model microorganism, thereby mimicking the effect of *Pa*KT. These conceptually new antimicrobial Abs which have the internal image of the active site of a KT, thus acting as antibiotics, have been designated "antibiobodies" or "killer Abs" (KAbs), and have been used to visualize specific KTRs on the surface of the presenting yeast cells. In particular, the reactivity against *C. albicans* varied according to the phase of growth, being maximal in budding cells and germ tubes, which are considered important virulence factors (Polonelli and Morace 1988; Polonelli et al. 1991).

By parenteral or intravaginal immunization with mAb KT4, as a new prophylactic experimental approach of "idiotypic vaccination", significant systemic and mucosal anticandidal immunoprotections were achieved, which were associated with rising titers of KAbs (Polonelli et al. 1993, 1994). Thus, the Id of KAbs could be assumed to be functionally equivalent to the active site of *Pa*KT, as the Id of mAb KT4 to the KT-binding site of *Pa*KTR. Intravaginal or intragastric inoculations of *Pa*KTR-bearing *C. albicans* were able to recall mucosal KAbs production in rats primarily immunized with mAb KT4 and also to elicit by themselves the production of KAbs in naïve animals. Significantly, natural polyclonal anti-receptor KAbs were also consistently found particularly in the vaginal fluid of women affected by vaginal candidiasis, who had, obviously, never been exposed to mAb KT4. All these KAbs were able to confer a significant anticandidal protection when passively transferred to naïve animals, thus suggesting that different KAbs elicited by Id vaccination or fungal infection may mimic the biological activity of *Pa*KT and may play a role in microbial infections, even though their clinical relevance still needs to be determined (Polonelli et al. 1996).

16.4.3 Antiidiotypic Monoclonal and Recombinant Antibiobodies

By using the same approach, i.e., by immunizing rats or mice with mAb KT4, a monoclonal (m-KAb, an IgM-designated mAb K10) and a recombinant (r-KAb, in the single-chain format, designated scFv H6) KAbs have been produced. M- and r-KAbs competed with *Pa*KT for the binding to mAb KT4 and *Pa*KTRs, allowed the visualization of the putative *Pa*KTRs on the *Candida* cell wall (mostly in the germ tubes and budding cells), exerted a fungicidal effect in vitro on the *Pa*KT-susceptible *C. albicans* reference strain and, significantly, they displayed a therapeutic effect in experimental animal models of systemic and mucosal candidiasis (Magliani et al. 1997a; Polonelli et al. 1997).

M- and r-KAbs could be produced indefinitely, and, as Abs, they demonstrated much more stability at different temperatures and pH than *Pa*KT, allowing their testing in vitro in well-defined experimental conditions against different microbial eukaryotic and prokaryotic pathogens and their use as therapeutic agents in different experimental animal models of infection. Furthermore, r-KAb has also been expressed in human commensal bacteria that secreted it in an active, therapeutic form (Beninati et al. 2000). When tested, both Abs displayed a wide *in vivo* and/or in vitro spectrum of antimicrobial activity including, in addition to *C. albicans* and other *Candida* spp., *P. carinii* (inhibition of attachment to epithelial lung cells and infectivity for nude rats), multidrug-resistant *Mycobacterium tuberculosis*, antibiotic-resistant gram-positive cocci, oral Streptococci, *A. fumigatus* (protection against invasive aspergillosis in allogeneic T-cell-depleted bone marrow transplanted mice), *Leishmania major*, *L. infantum*, and *Acanthamoeba castellanii*

(Magliani et al. 2005; Fiori et al. 2006). Thus, natural, monoclonal and recombinant Abs can directly cause microbicidal activity in vitro and therapeutic effects in animal models of infection against *Pa*KTR-bearing prokaryotic and eukaryotic pathogenic microorganisms, without requiring any other concomitant cell/factor or host's immune competence.

More recently, other candidacidal Abs have been described. R-KAbs have been produced that mimic the HM-1 KT by effectively inhibiting fungal β-glucan synthase activity (Selvakumar et al. 2006). Mycograb, a human recombinant Ab against fungal hsp90, which demonstrated synergy with antifungal agents such as amphotericin B and caspofungin, is currently suggested for the treatment of systemic candidiasis in combination therapy (Matthews and Burnie 2005). A mAb (C7), which reacts with the agglutinin-like cell surface protein 3 (Als3) component of a stress mannoprotein of $>$ 200 kDa from the cell wall of *C. albicans*, demonstrated antifungal and antitumor activities (Brena et al. 2007). Anti-β-glucan Abs, raised against a novel glyco-conjugate vaccine (β-glucan conjugated to a diphtheria toxoid), were candidacidal, and mediated *in vivo* protection against fungal lethal challenges (Torosantucci et al. 2005). Similar to what was observed with KAbs, all these candidacidal Abs showed an analogous wide spectrum of in vivo and/or in vitro fungicidal activity, including *A. fumigatus* and *C. neoformans*. This supports the existence of a family of microbicidal Abs that alone, even properly engineered, or in combination with current antimicrobial agents, could be useful in the treatment of relevant microbial infections (Magliani et al. 2005).

In the aim of obtaining less immunogenic, more stable, and safer molecules for consideration as potential therapeutic agents, KAb-derived killer mimotopes have been generated and investigated.

16.4.4 Antibiobody-Derived Killer Peptides

On the basis of the sequence of the variable region of scFv H6, peptides pertaining to its complementary determining regions (CDRs) and two residue-displaced decapeptides that reproduced its variable sequence have been synthesized, tested in vitro against *C. albicans*, and, when proved candidacidal, optimized through alanine scanning. This led to the selection, among other candidacidal peptides, of an engineered killer decapeptide (KP: AKVTMTCSAS), characterized by the highest in vitro activity, which appeared to be mediated by interaction with specific cell wall β-glucan KTRs (Polonelli et al. 2003).

Biological characteristics of KP strongly suggested its functional mimicry of *Pa*KT. In fact, KP demonstrated a strong in vitro activity against all the microbial pathogens previously shown to be sensitive to *Pa*KT and KAbs, and new ones have been added, such as *C. neoformans*, *Paracoccidioides brasiliensis*, phytopathogenic bacteria (*Pseudomonas syringae* and *Erwinia carotovora*) and fungi (*B. cinerea* and *Fusarium oxysporum*), owing to its ease of biotechnological manipulation, production and testing. Importantly, KP demonstrated a sufficient *in vivo* stability to be therapeutic against experimental infections in normal and immunocompromised animals, such as vaginal and systemic candidiasis, disseminated cryptococcosis, and paracoccidioidomycosis, as well as in *planta* (infections by *P. syringae* pv. *tabaci*) (Magliani et al. 2004a, b; Donini et al. 2005). Thus, KP could be considered as the first engineered killer mimotope exerting a wide-spectrum microbicidal activity presumably through interaction with KTRs or KTR-like molecules in the cell wall of target microbial cells, thus acting as a novel potent antimicrobial agent.

Besides retaining the same antimicrobial activity of the whole r-KAb from which it has been derived, KP exhibited additional relevant activities. In particular, it demonstrated an immunomodulatory activity, especially on dendritic cells, by improving their capacity to induce lymphocyte proliferation (Cenci et al. 2006), and a surprising antiviral activity. In fact, KP was able to specifically inhibit the replication of HIV-1 in endogenously and exogenously infected peripheral blood mononuclear cells (PBMCs) (Casoli et al. 2006). As a small synthetic decapeptide derived from physiological molecules, KP was devoid of immunogenicity and of any detectable cytotoxicity on PBMCs and in vitro cultured cell lines. Thus, it has been envisaged as a prototypal compound for the development of new wide-spectrum drugs, and also for the simultaneous treatment of HIV-1 and concomitant AIDS-related microbial opportunistic infections (Magliani et al. 2007).

16.5 Conclusions and Future Perspectives

As already discussed, new exciting perspectives should be envisaged on the basis of the potential industrial, biotechnological, and medical applications of the yeast killer phenomenon. In the field of infectious diseases, in particular, substantial advances in antimicrobial therapy and immunoprevention could result from the observation that anti-Id Abs can mimic KTs, and peptides derived from them can retain the microbicidal activity of the whole Ab. As for KTs, their killing activity is mediated by the interaction with specific KTRs, as sketched in Fig. 16.2, that are conserved by microbial cells through natural evolution and, as such, probably not rejectable by the microorganisms. Accordingly, no resistant mutant strains have been detected by screening KP activity against a wide *S. cerevisiae* nonessential gene deletion collection (Euroscarf, Frankfurt, Germany), including strains reported to be resistant to conventional antifungal drugs such as caspofungin and fluconazole (Conti et al. 2008). Thus, these killer molecules may constitute the basis for modeling new and safe wide-spectrum antimicrobial agents, which, as KP, could also have antiviral and immunomodulatory activities. Any new therapeutic option is always welcome, particularly in the field of infectious diseases. Their treatability is more and more challenged by the dramatically increasing incidence and prevalence of resistance to the available antimicrobial/antiviral drugs among epidemiologically relevant infectious agents and the emergence of new less susceptible/resistant agents (Cohen 2000). Although many questions on their potential

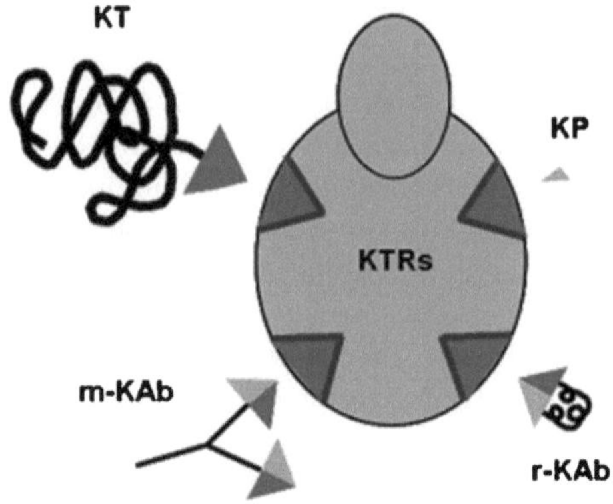

Fig. 16.2 Diagram depicting the interaction between cell wall killer toxin receptor (*KTR*) of *C. albicans*, as a representative microorganism, *Pichia anomala* killer toxin (*KT*), antiidiotypic monoclonal (m-KAb) and recombinant (r-KAb) antibodies its internal images and an Ab-derived peptide mimotope (*KP*)

clinical use still require active investigation, antigen/receptor-driven therapeutic approaches such as those already discussed show great promise in terms of enhanced specificity with lower side effects.

Furthermore, the detailed analysis of their mode of action can reveal a number of further potential targets against which antimicrobial agents could be potentially directed. As an example, besides being targeted by echinocandins (Morrison 2006), cell wall β-glucans have been suggested as target of some KTs, including *Pa*KT and HM-1. This observation is highly relevant, since these structures are not biosynthesized by mammalian cells, while they appear to be widespread vital components of microbial cell walls. As in idiotypic vaccination, both KT-neutralizing Abs (or their mimotopes) and KTRs could be used to produce candidate multivalent vaccines to elicit protective KAbs. The demonstration that a β-glucan-conjugate vaccine was protective against different fungal experimental infections, by eliciting fungicidal Abs, confirmed and strengthened this suggestion (Torosantucci et al. 2005). The use of "common" or "universal" antigens, such as β-glucans, KTRs or their mimotopes, as protective vaccines, introduces a new concept of transdisease immunoprophylaxis which might appear heretical according to immunological dogma, in that it is not pathogen-specific (Conti et al. 1998; Casadevall and Pirofski 2007).

Analogous relevant implications of the killer phenomenon could be envisaged in the field of phytopathogenic infectious agents, which have an enormous economic impact on agricultural production and storage, and pose serious problems for effective food production and distribution worldwide. Microbial spoilage can have detrimental effects on food and feeds, by also increasing the incidence of carcinogens, such as mycotoxins, which can affect human and animal health (Bryden 2007). Similar to what has been previously described for other antimicrobial peptides and plantibodies, the expression of KTs such as *U. maydis* KP4 (Clausen et al. 2000), as well as of killer peptides such as KP (Donini et al. 2005), could significantly increase the endogenous resistance of transgenic plants against relevant infectious phytopathogens, besides making available a potential new strategy for a rapid, and efficient large-scale production of antimicrobial molecules *in planta*. Once its environmental and health safety is determined, this genetic engineering could be applied to any commercial variety of plants, holding the promise of a novel resource for the biological control of relevant plant diseases as well as for scaling down herbicides, pesticides and preservatives in the agriculture and food industry. This could have undoubted benefits in terms of labor, safety, environmental hazard, and costs.

Thus, while future studies could confirm all the above-mentioned intriguing observations, the mere biological interest of the yeast killer phenomenon is going to be enriched.

References

Beninati C, Oggioni MR, Boccanera M, Spinosa MR, Maggi T, Conti S, Magliani W, De Bernardis F, Teti G, Cassone A, Pozzi G, Polonelli L (2000) Therapy of mucosal candidiasis by expression of an anti-idiotype in human commensal bacteria. Nat Biotechnol 18:1060–1064

Bennett AM, Norris AR, Limnander de Nieuwenhove A, Russell PJ (2002) Replication of a linear mini-chromosome with terminal inverted repeats from the *Kluyveromyces lactis* linear DNA plasmid k2 in the cytoplasm of *Saccharomyces cerevisiae*. Plasmid 48:13–23

Bleve G, Grieco F, Cozzi G, Logrieco A, Visconti A (2006) Isolation of epiphytic yeasts with potential for biocontrol of *Aspergillus carbonarius* and *A. niger* on grape. Int J Food Microbiol 108:204–209

Brena S, Omaetxebarría MJ, Elguezabal N, Cabezas J, Moragues MD, Pontón J (2007) Fungicidal monoclonal antibody C7 binds to *Candida albicans* Als3. Infect Immun 75:3680–3682

Bryden WL (2007) Mycotoxins in the food chain: human health implications. Asia Pac J Clin Nutr 16(Suppl 1):95–101

Buzzini P, Turchetti B, Vaughan-Martini AE (2007) The use of killer sensitivity patterns for biotyping yeast strains: the state of the art, potentialities and limitations. FEMS Yeast Res 7:749–760

Caprilli F, Prignano G, Latella C, Tavarozzi S (1985) Amplification of the killer system for differentiation of *Candida albicans* strains. Mykosen 28:569–573

Carrau FM, Neirotti E, Gioia O (1993) Stuck wine fermentations: effect of killer/sensitive yeast interactions. J Ferment Bioeng 76:67–69

Casadevall A, Pirofski LA (2007) Antibody-mediated protection through cross-reactivity introduces a fungal heresy into immunological dogma. Infect Immun 75:5074–5078

Casoli C, Pilotti E, Perno CF, Balestra E, Polverini E, Cassone A, Conti S, Magliani W, Polonelli L (2006) A killer mimotope with therapeutic activity against AIDS-related opportunistic microorganisms inhibits ex-vivo HIV-1 replication. AIDS 20:975–980

Cenci E, Pericolini E, Mencacci A, Conti S, Magliani W, Bistoni F, Polonelli L, Vecchiarelli A (2006) Modulation of phenotype and function of dendritic cells by a therapeutic synthetic killer peptide. J Leukoc Biol 79:40–45

Ciani M, Fatichenti F (2001) Killer toxin of *Kluyveromyces phaffii* DBVPG 6076 as a biopreservative agent to control apiculate wine yeasts. Appl Environ Microbiol 67:3058–3063

Clausen M, Kräuter R, Schachermayr G, Potrykus I, Sautter C (2000) Antifungal activity of a virally encoded gene in transgenic wheat. Nat Biotechnol 18:446–449

Cohen ML (2000) Changing patterns of infectious disease. Nature 406:762–727

Comitini F, De Ingeniis J, Pepe L, Mannazzu I, Ciani M (2004) *Pichia anomala* and *Kluyveromyces wickerhamii* killer toxins as new tools against *Dekkera/Brettanomyces* spoilage yeasts. FEMS Microbiol Lett 238:235–240

Conti S, Magliani W, Gerloni M, Salati A, Dieci E, Arseni S, Fisicaro P, Polonelli L (1998) A transphyletic anti-infectious control strategy based on the killer phenomenon. FEMS Immunol Med Microbiol 22:151–161

Conti S, Magliani W, Giovati L, Libri I, Maffei DL, Salati A, Polonelli L (2008) Screening of a *Saccharomyces cerevisiae* nonessential gene deletion collection for altered susceptibility to a killer peptide. New Microbiol 31:157–159

Donini M, Lico C, Baschieri S, Conti S, Magliani W, Polonelli L, Benvenuto E (2005) Production of an engineered killer peptide in *Nicotiana benthamiana* by using a potato virus X expression system. Appl Environ Microbiol 71:6360–6367

Druvefors U, Jonsson N, Boysen ME, Schnürer J (2002) Efficacy of the biocontrol yeast *Pichia anomala* during long-term storage of moist feed grain under different oxygen and carbon dioxide regimens. FEMS Yeast Res 2:389–394

Eiden-Plach A, Zagorc T, Heintel T, Carius Y, Breinig F, Schmitt MJ (2004) Viral preprotoxin signal sequence allows efficient secretion of green fluorescent protein by *Candida glabrata*, *Pichia pastoris*, *Saccharomyces cerevisiae*, and *Schizosaccharomyces pombe*. Appl Environ Microbiol 70:961–966

Fanti F, Conti S, Campani L, Morace G, Dettori G, Polonelli L (1989) Studies on the epidemiology of *Aspergillus fumigatus* infections in a university hospital. Eur J Epidemiol 5:8–14

Fiori PL, Mattana A, Dessì D, Conti S, Magliani W, Polonelli L (2006) In vitro acanthamoebicidal activity of a killer monoclonal antibody and a synthetic peptide. J Antimicrob Chemother 57:891–898

Fuentefria AM, Faganello J, Pazzini F, Schrank A, Valente P, Vainstein M (2007) Typing and patterns of cellular morphological alterations in *Cryptococcus neoformans* and *Cryptococcus gattii* isolates exposed to a panel of killer yeasts. Med Mycol 45:503–512

Gibbs MD, Reeves RA, Sunna A, Bergquist PL (2004) A yeast intron as a translational terminator in a plasmid shuttle vector. FEMS Yeast Res 4:573–577

Giga-Hama Y, Tohda H, Takegawa K, Kumagai H (2007) *Schizosaccharomyces pombe* minimum genome factory. Biotechnol Appl Biochem 46:147–155

Guyard C, Evrard P, Corbisier-Colson AM, Louvart H, Dei-Cas E, Menozzi FD, Polonelli L, Cailliez J (2001) Immuno-crossreactivity of an anti-*Pichia anomala* killer toxin monoclonal antibody with a *Williopsis saturnus* var. *mrakii* killer toxin. Med Mycol 39:395–400

Guyard C, Dehecq E, Tissier JP, Polonelli L, Dei-Cas E, Cailliez JC, Menozzi FD (2002) Involvement of [beta]-glucans in the wide-spectrum antimicrobial activity of *Williopsis saturnus* var. *mrakii* MUCL 41968 killer toxin. Mol Med 8:686–694

Heintel T, Zagorc T, Schmitt MJ (2001) Expression, processing and high level secretion of a virus toxin in fission yeast. Appl Microbiol Biotechnol 56:165–172

Izgü F, Altinbay D, Türeli AE (2007a) *In vitro* susceptibilities of *Candida* spp. to Panomycocin, a novel exo-beta-1,3-glucanase isolated from *Pichia anomala* NCYC 434. Microbiol Immunol 51:797–803

Izgü F, Altinbay D, Türeli AE (2007b) *In vitro* activity of panomycocin, a novel exo-beta-1, 3-glucanase isolated from *Pichia anomala* NCYC 434, against dermatophytes. Mycoses 50:31–34

Komiyama T, Shirai T, Ohta T, Urakami H, Furuichi Y, Ohta Y, Tsukada Y (1998) Action properties of HYI killer toxin from *Williopsis saturnus* var. *saturnus*, and antibiotics, aculeacin A and papulacandin B. Biol Pharm Bull 21:1013–1019

Lowes KF, Shearman CA, Payne J, MacKenzie D, Archer DB, Merry RJ, Gasson MJ (2000) Prevention of yeast spoilage in feed and food by the yeast mycocin HMK. Appl Environ Microbiol 66:1066–1076

Magliani W, Conti S, De Bernardis F, Gerloni M, Bertolotti D, Mozzoni P, Cassone A, Polonelli L (1997a) Therapeutic potential of antiidiotypic single chain antibodies with yeast killer toxin activity. Nat Biotechnol 15:155–158

Magliani W, Conti S, Gerloni M, Bertolotti D, Polonelli L (1997b). Yeast killer systems. Clin Microbiol Rev 10:369–400

Magliani W, Conti S, Salati A, Arseni S, Ravanetti L, Frazzi R, Polonelli L (2004a) Engineered killer mimotopes: new synthetic peptides for antimicrobial therapy. Curr Med Chem 11: 1793–800

Magliani W, Conti S, Salati A, Vaccari S, Ravanetti L, Maffei DL, Polonelli L (2004b) Therapeutic potential of yeast killer toxin-like antibodies and mimotopes. FEMS Yeast Res 5:11–18

Magliani W, Conti S, Frazzi R, Ravanetti L, Maffei DL, Polonelli L (2005) Protective antifungal yeast killer toxin-like antibodies. Curr Mol Med 5:443–452

Magliani W, Conti S, Maffei DL, Ravanetti L, Polonelli L (2007) Anti-idiotype-derived killer peptides as new potential tools to combat HIV-1 and AIDS-related opportunistic pathogens. Anti-Infective Agents Med Chem 6:263–272

Maia ML, Dos Santos JI, Viani FC, Larsson CE, Paula CR, Gambale W (2001) Phenotypic characterization of *Microsporum canis* isolated from cats and dogs. Mycoses 44:480–486

Marquina D, Santos A, Peinado JM (2002) Biology of killer yeasts. Int Microbiol 5:65–71

Matthews RC, Burnie JP (2005) Human recombinant antibody to HSP90: a natural partner in combination therapy. Curr Mol Med 5:403–411

Miyamoto M, Onozato N, Selvakumar D, Kimura T, Furuichi Y, Komiyama T (2006) The role of the histidine-35 residue in the cytocidal action of HM-1 killer toxin. Microbiology 152:2951–2958

Morace G, Archibusacci C, Sestito M, Polonelli L (1984) Strain differentiation of pathogenic yeasts by the killer system. Mycopathologia 84:81–85

Morace G, Manzara S, Dettori G, Fanti F, Conti S, Campani L, Polonelli L, Chezzi C (1989) Biotyping of bacterial isolates using the yeast killer system. Eur J Epidemiol 5:303–310

Morrison VA (2006) Echinocandin antifungals: review and update. Expert Rev Anti-Infect Ther 4:325–342

Passoth V, Fredlund E, Drufevors UA, Schnürer S (2006) Biotechnology, physiology and genetics of the yeast *Pichia anomala*. FEMS Yeast Res 6:3–13

Pérez F, Ramírez M, Regodón JA (2001) Influence of killer strains of *Saccharomyces cerevisiae* on wine fermentation. Antonie van Leeuwenhoek 79:393–399

Pettoello-Mantovani M, Nocerino A, Polonelli L, Morace G, Conti S, Di Martino L, De Ritis G, Iafusco M, Guandalini S (1995) *Hansenula anomala* killer toxin induces secretion and severe acute injury in the rat intestine. Gastroenterology 109:1900–1906

Polonelli L, Morace G (1986) Reevaluation of the yeast killer phenomenon. J Clin Microbiol 24:866–869

Polonelli L, Morace G (1988) Yeast killer toxin-like anti-idiotypic antibodies. J Clin Microbiol 26:602–604

Polonelli L, Archibusacci C, Sestito M, Morace G (1983) Killer system: a simple method for differentiating *Candida albicans* strains. J Clin Microbiol 17:774–780

Polonelli L, Castagnola M, Rossetti DV, Morace G (1985) Use of killer toxins for computer-aided differentiation of *Candida albicans* strains. Mycopathologia 91:175–179

Polonelli L, Lorenzini R, De Bernardis F, Morace G (1986) Potential therapeutic effect of yeast killer toxin. Mycopathologia 96:103–107

Polonelli L, Conti S, Campani L, Morace G, Fanti F (1989) Yeast killer toxins and dimorphism. J Clin Microbiol 27:1423–1425

Polonelli L, Conti S, Gerloni M, Magliani W, Castagnola M, Morace G, Chezzi C (1991) 'Antibiobodies': antibiotic-like anti-idiotypic antibodies. J Med Vet Mycol 29:235–242

Polonelli L, Menozzi MG, Campani L, Gerloni M, Conti S, Morace G, Chezzi C (1992) Anaerobic yeast killer systems. Eur J Epidemiol 8:471–476

Polonelli L, Lorenzini R, De Bernardis F, Gerloni M, Conti S, Morace G, Magliani W, Chezzi C (1993) Idiotypic vaccination: immunoprotection mediated by anti-idiotypic antibodies with antibiotic activity. Scand J Immunol 37:105–110

Polonelli L, De Bernardis F, Conti S, Boccanera M, Gerloni M, Morace G, Magliani W, Chezzi C, Cassone A (1994) Idiotypic intravaginal vaccination to protect against candidal vaginitis by secretory, yeast killer toxin-like anti-idiotypic antibodies. J Immunol 152:3175–3182

Polonelli L, De Bernadis F, Conti S, Boccanera M, Magliani W, Gerloni M, Cantelli C, Cassone A (1996) Human natural yeast killer toxin-like candidacidal antibodies. J Immunol 156: 1880–1885

Polonelli L, Séguy N, Conti S, Gerloni M, Bertolotti D, Cantelli C, Magliani W, Cailliez JC (1997) Monoclonal yeast killer toxin-like candidacidal anti-idiotypic antibodies. Clin Diagn Lab Immunol 4:142–146

Polonelli L, Magliani W, Conti S, Bracci L, Lozzi L, Neri P, Adriani D, De Bernardis F, Cassone A (2003) Therapeutic activity of an engineered synthetic killer antiidiotypic antibody fragment against experimental mucosal and systemic candidiasis. Infect Immun 71:6205–6212

Provost F, Polonelli L, Conti S, Fisicaro P, Gerloni M, Boiron P (1995) Use of yeast killer system to identify species of the *Nocardia asteroides* complex. J Clin Microbiol 33:8–10

Santos A, Marquina D (2004) Killer toxin of *Pichia membranifaciens* and its possible use as a biocontrol agent against grey mould disease of grapevine. Microbiology 150:2527–2534

Schlaich T, Urbaniak B, Plissonnier ML, Malgras N, Sautter C (2007) Exploration and Swiss field-testing of a viral gene for specific quantitative resistance against smuts and bunts in wheat. Adv Biochem Eng Biotechnol 107:97–112

Schmitt MJ, Breinig F (2006) Yeast viral killer toxins: lethality and self-protection. Nat Rev Microbiol 4:212–221

Séguy N, Cailliez JC, Polonelli L, Dei-Cas E, Camus D (1996) Inhibitory effect of a *Pichia anomala* killer toxin on *Pneumocystis carinii* infectivity to the SCID mouse. Parasitol Res 82:114–116

Selvakumar D, Miyamoto M, Furuichi Y, Komiyama T (2006) Inhibition of fungal beta-1,3-glucan synthase and cell growth by HM-1 killer toxin single-chain anti-idiotypic antibodies. Antimicrob Agents Chemother 50:3090–3097

Theisen S, Molkenau E, Schmitt MJ (2000) Wicaltin, a new protein toxin secreted by the yeast *Williopsis californica* and its broad-spectrum antimycotic potential. J Microbiol Biotechnol 10:547–550

Torosantucci A, Bromuro C, Chiani P, De Bernardis F, Berti F, Galli C, Norelli F, Bellucci C, Polonelli L, Costantino P, Rappuoli R, Cassone A (2005) A novel glyco-conjugate vaccine against fungal pathogens. J Exp Med 202:597–606

Van Vuuren HJJ, Jacobs CJ (1992) Killer yeasts in the wine industry: a review. Am J Enol Vitic 43:119–128

Yap NA, de Barros Lopes M, Langridge P, Henschke PA (2000) The incidence of killer activity of non-*Saccharomyces* yeasts towards indigenous yeast species of grape must: potential application in wine fermentation. J Appl Microbiol 89:381–389

Chapter 17
Trichothecenes as Toxin and Bioweapons: Prevention and Control

Avinash Ingle, Ajit Varma, and Mahendra Rai

17.1 Introduction

Mycotoxins are naturally occurring substances produced by fungi as secondary metabolites (Ingle et al. 2009). Many of these toxins are pathogenic to human beings and animals. It is estimated that more than 300 mycotoxins are produced by 350 species of fungi (Ingle et al. 2009). The trichothecene mycotoxins are a chemically related family of compounds that are produced by different species of *Fusarium*.

Species of the genus *Fusarium* are common plant pathogens occurring worldwide, mainly associated with cereal crops. *Fusarium* species can produce over 100 secondary metabolites, some of which can unfavorably affect human and animal health due to their immunosuppressive effects (Rabodonirina et al. 1994). The most important *Fusarium* mycotoxins, which can frequently occur at biologically significant concentrations in cereals, are fumonisins, zearalenone and trichothecenes [deoxynivalenol (DON), nivalenol (NIV) and T-2 toxin]. These compounds can occur naturally in cereals (Girish and Goyal 1986), either individually or as specific clusters of two or more of them, depending on the producing fungal species (or strain); of these, T-2 toxin and DON are of special importance, because these are causative agents in a variety of animal diseases and have been associated with human diseases. Wheat and barley are mainly subjected to contamination of T-2 toxin and related trichothecenes. The major *Fusarium* species producing these mycotoxins are *F. culmorum*, *F. graminearum*, *F. avenaceum*, *F. acuminatum*, *F. poae*, *F. solani*, *F. sporotrichioides*, *F. tricinctum* and *F. scripi*, etc. *F. sporotrichioides* and *F. graminearum* are known to produce T-2 toxin and DON respectively (Ueno et al. 1975; Edwards 2004).

Mycotoxins in general and trichothecenes in particular have been used as bioweapons in aerosol form ("yellow rain") to produce lethal and nonlethal casualties in

M. Rai (✉)
Department of Biotechnology, SGB Amravati University, Amravati, 444 602, Maharashtra, India
e-mail: mkrai123@rediffmail.com

M. Rai and A. Varma (eds.), *Mycotoxins in Food, Feed and Bioweapons*,
DOI: 10.1007/978-3-642-00725-5_17, © Springer-Verlag Berlin Heidelberg 2010

Laos (1975–1981), Cambodia (1979–1981), Afghanistan (1979–1981), and the Gulf War (1990–1991). The dissemination of T-2 toxin through aerosols was planned by Saddam Hussain during the Iraq-Iran war. These attacks were alleged to have occurred in remote forest areas, which made confirmation of attacks and recovery of the agent extremely difficult.

Trichothecenes (T-2 toxin and DON) and their derivatives enter the body through the skin and digestive or respiratory epithelium. They inhibit protein and nucleic acid synthesis. They mainly affect fast proliferating tissues like skin, germ cells, bone marrow and mucosal epithelia. In a successful attack with trichothecenes (T-2), the toxin(s) can be inhaled or ingested, or can penetrate the skin. If clothes are contaminated, the trichothecene remains for further toxin exposure. The most common symptoms are burning of the skin with severe pain, redness and pruritus, rash or blisters on the body, bleeding, vomiting, diarrhea, and dyspnea (Bunner et al. 1985; Wannemacher and Wiener 2001).

In India, an outbreak of the food-borne disease in humans in the Kashmir valley during the latter half of 1987 was identified as a distinct disease (trichothecene mycotoxicosis), caused by the consumption of mold-contaminated wheat products (Bhat et al. 1989). Since then, mycotoxins and other toxic metabolites produced by fungi have been a major topic of research.

17.2 Incidence in Agriculture

T-2 toxin and HT-2 toxin are mycotoxins produced by different species of the genus *Fusarium*, which are commonly found in various cereal crops (wheat, maize, barley, oats, and rye) (Rabie et al. 1986; Kim et al. 1993) and processed grains (malt, beer and bread). T-2 and HT-2 toxins often occur together in infected cereals. DON and NIV trichothecenes are the most important mycotoxins in cereal production; DON is actually the most diffuse mycotoxin in grains. The main DON producers *F. graminearum* and *F. culmorum* are responsible for two of the main diseases affecting crops worldwide: Fusarium head blight (FHB) in wheat and Fusarium ear rot (FER) in maize (Munkvold and Desjardins 1997). The fungi producing trichothecenes are soil fungi and are important plant pathogens which grow on the crop in the field. Natural occurrence has been reported in Asia, Africa, South America, Europe and North America (Krogh 1987). Natural levels range from near zero to 10 ppm with a few exceptions showing levels of 15–40 ppm. Toxin production is greatest with increased humidity and temperatures of 6–24°C (Paulo 2002).

17.3 Type of Trichothecenes

T-2 toxin is a member of the fungal metabolites known as the trichothecenes. The trichothecenes are divided into two groups: macrocyclic and nonmacrocyclic. T-2 belongs to the nonmacrocyclic group. This group is also subdivided into Type

A and Type B trichothecenes. Type A trichothecenes include T-2, HT-2 and DAS as mycotoxin components, while Type B trichothecenes include DON, NIV, 3- and 15-Acetyldeoxynivalenol (Ueno et al. 1975).

17.4 Chemical Nature of Trichothecenes

17.4.1 T-2 Toxin

Chemically T-2 toxin is known as - 4″,15-diacetoxy-3″-hydroxy-8″-(3-methyl-butyrlyloxy)-12,13-epoxytrichothec-9-ene (Fig. 17.1). The molecular formula of T-2 toxin is $C_{24}H_{34}O_9$, with a molecular weight of 466.58.

17.4.2 HT-2 toxin

Chemically this is l5-acetoxy-3″,4″-dihydroxy-8″-(3-methylbutyryloxy)-12″-epoxy-tricho-thec-9-ene (Fig. 17.2), with the molecular formula $C_{22}H_{32}O_8$ and a molecular weight of 424.54.

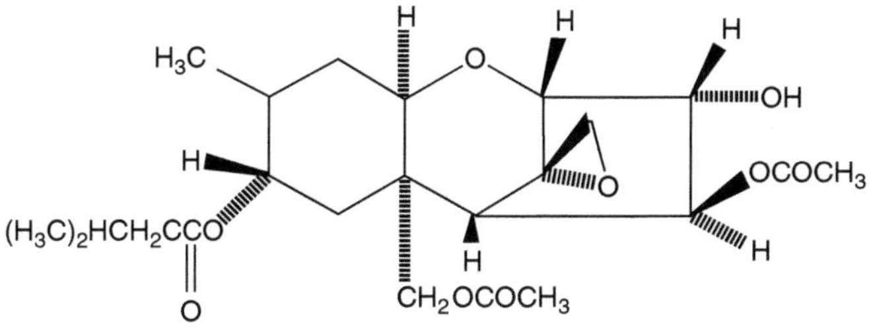

Fig. 17.1 Chemical structure of T-2 toxin

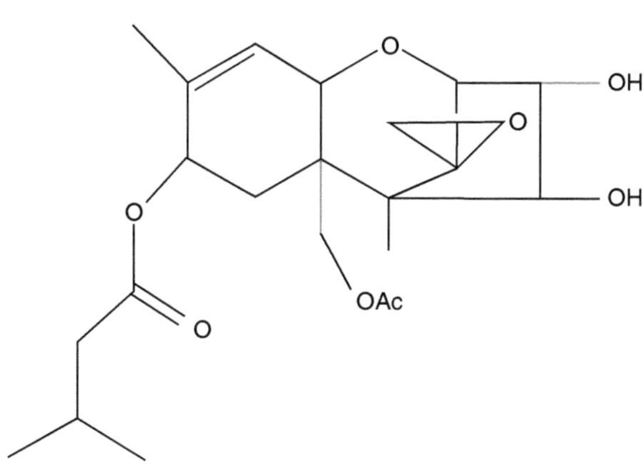

Fig. 17.2 Chemical structure of HT-2 toxin

17.4.3 DON

The chemical name of DON is (3″,7″,15-trihydroxy-12,13-epoxytricho-thec-9-en-8-one) (Fig. 17.3), with the molecular formula $C_{15}H_{20}O_6$, and a molecular weight of 296.32.

17.4.4 Nivalenol

Chemically this is 3″,4′,7″, 15-tetrahydroxy-12,13-epoxytrichothec-9-en-8-one) (Fig. 17.4), with the molecular formula $C_{15}H_{20}O_7$ and a molecular weight of 312.35.

T-2 toxin is rapidly metabolized to HT-2 toxin, which is a major metabolite *in vivo*, and therefore a common risk assessment for T-2 toxin and HT-2 toxin is necessary (Eriksen and Alexander 1998). Usually, the trichothecenes are very stable compounds, both during storage/milling and the processing/cooking of food, and they do not degrade at high temperature (Eriksen and Alexander 1998).

17.5 *In Vitro* Production of T-2 Toxin

In the recent past, research has been focused on different species of *Fusarium* and new techniques for production, isolation, detection and characterization of their mycotoxins. Schollenberger and his colleagues at the Institute of Animal Nutrition,

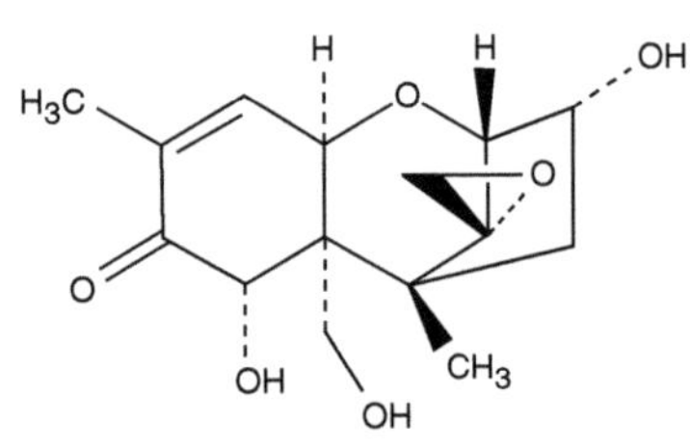

Fig. 17.3 Chemical structure of Deoxynivalenol

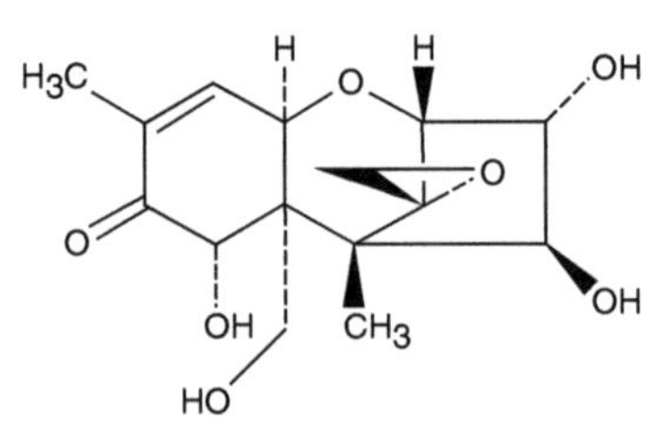

Fig. 17.4 Chemical structure of Nivalenol

University of Hohenheim, Germany are concentrating on German foodstuffs contaminated by mycotoxins (Schollenberger et al. 2005). A group led by Cervero at the Department of Biotechnology, Universidad Politécnica de Valencia, Spain is working on mycotoxins in commercially available corn-based food products in Spain (Cervero et al. 2007). The temperature, pH, moisture and growth period are the important physical parameters for the production of mycotoxins in *Fusarium* (Paulo 2002). *F. equiseti* and *F. graminearum* were successfully used for the production of T-2 toxin by Richardson et al. (1985), and identified by capillary gas chromatography–mass spectrometry.

Ueno et al. (1975) examined 12 T-2 toxin-producing isolates and four fusarenon-X-producing isolates of *Fusarium* species for their ability to produce trichothecene mycotoxins in shake culture and jar fermentation. The authors reported that *F. solani*, *F. sporotrichioides*, and *F. tricinctum* produced T-2 toxin in semisynthetic medium and *F. solani* M-1-1 produced the largest amount of the mycotoxins in a nutrient medium consisting of 5% glucose (or sucrose), 0.1% peptone, and 0.1% yeast extract in either shake culture or jar fermentation at 24–27°C for 5 days. None of the isolates produced significant amounts of fusarenon-X in shake cultures.

Radiolabelled T-2 toxin was synthesized biologically by *F. tricinctum* (NRRL-3299) on a solid rice medium in the presence of sodium acetate, mevalonic acid (Hagler 1981). Daniel et al. (1982) reported that strains of *Fusarium* produced high levels of T-2 toxin when cultured on certain media absorbed into vermiculite. Modified Gregory medium was nutritionally complex (2% soya meal, 0.5% corn steep liquor, 10% glucose) and, when inoculated with the appropriate fungal strain, yielded maximum T-2 toxin within 24 days of incubation at 19°C.

Torp and Langseth (1999) reported the production of trichothecenes and other metabolites by different isolates of *Fusarium* resembling *F. poae*. They found that all the isolates produced relatively large amounts of T-2 toxin (25–600 $\mu g\ g^{-1}$) and the amount of HT-2 toxin detected was on average 25% (5–50%) of the T-2 toxin. Moreover, low concentration of T-2 triol and T-2 tetraol were detected in most cultures.

Morin et al. (2000) studied the relationship between trichothecene production and pathogenicity for 29 isolates of *Fusarium tumidum*, a potential bioherbicide for gorse (*Ulex europaeus*) and broom (*Cytisus scoparius*) in New Zealand. The authors reported that T-2 tetraol and derivatives were produced in grain cultures while NIV and DON derivatives were not detected in any of the culture extracts.

Trichothecenes and other secondary metabolites can be easily produced by *F. culmorum* and *F. equiseti* on common laboratory media and a soil organic matter agar (Hestbjerg 2002). Recently, Jestoi et al. (2008) reported the *in vitro* and *in vivo* production of T-2 and HT-2 toxin by *F. sporotrichioides*. Murthy and his collaborators (2008) reported the presence of T-2 and HT-2 toxin in rice in Karnataka which were toxic to animals and responsible for many acute and chronic human diseases. They reported that rice is an important food commodity susceptible to fungal infection in field as well as storage causing significant loss in yield and quality. Hence rice samples grown under different agro-climatic conditions were screened for mold and toxin produced by *Fusarium* sp. Among the 64 *Fusarium* isolates, 17 were found to be toxigenic. HPLC analysis revealed the presence of DON, NIV, T-2 and HT-2 toxin; of these, DON was found in high levels, in the range 20–500 μg kg^{-1}.

17.6 Detection of Trichothecenes

Different mycotoxins such as NIV and fusarenon-X from *Fusarium nivale* and *F. episphaeria*, diacetylnivalenol from *F. oxysporum*, T-2 toxin, neosolaniol and diacetoxyscirpenol from *F. solani* were detected by thin layer chromatography with several solvent systems (Ueno et al 1973). Richardson et al. (1985) also reported that thin layer chromatography is a more effective method for the detection of trichothecenes such as T-2 toxin and DON produced by *Fusarium* than the gas-liquid chromatographic method.

A rapid and easy procedure for the detection of trichothecenes in plasma and urine was proposed by Rood et al. (1986). In their experiment, the trichothecene toxins were extracted using a Clin-Elut column, hydrolyzed to their corresponding parent alcohols and cleaned up with a silica cartridge followed by derivatization for gas chromatographic analysis. The detection of any of the parent alcohols in plasma or urine would indicate an exposure to trichothecenes. Recoveries in urine are between 78 and 119% at levels of 50–1,000 ng ml^{-1} and recoveries in plasma are between 80 and 116% at levels of 50–500 ng ml^{-1}. The limit of detection is better than 25 ppb.

Nikulin et al. (1996) reported the growth and toxin production of a highly toxic strain of *Fusarium sporotrichioides* on oat and wheat grains and on straw under experimental conditions, in which relative humidity (RH) of air was regulated. They found the production of different trichothecenes on all grains. Toxin production was measured by three biological toxicity tests (cytotoxicity test, dermotoxicity test, and yeast cell toxicity test), chemical analysis, and T-2 enzyme-linked immunosorbent (ELISA) assay. Cytotoxicity and production of trichothecenes were detected in all the samples. On oat and wheat grains, T-2 toxin, neosolaniol, and diacetoxyscirpenol were found, and on straw T-2 toxin, HT-2 toxin, neosolaniol, and T-2 tetraol were determined. In the T-2 ELISA assay, all material samples were found to contain T-2 toxin. The cytotoxicity test was found to be the most sensitive method for detecting biological toxicity of samples inoculated with fungus, while the T-2 ELISA assay and chemical analysis were about equally sensitive for detecting T-2 toxin in samples.

Yoshizawa et al. (2004) developed and tested an ELISA system for individual measurement of DON, NIV, T-2 and HT-2 toxin using monoclonal antibodies for 3,4,15-triacetyl-nivalenol, for both 3,4,15-triacetyl-nivalenol and 3,15-diacetyl-DON, and for acetyl-T-2 toxin. The assay system comprised three kits (designated the DON+NIV kit, the NIV kit, and the T-2+HT-2 kit). The practical performance of the ELISA system was assessed by assaying trichothecene mycotoxins in wheat kernels. They reported that the ELISA system meets all the requirements for use in a routine assay in terms of sensitivity. The ELISA system was found to be a useful alternative method to gas chromatography-mass spectrometry, liquid chromatography-mass spectrometry, or liquid chromatography-ultraviolet absorption for screening cereals and foods for trichothecene mycotoxin contamination.

Hoagland et al. (2008) developed a rapid ELISA technique for the detection of trichothecene produced by potential bioherbicidal fungus *Myrothecium verrucaria* (MV). The authors used commercially available ELISA plates for trichothecene detection with some modifications, possessing cross-reactivity with several trichothecene mycotoxins (e.g. verrucarin A, and J, roridin A, L-2, E and H), and tested for their ability to detect trichothecenes produced by a strain of MV in cell cultures and in plant tissues treated with purified roridin A, or ethyl acetate fractions of MV cultures. Evaluations of ELISA assays showed linear responses for standards of verrucarin A and roridin A over a concentration range of 0.2–20 ppb. Ethyl acetate or aqueous extractions were used to obtain samples from MV cultures and plant tissues for testing. ELISA is a sensitive and rapid assay method for quantifying trichothecenes produced by this bioherbicidal fungus and in certain plant tissues treated with trichothecenes.

17.7 Toxicity of T-2 Toxin

Poisoning with T-2 toxin may simply lead to vomiting, diarrhea, rejection of food, inflammation of the gastrointestinal tract, impairment of nerve cells, heart muscle, lymphatic system, testes and thymus, and formation of tissue necroses, or it may cause some severe infection such as alimentary toxic aleukia (ATA) (Joffe 1978; Bouaziz et al. 2006). ATA has been widely reported in the former U.S.S.R. since 1913 and has been attributed to the consumption of grain contaminated with toxigenic species of *Fusarium*, which have been shown to produce trichothecenes (T-2 toxin and other type A trichothecenes). The most severe outbreak occurred in the spring of 1944 in the Orenburg district, in which 10% of the population was affected and mortality rate was as high as 60% in some countries (Beardall and Miller 1994). Clinical features of the disease include leukopenia, agranulocytosis, bleeding from the nose, throat, gums, necrotic angina, a hemorragic rash, sepsis, exhaustion of the bone marrow, fever (Beardall and Miller 1994).

Pandey and his coworkers (2005) of the Department of Veterinary Microbiology, Nagpur Veterinary College, Nagpur, India reported the acute intoxication in humans due to toxicity of T-2 toxin after ingestion of contaminated food such as meat, eggs and milk. In their study they proved that mycotoxins can be passed through the food chain to humans. They selected 20 day-old broiler chicks, and divided them into two equal groups A and B. Group A served as control which received feed free of toxin, while group B received feed containing T-2 toxin at 4 ppm for 28 days. Broilers from both the groups were slaughtered at 28 days. Liver and breast muscle of the chicks from these groups were collected and processed for the detection of T-2 toxin. The result of this study indicated presence of T-2 toxin and its metabolites in the liver, while the breast muscle revealed presence of only T-2 metabolites, thus confirming the pathway of T-2 toxin in the human food chain, which substantiates the reports of human illness associated with T-2 contaminated foodstuffs.

T-2 toxin is the most toxic of the *Fusarium* trichothecenes, though less widely distributed than DON. In pigs, clinical signs of T-2 toxicosis include emesis, posterior paresis, lethargy and frequent defecation. At natural levels of contamination in the diet T-2 toxin causes reduced feed intake and animal performance. At high concentrations in the diet it produces diarrhea, emesis and feed refusal. T-2 toxicosis in poultry causes oral lesions, reduced feed consumption and growth rate in young animals and reduced egg production in laying hens (Sokolovic et al. 2008). Sokolovic et al. (2008) also reported the mechanism of action of T-2 toxin in poultry. They found that after exposure of T-2 toxin by the oral, dermal or inhalation route, various animal organs and tissues can be severely affected. So far, toxic effects have been evidenced in the cells of fungi, protozoa, insects, molds, plants, and different cell cultures. In poultry, the toxic effects of T-2 toxin can be classified as genotoxic and cytotoxic, immunomodulatory effects, effects on the cells of the digestive system and liver, effects on the nervous system and skin and impairment of poultry performance (Sokolovic et al. 2008).

The impact on poultry production becomes important at dietary concentrations above 2 mg kg^{-1}. In ruminants T-2 toxicosis results in a wide range of responses, such as feed refusal, leukopenia, depression, diarrhea, coagulopathy, enteritis, and posterior ataxia. Reduction of humoral immunity is a common effect in pigs, poultry and ruminants exposed to low concentrations of T-2 toxin in the diet. This increases susceptibility to other diseases and the effects of poor management practices. Most of the clinical signs caused by the ingestion of T-2 toxin are also observed with diacetoxyscirpenol and, to a slightly lower extent, with HT-2 toxin (Prelusky et al. 1994).

17.8 Role of Trichothecenes in War

"Biological warfare" (BW) is defined as the "employment of biological agents to produce casualties in man or animals or damage to plants" (NATO 1996). Trichothecene-producing fungi are plant pathogens and invade various agricultural products and plants. Since *Fusarium* and other related fungi infect important foodstuff, they have been associated worldwide with intoxication of humans and animals. Thus, these fungi have potential as biological weapons (Wannemacher and Wiener 2001).

The delivery of trichothecene mycotoxins can be made as dusts, droplets, aerosols or smoke from aircraft, rockets, missiles, artillery, mines or portable sprayers. Because of their antipersonnel properties, ease of large-scale production, and apparent proven delivery by various aerial dispersal systems, the trichothecene mycotoxins (especially T-2 toxin) have an excellent potential for weaponization (Wannemacher and Wiener 2001). When delivered at low doses, trichothecene causes skin, eye, and gastrointestinal problems. In nanogram amounts, they (T-2 toxin, in particular) cause severe skin irritation (erythema, edema, and necrosis) (Ueno 1989; Wannemacher et al. 1985, 1991). Skin vesication has been observed in a number of humans exposed to yellow rain attacks (Ueno 1989). T-2 toxin is about 400-fold more powerful (50 ng vs 20 μg) than mustard in producing skin injury. Lower-microgram quantities of trichothecene mycotoxins cause severe eye irritation, corneal damage, and impaired vision (Watson et al. 1984; Bunner et al. 1985; Stahl et al. 1985; Ueno 1989). Emesis and diarrhea have been observed at amounts that are one-fifth to one-tenth the lethal doses of trichothecene mycotoxins (Bunner et al. 1985).

From 1974 to 1981, toxic agents were used by the Soviet Union and its client states in such cold war sites as Afghanistan, Laos, and Kampuchea (Cambodia). Aerosol and droplet clouds were produced by delivery systems in the Soviet arsenal such as aircraft spray tanks, aircraft-launched rockets, bombs (exploding cylinders), canisters, a Soviet hand-held weapon (DH-10), and booby traps. Aircraft used for delivery included L-19s, AN-2s, T-28s, T-41s, MiG-21s (in Laos) and Soviet MI-24 helicopters (in Afghanistan and Laos) (Wannemacher and Wiener 2001).

There are reports of alleged use of trichothecenes in the 1964 Egyptian (or Russian) attacks on Yemeni Royalists in Yemen (Ricaud 1983) and in combination with mustard gas during chemical warfare attacks in the Iran-Iraq War (1983–1984) (Ember et al. 1984). According to European sources, Soviet-Cuban forces in Cuba are said to have been equipped with mycotoxins, and a Cuban agent is said to have died of a hemorrhagic syndrome induced by a mycotoxin agent (Seagrave 1981).

17.9 Mode of Action of T-2 and HT-2 Toxin

17.9.1 Effects on DNA and RNA Synthesis

T-2 toxin and DON inhibit the synthesis of DNA and RNA both *in vivo* [0.75 mg kg^{-1} body weight (bw) single or multiple doses] and *in vitro* (>0.1–1 ng ml^{-1}) (Rosenstein and Lafarge-Frayssinet 1983). DON also inhibits the synthesis of DNA and RNA and protein synthesis at the ribosomal level. The toxin has a hemolytic effect on erythrocytes. An acute dose of DON can induce vomiting in pigs, whereas at lower concentrations in the diet it reduces growth and feed consumption. Both effects, which are also seen with other trichothecene toxins, are thought to be mediated by affecting the serotonergic activity in the CNS or via peripheral actions on serotonin receptors (Eriksen and Alexander 1998).

17.9.2 Effects on Protein Synthesis

T-2 toxin inhibited the activity of peptidyl transferase, and consequently the protein synthesis in the initiation phase (Beasley 1989). T-2 toxin inhibited protein synthesis both *in vitro* (0.01 ng ml^{-1} in suspensions of rat hepatocytes gave 75% inhibition) and *in vivo* (WHO 1990). *In vivo* inhibition of synthesis of proteins has been demonstrated in cells from bone marrow, spleen and thymus (0.75 mg kg^{-1} bw single dose in mice) (Rosenstein and Lafarge-Frayssinet 1983; Thompson and Wannemachter 1990; WHO 1990).

17.9.3 Effects on Membranes and Lipid Peroxidation

T-2 toxin affected the permeability of cell membranes *in vitro* at concentrations of 0.4 pg ml^{-1} (Bunner and Morris 1988), caused changes in the phospholipid turnover in bovine platelets (Grandoni et al. 1992) and hemolysis of erythrocytes *in vitro* (Rizzo et al. 1992). One single oral dose of 3.6 mg kg^{-1} bw increased lipid peroxides in the liver of rats (Rizzo et al. 1994). Ascorbic acid, a-tocopherol and selenium (Rizzo et al. 1992, 1994), as well as glutathione precursors (Fricke and Jorge 1991), have a protective effect against lipid peroxidation induced by T-2 toxin (Bouaziz et al. 2006).

17.9.4 Effect on Nutrients

The effect of T-2 toxin on intestinal absorption of monosaccharides was studied in rats. The absorption of 3-*O*-methylglucose was reduced by one- to three-fold after injection of T-2 toxin into the lumen or after intravenous injection. Jejunal function was impaired by specific damage to the active transport and diffusional movement of monosaccharides (Kumagai and Shimizu 1988).

17.9.5 Alterations of Cellular Membranes

At a concentration of T-2 toxin of 20 μg ml^{-1}, no entry of [^{14}C] sucrose or [^{3}H] inulin was observed in bovine erythrocytes *in vitro*. Very little radiolabel was bound to bovine erythrocytes, and the binding was independent of the T-2 toxin concentration. The toxin had no effect on the entrapment of sucrose or inulin. Carrier erythrocytes retained 85% of [^{14}C] sucrose and only 18% of [^{3}H] T-2 toxin. Thus, T-2 toxin diffused from carrier cells more rapidly than sucrose. The authors concluded that the interaction of T-2 toxin with bovine erythrocytes was minimal and intercalation with the inner bilayer was unlikely, because the increase in cell volume that would have resulted did not occur (DeLoach et al. 1987).

17.9.6 Apoptosis of Human Lymphocytes In Vitro

Apoptosis of human peripheral lymphocytes was observed after exposure *in vitro* to T-2 toxin at a concentration of 0.1, 1, 10, or 100 ng ml^{-1} of culture medium. A concentration- and time-dependent apoptotic response was observed that was inhibited by chelating intracellular calcium with BAPTA-AM, a chelator activated by cytosolic esterases. No response was seen with T-2 toxin at 0.01 or 0.1 ng ml^{-1} after up to 5 days of incubation. Increased apoptotic cell counts were observed after 3 or 5 days of incubation with concentrations $\geq$1 ng ml^{-1}. Apoptosis was observed in all lymphocyte types (Yoshino et al. 1997).

17.10 Prevention and Control of Trichothecenes

The hazards of mycotoxins in general and trichothecenes in particular to humans and animals are now well-recognized. There has been great concern in recent years about the control of these toxins, and prevention of mycotoxin contamination of grain is the main goal of food and agricultural industries throughout the world. But unfortunately under fast-changing environmental conditions the contamination of various cereal grains with *Fusarium* and mycotoxins can not be avoided. While certain treatments have been found to reduce concentrations of specific mycotoxins, no single method has been developed that is equally effective for the wide variety of mycotoxins. The following methods can be helpful in controlling the contamination of mycotoxins such as trichothecenes.

17.10.1 Prevention of Mold Growth

The fact that *Fusarium* species secretes trichothecenes (T-2, HT-2 and DON) is well-known. The *Fusaria* secretes trichothecenes in stored food and food products, and causes many severe infections, as discussed in this chapter. In stored grain, mold damage may be prevented mainly by drying of grain, controlled atmosphere storage and chemical treatment.

17.10.1.1 Drying and Cooking of Grains

It is well known that dry grains can be stored for longer periods, safe from insects and molds due to lack of moisture for their development. The average Indian farmer performs drying of grain conventionally under direct sun light (Girish and Goyal 1986). The most widely used indigenous practice of grain-drying is to spread threshed grains in thin layers on a plain surface in open sunlight. Some farmers have been using concrete floors for drying which is quicker than a mud floor. Exposure of aflatoxin-contaminated groundnut oil to sunlight destroyed about 99% of the aflatoxins (Bilgrami and Sinha 1984). Other methods of grain-drying include mechanical drying, infrared, microwave and solar energy drying. Research is also being conducted into the use of these methods (Girish and Goyal 1986). Sugita-Konishi et al. (2006) reported that proper cooking was one of the best methods of detoxifying the DON in food. Their study showed the retention in the DON level in noodles and bread made from naturally contaminated wheat flour after proper cooking and boiling.

17.10.1.2 Controlled Atmosphere Storage

Environmental conditions can be controlled by the traditional method of underground storage. The main advantage of underground storage lies behind the principle of grain-cooling and depleting the oxygen content to the desired level at which microbes and insects cannot grow. Air-tight storage also works on the same principle where the depletion of oxygen by grain respiration manipulates disinfection by inhibiting aerobic fungi, eliminating mycotoxin production, and conserving desirable quality factors in the grain. For preservation of grains, natural cooling is another effective method of preservation. The low temperature prevents the microflora, as most of them are thermophilic molds and grow at relatively high temperatures.

17.10.1.3 Chemical Treatment

Chemical control of fungal deterioration to stored grain is not used for grain intended for food and feed, but is restricted to the treatment of grain for seed purposes only. Experiments conducted by the Indian Grain Storage Institute showed that "grain treat" (mixture of propionic acid, acetic acid and benzoic acid) against *Aspergillus* and *Penicillium* spp. did not produce effective results in maize. Seed treatment with Bavistin and TMTD (trimethyl thiuram disulfide) at 0.25% concentration gave 100% protection against *Aspergillus* and *Penicillium* spp. to wheat grain in 1-year storage (Girish and Goyal 1986).

17.11 Conclusion and Future Directions

Trichothecenes (T-2, HT-2, DON and NIV) are produced by different *Fusarium* species in significant quantity. These toxins are found to be highly toxic to human beings and animals and cause severe infection. They are the main source of contamination of food and feed. Trichothecenes, in particular T-2 toxin, if used as a bio-weapon in war, may cause severe health hazards to livestock and humans. There is urgent need to develop rapid molecular methods and assays for the detection of toxigenic fungi. Prevention and control of the production and contamination of toxins in food and food products can be achieved by the early detection of these toxins through the use of commercially available different assays and kits such as ELISA. Retention in the level and detoxification of trichothecenes can also be managed to a large extent by good practices by using different methods such as drying of grains, ensuring proper atmospheric conditions, and by chemical treatment.

References

Beardall JM, Miller JD (1994) Diseases in humans with mycotoxins as possible causes. In: Miller JD, Trenholm HL (eds) Mycotoxins in grain. Compounds other than aflatoxins. Eagan, St. Paul, MN, USA, pp 487–539

Beasley VR (ed) (1989) Trichothecene mycotoxicosis: pathophysiologic effect. CRC, Boca Raton

Bilgrami KS, Sinha KK (1984) Mycotoxin contamination in food and its control. Indian Rev Life Sci 4:19–36

Bhat RV, Beedu SR, Ramakrishna Y, Munshi KL (1989) Outbreak of trichothecene mycotoxicosis associated with consumption of mould-damaged wheat production in Kashmir Valley, India. Lancet 7:35–37

Bouaziz C, Abid-Essefi S, Bouslimi A, El-Golli E, Bacha H (2006) Cytotoxicity and related effects of T-2 toxin on cultured Vero cells. Toxicon 48:343–352

Bunner DL, Upshall DG, Bhatti AR (1985) Toxicology data on T-2 toxin. In: Report of Focus Officers Meeting on Mycotoxin Toxicity, September 23–24, 1985. Defense Research Establishment at Suffield, Alberta, Canada

Bunner DL, Morris ER (1988) Alteration of multiple cell membrane functions in L-6 myoblasts by T-2 toxin: an important mechanism of action. Toxicol Appl Microbiol 92:113–121

Cervero MC, Castillo MA, Montes R, Hernández E (2007) Determination of trichothecenes, zearalenone and zearalenols in commercially available corn-based foods in Spain. Revista Iberoamericana de Micología 24:52–55

Daniel C, Eugene B, Smalley S, Caldwell RW (1982) New process for T-2 toxin production. Appl Environ Microbiol 44:371–375

DeLoach JR, Andrews K, Nagi A (1987) Interaction of T-2 toxin with bovine erythrocyte, effects on cell lysis, permeability and entrapment. Toxicol Appl Pharmacol 88:123–131

Edwards SG (2004) Influence of agricultural practices on *Fusarium* infection of cereals and subsequent contamination of grains by trichothecene mycotoxin. Toxicol Lett 153:29–35

Ember LR, Sorenson WG, Lewis DM (1984) Charges of toxic arms use by Iraq escalate. Chem Eng News 62:16–18

Eriksen GS, Alexander J (eds) (1998) *Fusarium* toxins in cereals — a risk assessment. Nordic Council of Ministers, Copenhagen, pp 7–44

Fricke RF, Jorge J (1991) Methylthiazolidine-4-carboxylate for treatment of acute T-2 toxin exposure. J Appl Toxicol 11:135–140

Girish GK, Goyal RK (1986) Prevention and control of mycotoxins in food grains in India. Bull Grain Technol 24:157–177

Grandoni KM, Gentry PA, Holub BJ, Yagen B (1992) Trichothecene mycotoxins inhibit phosphoinositide hydrolysis in bovine platelets stimulated with platelet activating factor. Toxicology 72:5160–5170

Hagler WM (1981) Biosynthesis of radiolabeled T-2 toxin by *Fusarium tricinctum*. Appl Environ Microbiol 41:1049–1051

Hestbjerg H (2002) Production of trichothecenes and other secondary metabolites by *Fusarium culmorum* and *Fusarium equiseti* on common laboratory media and a soil organic matter agar: an ecological interpretation. J Agric Food Chem 50:7593–7599

Hoagland RE, Weaver MA, Boyette CD (2008) Enzyme-linked immunosorbent assay detection of trichothecenes produced by the bioherbicide *Myrothecium verrucaria* in cell cultures, extracts, and plant tissues. Commun Soil Sci Plant Anal 39:3057–3073

Ingle AP, Karwa AS, Rai MK, Gherbawy Y (2009) *Fusaria*: Mycotoxin production and species differentiation by molecular markers. In: Gherbawy Y, Mach RL, Rai MK (eds) Current advances in molecular mycology. Nova Science Publishers, Hauppage, NY, USA, pp. 85–107

Jestoi M N, Paavanen-Huhtala S, Parikka P, Yli-Mattila T (2008) *In vitro* and *in vivo* mycotoxin production of *Fusarium* species isolated from Finnish grains. Arch Phytopathol Plant Prot 41:545–558

Joffe AZ (1978) *Fusarium poae* and *F. sporotrichioides* as principal causal agents of alimentary toxic aleukia. In: Wyllie TD, Morehouse LG (eds) Mycotoxic fungi, mycotoxins, mycotoxicoses: an encyclopaedic handbook, vol. 3. Marcel Dekker, New York, pp 21–86

Kim JC, Kang HJ, Lee DH, Lee YW, Yoshizawa T (1993) Natural occurrence of *Fusarium* mycotoxins (trichothecenes and zearalenone) in barley and corn in Korea. Appl Environ Microbiol 59:3798–3802

Krogh P (1987) Mycotoxins in food. Academic, San Diego

Kumagai S, Shimizu T (1988) Effects of Fusarenon-X and T-2 toxin on intestinal absorption of monosaccharide in rats. Arch Toxicol 61:489–495

Morin L, Gianotti, AF, Lauren DR (2000) Trichothecene production and pathogenicity of *Fusarium tumidum*, a candidate bioherbicide for gorse and broom in New Zealand. Mycol Res 104:993–999

Munkvold GP, Desjardins AE (1997) Fumonisins in maize: can we reduce their occurrence? Plant Dis 81:556–565

Murthy KK, Rati ER, Manonmani HK (2008). Incidence of *Fusarium* toxins in rice from Karnataka, India. Res J Toxin 1:1–7

Nikulin M, Lappalainen S, Pasanen AL, Laamanen I, Veijalainen P, Berg S, Hintikka EL (1996) Comparison of detection methods for trichothecenes produced by *Fusarium sporotrichioides* on fodder and grains at different air humidities. Nat Toxins 4:117–121

North Atlantic Treaty Organization (NATO) (1996) NATO handbook on the medical aspects of NBC defensive operations. Part II–Biological. NATO Amed P-6(B) Anonymous

Pandey VV, Kurkure NV, Kalorey DR, Bhandarkar AG (2005) Detection of residual T2 toxin in liver and muscle of experimentally toxicated broilers. XVIIth European Symposium on the Quality of Poultry Meat, pp 170–171

Paulo D (2002) Production of fumonisins by strains of *Fusarium moniliforme* according to temperature, moisture and growth period. Brazil J Microbiol 33:27–36

Prelusky DB, Rotter BA, Rotter RG (1994) Toxicology of mycotoxins. In: Miller JD, Trenholm HL (eds) Mycotoxins in grain. Compounds other than aflatoxins. Eagan, St. Paul, MN, USA, pp 359–404

Rabie CJ, Sydenham EW, Thiel PG, Lubben A, Marasas WFO (1986) T-2 toxin production by *Fusarium acuminatum* isolated from oats and barley. Appl Environ Microbiol 52:594–596

Rabodonirina M, Piens MA, Monier MF, Gueho E, Fiere D, Mojon M (1994) *Fusarium* infections in immunocompromised patients: case report and literature review. Eur J Clin Microbiol Infect Dis 13:152–161

Ricaud D (1983) Les Recherche de Défense contre les Armés Biologique et Chimiques. École Polytechique, Paris, France, ISBN 2-7170–0738–5

Richardson KE, Hagler WM Jr, Campbell CL, Hamilton PB (1985) Production of zearalenone, T-2 toxin, and deoxynivalenol by *Fusarium* spp. isolated from plant materials grown in North Carolina. Mycopathologia 90:155–160

Rizzo A, Atroshi R, Hirvi T, Saloniemi H (1992) The hemolytic activity of deoxynivalenol and T-2 toxin. Nat Toxins 1:106–110

Rizzo A, Atroshi F, Ahotupa M, Sankari S, Elovaara E (1994) Protective effect of antioxidants against free radical-mediated lipid peroxidation induced by DON or T-2 toxin. J Vet Med 41:81–90

Rood HD, Swanson SP, Buck WB (1986) Rapid screening procedure for the detection of trichothecenes in plasma and urine. J Chromatogr 378:375–383

Rosenstein Y, Lafarge-Frayssinet C (1983) Inhibitory effect of *Fusarium* T-2 toxin on lymphoid DNA and protein synthesis. Toxicol Appl Pharmacol 70:283–288

Schollenberger M, Drochner W, Rüfle M, Suchy S, Terry-Jara H, Müller H (2005) Trichothecene toxins in different groups of conventional and organic bread of the German market. J Food Comp Anal 18:69–78

Seagrave S (1981) Yellow rain: a journey through the terror of chemical warfare. M Evans, New York

Sokolovic M, Garaj-Vrhovac V, Simpraga B (2008) T-2 toxin: incidence and toxicity in poultry. Arh Hig Rada Toksikol 59:43–52

Stahl CJ, Green CC, Farnum JB (1985) The incident at Tuol Chrey: pathological and toxicological examination of a casualty after chemical attack. J Forensic Sci 30:317–337

Sugita-Konishi Y, Park BJ, Kobayashi-Hattori K, Tanaka T, Chonan T, Yoshikawa K, Kumagai S (2006) Effect of cooking process on the deoxynivalenol content and its subsequent cytotoxicity in wheat products. Biosci Biotechnol Biochem 70:1764–1768

Thompson WL, Wannemachter RW (1990) *In vivo* effects of T-2 mycotoxin on synthesis of proteins and DNA in rat tissues. Toxicol Appl Pharmacol 105:483–491

Torp M, Langseth W (1999) Production of T-2 toxin by a *Fusarium* resembling *Fusarium poae.* Mycopathologia 147:89–96

Ueno Y, Sato N, Ishii K, Sakai K, Tsunoda H, Enomoto M (1973) Biological and chemical detection of trichothecene mycotoxins of *Fusarium* species. Appl Microbiol 25:699–704

Ueno Y, Sawano M, Ishii K (1975) Production of trichothecene mycotoxins by *Fusarium* species in shake culture. Appl Microbiol 30:4–9

Ueno Y, (1989) Trichothecene mycotoxins: mycology, chemistry, and toxicology. Adv Nutr Res 3:301–353

Yoshino N, Takizawa M, Tashiro F, Honda M, Ueno Y (1997) T-2 toxin induced apoptosis in human peripheral blood lymphocytes in vitro. Res Commun Biochem Cell Mol Biol 1:218–227

Yoshizawa T, Kohno H, Ikeda K, Shinoda T, Yokohama H, Morita K, Kusada O, Kobayashi Y (2004) A practical method for measuring deoxynivalenol, nivalenol, and T-2+HT-2 toxin in foods by an enzyme-linked immunosorbent assay using monoclonal antibodies. Biosci Biotechnol Biochem 68:2076–2085

Wannemacher RW, Bunner DL, Pace JG, Neufeld HA, Brennecke LH, Dinterman RE (1985) Dermal toxicity of T-2 toxin in guinea pigs, rats, and cynomolgus monkeys. In: Lacey J (ed) Trichothecenes and other mycotoxins. Wiley, Chichester, pp 423–432

Wannemacher RW, Bunner DL, Neufeld HA (1991) Toxicity of trichothecenes and other related mycotoxins in laboratory animals. In: Smith JE, Henderson RS (eds) Mycotoxins and animal foods. CRC, Boca Raton, pp 499–552

Wannemacher RW, Wiener SL (2001) Trichothecene mycotoxins. In: Medical aspects of chemical and biological warfare, chap. 34. US Department of Army, Washington DC, pp 655–676

Watson SA, Mirocha CJ, Hayes AW (1984) Analysis for trichothecenes in samples from Southeast Asia associated with "yellow rain." Fundam Appl Toxicol 4:700–717

WHO (1990) Selected mycotoxins: ochratoxins, trichothecenes, ergot. Environmental Health Criteria 105. World Health Organization, Geneva

Chapter 18
Risk Assessment of Ochratoxin A (OTA)

Diana Ringot and Abalo Chango

18.1 Introduction

Ochratoxin A (OTA) is a mycotoxin produced by fungi of two genera: *Penicillium* and *Aspergillus*. OTA is a worldwide spread mycotoxin that contaminates various food commodities and with harmful effects to animals and humans. OTA has been shown to be nephrotoxic, hepatotoxic, teratogenic and immunotoxic to several species of animals and to cause kidney and liver tumours in mice and rats. The genotoxic status of OTA is still controversial because contradictory results were obtained in various microbial and mammalian tests, notably regarding the formation of DNA adducts. In humans, intake of high amount of OTA has been linked to Balkan endemic nephropathy (BEN), a chronic nephropathy described in several rural regions of Bulgaria, Romania, Serbia, Croatia and Bosnia. Risk analysis of a toxicant (such as OTA) is made up of three parts: risk assessment, risk management and risk communication. The risk assessment of the food-borne OTA is the scientific evaluation of its known or potential adverse health effects resulting from human exposure throughout foods. It provides a qualitative and quantitative estimation of the severity and probability of harm resulting from exposure to these hazards. The "hazard" is defined as the intrinsic property of a biological, chemical or physical agent (in this case, OTA) to cause adverse health effects under specific conditions. This definition implies some certainties that, under similar conditions, the agent will cause similar adverse health effects. "Risk" is defined as the estimated probability of an adverse health effect occurring in humans as a result of exposure to a biological, chemical or physical agent in food. Risk assessment involves a complete toxicological assessment, an epidemiological assessment, an exposure assessment and a risk characterization. Two international organizations,

D. Ringot (✉)
Institut Polytechnique Lasalle, Beauvais, Rue Pierre Waguet, BP 30313, Beauvais, Cedex 60026, France
e-mail: diana.ringot@lasalle-beauvais.fr

M. Rai and A. Varma (eds.), *Mycotoxins in Food, Feed and Bioweapons*,
DOI: 10.1007/978-3-642-00725-5_18, © Springer-Verlag Berlin Heidelberg 2010

the Joint FAO/WHO Expert Committee on Food Additives (JECFA) and the Scientific Committee for Food of the European Commission (SCF) assessed the risk for the human associated with the consumption of foods containing OTA (SCF 1996, 1998; WHO/FAO 1991, 1996, 2001). More recently, a workshop on "Ochratoxin A in Food" organised by the International Life Science Institute Europe in Baden (Hazel and Walker 2005) provided an updated view on the recent research on OTA. This chapter offers a concise review of these different issues and of other new information on OTA risk assessment.

18.2 Toxicological Assessment of OTA

Toxicological assessment is the first step in risk assessment. In the case of mycotoxins, it consists of an identification of the various food mycotoxins and a study of their toxic effects on animals and humans (i.e. acute toxicity, chronic toxicity, carcinogenicity, immunotoxicity, teratogenicity, etc.). The end point of this step is the calculation of a "no-observed-adverse-effect-level" [NOAEL, mg kg^{-1} of body weight (bw) per day, namely the greatest concentration or amount of mycotoxin that does not cause detectable adverse effects], or of a "lowest-observed-adverse-effect-level" [LOAEL, mg kg^{-1} of bw per day, namely the lowest concentration or amount of mycotoxin that causes a detectable adverse effect].

18.2.1 Acute Toxicity

Kuiper-Goodman and Scott (1989) reviewed the data on the acute toxicity of OTA. A comparison of lethal dose (LD_{50}) values in different species and using different routes of exposition is presented in Table 18.1. These results indicate that dogs and pigs are the most sensitive species, whereas rats and mice are the least sensitive. Acute LD of OTA produced haemorrhages and necrosis of liver, kidney and lymphoid organs. As is the case with many xenobiotics, neonatal rats are more

Table 18.1 LD50 values of OTA in several animal species. Based on data from Kuiper-Goodman and Scott (1989) and WHO/FAO (2001)

Species	LD_{50} values (mg kg^{-1} bw)		
	Oral	Intraperitoneally (i.p.)	Intravenous (i.v.)
Mouse	46–58.3	22–40.1	25.7–33.8
Rat	20–30.3	12.6	12.7
Rat neonate	3.9		
Dog	0.2		
Pig	1–6		
Chicken	3.3		

sensitive than adult rats. No LD_{50} values were observed in ruminants; however, high doses, in the range of 2.0 mg kg^{-1} bw, seemed to cause clinical depression, reduced bw and dehydration in calves (Marquardt and Frohlich 1992).

18.2.2 Short-term to 90-Day Studies

OTA has been demonstrated as a nephrotoxic agent in all mammalian species tested so far. The results of short-term studies in rodents, dogs and pigs have been reviewed by Kuiper-Goodman and Scott (1989). Moreover, Munro et al. (1974) showed that OTA was involved in the karyomegaly of the proximal convoluted tubules cells. Similar results were obtained in 16-day and 91-day gavage studies on the Fisher rat (F344/N) (NTP 1989). Berndt and Hayes (1979) found that the Sprague-Dawley rat was more sensitive than the Wistar rat and that the males were more sensitive than females. Dogs and pigs are also very sensitive to the nephrotoxic effect of orally administrated OTA (Kitchen et al. 1977; Elling 1979; Krogh et al. 1988).

18.2.3 Long-term Studies on Toxicity and Carcinogenicity

The toxicological status of OTA has been reviewed several times and detailed monographs have been published by the International Agency for Research on Cancer (IARC 1993) and by the JECFA (WHO/FAO 1991, 1996, 2001). These reports presented the nephrotoxic and carcinogenic effects of OTA. The toxin is also a teratogenic and immunotoxic compound, affecting both the humoral and cell-mediated immunity. Evaluation of the OTA nephrotoxic and carcinogenic properties has been the major focus of large numbers of studies. The first studies on OTA carcinogenicity were performed on rats (Purchase and Van der Watt 1971), trouts (Doster et al. 1971) and mice by oral or intraperitoneal administration (Dickens and Waynforth 1968; Kanisawa and Suzuki 1978). The first evaluations by the IARC (1976, 1983) concluded that the evidence for carcinogenicity was limited. Later on, several studies confirmed the carcinogenic potency of the toxin in mice (Kanisawa 1984; Bendele et al. 1985a) and in rats (Boorman 1989). In this regard, the potency varies markedly with species and sex, with the mouse being much less sensitive than the rat and with male rats being more sensitive than females. In addition, the pig is remarkably more sensitive than the rat with regard to nephropathy. The LOAEL and NOAEL are presented in Table 18.2. On the basis of these results, IARC (1993) evaluated the experimental evidence for carcinogenicity as sufficient and classified OTA as "possibly carcinogenic to humans" (group 2B) but not yet "probably carcinogenic to humans (group 2A)". However, more recent studies have brought additional data in favour of a classification into group 2A (Petzinger and Ziegler 2000). It is not yet clear whether OTA acts as a direct genotoxic mutagen or

Table 18.2 Several long-term studies on the OTA nephrotoxicity and carcinogenicity (Based on data from WHO/FAO 2001)

Species	Effect	Duration of studies	LOAEL ($\mu g\ kg^{-1}$ bw day^{-1})	NOAEL ($\mu g\ kg^{-1}$ bw day^{-1})
Mouse: male	Kidney tumours	2 years	4,400	130
Rat: male	Karyomegaly of cells of proximal tubule	90 days	15	Not demonstrated
	Kidney tumours	2 years	70	21
Pig	Impaired renal function	90 days	8	Not demonstrated
	Progressive nephropathy	2 years	40	8

whether its carcinogenicity is related to indirect mechanisms, such as induction of cytotoxicity and increased cellular proliferation as a consequence of tissue injury (for a review see Ringot et al. 2006). Nevertheless, OTA leads to gene and chromosomal mutations (genotoxicity), induction of cell death (cytotoxicity) and induction of altered expression of genes (epigenesis). These three types of effects are described below.

18.2.4 Genotoxicity, Cytotoxicity and Epigenesis

Concerning *gene mutation*, in the first instance, OTA was regarded as non-mutagenic because most standard bacterial cell assays gave negative results (for reviews see IARC 1993). In the bacterial Ames test system for gene mutation, only two studies showed positive results, when conditioned medium from OTA-exposed hepatocytes was used as the test substances (Hennig et al. 1991) or when kidney microsomal fractions were used as the metabolic activation system (Obrecht-Pflumio et al. 1999). A smaller number of experiments have been performed in mammalian culture systems. Gene mutations were not induced in mouse lymphoma cells (Bendele et al. 1985b) or in C3H mouse mammary cells (Umeda et al. 1977). By contrast, in stably transfected murine fibroblast NIH/3T3 cell lines stably expressing human cytochrome P450, OTA increased the mutation frequency at a concentration of 25 $\mu g\ l^{-1}$ (De Groene et al. 1996).

Contradictory results were found concerning the induction of *chromosomal aberrations*. Negative effects on sister chromatid exchange (SCE) frequency were found in human peripheral blood (HPBL) cells (Cooray 1984) and in medullar cells from OTA-treated Chinese hamsters (Bendele et al. 1985b). By contrast, Föllmann et al. (1995) clearly showed an induction of SCE by OTA in primary cultures of porcine urinary bladder cells. OTA was also found to induce micronuclei formation in cytokinesis-blocked binucleated ovine seminal vesicles (Degen et al. 1997), in Syrian hamster embryo fibroblasts (SHE) (Dopp et al. 1999), in the HepG2 human-derived hepatoma cell line (Ehrlich et al. 2002) and in rat and human kidney cells (Robbianno et al. 2004). In addition, the control mechanisms for proper separation of chromosomes during cell division may be impaired, leading to chromosomal aberrations and thus to phenotypical and karyotypical cell conversion without general damage (Gekle et al. 1998).

It has been demonstrated that OTA causes *single-strand DNA breaks in vitro* in mouse spleen cells in culture (Creppy et al. 1985), in Madine-Darby canine kidney MDCK cells (Lebrun and Föllmann 2002), as well as in liver and kidney cells of rats that have ingested OTA (Kane et al. 1986). Several investigators have found that OTA causes *inhibition of scheduled DNA* synthesis in cultured CHO cells, rat fibroblasts (Stetina and Votava 1986), mouse L cells (Jeffery et al. 1984) and HPBL cells (Cooray 1984). *DNA repair* as a result of unscheduled DNA synthesis was observed in SOS DNA repair assays conducted in the *Escherichia coli* strain PQ37 (Malaveille et al. 1994) and in studies with primary cultures of mouse and rat hepatocytes (Mori et al. 1984), porcine urinary bladder epithelial cells (Dörrenhaus and Föllmann 1997) and human urothelial cells (Föllmann et al. 1997).

In recent years, the debate has been controversial with regard to the involvement of OTA in *DNA adducts* formation (Ringot et al. 2006).

In acute toxicity studies, OTA-induced *cell death* has been reported *in vivo* in rat renal tubules (Albassam et al. 1987), in mouse embryo (Wei and Sulik 1993) and mouse liver (Atroshi et al. 2000) and *in vitro*, in various cell types, such as human lymphocytes (Seegers et al. 1994), HL-60 human promyelotic leukaemia cells (Ueno et al. 1995), MDCK cells (Gekle et al. 1998, 2000; Schwerdt et al. 1999), Chinese hamster lung fibroblasts (V79 cells), and African green monkey kidney fibroblast cells (CV-1 cells) (Kamp et al. 2005). In food, the amount of the toxin is small (in the range of nanomols). At this level, OTA have been shown to facilitate apoptosis by causing the reduction of protein synthesis (Ueno et al. 1995), an increase of caspase-3 activity (a protease leading to the activation of DNA ladder nuclease), DNA fragmentation and chromatin condensation (Schwerdt et al 1999; Gekle et al. 2000). Schwerdt et al. (2003) studied the role of the mitochondrial function on the caspase-3 activity in MDCK-C7 cells. They showed that the apoptosis induction by OTA was not dependent on a decrease of the mitochondrial potential or on a subsequent cytochrome C release.

OTA, like many environmental xenobiotics, can act through non-genotoxic mechanisms that *epigenetically* control gene expression leading to a disease state. A detailed description of these mechanisms is presented in a previously published paper (Ringot et al. 2006).

18.2.5 Teratogenicity

At relative large doses administered intraperitoneally, OTA was shown to be a teratogen in mice (Hayes et al. 1974; Arora 1982), rats (Still et al. 1971; Brown et al. 1976; Szczech and Hood 1981; Mayura et al. 1998; Abdel-Wahhab et al. 1999; Wangikar et al. 2004a), hamsters (Hood et al. 1976) chicken (Edrington et al. 1995; Lalithakunjamma and Krishnan 1997), and rabbit (Wangikar et al. 2004b, 2005), but not in pig (Kuiper-Goodman and Scott 1989). Thus, sufficient experimental evidence exists in the scientific literature to classify OTA as a teratogen, affecting the nervous system, the skeletal structures and the immune system of research animals. In addition, the prenatal exposure of Sprague-Dawley dams may induce immunosuppression in pups, whereas a perinatal short-term exposure via the milk stimulates the proliferative responses of lymphocytes to polyclonal activation (Thuvander et al. 1996).

18.2.6 Immunotoxicity

The OTA immunotoxic potential has been studied in different experimental models, after single and repeated administration of different doses and using various routes of administration. Therefore, the data available are often contradictory and difficult to interpret. Several studies showed that OTA suppresses the antibody response in a number of species (Haubeck et al. 1981; Harvey et al. 1992; Stormer and Lea 1995; Müller et al. 1999). In relation to the humoral immunity, OTA induces a dose-dependent regression of IgG-, IgA- and IgM-immunoglobulins (Müller et al. 1995; Dortant et al. 2001). Weidenbach et al. (2001) found that, among early response cytokines, tumour necrosis factor α (TNFα) and another pro-inflammatory cytokine, interleukin 6 (IL-6), are released in significant amounts from a blood-free perfused rat liver, following OTA exposure. Subsequently, Petzinger and Weidenbach (2002) demonstrated the immunostimulatory and modulating effect of OTA in the liver, this toxin representing a receptor-dependent trigger of a signal cascade leading to TNFα secretion by the Kupffer cells and parenchymal cells. Concerning the cell-mediated immunity, the proliferative response of B cells is not modified by OTA in rats (Dortant et al. 2001) or mice (Thuvander et al. 1995). However, in vitro, human B lymphocytes are inhibited (Lea et al. 1989). The proliferative response of T lymphocytes is decreased in pigs (Harvey et al. 1992) and broiler (Elissalde et al. 1994), as well as in in vitro studies (Lea et al. 1989). T cytotoxic lymphocyte (CTL) activity was not affected in mice (Luster et al. 1987). A decrease in the natural killer cells (NK) activity was observed in mice (Luster et al. 1987) and rats (Alvarez et al. 2004), whereas no effect was found in other studies with rodents (Thuvander et al. 1995; Dortant et al. 2001). OTA was also reported to affect macrophage bactericide activity in rodents (Boorman et al. 1984; Müller et al. 1995).

18.2.7 Nephrotoxicity

The kidney is the main target organ for OTA toxicity. OTA is nephrotoxic to various animal species (mammals and avian) (Kuiper-Goodman and Scott 1989) with the exception of adult ruminants (Ribelin et al. 1978). Swine, rats and dogs are animal species known to be sensitive to the nephrotoxic effects of the toxin. Renal function and morphology are greatly affected at high doses of OTA, as indicated by increased kidney weight, urine volume and blood urea nitrogen (Hatey and Galtier 1977). The NOAELs for changes in renal function depend on the species and on the parameter tested.

Various groups of investigators showed that the nephrotoxic effect of OTA is due to its action on the organic anion transport (OAT) system, located in basolateral and brush border membranes of the proximal tubule cells of the nephron and also involved in the absorption/reabsorption and excretion of the toxin in kidney (WHO/FAO 2001). The middle (S2) and terminal (S3) segments of the proximal tubule of isolated nephrons were found to be the most sensitive to the toxic effect of OTA (Jung and Endou 1989).

Several authors have reported decrease in the activity of several urinary kidney enzymes in rodents, such as muramidase, alkaline phosphatase, alanine peptidase, lactate dehydrogenase, glutamate dehydrogenase, and phosphoenolpyruvate carboxykinase (Ngaha 1985; Endou et al. 1986; Meisner and Krogh 1986). A late event was the urinary increase of *N*-acetyl-β-Glucosidase, a lysosomal enzyme, indicating the active regeneration and exfoliation of necrotic proximal convoluted tubular cells (Stonard et al. 1987). Pigs are very sensitive to the effect of OTA on renal enzyme activities. In pig kidney, a dose-related decrease in the activity of cytosolic phosphoenolpyruvate carboxykinase and of the γ-glutamyl transpeptidase was accompanied by a decrease in the kidney function (Meisner and Krogh 1986). When OTA was administered intravenously by a single injection of 3 μmol kg^{-1} bw or by six injections of 1.2 μmol kg^{-1} bw in male Wistar rats, it upset the pH homeostasis, leading to the alkalinization in the interstitium of the renal papilla, in addition to impairment of urinary acidification (Kuramochi et al. 1997). The histological examination of rat kidney showed karyomegaly and abnormal mitosis in tubule cells and a limited degeneration of interstitial tissues (Mantle et al. 1991; Levin 1998; Maaroufi et al. 1999). In pigs, major kidney lesions have been observed in the epithelial cells of the proximal tubules (Stoev et al. 1998).

18.3 Toxic Effects of OTA on Farm Livestock

Pigs are generally considered to be the most sensitive farm animal species for the nephrotoxic effect of OTA. Porcine nephropathy is a major renal disease in swine known to be naturally occurring in certain European countries such as Denmark, Hungary, Bulgaria and Poland (Krogh et al. 1973). Porcine nephropathy is a chronic renal disorder characterised by very low mortality, because of the short finishing period of bacon pig. Renal lesions in pig include degeneration of the proximal tubules, interstitial fibrosis and hyalinization of the glomeruli. The disease is endemic and has been associated with bad weather conditions, and a positive correlation has been observed between its prevalence rates and the consumption of mycotoxin (OTA and citrinin-CIT)-contaminated feed (Hald 1991). Stoev et al. (2002b) showed that a high OTA concentration in diet (800 μg kg^{-1}) has been quite well tolerated in pig, and they suggested that synergistic effects between OTA and other mycotoxins (i.e. penicillic acid) could be responsible for the porcine nephropathy in Bulgaria.

Chicken is also a sensitive species and OTA is the most important cause of poultry nephropathy (Elling et al. 1975; Huff et al. 1975; Stoev et al. 2002a). Moreover, primary hepatic degeneration can occur in ducklings (Theron et al. 1966). In a study with broiler chicken fed an OTA-contaminated diet at 2.5 mg kg^{-1}, Gentles et al. (1999) showed a significant reduction in weight gain, an increase in relative kidney weight associated with increase in serum uric acid and decrease in total protein and albumin.

Cows, sheeps and goats are considered to be less sensitive to the toxicity of OTA due to its hydrolysis into the less toxic Otα by the protozoa in the rumen. However, polyuria, enteritis and tubular degeneration have been observed in preruminant calves fed an OTA contaminated diet at 0.1–2 mg kg^{-1} bw for 30 days (Pier et al. 1976).

18.4 Toxic Effects of OTA in Humans

OTA has received considerable attention because of its deleterious effects on human health. OTA accumulates in various organs in the body; kidney is the main target where it exerts toxic and carcinogenic effects (Kuiper-Goodman and Scott 1989). Three forms of human renal disease appear to be caused, at least in part, by enhanced exposure to OTA: BEN, chronic interstitial nephritis and karyomegalic interstitial nephritis.

BEN is an acquired, environmental disease of the entire urinary tract, with long latency (Bozic 1994). The disease occurs in Romania, Bulgaria, Croatia, Bosnia and Herzegovina and Yougoslavia, in certain rural areas along river valleys belonging to the Danube river basin. BEN affects all age groups, but rarely adolescents and children (Bozic et al. 1995). BEN tends to occur more frequently in females, but studies do not offer a statistically significant difference; the sexual distribution observed could be caused by a greater prevalence of females in the observed population (Plestina 1992). Tubulointerstitial chronic nephritis, urotheliomas and renal carcinoma occur in patients with BEN (Castegnaro and Chernozemsky 1987). Several hypotheses on the aetiology of BEN considered genetic and environmental factors such as heavy metals, silica, bacteria, leptospira, viruses, aristolochic acid (alkaloid from *Aristolochia fangchi*), pliocene lignites and fungal mycotoxins (OTA and CIT) (Pfohl-Leszkowicz et al. 2002). Because of certain similarities of this disease to OTA-induced porcine nephropathy (Krogh 1974), it has been suggested that OTA has a causal role in the induction of BEN. However, the physiopathological role of OTA in this disease has not yet been completely determined.

Several studies indicated the presence of OTA residues in local foodstuffs as well as in the blood of patients with nephropathy, and correlations have been observed between OTA levels in food and sera and BEN incidence in endemic and non-endemic areas (Krogh et al. 1977; Hult et al. 1982; Petkova-Bocharova et al. 1988; Radovanovic 1989; Nikolov et al. 1996). The possibility, however, still remains that the increased OTA serum levels in patients and other subject from endemic areas is the result of kidney damage rather than the cause of it.

The occurrence of OTA in human sera from individuals with and without nephropathy has also been studied in countries where BEN has not been detected such as Canada, Tunisia, Algeria, Turkey and Italy. In Canada, no difference has been found in the frequency of OTA-positive samples between non-affected and affected patients (Kuiper-Goodman et al. 1993). In Algeria, a higher incidence of OTA-positive samples has been found in the group of subjects suffering from different nephropathies compared to the general population (Khalef et al. 1993). In Tunisia and Turkey, a higher mean concentration of OTA has been observed in serum samples collected from patients with nephropathy (Maaroufi et al. 1995; Özçelik et al. 2001).

Moreover, in the past 15 years, direct evidence of general human population exposure to OTA has been obtained in many countries. Approximately 50% of the European population appears to be exposed to OTA and the mean serum concentration of OTA in the general population does not exceed 1.0 ng ml^{-1} (Peraica et al. 1999). From these studies, it is apparent that there are fluctuations in human serum OTA levels, which probably reflect local, seasonal or yearly fluctuations in the levels of OTA in foods. In addition, sex-specific differences in OTA blood level have been found in Switzerland, in the south of the Alps, with a higher mean concentration in men than in women (Zimmerli and Dick 1995). The authors hypothesised that the difference is due to different dietary habits, because men consume considerably more wine, beer and coffee. Sex-specific differences, however, were not found in the north of the Alps (Zimmerli and Dick 1995) or in any other country in which such investigation was undertaken such as Sweden (Breitholtz et al. 1991), Hungary (Tapai et al. 1997), Canada (Scott et al. 1998), Spain (Jiménez et al. 1998) and Japan (Ueno et al 1998).

18.5 Endpoint of OTA Hazard Characterization – Provisional Tolerable Daily Intake

Hazard characterization is the extrapolation phase of risk assessment. Its aim is to make a predictive characterization of the hazard to humans, based on animal studies (species extrapolation) under low exposure conditions (extrapolation from high to low dose). The endpoint of hazard characterization is the estimation of a "safe dose" such as a provisional tolerable daily intake (PTDI). To derive a tolerable daily intake (TDI) for humans, it has been common practice to divide the NOAEL by a safety factor of 100 when extrapolating from animals to humans. When there are significant irreversible effects for which thresholds have been established (as is the case for non-genotoxic carcinogens) or insufficient data, additional uncertainty factors may be added.

The JECFA, on the basis of the nephrotoxicity of OTA, proposed a provisional tolerable weekly intake (PTWI) for OTA of 0.1 μg kg^{-1} bw, corresponding to a PTDI of 14 ng kg^{-1} bw day^{-1} (WHO/FAO 1991, 1996). This assessment has been based on the LOAEL for kidney damage in pig and a safety factor of 500. This PTWI was reconfirmed at the 56th meeting of JECFA (WHO/FAO 2001). JECFA also noted that a "no effect level" had not been observed for this endpoint.

However, on the basis of carcinogenicity data, the Working Group of the Nordic Council of Ministers proposed a maximum TDI of 5 ng kg^{-1} bw of toxin (Olsen et al. 1991). A similar PTDI (1.2–5.7 ng kg^{-1} bw) has been established by the Canadian authority (Kuiper-Goodman 1996), based on the carcinogenic effects and other toxic properties of OTA, and using both a large safety factor of 5,000 and a model-based approach in extrapolations. In 1998, taking into account the OTA exposure data of the European population, the SCF suggested that it was prudent to reduce exposure to OTA as much as possible, "ensuring that exposures are towards the lower end of the range of TDI of 1.2–14 ng kg^{-1} bw day^{-1}, which have been estimated by other bodies, e.g. below 5 ng kg^{-1} bw day^{-1}" (SCF 1998).

Both JECFA and SCF recognised that the new data raised further questions about the mechanism by which OTA causes nephrotoxicity and renal carcinogenicity, and recommended a further review of OTA. In addition, JECFA recommended the improvement in food sampling procedures and pointed out that further surveys on OTA food contamination outside Europe were needed for the assessing of OTA intake.

Accordingly, on April 2006, SCF (2006) concluded on the new opinion on OTA in foods. Considering the lack of evidence for the existence of DNA adducts, the TWI has been calculated using a threshold-based approach. On the basis of the LOAEL of 8 μg kg^{-1} bw day^{-1} for the renal toxicity in pigs (the most sensitive animal species), and applying an uncertainty factor of 450, TWI calculated value is 120 ng kg^{-1} bw.

18.6 Exposure Assessment of Animal and Human to OTA

Exposure assessment is an important step in the risk assessment process of OTA and consists of the quantitative evaluation of its likely intake by consumers, humans or animals via food or feed respectively. The estimation of dietary intake of OTA requires information on the contamination level of foods and feeds and on the consumption of relevant food and feed.

18.6.1 OTA Occurrence in Feed and its Intake by Farm Livestock

As seen above, at high levels in feed, OTA may cause illness in farm animals through the development of animal ochratoxicosis. At lower levels in feed, OTA may have no apparent effect on livestock production, but their residues and related substances might move up the food chain. This indirect intake of OTA from the consumption of animal food products may pose a health hazard to humans (Kuiper-Goodman 1991). OTA occurrence in animal feed is mainly a consequence of inadequate drying of cereals before storage and of wrong post-harvest storage conditions, leading to peaks in mould contamination and ochratoxinogenesis. Substantial surveillance data have been generated by EU Member States in the frame of two large projects, the SCOOP 1 and 2 projects (EC 1998, 2002). These studies aimed at assessing human OTA exposure. In the SCOOP 2, only 4% (61/1,500) of cereal (wheat, oat, rye, maize and barley) samples have been identified as OTA-positive, with a mean value of 1 $\mu g\ kg^{-1}$, the highest value reported being 27 $\mu g\ kg^{-1}$ in a rye sample from Sweden. By contrast, earlier data obtained in 1997 in the UK on farm-stored grain samples showed 15% OTA-positive samples (45/306), with a mean value of 1 $\mu g\ kg^{-1}$ and the highest value of 17.8 $\mu g\ kg^{-1}$ (in a sample of barley) (Scudamore et al. 1999). Data from Hungary (Rafai et al. 2000) reported 2.5–18% OTA-positive cereal samples, with average levels from 76 to 350 $\mu g\ kg^{-1}$ were also significantly higher than the data in the SCOOP report. Jonsson and Pettersson (1999) have studied the effect of preservation methods on the OTA level in grain during farm storage. OTA ($>1\ \mu g\ kg^{-1}$) was detected in 28% and 15% of samples from cereals dried with ambient air and heated air respectively. Cereals from other preservation methods, such as acid treatment and air-tight storage, did not contain detectable levels of OTA. Accordingly, Garaleviciene et al. (2003) found a high prevalence of OTA in final Lithuanian mixed feed for pigs (92%) and poultry (93%) produced in 1999, the highest OTA level determined being 68 $\mu g\ kg^{-1}$ feed.

The exposure of farm animals to OTA is difficult to assess because of its heterogeneous distribution in contaminated feed and a large variety of diet composition. For pigs, an alternative method for OTA monitoring in feed is based on the relationship of exposure to tissue and serum levels, which reflects the toxin content of the ingested feed (Hult et al. 1984). Holmberg et al. (1990) analysed OTA pig blood over 6 years to study OTA contamination of cereal grains in Sweden. OTA was found in 18% of the samples ($\geq 2\ ng\ ml^{-1}$ blood), with a yearly incidence variation between 11 and 35%. These authors observed an increase in contamination during years with wet harvest conditions and after prolonged storage of cereals. Curtui and Gareis (2001), in a Romanian study on OTA levels in blood serum collected from slaughtered pigs, found an OTA incidence of 98%, indicating a high prevalence of feed contamination. This approach is unfortunately not pertinent for animals with a high plasma clearance such as poultry and fish, or for animals with low absorption, such as ruminants (EFSA 2004).

18.6.2 OTA Occurrence in Food and Human Dietary Exposure

JECFA (WHO/FAO 2001) published survey data on the distribution and level of OTA contamination in food commodities. Results concerning 23,167 samples from various commodities have been provided by various countries, with 85% of the samples originating from Europe (Croatia, Denmark, Finland, France, Germany, Italy, the Netherlands, Norway, Spain, Sweden, Switzerland and the United Kingdom), 7% from South America (Brazil and Uruguay), 6% from North America (Canada and USA), 1% from Asia (Dubai and Japan) and 1% from Africa (Sierra Leone and Tunisia). The prevalence of contaminated samples varied between commodities and the concentrations of OTA in the various commodities were highly variable. The weighed mean concentrations of OTA in cereals, cereals products, cocoa and chocolate, coffee, dried vine fruits, grape juice, wine, pig kidney and other meat products are shown in Table 18.3.

Since most samples (85%) originated from the European countries, it was difficult to evaluate the geographical distribution of OTA. Moreover, some data (Brazil, Uruguay) could not be used since they were obtained using inadequate analytical methods. Therefore, JECFA decided to focus on the European data. According to the recommendation of WHO/FAO, the data on food consumption obtained from the GEMS/Food database and from the report of the European Commission (EC 1998) were used to estimate the mean total intake of OTA. With this approach, the mean total intake of OTA has been estimated to be 45 ng kg^{-1} bw $week^{-1}$, assuming a bw of 60 kg. Cereals and wine contributed to about 25 and 10 ng kg^{-1} bw $week^{-1}$, respectively, whereas grape juice and coffee each contributed by 2–3 ng kg^{-1} bw $week^{-1}$. Other food products (dried fruits, beer, tea, milk, cocoa, poultry) each contributed by less than 1 ng kg^{-1} bw $week^{-1}$. Pig meat and products contributed by about 1.5 ng kg^{-1} bw $week^{-1}$, but this value was considered as overestimated, because most of the submitted data were for pig kidney and liver, whereas food consumption calculation was based on pig meat.

Table 18.3 Weighed mean concentrations of OTA in various commodities (Based on data from WHO/FAO 2001)

Commodity	No. of samples	OTA weighted mean concentrations (μg kg^{-1})
Cereals	2,700	0.94
Cereal products	1,500	0.19
Cocoa and chocolate	270	0.18
Coffee, green and roasted	1,900	0.86
Dried vine fruits	860	2.3
Grape juice	68	0.44
Pig kidney	380	0.12
Meat products (liver, sausages, meat)	810	0.052
Wine	260	0.1

Since a lot of new data have recently been generated by European countries and presented are collectively, it is worth re-evaluating the value of OTA intake (Jorgensen 2005).

In this European study, data from participants were collected during the period November 1999–September 2000. All 13 participating countries (Italy, Denmark, Finland, France, Germany, Greece, Ireland, Norway, Portugal, Spain, Sweden, the Netherlands and the United Kingdom) have sent results on OTA occurrence in food. Moreover, six participants (Germany, Italy, Norway, Spain, Sweden and UK) have sent results on serum/blood and four (Germany, Italy, Norway and Sweden) on breast milk. Results on other human fluids (urine and amniotic fluid) and tissues (placenta and funiculum) have been provided by three countries (Denmark, Italy and UK). All participating countries provided results on several food matrices such as cereals, green coffee, processed coffee, cocoa and derived products, dried fruits, spices and meat product.

In order to obtain the overall European scenario of *the exposure to OTA*, through the combination of OTA occurrence data in food products and consumption data, various approaches were followed – by country, by food commodity and by groups of population. On the basis of the provided data, different estimates of *dietary intakes* from each food commodity, as derived by the combination of the above sets of data, were calculated. The intake estimates have been calculated as referred both to person and per kg of bw, the latter being calculated on the bw values as sent by participants.

The contribution to dietary intake for all population of each food group of commodities in participating was also calculated. Cereals resulted the main contributors (50%) followed by wine (13%), coffee (10%), spices (8%), others (6%), beer (5%), cocoa (4%), dried fruits (3%) and meat (1%).

Another approach to assessing the exposure to mycotoxins applies to data on *biological fluids*. A total of 2,712 items of data concerning OTA occurrence in serum and plasma was provided by six countries (Germany, Italy, Norway, Spain, Sweden and UK). The considered biological fluids and tissues were serum plasma, urine, milk, amniotic fluid, placenta and funiculum. The range of OTA concentration in individual serum/plasma samples ranged from 0.11 $\mu g\ l^{-1}$ (Germany) to 5.58 $\mu g\ l^{-1}$ (Spain). The weighed mean for all individual values for adult European population was 0.34 $\mu g\ l^{-1}$. OTA levels in serum/plasma samples were used to calculate the estimate daily intake through the Klaassen equation. According to Hagelberg et al. (1989) the bioavailability of OTA is around 50% and plasma clearance is calculated by considering renal filtration as the only route of elimination. The *estimated daily intake by Klaassen equation* ranged from 0.41 ng kg^{-1} bw day $^{-1}$ (Sweden) to 2.34 ng kg^{-1} bw day $^{-1}$ (Spain) and the mean value was 0.67 ng kg^{-1} bw day $^{-1}$.

Data on OTA contamination of *human milk* (324 data) were provided by Germany, Italy, Norway and Sweden. The overall weighed means for all individual values and for all positive values, approximately indicating an average European level, were 0.09 and 0.18 $\mu g\ l^{-1}$. On the basis of these data, *OTA dietary intake for babies from human milk* was calculated, on the basis of 600 ml milk consumption

(100 ml suck^{-1}) and on the mean OTA level in milk. The average calculated OTA intake for babies from human milk ranged from 1.00 ng kg^{-1} bw day $^{-1}$ (Norway) to 24.00 ng kg^{-1} bw day $^{-1}$ (Italy).

18.7 Risk Characterisation

Risk characterization is the qualitative and/or quantitative estimation, including the attendant uncertainties, of the severity and probable occurrence of known and potential adverse health effects on an exposed population. It is based on hazard identification, hazard characterization and exposure assessment. Risk characterization can be the establishment of levels of daily exposure at which the risk is insignificant over a lifetime (i.e. exposure needs to be below the TDI or other measure of safe dose). As well as considering the average population, risk characterization also needs to consider those groups that are most vulnerable to exposure, such as children, and other groups for which there may be some differences in bioavailability, metabolism or genetic disposition, such as the elderly. In this regard, the adequacy of a tenfold safety factor to address differences in human susceptibility arising from human variability needs to be examined. The risk characterization of OTA is difficult to assess and only a small amount of data have been published on this subject. Several factors make this process difficult: the controversial levels of tolerable intakes proposed by the various evaluators, the variability of the contamination of food commodities over space and time, the uncertainty of sampling and of analytical methods, the limited data on OTA contamination of food over the world, the utilization of food consumption data, and the limited information on OTA level in human fluids (blood, plasma, milk, urine).

The dietary exposure assessment performed by the European SCF on the bases of deterministic estimates performed by eight European member states (EC 1998) showed that the mean exposure was in the range of 5–32 ng kg^{-1} bw week^{-1}. The exposure of European population seemed to be in most cases quite below the TDI value for OTA (5 ng kg^{-1} bw day $^{-1}$) suggested by the SCF (1998). A second dietary OTA exposure assessment was performed by JECFA (2001) on the basis of the international mean values for both consumption and contamination levels. Accordingly, the average exposure was calculated at 45 ng kg^{-1} bw week^{-1} and was largely below the PTWI value (100 ng kg^{-1} bw week^{-1}) proposed by JECFA. Another JECFA dietary exposure calculation combined the distribution of French consumption data with the mean values of international data on occurrence of OTA in foods. As a result, the estimated dietary exposure at the 95th percentile was around 92 ng kg^{-1} bw week^{-1}. These values are quite below the PTWI proposed by JECFA. Moreover, based on the OTA contamination data for cereal OTA, JECFA noted that the intake of OTA by 95th percentile consumers of cereal might approach the PTWI from this source alone. Hence, for MRLs in cereals of 5 μg kg^{-1}, the estimated intake of cereal consumers on a European diet was about 84 ng kg^{-1} bw week^{-1}. The Committee has been unable to estimate the risk for nephrotoxicity if the PTWI is exceeded. Nevertheless, some countries seemed to be concerned to a more relevant contamination especially if specific group of consumers are considered (i.e. babies, children, pregnant women).

More recently, Tressou et al. (2004) used three statistical methodologies to evaluate the food exposure to OTA of different groups of the French population and to qualify the impact of new MRLs in cereals and wine on the dietary intake. The results of this national French survey, called INCA and realised by CREDOC-AFSSA-DGAL, were used for the consumption data for two groups of population: adults (over 15 years old) and children (3–14 years old). Contamination data are issued from several control and survey studies realised by the Ministry of Agriculture, Ministry of Economy and Finances, National Office of Wines and National Institute of Agronomical Research. They showed that, for the total adult population, the risk of surpassing the international PTWI for adults is at maximum 4.76%, whereas for children the probability of exceeding the PTWI is three times higher. However, the risk of exceeding the PTWI is significantly reduced for adults when applying a MRL of 5 $\mu g\ kg^{-1}$ on cereals (max. 1.36% vs max.4.76%), whereas for children this reduction does not appear to be statistically significant. In addition, the recently fixed MRL on wine does not have a significant impact on adult consumers. This is explained essentially by the fact that cereals are the main contributor.

18.8 Conclusions

As far as presented above, OTA is a mycotoxin spread worldwide that contaminates various food commodities and with harmful effects on animals and humans. The OTA toxicological status is still controversial, especially its genotoxicity. To complete the OTA hazard characterization, further data are needed to clarify the genotoxic status of this toxin, to explore its role in human chronic renal disease and to evaluate its synergistic or additive effects with other food toxins and pollutants. To improve the OTA risk assessment further research should be focused on appropriate sampling and analytical methods, on the collection of relevant contamination data at an international level, on the estimation of potential biomarkers for the exposure assessment, and on the development of appropriate probabilistic models of assessment. A further JECFA evaluation of OTA will permit an updated risk characterization of this toxin.

References

Abdel-Wahhab M, Nada SA, Arbid MS (1999) Ochratoxicosis: prevention of developmental toxicity by L-methionine in rats. J Appl Toxicol 19:7–12

Albassam M, Yong SI, Bhatnagar R, Sharma AK, Prior MG (1987) Histopathologic and electron microscopic studies on the acute toxicity of ochratoxin A in rats. Vet Pathol 24:427–435

Alvarez L, Gil AG, Ezpeleta O, Garcia-Jalon JA, Lopez de Cerain A (2004) Immunotoxic effects of ochratoxin A in Wistar rats after oral administration. Food Chem Toxicol 42:825–834

Arora RG (1982) Mycotoxin-induced effects on the prenatal development in mice: a teratopathological and autoradiographic study of aflatoxin B1, ochratoxin A and zearalenone. PhD Dissertation, Swedish University of Agricultural Sciences, Uppsala, Sweden

Atroshi F, Biese I, Saloniemi H, Ali-Vehmas T, Saari S, Rizzo A, Veijalainen P (2000) Significance of apoptosis and its relationship to antioxidants after ochratoxin A administration in mice. J Pharm Pharm Sci 3:281–291

Bendele AM, Carlton WW, Krogh P, Lillehoj EB (1985a) Ochratoxin A carcinogenesis in the (C57BL/6J X C3H)F1 mouse. J Natl Cancer Inst 75:733–742

Bendele AM, Neal SB, Oberly TJ, Thompson CZ, Bewsey BJ, HillL E, Rexroat MA, Carlton WW, Probst GS (1985b) Evaluation of ochratoxin A for mutagenicity in a battery of bacterial and mammalian cell assays. Food Chem Toxicol 23:911–918

Berndt WO, Hayes AW (1979) *In vivo* and *in vitro* changes in renal function caused by ochratoxin A in the rat. Toxicology 12:5–17

Boorman GA, Hong HL, Dieter MP, Hayes HT, Pohland AE, Stack M, Luster MI (1984) Myelotoxicity and macrophage alteration in mice exposed to ochratoxin A. Toxicol Appl Pharmacol 72:304–312

Boorman GA (1989) NTP technical report 358 on the toxicology and carcinogenesis studies of ochratoxin A. US Department of Health and Human Services, National Institutes of Health

Bozic Z (1994) Epidemiologija I etiologija urogenitalong carcinoma. Panonsko-balkanska endemicna uropatija. In: Novak R, Tucak A (eds) Uroloska onkologija. Zagreb, Medicina naklada, pp 9–10

Bozic Z, Duancic V, Belicza M, Kraus O, Skljarov I (1995) Balkan endemic nephropathy: still a mysterious disease. Eur J Epidemiol 11:235–238

Breitholtz A, Olsen M, Dahlback A, Hult K (1991) Plasma ochratoxin A levels in three Swedish populations surveyed using an ion-pair HPLC technique. Food Addit Contam 8:183–192

Brown MH, Szczech GM, Purmalis BP (1976) Teratogenic and toxic effects of ochratoxin A in rats. Toxicol Appl Pharmacol 37:331–338

Castegnaro M, Chernozemsky I (1987) Endemic nephropathy and urinary tract tumors in the Balkans. Cancer Res 47:3608–3609

Cooray R (1984) Effects of some mycotoxins on mitogen-induced blastogenesis and SCE frequency in human lymphocytes. Food Chem Toxicol 22:529–534

Creppy EE, Kane A, Dirheimer G, Lafarge-Frayssinet C, Mousset S, Frayssinet C (1985) Genotoxicity of ochratoxin A in mice: DNA single-strand break evaluation in spleen, liver and kidney. Toxicol Lett 28:29–35

Curtui VG, Gareis M (2001) A simple HPLC method for the determination of the mycotoxins ochratoxin A and B in blood serum of swine. Food Addit Contam 18:635–643

De Groene EM, Hassing IG, Blom MJ, Seinen W, Fink-Gremmels J, Horbach GJ (1996) Development of human cytochrome P450-expressing cell lines: application in mutagenicity testing of ochratoxin A. Cancer Res 56:299–304

Degen GH, Gerber MM, Obrecht-Pflumio S, Dirheimer G (1997) Induction of micronuclei with ochratoxin A in ovine seminal vesicle culture cells. Arch Toxicol 71:365–371

Dickens F, Waynforth HB (1968) Studies on the carcinogenesis of lactones and related substances. Report on the British Empire Cancer Campaign 46:108

Dopp E, Muller J, Hahnel C, Schiffmann D (1999) Induction of genotoxic effects and modulation of the intracellular calcium level in Syrian hamster embryo (SHE) fibroblasts caused by ochratoxin A. Food Chem Toxicol 37:713–721

Dörrenhaus A, Föllmann W (1997) Effects of ochratoxin A on DNA repair in cultures of rat hepatocytes and porcine urinary bladder epithelial cells. Arch Toxicol 71:709–713

Dortant PM, Peters-Volleberg GW, Van Loveren H, Marquardt RR, Speijers GJ (2001) Age-related differences in the toxicity of ochratoxin A in female rats. Food Chem Toxicol 39:55–65

Doster RC, Sinnhuber RO, Wales JH, Lee DJ (1971) Acute toxicity and carcinogenicity of ochratoxin A and B derivatives in rainbow trout (*Salmo garidneri*). Food Cosmet Toxicol 12:499–505

European Commission (EC) (1998) Report on task. 3.2.7 for scientific cooperation (SCOOP) assessment of dietary intake of ochratoxin A by the population of EU member states. European Commission, Directorate-General for Industry, Brussels

European Commission (EC) (2002) Report on task. 3.2.7 for scientific cooperation (SCOOP), assessment of dietary intake of ochratoxin A by the population of EU member states. European Commission, Directorate-General Health and Consumer Protection, Brussels

Edrington TS, Harvey RB, Kubena LF (1995) Toxic effects of aflatoxin B1 and ochratoxin A, alone and in combination, on chicken embryos. Bull Environ Contam Toxicol 54:331–336

European Food Security Agency (EFSA) (2004) Opinion of the Scientific Panel on contaminants in food chain on a request from the Commission related to ochratoxin A (OA) as undesirable substance in animal feed. EFSA J 101:1–36

Ehrlich V, Darroudi F, Uhl M, Steinkellner H, Gann M, Majer BJ, Eisenbauer KS (2002) Genotoxic effects of ochratoxin A in human-derived hepatoma (HepG2) cells. Food Chem Toxicol 40:1085–1109

Elissalde MH, Ziprin RL, Huff WE, Kubena LF, Harvey RB (1994) Effects of ochratoxin A on *Salmonella*-challenged broiler chicks. Poult Sci 73:1241–1248

Elling F, Hald B, Jacobsen C, Krogh P (1975) Spontaneous toxic nephropathy in poultry associated with ochratoxin A. Acta Pathol Microbiol Scand A 83:739–741

Elling F (1979) Ochratoxin A-induced mycotoxic porcine nephropathy: alterations in enzyme activity in tubular cells. Acta Pathol Microbiol Scand 87:237–243

Endou H, Koseki C, Yamada H, Obara T (1986) Evaluation of nephrotoxicity using isolated nephron segments. Dev Toxicol Environ Sci 14:207–216

Föllmann W, Hillebrand IE, Creppy EE, Bolt HM (1995) Sister chromatid exchange frequency in cultured isolated porcine urinary bladder epithelial cells (PUBEC) treated with ochratoxin A and alpha. Arch Toxicol 69:280–286

Föllmann W, Flieger A, Dörrenhaus A (1997) Induktion von DNA Reparatursynthese durch Ochratoxin A in vitro. In: Märtlbauer E, Usleber E (eds). Proceedings 19th Mykotoxin-Workshop, Munich, pp 78–83

Garaleviciene D, Petterson H, Agnedal M (2003) Occurrence of trichothecenes, zearalenone and ochratoxin A in cereals and mixed feed from central Lithuania. Mycotox Res 18:77–89

Gekle M, Gassner B, Freudinger R, Mildenberger S, Pfaller W, Schramek H (1998) Characterization of an ochratoxin A-dedifferentiated and cloned renal epithelial cell line. Toxicol Appl Pharmacol 152 282–291

Gekle M, Schwerdt G, Freudinger R, Mildenberger S, Wilflingseder D, Pollack V, Dander M, Schramek H (2000) Ochratoxin A induces JNK activation and apoptosis in MDCK-C7 cells at nanomolar concentrations. J Pharmacol Exp Ther 293:837–844

Gentles A, Smith EE, Kubena LF, Duffus E, Johnson P, Thompson J, Harvey RB, Edrington TS (1999) Toxicological evaluations of cyclopiazonic acid and ochratoxin A in broilers. Poult Sci 78:1380–1384

Hagelberg S, Hult K, Fuchs R (1989) Toxicokinetics of ochratoxin A in several species and its plasma-binding properties. J Appl Toxicol 9:91–96

Hald B (1991) Porcine nephropathy in Europe. IARC Sci Publ 115:49–56

Harvey RB, Elissalde MH, Kubena LF, Weaver EA, Corrier DE, Clement BA (1992) Immunotoxicity of ochratoxin A to growing gilts. Am J Vet Res 53:1966–1970

Hatey F, Galtier P (1977) Toxicité à court terme de l'ochratoxine A chez le rat; quelques manifestations biochimiques de l'intoxication. Ann Rech Vét 8:7–12

Haubeck HD, Lorkowski G, Kölsch E, Röschenthaler R (1981) Immunosuppression by ochratoxin A and its prevention by phenylalanine. Appl Environ Microbiol 41:1040–1042

Hayes AW, Hood RD, Lee HL (1974) Teratogenic effects of ochratoxin A in mice. Teratology 9:93–97

Hazel C, Walker R (2005) Ochratoxin A in food, recent development and significance. Proceedings of a workshop held on 29, 30 June and 1 July 2005 in Baden, Food Addit Contam 22 (Suppl 1):1–107
Hennig A, Fink-Gremmels J, Leistner L (1991) Mutagenicity and effects of ochratoxin A on the frequency of sister chromatid exchange after metabolic activation. IARC Sci Publ 115:255–260
Holmberg T, Hagelberg S, Lundeheim N, Thafvelin B, Hult K (1990) Ochratoxin A in swine blood used for evaluation of cereal handling procedures. Zentralbl Veterinarmed B 37:97–105
Hood RD, Naughton MJ, Hayes AW (1976) Prenatal effects of ochratoxin A in hamsters. Teratology 13:11–14
Huff WE, Wyatt RD, Hamilton PB (1975) Nephrotoxicity of dietary ochratoxin A in broiler chickens. Appl Microbiol 30:48–51
Hult K, Plestina R, Habazin-Novak V, Radic B, Ceovic S (1982) Ochratoxin A in human blood and Balkan endemic nephropathy. Arch Toxicol 51:313–321
Hult K, Rutqvist L, Holmberg T, Thafvelin B, Gatenbeck S (1984) Ochratoxin A in blood of slaughter pigs. Nord Vet Med 36:314–316
IARC (1976) IARC monographs on the evaluation of carcinogenic risk of chemicals to humans. Some naturally occurring substances. International Agency for Cancer, Lyon, 10:191–197
IARC (1983) IARC monographs on the evaluation of carcinogenic risk of chemicals to humans. Some food additives, feed additives and naturally occurring substances. International Agency for Cancer, Lyon, 31:191–206
IARC (1993) IARC monographs on the evaluation of carcinogenic risks of chemicals to humans. Some naturally occurring substances: food items and constituents, heterocyclic aromatic amines and mycotoxins. International Agency for Cancer, Lyon, 56:489–521
Jeffery FH, Morton JG, Miller JK (1984) Effects of some clinically significant mycotoxins on the incorporation of DNA, RNA and protein precursors in cultured mammalian cells. Res Vet Sci 37:30–38
Jiménez AM, López de Cerain A, González Peñas E, Bello J, Betbeder AM, Creppy EE (1998) Exposure to ochratoxin A in Europe: comparison with a region of northern Spain. J Toxicol Toxin Rev 17:479–491
Jonsson N, Pettersson H (1999) Evaluation of different conservation method for grain-based on analysis of hygiene quality. Swedish Institute of Agricultural and Environment Engeneering, Report 263
Jorgensen K (2005) Occurrence of ochratoxin A in commodities and processed foods - a review of EU occurrence data. Food Addit Contam 22(Suppl 1):26–30
Jung KY, Endou H (1989) Nephrotoxicity assessment by measuring cellular ATP content. II Intranephron site of ochratoxin A nephrotoxicity. Toxicol Appl Pharmacol 100:383–390
Kamp HG, Eisenbrand G, Schlatter J, Würth K, Janzowski C (2005) Ochratoxin A: induction of (oxidative) DNA damage, cytotoxicity and apoptosis in mammalian cell lines and primary cells. Toxicology 206:413–425
Kane A, Creppy EE, Roth A, Röschenthaler R, Dirheimer G (1986) Distribution of the [3H]-label from low doses of radioactive ochratoxin A ingested by rats, and evidence for DNA single-strand breaks caused in liver and kidneys. Arch Toxicol 58:219–224
Kanisawa M (1984) Synergistic effects of citrinin on hepatorenal carcinogenesis of ochratoxin A in mice. Dev Food Sci 7:245–254
Kanisawa M, Suzuki S (1978) Induction of renal and hepatic tumors in mice by ochratoxin A, a mycotoxin. Gann 69:599–600
Khalef A, Zidane C, Charef A, Gharbi A, Tadjerouna M, Betbeder AM, Creppy EE (1993) Ochratoxicose humaine en Algerie. In: Creppy EE, Castegnaro M, Irheimer G (eds) Human ochratoxicosis and its pathologies, vol 231. Colloque INSERM/John Libbey Eurotext Ltd., Montroge, France, pp 123–127
Kitchen DN, Carlton WW, Tuite J (1977) Ochratoxin A and citrinin induced nephrosis in beagle dogs. I. Clinical and clinicopathological features. Vet Pathol 14:154–172

Krogh P (1974) Mycotoxin porcine nephropathy: a possible model for Balkanic endemic nephropathy. In: Puchlev A, Dinev I, Milev B, Doichinov D (eds) Endemic nephropathy. Publishing House of Bulgarian Academy of Sciences, Sofia, pp 266–270

Krogh P, Hald B, Pedersen EJ (1973) Occurrence of ochratoxin A and citrinin in cereals associated with mycotoxic porcine nephropathy. Acta Pathol Microbiol Scand B 81:689–695

Krogh P, Hald B, Plestina R, Ceovic S (1977) Balkan (endemic) nephropathy and foodborne ochratoxin A: preliminary results of a survey of foodstuffs. Acta Pathol Microbiol Scand [B] 85:238–240

Krogh P, Gyrd-Hansen, Hald B, Larsen S, Nielsen JP, Smith M, Ivanoff C, Meisner H (1988) Renal enzyme activities in experimental ochratoxin A-induced porcine nephropathy: diagnostic potential of phosphoenolpyruvate carboxykinase and gamma-glutamyl transpeptidase activity. J Toxicol Environ Health 23:1–14

Kuiper-Goodman T (1991) Risk assessment to humans of mycotoxins in animal-derived food products. Vet Hum Toxicol 33:325–333

Kuiper-Goodman T (1996) Risk assessment of ochratoxin A: an update. Food Addit Contam 13:53–57

Kuiper-Goodman T, Scott PM (1989) Risk assessment of the mycotoxin ochratoxin A. Biomed Environ Sci 2:179–248

Kuiper-Goodman T, Ominsky K, Marquardt RR, Malcom S, McMullen E, Lombaert GA, Morton T (1993) Estimating human exposure to ochratoxin A in Canada. In: Creppy EE, Castegnaro M, Dirheimer G (eds) Human ochratoxicosis and its pathologies, vol 231. Colloque INSERM/ John Libbey Eurotext Ltd., Montroge, France, pp 167–174

Kuramochi G, Gekle M, Silbernagl S (1997) Derangement of pH homeostasis in the renal papilla: ochratoxin A increases pH in vasa recta blood. Nephron 76:472–476

Lalithakunjamma CR, Krishnan Nair M (1997) Pathomorphology of the combined effect of ochratoxin A and citrinin in chicken embryo. Indian Vet J 74:22–25

Lea T, Steien K, Stormer FC (1989) Mechanism of ochratoxin A-induced immunosuppression. Mycopathologia 107:153–159

Lebrun S, Föllmann W (2002) Detection of ochratoxin A-induced DNA damage in MDCK cells by alkaline single cell electrophoresis (comet assay). Arch Toxicol 75:734–741

Levin S (1998) Apoptosis, necrosis, or oncosis: what is your diagnosis? A report from the Cell Death Nomenclature Committee of the Society of Toxicologic Pathologists. Toxicol Sci 41:155–156

Luster MI, Germolec DR, Burleson GR, Jameson CW, Ackermann MF, Lamm KR, Hayes HT (1987) Selective immunosuppression in mice of natural killer cell activity by ochratoxin A. Cancer Res 47:2259–2263

Maaroufi K, Achour A, Betbeder AM, Hammami M, Ellouz F, Creppy EE, Bacha H (1995) Foodstuffs and human blood contamination by the mycotoxin ochratoxin A: correlation with chronic interstitial nephropathy in Tunisia. Arch Toxicol 69:552–558

Maaroufi K, Zakhama A, Baudrimont I, Achour A, Abid S, Ellouz F, Dhouib S, Creppy EE, Bacha H (1999) Karyomegaly of tubular cells as an early stage marker of nephrotoxicity induced by ochratoxin A in rats. Hum Exp Toxicol 18:410–415

Malaveille C, Brun G, Bartsch H (1994) Structure-activity studies in *E. coli* strains on ochratoxin A (OTA) and its analogues implicate a genotoxic free radical and a cytotoxic thiol derivative as reactive metabolites. Mutat Res 307:141–147

Mantle PG, McHugh KM, Adatia R, Gray T, Turner DR (1991) Persistent karyomegaly caused by *Penicillium* nephrotoxins in the rat. Proc Biol Sci 246:251–259

Marquardt RR, Frohlich AA (1992) A review of recent advances in understanding ochratoxicosis. J Anim Sci 70:3968–3988

Mayura K, Abdel-Wahhab MA, McKenzie KS, Sarr AB, Edwards JF, Naguib K, Phillips TD (1998) Prevention of maternal and developmental toxicity in rats via dietary inclusion of common aflatoxin sorbents: potential for hidden risks. Toxicol Sci 41: 175–182

Meisner H, Krogh P (1986) Phosphoenolpyruvate carboxykinase as a selective indicator of ochratoxin A-induced nephropathy. Dev Toxicol Environ Sci 14:199–206

Mori H, Kawai K, Ohbayashi F, Kuniyasu T, Yamazaki M, Hamasaki T, Williams GM (1984) Genotoxicity of a variety of mycotoxins in the hepatocyte primary culture/DNA repair test using rat and mouse hepatocytes. Cancer Res 44:2918–2923

Müller G, Kielstein P, Köhler H, Berndt A, Rosner H (1995) Studies of the influence of ochratoxin A on immune and defense reactions in the mouse model. Mycoses 38:85–91

Müller G, Kielstein P, Rosner H, Berndt A, Heller M, Köhler H (1999) Studies of the influence of ochratoxin A on immune and defense reactions in weaners. Mycoses 42:495–505

Munro IC, Moodie CA, Kuiper-Goodman T, Scott PM, Grice HC (1974) Toxicological changes in rats fed graded dietary levels of ochratoxin A. Toxicol Appl Pharmacol 28:180–188

Ngaha EO (1985) Biochemical changes in the rat during experimentally induced acute ochratoxicosis. Enzyme 33:1–8

Nikolov IG, Petkova-Bocharova D, Castegnaro M, Pfohl-Leskowicz A, Gill C, Day N, Chernozemsky IN (1996) Molecular and epidemiological approaches to the etiology of urinary tract tumors in an area with Balkan endemic nephropathy. J Environ Pathol Toxicol Oncol 15:201–207

NTP (1989) National Toxicology Program. Technical report on the Toxicology and Carcinogenesis studies of ochratoxin A (CAS 303–47–9) in F344/N rats (Gvaavge studies).In: Boorman G (ed) NIH publication No. 82–2813, US Department of Health and Human Services, National Institute of Health, Research Triangle Park, NC, USA

Obrecht-Pflumio S, Chassat T, Dirheimer G, Marzin (1999) Genotoxicity of ochratoxin A by *Salmonella* mutagenicity test after bioactivation by mouse kidney microsomes. Mutat Res 446:95–102

Olsen M, Thorup I, Knudsen I, Larsen J-J, Hald B, Olsen J (1991) Health evaluation of ochratoxin A in food products. In: Nordiske seminar-og Arbejds-rapporter, vol 545. Nordic Council of Ministers, Copenhagen

Özçelik N, Kosar A, Soysal D (2001) Ochratoxin A in human serum samples collected in Isparta, Turkey from healthy individuals and individuals suffering from different urinary disorders. Toxicol Lett 121:9–13

Peraica M, Domijan AM, Fuchs R, Lucic A, Radic B (1999) The occurrence of ochratoxin A in blood in general population of Croatia. Toxicol Lett 110:105–112

Petkova-Bocharova T, Chernozemsky IN, Castegnaro M (1988) Ochratoxin A in human blood in relation to Balkan endemic nephropathy and urinary system tumors in Bulgaria. Food Addit Contam 5:299–301

Petzinger E, Ziegler K (2000) Ochratoxin A from a toxicological perspective. J Vet Pharmacol Therap 23:91–98

Petzinger E, Weidenbach A (2002) Mycotoxins in the food chain: the role of ochratoxins. Livest Prod Sci 76:245–250

Pfohl-Leszkowicz A, Petkova-Bocharova T, Chernozemsky IN, Castegnaro M (2002) Balkan endemic nephropathy and associated urinary tract tumours: a review on aetiological causes and the potential role of mycotoxins. Food Addit Contam 19:282–302

Pier AC, Cysewski SJ, Richard JL, Baetz AL, Mitchell L (1976) Experimental mycotoxicoses in calves with aflatoxin, ochratoxin, rubratoxin, and T-2 toxin. Proc Annu Meet U S Anim Health Assoc 80:130–148

Plestina R (1992) Some features of Balkan endemic nephropathy. Food Chem Toxicol 30: 177–181

Purchase IF, Van der Watt JJ (1971) The long-term toxicity of ochratoxin A to rats. Food Cosmet Toxicol 9:681–682

Radovanovic Z (1989) Aetiology of Balkan nephropathy: a reappraisal after 30 years. Eur J Epidemiol 5:372–377

Rafai P, Bata A, Jakab L, Vanyi A (2000) Evaluation of mycotoxin-contaminated cereals for their use in animal feeds in Hungary. Food Addit Contam 17:799–808

Ribelin WE, Fukushima K, Still PE (1978) The toxicity of ochratoxin to ruminants. Can J Comp Med 42:172–176

Ringot D, Chango A, Schneider YJ, Larondelle Y (2006) Toxicokinetics and toxicodynamics of ochratoxin A, an update. Chem-Biol Interact 159:18–46

Robbiano L, Baroni D, Carrozzino R, Mereto E, Brambilla G (2004) DNA damage and micronuclei induced in rat and human kidney cells by six chemicals carcinogenic to the rat kidney. Toxicology 204:187–195

SCF (1996) Opinion on ochratoxin A and patulin expressed on 23rd September 1994. Reports of the Scientific Committee on Food, 35th Series, Luxembourg

SCF (1998) Opinion on ochratoxin A expressed on 17 September 1998. European Commission, Bruxelles. http://europa.eu.int/comm/food/fs/sc/scf/out14_en.html. Accessed October 2007

SCF (2006) Opinion of scientific panel on contaminants in the food chain on a request from the commission related to ochratoxin A in food. EFSA J 365:1–56

Schwerdt G, Freudinger R, Silbernagl S, Gekle M (1999) Ochratoxin A-binding proteins in rat organs and plasma and in different cell lines of the kidney. Toxicology 135:1–10

Schwerdt G, Freudinger R, Schuster C, Silbernagl S, Gekle M (2003) Inhibition of mitochondria prevents cell death in kidney epithelial cells by intra- and extracellular acidification. Kidney Int 63:1725–1735

Scott PM, Kanhere SR, Lau BP, Lewis DA, Hayward S, Ryan JJ, Kuiper-Goodman T (1998) Survey of Canadian human blood plasma for ochratoxin A. Food Addit Contam 15:555–562

Scudamore KA, Patel S, Breeze V (1999) Surveillance of stored grain from the 1997 harvest in the United Kingdom for ochratoxin A. Food Addit Contam 16:281–290

Seegers JC, Lottering ML, Garlinski PJ (1994) The mycotoxin ochratoxin A causes apoptosis-associated DNA degradation in human lymphocytes. Med Sci Res 22:417–419

Stetina R, Votava (1986). Induction of DNA single-strand breaks and DNA synthesis inhibition by patulin, ochratoxin A, citrinin and aflatoxin B1 in cell lines CHO and AWRF. Folia Biol 32:128–144

Still PE, Macklin AW, Ribelin WE, Smalley EB (1971) Relationship of ochratoxin A to foetal death in laboratory and domestic animals. Nature 31:563–564

Stoev SD, Hald B, Mantle PG (1998) Porcine nephropathy in Bulgaria: a progressive syndrome of complex or uncertain (mycotoxin) aetiology. Vet Rec 142:190–194

Stoev SD, Daskalov H, Radic B, Domijan AM, Peraica M (2002a) Spontaneous mycotoxic nephropathy in Bulgarian chickens with unclarified mycotoxin aetiology. Vet Res 33:83–93

Stoev SD, Paskalev M, MacDonald S, Mantle PG (2002b) Experimental one-year ochratoxin A toxicosis in pigs. Exp Toxicol Pathol 53:481–487

Stonard MD, Gore CW, Oliver GJ, Smith IK (1987) Urinary enzymes and protein patterns as indicators of injury to different regions of the kidney. Fundam Appl Toxicol 9:339–351

Stormer FC, Lea T (1995) Effects of ochratoxin A upon early and late events in human T-cell proliferation. Toxicology 95:45–50

Szczech GM, Hood RD (1981) Brain necrosis in mouse fetuses transplacentally exposed to the mycotoxin ochratoxin A. Toxicol Appl Pharmacol 57:127–137

Tapai K, Teren J, Mesterhazy A (1997) Ochratoxin A in the sera of blood donors and ill persons. Cereal Res Commun 25:307–308

Theron JJ, van der Merwe KJ, Liebenberg N, Joubert HJ, Nel W (1966) Acute liver injury in ducklings and rats as result of ochratoxin poisoning. J Pathol Bacteriol 91:521–529

Thuvander A, Breitholtz-Emanuelson A, Olsen M (1995) Effects of ochratoxin A on the mouse immune system after subchronic exposure. Food Chem Toxicol 33:1005–1011

Thuvander A, Funseth E, Breitholtz-Emanuelsson A, Hallen IP, Oskarsson A (1996) Effects of ochratoxin A on the rat immune system after perinatal exposure. Nat Toxins 4:141–147

Tressou J, Leblanc JCH, Feinberg M, Bertail P (2004) Statistical methodology to evaluate food exposure to a contaminant and influence of sanitary limits: application to ochratoxin A. Regul Toxicol Pharmacol 40:252–263

Ueno Y, Maki S, Lin J, Furuya M, Sugiura Y, Kawamura O (1998) A 4-year study of plasma ochratoxinA in a selected population in Tokyo by immunoassay and immunoaffinity column linked HPLC. Food Chem Toxicol 36:445–449

Ueno Y, Umemori K, Niimi E, Tanuma S, Nagata S, Sugamata M, Ihara T, Sekijima M, Kawai K, Ueno I, Tashiro F (1995) Induction of apoptosis by T-2 toxin and other natural toxins in HL-60 human promyelotic leukaemia cells. Nat Toxins 3:129–137

Umeda M, Tsutsui T, Saito M (1977) Mutagenicity and inducibility of DNA single-strand breaks and chromosome aberrations by various mycotoxins. Gann 69:619–625

Wangikar PB, Dwivedi P, Sinha N (2004a) Effect in rat of simultaneous prenatal exposure to ochratoxin A and aflatoxin B1. I. Maternal toxicity and fetal malformations. Birth Defects Res Part B: Dev Reprod Toxicol 71:343–351

Wangikar PB, Dwivedi P, Sinha N (2004b) Teratogenic effects of ochratoxin A in rabbits. World Rabbit Sci 12:159–171

Wangikar PB, Dwivedi P, Sinha N, Sharma AK, Telang AG (2005) Teratogenic effects in rabbits of simultaneous exposure to ochratoxin A and aflatoxin B1 with special reference to microscopic effects. Toxicology 215:37–47

Wei X, Sulik KK (1993) Pathogenesis of craniofacial and body wall malformations induced by ochratoxin A in mice. Am J Med Genet 47:862–871

Weidenbach A, Schuh K, Failing K, Petzinger E (2001) Ochratoxin A-induced TNF-alpha release from the isolated blood-free perfused liver. Mycotoxin Res 16A:189–193

World Health Organization/Food and Agriculture Organization (WHO/FAO) (1991) Ochratoxin A, toxicological evaluation of certain food additives and contaminants. WHO Food Addit Ser 28, Geneva, pp 365–417

World Health Organization/Food and Agriculture Organization (WHO/FAO) (1996) Ochratoxin A, Toxicological evaluation of certain food additives and contaminants, WHO Food Addit Ser 35, Geneva, pp 363–376

World Health Organization/Food and Agriculture Organization (WHO/FAO) (2001) Ochratoxin A. In: Safety Evaluation of Certain Mycotoxins in Food. WHO Food Addit Ser 47, Geneva, pp 281–415

Zimmerli B, Dick R (1995) Determination of ochratoxin A at the ppt level in human blood, serum, milk and some foodstuffs by high-performance liquid chromatography with enhanced fluorescence detection and immunoaffinity column cleanup: methodology and Swiss data. J Chromatogr B Biomed Appl 666:85–99

Chapter 19
Natural Aflatoxin Inhibitors from Medicinal Plants

Mehdi Razzaghi-Abyaneh, Masoomeh Shams-Ghahfarokhi, Mohammad-Bagher Rezaee, and Shohei Sakuda

19.1 Introduction

Aflatoxins (AF) are real public health hazards due to their potential carcinogenic, teratogenic, and mutagenic activities toward humans and a wide range of animal species (Allameh and Razzaghi-Abyaneh 2001; Bennett and Klich 2003). They are a major group of polyketide mycotoxins produced mainly by specific members of *Aspergillus* section *Flavi*. AF-producing fungi, especially *Aspergillus flavus* and *Aspergillus parasiticus*, have worldwide distribution and they are able to contaminate a wide range of substrates including cereal grains, oilseeds, and tree nuts under favorable conditions of temperature and relative humidity (Bennett and Klich 2003). The importance of these toxins, especially aflatoxin B_1 (AFB_1), the most potent natural hepatocarcinogen, led to significant advances in the field of secondary metabolism, from biochemistry to genetics and control strategies. Genetic studies of AF biosynthesis in the major producers, *A. flavus* and *A. parasiticus*, led to the identification of at least 25 clustered genes within a 70 kb DNA region responsible for the enzymatic conversions in the AF biosynthetic pathway (Yu et al. 2005). The role of regulatory elements, nutritional and environmental factors and fungal development in AF formation has been studied with special focus on microarray technology using expressed sequence tags of *A. flavus* and *A. parasiticus* (Yu et al. 2004). Despite these promising data, concerns over AF contamination concerns are still far from being solved, due to our low level of understanding about signal transduction pathways underlying toxin formation by producing fungi, and about the dynamics of toxigenic fungus-host plant interactions during the infection process. Since the discovery of AF in the early 1960s, a large number of chemicals have been screened with the aim of finding AF biosynthesis inhibitors (Zaika and Buchanan 1987; Razzaghi-Abyaneh et al. 2006a).

M. Razzaghi-Abyaneh (✉)
Department of Mycology, Pasteur Institute of Iran, Tehran 13164, Iran
e-mail: mrab442@pasteur.ac.ir; mrab442@yahoo.com

M. Rai and A. Varma (eds.), *Mycotoxins in Food, Feed and Bioweapons*,
DOI: 10.1007/978-3-642-00725-5_19,

Most effective chemicals identified suffer from large limitations, including human and animal toxicity at effective concentrations, induction of resistance in toxigenic fungi, high costs for safe experiments, and undesirable effects on non-target organisms sharing the ecosystem. So, researchers focused on evaluating natural sources including plants and microorganisms for effective and safer inhibitors of AF biosynthesis. In this area, screening of a large number of plant essential oils (EOs) and extracts (Es) led to the identification of several inhibitory plants from diverse family groups with potential activity against fungal growth and/or AF production (Table 19.1). Recent advances in drug discovery from medicinal plants led to the isolation and characterization of pharmaceutically active compounds with potential activities against cancer, HIV/AIDS, malaria, pain, etc. (Balunas and Kinghorn 2005). Along with this interesting progress toward marketing plant-originated compounds, effective components of some medicinal plants which strongly inhibit AF formation in toxigenic fungi have now been successfully isolated and identified (Table 19.1).

In this chapter, plants with potential inhibitory activity toward AF biosynthesis are discussed in detail, with special reference to their effective components and modes of action. A short review of new findings in AF research areas is presented as an initial step for better understanding of further descriptions. Then, selected AF-inhibitory plants are described. Finally, we discuss a practical approach to isolate dillapiol, a pathway-specific AF inhibitor, from a medicinal plant named *Anethum graveolens* L. as a successful example of a simple low-cost method for characterizing novel AF inhibitors from natural sources. Special consideration is given to the mode of action of dillapiol at cellular and molecular levels. With a better understanding of natural inhibitors of AF biosynthesis, researchers will find suitable tools for further evaluation of regulatory mechanisms of AF biosynthesis by producing fungi.

19.2 AF-Producing Fungi

AF are polyketide-derived mycotoxins produced in a complex pathway in *A. flavus* and *A. parasiticus* by involving about 29 genes catalyzed in at least 23 enzymatic reactions (Yu et al. 2005). After the initial description of *A. flavus* as an AF-producing fungus in the 1960s, about ten other *Aspergillus* species have been recognized as producing AF in various amounts (For an excellent review, see Frisvad et al. 2007).

A. flavus and *A. parasiticus*, the most important AF producers, are frequently isolated from soil and various agricultural commodities in most regions of the world (Payne and Brown 1998; Razzaghi-Abyaneh et al. 2006b). Infective propagules (conidia, mycelia, and sclerotia) are easily distributed by wind or by insects. Under favorable environmental conditions, the propagules grow on plants and other materials and produce carcinogenic AF. The exact role of AF in producing fungi is not clear. It has been estimated that the genes for AF biosynthesis have persisted for more than 100 million years (Cary and Ehrlich 2006). It is believed that AF are produced as a fungal defense response to stressful conditions, for protecting the fungus from UV damage, as virulence factors, or as a part of defense mechanisms to protect the fungus from predators in the environment (Cary and Ehrlich 2006).

19.3 Control Strategies

The strategies for preventing AF contamination are generally divided into two categories including pre and post-harvest controls (For an excellent review, see Kabak et al. 2006). Pre-harvest control strategies are mainly performed through appropriate field management practices (crop rotation, irrigation, soil cultivation, etc.), producing AF-resistant crop varieties, and application of biological (inhibitory fungi and bacteria), natural, and chemical agents. Post-harvest control strategies are based on improvement of drying and storage conditions, by use of biological, natural, and chemical agents and irradiation. With regard to biocontrol methods, several metabolites from *Streptomyces* spp. (dioctatin A, aflastatin A, blasticidin A), *Achromobacter xylosoxidans* [cyclo(L-leucyl-L-propyl)] and *Bacillus subtilis* (bacillomycin D, iturin A, etc.) have been introduced as potent inhibitors of AF biosynthesis in laboratory conditions (see in Holmes et al. 2008). Sakuda et al. (2006) examined the inhibition of AF by dioctatin A in a row of peanuts model system and suggested that this compound may be effective in preventing AF contamination of peanuts during storage. Many EOs and Es from a wide variety of plant species were also reported to inhibit toxigenic fungal growth and/or AF production. Likewise, complementary studies have led to the isolation and characterization of several plant metabolites as strong inhibitors of AF biosynthesis.

19.4 Plants as AF Inhibitors

Herbal medicine has a history of thousands of years (Samuelsson 2004). Medicinal plants are widely distributed in the world and are rich sources of useful metabolites such as flavonoids, alkaloids, terpenoids, terpenes, aromatic compounds (aldehydes, alcohols, phenols, etc.) etc. In nature, these metabolites play an important role in the protection of plants from diverse groups of phytopathogenic microorganisms, insects, and herbivores (Bakkali et al. 2008). Since plants are generally regarded as safe (Samuelsson 2004), their formulation as natural antimicrobial preservatives may be an attractive way to controlling microbial food contamination. In this way, researchers have focused on the potential of plants and their metabolites to inhibit toxigenic fungus growth and/or AF production as a useful tool for controlling AF contamination of crops and agricultural commodities. As described in Table 19.1, a large number of plants and their effective components classified in 52 genera from 28 families were found to be potent inhibitors of AF biosynthesis.

Table 19.1 Medicinal plants and their active components inhibitory to aflatoxin (AF) production

Family	Species	Common name	Part used	Effective component (s)	Inhibitory concentration	Reference
Agavaceae	*Agave asperrima*	Agave	Flower	Mathanolic and aqueous extracts[a]	<2 mg ml^{-1}	Sánchez et al. (2005)
	Agave striata	Agave	Flower	Mathanolic and aqueous extracts[a]	<2 mg ml^{-1}	Sánchez et al. (2005)
Alliaceae	*Allium cepa*	Welsh onion	Bulb	Ethanolic extract[a]	<10 mg ml^{-1}	Fan and Chen (1999)
	Allium cepa	Onion	Bulb	Aqueous extract[a]	NA	Bilgrami et al. (1992)
	Allium sativum	Garlic	Bulb	Aqueous extract[a]	NA	Bilgrami et al. (1992)
Amarantaceae	*Chenopodium ambrosioides* Linn.	Goose foots	Leaves	Essential oil[a]	<100 μg ml^{-1}	Kumar et al. (2007)
Anacardiaceae	*Pistacia vera* L.	Pistachio	kernel	Caffeic acid	12 mM	Molyneux et al. (2007)
	Pistacia vera L.	Pistachio	kernel	3,4-digalloyl quinic acid	12 mM	Molyneux et al. (2007)
	Pistacia vera L.	Pistachio	kernel	Quinic acid	12 mM	Molyneux et al. (2007)
	Pistacia vera L.	Pistachio	kernel	Chlorogenic acid	12 mM	Molyneux et al. (2007)
Apiaceae	*Ammi visnaga*	Toothpick weed	Leaves	Aqueous extract[a]	4–10 mg ml^{-1}	Mahmoud (1999)
	Trachyspermum copticum L.	Carom seed	Seed	Essential oil[a]	450 ppm	Rasooli et al. (2008)
	Coriandrum sativum L.	Coriander		Essential oil[a]	~1%	Tantaoui-Elaraki and Beraoud (1994)
	Cuminum cyminum L.	Cumin	Seed	Essential oil[a]	1–10%	Tantaoui-Elaraki and Beraoud (1994)
						Mabrouk and El-Shayeb (1980)
						Hasan and Mahmoud (1993)
	Anethum graveolens L.	Dill	Leaves	Dillapiol	0.15 μM	Razzaghi-Abyaneh et al. (2007)
	Petroselinum crispum	Parsley	Seed	Apiol	0.24 μM	Razzaghi-Abyaneh et al. (2007)
	Petroselinum crispum	Parsley	Seed	Myristicin	3.55 μM	Razzaghi-Abyaneh et al. (2007)

	Ammi visnaga (L.) Lam.	Toothpick weed	Seed	Xanthotoxin	<0.1 mM	Mabrouk and El-Shayeb (1992)
	Ammi visnaga (L.) Lam.	Toothpick weed	Seed	Bergapten	<0.1 mM	Mabrouk and El-Shayeb (1992)
	Ammi visnaga (L.) Lam.	Toothpick weed	Seed	Psoralene	<0.1 mM	Mabrouk and El-Shayeb (1992)
	Ammi visnaga (L.) Lam.	Toothpick weed	Seed	Khellin	<0.1 mM	Mabrouk and El-Shayeb (1992)
	Ammi visnaga (L.) Lam.	Toothpick weed	Seed	Visnagin	<0.1 mM	Mabrouk and El-Shayeb (1992)
	Pimpinella anisum	Anise	Seed	Essential oil[a]	500 $\mu g\ g^{-1}$	Bluma and Etcheverry (2008)
Araceae	*Amorphophallus campanulatus* Blume	Anto Apong-apong	Leaves	Aqueous extract[a]	1.5–8.0 $mg\ ml^{-1}$	Prasad et al. (1994)
Asteraceae	*Polymnia sanchifolia*	Yacon	Leaves	Aqueous extract[a]	50–200 $\mu g\ ml^{-1}$	Pinto et al. (2001)
	Polymnia sanchifolia	Yacon	Leaves	Ethanolic extract[a] Methanolic extract[a] Hexanic extract[a]	50–350 $\mu g\ ml^{-1}$ 25–100 $\mu g\ ml^{-1}$ 100 $\mu g\ ml^{-1}$	Gonçalez et al. (2003)
	Xanthium pungens	Cocklebur	Leaves	Extract[a]	4–10 $mg\ ml^{-1}$	Mahmoud (1999)
	Artemisia vulgaris L.	Mugwort	Aerial parts	Essential oil[a]	~1%	Tantaoui-Elaraki and Beraoud (1994)
	Matricaria recutita L.	German Chamomile	Flower heads	(E)-spiroether (Z)-spiroether	2.8 μM 20.8 μM	Yoshinari et al. (2008)
Caprifoliaceae	*Lonicera* sp.	Honeysuckle	Flower and root stem	Extract[a]	2% (v/v)	Bahk and Marth (1983)
Clusiaceae	*Garcinia indica*	Kokum	Fruit rinds	Chloroformic extract[a]	500–3,000 ppm	Selvi et al. (2003)
	Garcinia cowa	Garcinia	Fruit rinds	Hexane and chloroformic extracts[a]	2,000 ppm	Joseph et al. (2005)
	Garcinia pedunculata	Garcinia	Fruit rinds	Hexane and chloroformic extracts[a]	2,000 ppm	Joseph et al. (2005)

(*continued*)

Table 19.1 (continued)

Family	Species	Common name	Part used	Effective component (s)	Inhibitory concentration	Reference
Cyperaceae	*Cyperus rotundus*	Papyrus sedges	Leaves	Aqueous extract[a]	4.2% (v/v)	Masood and Ranjan (1991)
Euphorbiaceae	*Euphorbia hirta* L.	Spurge	Leaves	Aqueous extract[a]	4.2% (v/v)	Masood and Ranjan (1991)
Fabaceae	*Glycyrrhiza glabra* L.	Liquorice	Root	Extract[a]	2% (v/v)	Bahk and Marth (1983)
	Arachis hypogaea	Peanut	Seed	Ferulic acid[a]	0.5–1 mM	Chipley and Uraih (1980) Fajardo et al. (1995)
	Arachis hypogaea	Peanut	Seed	Vanillic acid[a]	~1 mM	Fajardo et al. (1995)
	Arachis hypogaea	Peanut	Seed	p-coumaric acid[a]	1 mM	Fajardo et al. (1995)
	Arachis hypogaea	Peanut	Shell	Eriodictyol[a]	~0.05 mM	Delucca et al. (1987)
	Arachis hypogaea	Peanut	Shell	5,7-dihydroxychromone	<100 μM	Delucca et al. (1987)
	Glycine max L.	Soybean	Seed	Phytic acid	10 mM	Gupta and Venkitasubramanian (1975)
	Lablab purpureus	Hyacinth bean	Seed	α-amylase inhibitor-like protein[a] (AILP)	NA	Fakhoury and Woloshuk (2001)
	Lupinus albus	Lupin	Leaves	Aqueous extract[a]	4–10 mg ml^{-1}	Mahmoud (1999)
Juglandaceae	*Juglans regia* L.	Common walnut	Husks	1,4-naphtoquinone[a]	>100 ppm	Mahoney et al. (2000)
	Juglans regia L.	Common walnut	Husks	Juglone[a]	>100 ppm	Mahoney et al. (2000)
	Juglans regia L.	Common walnut	Husks	2-Methyl-1,4-naphtoquinone[a]	>100 ppm	Mahoney et al. (2000)
	Juglans regia L.	Common walnut	Husks	Plumbagin[a]	>100 ppm	Mahoney et al. (2000)
	Juglans regia L.	Common walnut	Seed coat	Tannic acid	2 mM	Mahoney and Molyneux (2004)
	Juglans regia L.	Common walnut	Seed coat	Gallic acid	12 mM	Mahoney and Molyneux (2004)

	Juglans regia L.	Common walnut	Seed coat	Ellagic acid		12 mM	Mahoney and Molyneux (2004)
Lamiaceae	*Hyptis suaveolens*	Bushmint	Leaves	Powder		5–10% (w/w)	Krishnamurthy and Shashikala (2006)
	Ocimum sp.	Basil	Leaves	Powder		NA	Awuah (1996)
	Rosmarinus officinalis	Rosmary	Aerial parts	Essential oil[a]		450 ppm	Rasooli et al. (2008)
	Mentha x-piperita L.	Peppermint	Leaves	Essential oil[a]		≥5%	Mabrouk and El-Shayeb (1980)
	Satureja hortensis L.	Savory	Leaves	Carvacrol[a]		~0.5 mM	Razzaghi-Abyaneh et al. (2008)
	Satureja hortensis	Summer savory	Leaves	Thymol[a]		~0.7 mM	Razzaghi-Abyaneh et al. (2008)
	Thymus eriocalyx	Thyme	Leaves	Essential oil[a]		5 μl ml^{-1}	Rasooli and Razzaghi-Abyaneh (2004)
	Thymus x-prolock	Thyme	Leaves	Essential oil[a]		5 μl ml^{-1}	Rasooli and Razzaghi-Abyaneh (2004)
	Thymus vulgaris	Thyme	Leaves	Carvacrol[a]		~0.5 mM	Razzaghi-Abyaneh (unpublished data)
	Thymus vulgaris	Thyme	Leaves	Thymol[a]		~0.6 mM	Razzaghi-Abyaneh (unpublished data)
	Ocimum basilicum	Sweet basil	Leaves	Essential oil[a]		5% (v/v)	Atanda et al. (2007)
	Hedeoma multiflora	Mountain thyme	Aerial parts	Essential oil[a]		500 μg g^{-1}	Bluma and Etcheverry (2008)
	Origanum vulgare L.	Oregano pot marjoram	Aerial parts	Essential oil[a]		500 μg g^{-1}	Bluma and Etcheverry (2008)
	Minthostachys verticillata	Peperina	Aerial parts	Essential oil[a]		500 μg g^{-1}	Bluma and Etcheverry (2008)
Lauraceae	*Cinnamomum zeylanicum*	Cinnamon	Leaves	Essential oil[a]		200–250 ppm	Bullerman et al. (1977)
	Cinnamomum cassia	Cassia	Leaves	Essential oil		5% (v/v)	Atanda et al. (2007)
	Laurus nobilis L.	Bay laurel	Leaves	Essential oil		5% (v/v)	Atanda et al. (2007)

(continued)

Table 19.1 (continued)

Family	Species	Common name	Part used	Effective component (s)	Inhibitory concentration	Reference
Malvaceae	*Gossypium* L.	Cotton	Leaves	Nonanol[a]	>50 μl	Zeringue and McCormick (1990)
	Gossypium L.	Cotton	Leaves	Limonene[a]	>50 μl	Zeringue and McCormick (1990)
	Gossypium L.	Cotton	Leaves	Camphene[a]	>100 μl	Zeringue and McCormick (1990)
	Gossypium hirsutum L.	Cotton	Carpel	Methyl jasmonate	NA	Zeringue (2002)
	Gossypium L.	Cotton	Leaves	Nonanol[a]	0.25 μl ml^{-1}	Greene-McDowelle et al. (1999)
	Gossypium L.	Cotton	Leaves	Camphene[a]	0.25 μl ml^{-1}	Greene-McDowelle et al. (1999)
	Gossypium L.	Cotton	Leaves	Limonene[a]	0.25 μl ml^{-1}	Greene-McDowelle et al. (1999)
Meliaceae	*Azadirachta indica* A. Juss	Neem	Seed oil	Volatile compounds[a] including trans-2-heptenal	NA	Zeringue et al. (2001)
	Azadirachta indica A. Juss	Neem margosa	Leaves	Aqueous extract	10–50% (v/v)	Bhatnagar and McCormick (1988) Ghorbanian et al. (2008) Allameh et al. (2001)
	Azadirachta indica A. Juss	Neem margosa	Seed	Aqueous extract	>10% (v/v)	Razzaghi-Abyaneh et al. (2005)
Monimiaceae	*Peumus boldus* Molina	Boldo	Leaves	Essential oil[a]	500 μg g^{-1}	Bluma and Etcheverry (2008)
Myrtaceae	*Syzygium aromaticum* L.	Clove	Leaves	Essential oil[a]	200–250 ppm	Bullerman et al. (1977) Hasan and Mahmoud (1993)
					>0.1% (v/v)	Mabrouk and El-Shayeb (1980) Juglal et al. (2002)

Ochnaceae	*Ouratea spectabilis*	Ouratea	Leaves	6,6″-bigenkwanin	10 μg ml^{-1}	Gonçalez et al. (2001)
	Ouratea spectabilis	Ouratea	Leaves	6,6″-tetradimethoxybigenkwanin	10 μg ml^{-1}	Gonçalez et al. (2001)
	Ouratea multiflora	Ouratea	Leaves	Amentoflavone	10 μg ml^{-1}	Gonçalez et al. (2001)
	Ouratea parviflora	Ouratea	Leaves	7,7″-dimethoxyagastisflavone	10 μg ml^{-1}	Gonçalez et al. (2001)
Oleaceae	*Olea europaea* L.	Olive	Cake	Vanillic acid	0.2 mg ml^{-1}	Aziz et al. (1998)
	Olea europaea L.	Olive	Cake	Caffeic acid	0.2 mg ml^{-1}	Aziz et al. (1998)
	Olea europaea L.	Olive	Cake	*p*-coumaric acid	0.3 mg ml^{-1}	Aziz et al. (1998)
	Olea europaea L.	Olive	Callus	Ethanolic extract	~1%	Paster et al. (1988)
	Olea europaea L.	Olive	Callus	Caffeic acid	10 mM	Paster et al. (1988)
	Olea europaea L.	Olive	Callus	*O*-coumaric acid	10 mM	Paster et al. (1988)
Papaveraceae	*Argemone mexicana*	Mexican poppy	Leaves	Aqueous extract	4.2% (v/v)	Masood and Ranjan (1991)
Piperaceae	*Piper nigrum* L.	Black pepper	Fruit	Essential oil[a]	>100 ppm	Madhyastha and Bhat (1984)
	Piper longum	Long pepper	Fruit	Piperoctadecalidine[a]	<20 mM	Lee et al. (2002)
	Piper longum	Long pepper	Fruit	Piperine[a]	<25 mM	Lee et al. (2002)
	Piper longum	Long pepper	Fruit	Piperlongumine[a]	<6 mM	Lee et al. (2002)
	Piper longum	Long pepper	Fruit	Pipernonaline[a]	<2 mM	Lee et al. (2002)
	Piper nigrum L.	Black pepper	Seed	Essential oil[a]	1–10%	Tantaoui-Elaraki and Beraoud (1994)
					10%	Mabrouk and El-Shayeb (1980)
Poaceae	*Zea mays* L.	Maize	Seed	Eugenol	<1 μg g^{-1}	Bilgrami et al. (1992)
	Zea mays L.	Maize	Seed	Hexanal[a]	0.25 μl ml^{-1}	Wright et al. (2000)
	Zea mays L.	Maize	Seed	*n*-decyl aldehyde[a]	0.25 μl ml^{-1}	Wright et al. (2000)
	Zea mays L.	Maize	Silk	Volatile compounds[a]	NA	Zeringue (2000)
	Zea mays L.	Maize	Seed	4-acetyl-benzoxazolin-2-one[a]	~0.1 mM	Miller et al. (1996)
	Zea mays L	Maize	Seed	6-methoxy-benzoxazolin-2-one[a]	~0.1 mM	Miller et al. (1996)
	Zea mays L.	Maize	Seed	Nonyl aldehyde	NA	Zeringue et al. (1996)

(*continued*)

Table 19.1 (continued)

Family	Species	Common name	Part used	Effective component (s)	Inhibitory concentration	Reference
	Cymbopogon sp.	Lemon grass	Leaves	Essential oil[a]	1–5%	Thanaboripat et al. (2004)
	Zea mays L.	Maize	Seed	Trans-cinnamic acid	30 mM	Nesci et al. (2007)
	Zea mays L.	Maize	Seed	Ferulic acid	30 mM	Nesci et al. (2007)
	Cymbopogon citratus L.	Lemon grass	Aerial parts	Essential oil[a]	0.1 mg ml^{-1}	Paranagama et al. (2003)
Rosaceae	*Prunus dulcis*	Almond	kernel	Vanillic acid	12 mM	Molyneux et al. (2007)
	Prunus dulcis	Almond	kernel	4-hydroxybenzoic acid	12 mM	Molyneux et al. (2007)
	Prunus dulcis	Almond	kernel	Protocatechuic acid	12 mM	Molyneux et al. (2007)
	Prunus dulcis	Almond	Seed	Catechin	12 mM	Molyneux et al. (2007)
Rutaceae	*Citrus sinensis* L.	Orange	Peel	Powder	5–10% (w/w)	Krishnamurthy and Shashikala (2006)
	Citrus medica L.	Citron	Peel	Powder	5–10% (w/w)	Krishnamurthy and Shashikala (2006)
	Citrus sinensis L.	Orange	Fruit	6,7-dimethoxycoumarin	ND	Mohanlall and Odhav (2006)
	Citrus sinensis L.	Orange	Peel	Essential oil[a]	3,000 ppm	Alderman and Marth (1976)
	Citrus x-limon L.	Lemon	Peel	Essential oil[a]	2,000 ppm	Alderman and Marth (1976)
Solanaceae	*Solanum nigrum* L.	Black nightshade	Leaves	Aqueous extract	4.2% (v/v)	Masood and Ranjan (1991)
	Capsicum annum	Chilli pepper (red chilli)	Pods	Capsanthin[a] Capsaicin[a]	0.2–1 mg ml^{-1} 1 mg ml^{-1}	Masood et al. (1994)
	Withania somnifera	Ashwagandha	Leaves	Powder	5–10% (w/w)	Krishnamurthy and Shashikala (2006)
Zingiberaceae	*Zingiber officinale*	Ginger	Root	Essential oil[a]	≥5%	Mabrouk and El-Shayeb (1980)
	Zingiber officinale	Ginger	Root	Extract[a]	2% (v/v)	Bahk and Marth (1983)
	Aldehydes			Cinnamaldehyde[a]	150 ppm	Bullerman et al. (1977)

Miscellaneous (Pure compounds, different families)			200 ppm	Mahmoud (1994)
		Syringaldehyde	2–4 mM	Hua et al. (1999)
	Phenolics	Acetosyringone	2–4 mM	Hua et al. (1999)
				Hua (2001)
		Curcumin[a]	~1%	Tantaoui-Elaraki and Beraoud (1994)
		Eugenol[a]	125 ppm	Bullerman et al. (1977)
			0.75 mM	Jayashree and Subramanyam (1999)
		Vanillylacetone	<5 mM	Kim et al. (2004 a, b)
	Alkaloids	Piperine[a]	≥100 ppm	Lee et al. (2002)
	Phenylpropanoids	Sinapinic acid	4 mM	Hua et al. (1999)
	Terpenoids	Canthaxanthin	0.43 mM	Norton (1997)
		α-carotene	~6 μM	Norton (1997)
		β-carotene	0.12 mM	Norton (1997)
		β-cryptoxanthin	0.01 mM	Norton (1997)
		α-lonone	0.4 μM	Norton (1997)
		β-lonone	~37 μM	Norton (1997)
		Lutein	1.1 μM	Norton (1997)
		Lycopene	0.24 mM	Norton (1997)
		Zeaxanthin	0.06 mM	Norton (1997)
		Geraniol[a]	500 ppm	Mahmoud (1994)
		Citronellol[a]	500 ppm	Mahmoud (1994)
		Nerol[a]	500 ppm	Mahmoud (1994)
	Flavonoids	Cyanidin	~1.7 mM	Norton 1999
		Delphinidin	~0.5 mM	Norton (1999)
		Kaempferol	4 mM	Norton (1999)
		Luteolin	6 mM	Norton (1999)
		Malvidin	2 mM	Norton (1999)
		Pelargonidin	1 mM	Norton (1999)
		Peonidin	2 mM	Norton (1999)

[a]AF inhibitory components with growth inhibitory activity toward aflatoxigenic fungi tested

We describe here the most important plant-derived AF inhibitory compounds with special reference to their mode of action.

19.4.1 Phenolics from Pistacia vera L

From the work of Molyneux et al. (2007), it has been shown that some constituents of *Pistacia vera* L. (pistachio) kernel including caffeic acid, 3,4-digalloyl quinic acid, quinic acid, and chlorogenic acid inhibit AF production in *A. flavus* by ~90% at a concentration of 12 mM. The authors suggested that these phenolic compounds may suppress AF biosynthesis by relieving the stimulatory effects of oxidative stress on the fungus. So, producing genetically engineered pistachios with high levels of such compounds could limit nut contamination to acceptable levels.

19.4.2 Flavonoids and Coumarins from Ammi visnaga L.

Mabrouk and El-Shayeb (1992) reported that some flavonoids (khellin and visnagin) and coumarins (xanthotoxin, bergapten, and psoralene) isolated from the seeds of *Ammi visnaga* L. inhibited AF production in *A. flavus* by an IC_{50} about 0.1 mM. Since the role of cAMP in AF production has now been established (Roze et al. 2004), AF suppression by the above flavonoids may be attributed to their inhibitory effects on cAMP-phosphodiesterase as confirmed by Duarte et al. (1998). For the coumarins tested, the authors claimed that they may competitively inhibit the enzymes involved in AF biosynthetic pathway as a result of structural similarities with AF.

19.4.3 Phenylpropanoids from Anethum graveolens L. and Petroselinum crispum (Mill.)

Razzaghi-Abyaneh et al. (2007) reported the isolation of a phenylpropanoid compound named dillapiol from leaf EO of *A. graveolens* L. (dill) as specific inhibitor of AFG_1 production by *A. parasiticus* without any obvious effect on fungal growth and AFB_1 synthesis. The compound inhibited AFG_1 with an IC_{50} equal to 0.15 μM. Two other phenylpropanoids i.e., apiol and myristicin, isolated from seed EO of *Petroselinum crispum* (Parsley), showed inhibitory effects on AFG_1 production with IC_{50} values of 0.24 and 3.55 μM respectively (Razzaghi-Abyaneh et al. 2007). The authors proposed that these phenylpropanoids may inhibit AFG_1 biosynthesis via inhibition of cypA, a cytochrome P-450-dependent monooxygenase involved in conversion of *O*-methylsterigmatocystin (OMST) to AFG series in AF biosynthetic pathway (for more details, see Sect. 20.5).

19.4.4 Phenolics and Flavonoids from *Arachis hypogaea* L.

Various compounds isolated from *Arachis hypogaea* (peanut) seed and shell including ferulic acid (0.5–1 mM), vanillic acid (~1 mM), p-coumaric acid (1 mM), eriodictyol (~0.05 mM), and 5,7-dihydroxychromone (<100 μM) were shown to efficiently inhibit AF production by *A. flavus* and *A. parasiticus* in synthetic sucrose-yeast extract and Adye and Mateles (A&M) culture media (Chipley and Uraih 1980; Delucca et al. 1987; Fajardo et al. 1995). All compounds inhibited fungal growth weakly at concentrations used. Since ferulic acid is an antioxidant in the sense that it is reactive toward free radicals such as reactive oxygen species, it may inhibit AF production by suppressing oxidative stress response to toxigenic fungus. The presence of frolic acid in plant cell walls as well as in the seeds of some plants including rice, wheat, and oats makes it a promising target for enhancing through genetic engineering in susceptible crops, with the aim of creating more resistant varieties to AF contamination on susceptible crops.

19.4.5 Phenolics from *Satureja hortensis* L.

The phenolic compounds, thymol and carvacrol, were successfully isolated from leaf EO of *Satureja hortensis* L. (Razzaghi-Abyaneh et al. 2008). Microbioassay on cell culture microplates contained potato-dextrose broth medium and subsequent analysis of cultures with HPLC technique revealed that both carvacrol and thymol were able to effectively inhibit fungal growth as well as AFB_1 and AFG_1 production in a dose-dependent manner at all two-fold concentrations from 0.041 to 1.32 mM. The IC_{50} values for growth inhibition were calculated as 0.79 and 0.86 mM for carvacrol and thymol, while for AFB_1 and AFG_1, it was reported as 0.50 and 0.06 mM for carvacrol and 0.69 and 0.55 mM for thymol. The same compounds isolated from the leaf EO of another plant named *Thymus vulgaris* showed AF inhibition by *A. parasiticus* (Razzaghi-Abyaneh et al. 2008). The authors concluded that carvacrol and thymol may be useful for controlling AF contamination of susceptible crops in the field. Since these phenolics have potent antioxidant activity, they may inhibit AF production in *A. parasiticus* through suppression of oxidative stress response to fungus. Another possibility is the inhibition of *nor-1* expression, an important gene in AF biosynthetic pathway, as indicated for a phenolic compound, acetosyringone (Hua 2001).

19.4.6 Alkaloids from *Piper longum* L.

Lee et al. (2002) reported that some alkaloids isolated from *Piper longum* seed, including piperine (<25 mM), piperlongumine (<6 mM), pipernonaline (<2 mM) and piperoctadecalidine (<20 mM), inhibited AF production in *A. flavus* grown for 7 days on potato dextrose agar by 75, 96, 92 and 100% respectively. Since piperine has been reported to inhibit human CYP3A4 enzyme (Bhardwaj et al. 2002), it may suppress AF production by affecting cytochrome P-450 dependent monooxygenases involved in AF biosynthesis pathway by a similar manner.

19.4.7 Phenolics from Prunus dulcis (Mill.)

Molyneux et al. (2007) reported that some phenolic ingredients of *Prunus dulcis* (Almond) kernel including vanillic acid, catechin, 4-hydroxybenzoic acid, and protocatechuic acid were inhibited AF production in *A. flavus* by 85.6, 69.0, 76.4 and 69.3%, respectively at 12 mM concentration. By adding organic peroxide, *tert*-butyl hydroperoxide (*t*-BuOOH), in a concentration of 100 μM to *A. flavus* cultures for inducing oxidative stress, a 34% increase in AF production was observed after incubation for 5 days. So the authors concluded that the phenolics tested may inhibit AF production by suppressing oxidative stress response to the fungus.

19.4.8 Spiroethers from Matricaria recutita L.

From the work of Yoshinari et al. (2008), it was concluded that (E)- and (Z)-spiroethers isolated from *Matricaria recutita* (German chamomile) EO specifically inhibited AFG_1 production by *A. parasiticus* with IC_{50} values of 2.8 and 20.8 μM respectively, without affecting fungal growth. The authors suggested that the spiroethers may inhibit AFG_1 production by a similar mechanism with dillapiol i.e. inhibiting cypA, a cytochrome P-450 monooxygenase essential for AFG group production in AF biosynthesis pathway.

19.5 A Practical Approach to Isolation and Characterization of AF Inhibitors from Medicinal Plants: Dillapiol from *Anethum graveolens* L.

The history of medicinal and herbal knowledge in Iran goes back to 100 years ago when the ancient inhabitants mentioned the names and properties of thousands of healing plants in Avesta, the holy book of the Zoroastrians. It has been estimated that more than 7,500 plant species are grown in Iran out of which about 1,800 species are considered medicinal (Adhami et al. 2007). Among medicinal plants, the members of the Apiaceae family (Umbelliferae) have gained increasing interest because they are composed of different bioactive chemicals such as monoterpens, flavonoids, furanocumarins, phenolic acids and phenylpropanoids (Crowden et al. 1969). *Anethum graveolens* L. (dill, CAS No. 8006–75–5) is a short-lived annual herb of Apiaceae family, cultivated as a native plant in southwest and central Asia including Iran. Different biological activities of *A. graveolens* L. contribute to major constituents of the whole plant including monoterpenes, flavonoids, furanocumarins and phenylpropanoids (Crowden et al. 1969). Recently, we demonstrated a novel biological activity for an Iranian variety of *A. graveolens* L. as pathway-specific inhibition of AFG_1 biosynthesis by *A. parasiricus* (Razzaghi-Abyaneh et al. 2007).

19.5.1 Culture Conditions for Analysis of *A. parasiticus* Growth and AF Production

Anethum graveolens L. was collected from the National Botanical Garden of Iran (NBGI) during June 2006. Essential oil of plant leaves (LEO) was prepared by hydrodistillation of sterilized leaves using a Clevenger-type apparatus.

Difco Potato Dextrose liquid medium was added to 24-well flat-bottom microplates at 5 ml per well and then incubated with spore suspension of *A. parasiticus* NRRL 2,999 (1×10^6 spores per well) prepared in distilled water containing 0.01% Tween 80. Different concentrations of the LEO of *A. graveolens* L. and its silica gel column fractions prepared in MeOH (final concentration of 1%) were added to the test wells. For purified dillapiol, serial twofold concentrations of 0.031–64 μM prepared in MeOH were treated in the same manner except that 6-well flat-bottom microplates were used. Microplates were incubated statically for 3 days at 28°C.

The culture broth was separated from the fungal mat by filtration. After thoroughly washing with distilled water, the mycelia were transferred into 1.8 ml microtubes and dried at 80°C until a constant weight was obtained. Mycelial weight was calculated by subtracting the weight of a 1.8-ml microtube without the mycelia from the total weight. The amounts of AFB_1 and AFG_1, which were the major detectable AF in the culture filtrates under test conditions, were analyzed by HPLC [column: Cosmosil 5Ph-AR-300 Waters; 4.6 mm × 150 mm; an isocratic elution of water/acetonitrile/methanol (60: 25: 15, v/v/v); flow rate: 1.0 ml min^{-1}; detection: 365 nm; the retention times of AFB_1 and AFG_1 were 5.0 and 6.3 min, respectively].

19.5.2 Isolation of Dillapiol from Essential Oil of *A. graveolens* L.

A glass column (50 × 2.5 cm) was packed with silica gel (Wakogel; 75–150 μm particle size) using *n*-hexane. One gram of LEO of *A. graveolens* L. was directly loaded on top of the gel and eluted stepwise with *n*-hexane (500 ml), *n*-hexane/ethyl acetate (in three steps of 95:5, 90:10 and 80:20, v/v, each of 500 ml) and ethyl acetate (500 ml), successively. The flow rate was 5 ml min^{-1}, and 5 fractions were collected in amounts of 500 ml each (Table 19.2). All fractions were condensed by an IWAKI REN-1 series rotary evaporator.

Table 19.2 General characteristics of silica gel column fractions of leaves essential oil (LEO) prepared from *Anethum graveolens* L

Fraction number	Elution solvent[a]	Yield (mg g^{-1})	AFB_1[b]	AFG_1[b]
1	*n*-hexane	576.0	–	–
2	*n*-hexane/ethyl acetate (95:5, v/v)	232.3	–	+
3	*n*-hexane/ethyl acetate (90:10, v/v)	29.1	–	–
4	*n*-hexane/ethyl acetate (80:20, v/v)	32.5	–	–
5	Ethyl acetate	32.5	–	–

[a]The column was washed in each step with 500 ml of each solvent
[b]The results are from visual comparison of equal amounts of treated samples and non-treated controls on TLC plates (+ inhibited; – not inhibited)

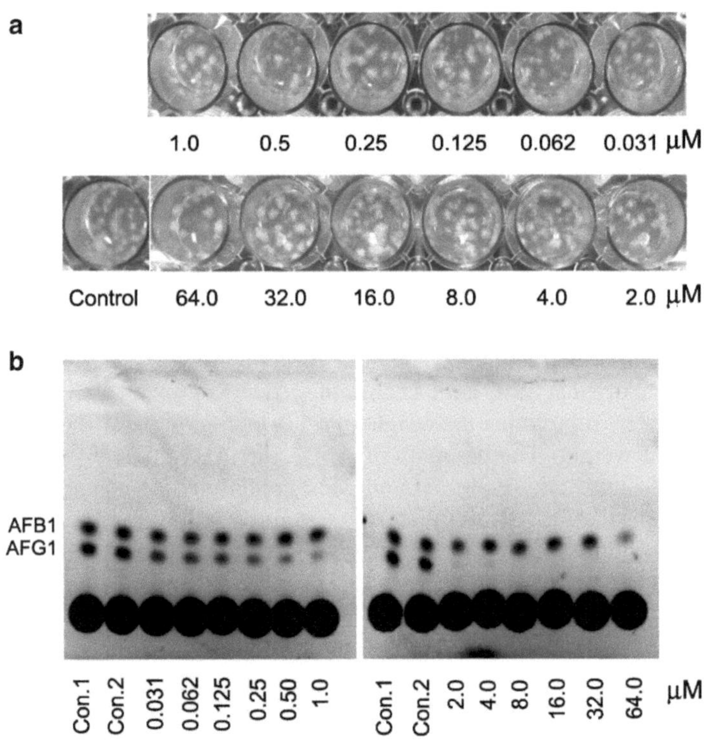

Fig. 19.1 Effect of dillapiol on *A. parasiticus* growth and aflatoxin production. **a** Microbioassay results of fungal growth in 24-well flat-bottom microplates in treated and non-treated (control) fungus show no inhibition of fungal growth at all concentrations used. **b** TLC analysis of cultures indicates dose-dependent inhibition of aflatoxin G_1 production without any effect on aflatoxin B_1 synthesis

The fraction of *n*-hexane/ethyl acetate (95:5, v/v) was purified by HPLC (column:CapCell-Pak C_{18}, 10 mm × 250 mm, UG 120 A°, 5 μm; an isocratic elution of 60% acetonitrile in water; flow rate: 3 ml min^{-1}; detection: 254 nm) to obtain dillapiol (retention time: 12.5 min; yield: 45 mg). Dillapiol ($C_{12}H_{14}O_4$; 222.23 g mol^{-1}): FAB-MS (3-nitrobenzyl alcohol matrix) m/z 223 $(M + H)^+$; δ_H ($CDCl_3$, 500 MHz): 6.34 (1H, s, H-6), 5.90 (1H, m, H-2′), 5.87 (2H, s, O-CH_2-O), 5.05–5.01 (2H, m, H-3′), 4.00 (3H, s, 3-OCH_3), 3.74 (3H, s, 2-OCH_3), 3.29 (2H, dt, $J = 6.5$ Hz, 1.5 Hz, H-1′); δc ($CDCl_3$, 500 MHz): 144.6 (C4), 144.3 (C2), 137.6 (C3), 137.4 (C2′), 135.9 (C5), 126.0 (C1), 115.5 (C3′), 102.7 (C6), 101.1 (O-CH_2-O), 61.3 (2-OCH_3), 59.9 (3-OCH_3), 33.9 (C1′).

As illustrated in Fig. 19.1a, dillapiol did not affect the fungal growth in micro-bioassay at all concentrations used. The qualitative TLC results of dillapiol-treated cultures showed that this compound was able to strongly inhibit AFG_1 production in a dose-dependent manner without affecting AFB_1 synthesis (Fig. 19.1b). Based on

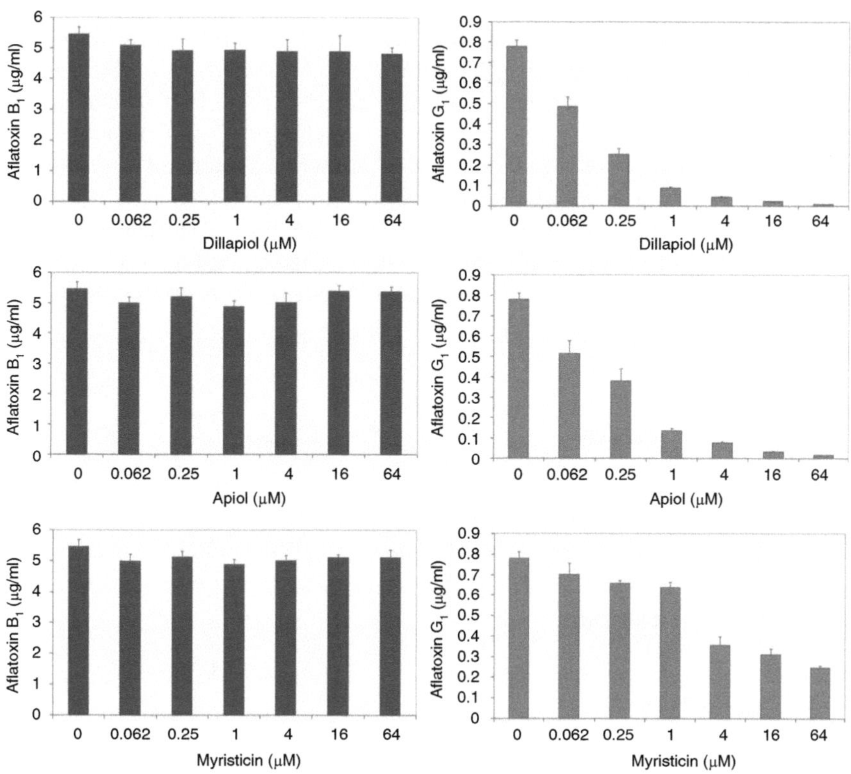

Fig. 19.2 HPLC results of the effect of dillapiol, parsley apiol and myristicin on aflatoxins B_1 and G_1 production by *A. parasiticus* show dose-dependent inhibition only for aflatoxin G_1. All compounds were not able to inhibit aflatoxin B_1 production. Data are the mean ± SEM obtained from two separate experiments in two triplicate sets each. *Error bars* indicate SEM. *Asterisks* show statistically significant differences with a control (ANOVA, $P < 0.05$)

the HPLC quantization, AFG_1 inhibition was reported between 32.67% and 98.95% at dillapiol concentrations of 0.031–64 μM after comparing with untreated controls (Fig. 19.2). The IC_{50} value of dillapiol for AFG_1 inhibition was measured as 0.146 μM based on the logarithmic calculations of the quantitative data.

19.5.3 Mode of Action of Dillapiol

A plausible mode of action of dillapiol for specific inhibition of AFG_1 is shown in Fig. 19.3. In recent years, researchers have clarified more about the role of enzymes catalyze oxidative reactions such as cytochrome P-450 monooxygenases in AF biosynthetic pathway, especially in the final steps where the separation of AF B and G series occurs (Yu et al. 2000). Yu et al. (1998) showed that one of these monooxygenases, OrdA, is responsible for conversion of the precursor, OMST to AFB_1 and AFB_2. Complementary results of Udwary et al. (2002) showed that OrdA was responsible for two oxidation reactions on OMST that resulted in the formation of two intermediates, 11-hydroxyOMST and an open chain butenyl carboxylic acid. The recent work of Ehrlich et al. (2004) clarified that an early gene in AF gene cluster, *cypA*, encodes a cytochrome P-450 monooxygenase required for formation of AF G-group. An attractive feature of the biological activities of dillapiol is its ability to inhibit aldrin epoxidase as a cytochrome P-450 monooxygenase enzyme. Based on the data from MacRae and Towers (1984) on the mode of action of natural monooxygenase inhibitors, the inhibitory effect of dillapiol on AFG_1 production could be attributed to formation of stable

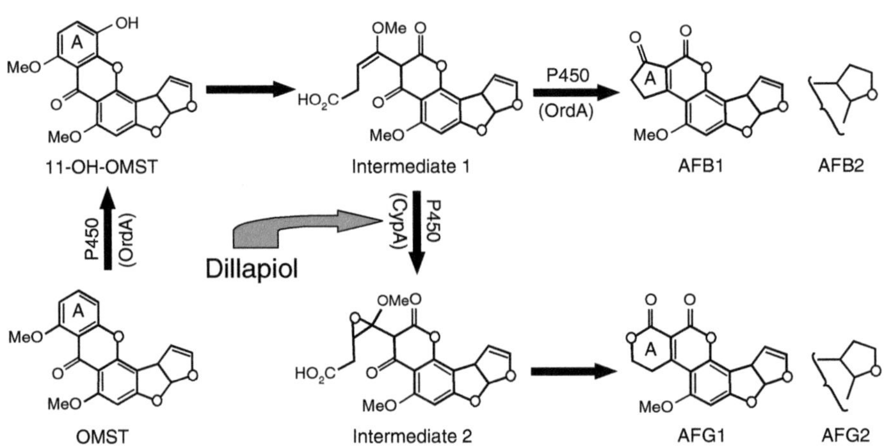

Fig. 19.3 Plausible scheme of inhibition of aflatoxin G_1 biosynthesis by dillapiol. It is proposed that dillapiol specifically interacts with *cypA* gene encoded, a cytochrome P-450 monooxygenase enzyme (cypA) responsible for two-step converting of an onen-ring intermediate (No. 1) to aflatoxin G_1

adducts with P-450 monooxygenases through binding of its methylenedioxyphenyl group to the hemi moiety of the enzyme. With respect to lower inhibitory activity of myristicin, with one methoxy group compared with dillapiol and apiol which have two methoxy groups, it seems that the presence and position of a methoxy group on the benzene ring of these compounds are important factors affecting their inhibitory activity toward AFG_1.

Taken together, study of the effect of dillapiol on genes with unknown function in AF biosynthesis that are responsible for production of oxidoreductases such as *ordB*, *norA,* and *norB* may be useful for further clarification of the conclusive pattern of AF B and G series biosynthesis.

19.6 Conclusion and Future Perspectives

Within hundreds of compounds that influence AF biosynthesis, natural inhibitors originated from plants have received major considerations with regard to their safety for use in biological systems. The plant EOs have been considered as rich sources of inhibitory compounds affecting toxigenic fungus growth and/or its AF productivity. So, many researchers are involved in screening a large number of medicinal plants to find novel AF inhibitors. Since plant preparations including oils and extracts are composed of a wide range of structurally different compounds with diverse biological activities, it is necessary to isolate their effective constituents affecting AF production. In recent years, several plant-originated compounds such as alkaloids, terpenoids, flavonoids and phenylpropanoids, etc., have been successfully characterized as AF-production inhibitors in vitro (Holmes et al. 2008). The most important concerns with regard to these inhibitors are the lack of information about their effectiveness in pre- or post-harvest conditions and their unknown or poorly understood mechanisms of action. Although many inhibitors of AF biosynthesis are not examined on stored seeds, they may be useful for studying the complex regulatory pathway of AF formation. Good examples of such inhibitors are dillapiol from an aromatic plant, e.g., *Anethum graveolens* L. (Razzaghi-Abyaneh et al. 2007), and spiroethers from *Matricaria recutita* L. (Yoshinari et al. 2008). It has been shown that dilapiol and spiroethers, potent inhibitors of AFG_1 production, exert their effect probably through interaction with cypA, a cytochrome P-450-dependent monooxygenase enzyme, involved in the later stages of AF biosynthesis in *A. parasiticus*. The availability of *A. flavus* genome sequence data will provide us with a unique opportunity to elucidate the targets of AF production inhibitors in order to understand about AF biosynthesis regulatory mechanisms and host-aflatoxigenic fungus interactions during plant infection.

Acknowledgements The authors wish to thank Prof. Hiromichi Nagasawa, the head of Applied Biological Chemistry Department, Graduate School of Agricultural and Life Sciences, Tokyo University, Japan for kind providing laboratory facilities for research on medicinal plants.

References

Adhami HR, Farsam H, Mesgarpour B (2007) Herbal medicine in Iran. Herbal Gram 74:34–43

Alderman GG, Marth EH (1976) Inhibition of growth and aflatoxin production of *Aspergillus parasiticus* by *Citrus* oils. Z Lebensm Unters Forsch 160:353–358

Allameh A, Razzaghi-Abyaneh M (2001) Mycotoxins. Emam Hossein University Publication, Tehran, Iran (678 pages in Persian)

Allameh A, Razzaghi-Abyaneh M, Shams M, Rezaee MB, Jaimand K (2001) Effects of neem leaf extract on production of aflatoxins and activities of fatty acid synthetase, isocitrate dehydrogenase and glutathione S-transferase in *Aspergillus parasiticus*. Mycopathologia 154:79–84

Atanda OO, Akpan I, Oluwafemi F (2007) The potential of some spice essential oils in the Control of *A. parasiticus* CFR 223 and aflatoxin production. Food Control 18:601–607

Awuah RT (1996) Possible utilization of plant products in grain storage. In: Cardwell KF (ed) Proceedings of the workshop on mycotoxins in food in Africa. International Institute of Tropical Agriculture, Cotonou, Benin, 6–10 November 1995, p 32

Aziz NH, Farag SE, Mousa LAA, Abo-Zaid MA (1998) Comparative antibacterial and antifungal effects of some phenolic compounds. Microbios 93:43–54

Bhardwaj RK, Glaeser H, Becquemont L, Klotz U, Gupta SK, Fromm MF (2002) Piperine, a major constituent of black pepper, inhibits human P-glycoprotein and CYP3A4. J Pharmacol Exp Ther 302:645–650

Bahk J, Marth EH (1983) Aflatoxin production is inhibited by selected herbal drugs. Mycopathologia 83:129–134

Bakkali F, Averbeck S, Averbeck D, Idaomar M (2008) Biological effects of essential oils — a review. Food Chem Toxicol 46:446–475

Balunas MJ, Kinghorn AD (2005) Drug discovery from medicinal plants. Life Sci 78:431–441

Bennett JW, Klich M (2003) Mycotoxins. Clin Microbiol Rev 16:497–516

Bhatnagar D, McCormick SP (1988) The inhibitory effect of neem (*Azadirachta indica*) leaf extracts on aflatoxin synthesis in *Aspergillus parasiticus*. J Am Oil Chem Soc 65:1166–1168

Bilgrami KS, Sinha KK, Sinha AK (1992) Inhibition of aflatoxin production and growth of *Aspergillus flavus* by eugenol, onion and garlic extracts. Indian J Med Res 96:171–175

Bluma R, Etcheverry M (2008) Application of essential oils in maize grain: impact on *Aspergillus* Section *Flavi* growth parameters and aflatoxin accumulation. Food Microbiol 25:324–334

Bullerman LB, Lieu FY, Seier SA (1977) Inhibition of growth and aflatoxin production by cinnamon and clove oils, cinnamic aldehyde and eugenol. J Food Sci 42:637–646

Cary JW, Ehrlich KC (2006) Aflatoxigenicity in *Aspergillus*: molecular genetics, phylogenetic relationships and evolutionary implications. Mycopathologia 162:167–177

Chipley JR, Uraih N (1980) Inhibition of *Aspergillus* growth and aflatoxin release by derivatives of benzoic acid. Appl Environ Microbiol 40:352–357

Crowden RK, Harborne JB, Heywood VH (1969) Chemosystematics of the Umbelliferae — a general survey. Phytochemistry 8:1963–1984

Delucca A, Palmgren MS, Daigle DJ (1987) Depression of aflatoxin production by flavonoid-type compounds from peanut shells. Phytopathology 77:1560–1563

Duarte J, Lugnier C, Torres AI, Perez-Vizcaino F, Zarzuelo A, Tamargo J (1998) Effects of visnagin on cyclic nucleotide phosphodiesterases and their role in its inhibitory effects on vascular smooth muscle contraction. Gen Pharmacol 32:71–74

Ehrlich KC, Chang P-K, Yu J, Cotty PJ (2004) Aflatoxin biosynthesis cluster gene *cyp*A is required for G aflatoxin formation. Appl Environ Microbiol 70:6518–6524

Fajardo JE, Waniska RD, Cuero RG, Pettit RE (1995) Phenolic compounds in peanut seeds-enhanced elicitation by chitosan and effects on growth and aflatoxin B_1 production by *Aspergillus flavus*. Food Biotechnol 9:59–78

Fakhoury AM, Woloshuk CP (2001) Inhibition of growth of *Aspergillus flavus* and fungal α-amylases by a lectin-like protein from *Lablab purpureus*. Mol Plant Microbe Interact 14:955–961

Fan JJ, Chen JH (1999) Inhibition of aflatoxin-producing fungi by Welsh onion extracts. J Food Prot 62:414–417

Frisvad JC, Skouboe P, Samson RA (2007) Taxonomic comparison of three different groups of aflatoxin producers and a new efficient producer of aflatoxin B_1, sterigmatocystin and 3-O-methylsterigmatocystin, *Aspergillus rambellii* sp. nov. Sys Appl Microbiol 28:442–453

Gonçalez E, Felicio JD, Pinto MM (2001) Biflavonoids inhibit the production of aflatoxin by *Aspergillus flavus*. Braz J Med Biol Res 34:1453–1456

Gonçalez E, Felício JD, Pinto MM, Rossi MH, Medina C, Fernández MJB, Simoni IC (2003) Inhibition of aflatoxin production by *Polymnia sonchifolia* and its in vitro cytotoxicity. Arq Inst Biol 70:139–143

Greene-McDowelle DM, Ingber B, Wright MS, Zeringue HJ, Bhatnagar D, Cleveland TE (1999) The effects of selected cotton-leaf volatiles on growth, development and aflatoxin production of *Aspergillus parasiticus*. Toxicon 37:883–893

Gupta SK, Venkitasubramanian TA (1975) Production of aflatoxin on soybeans. Appl Environ Microbiol 29:834–836

Holmes RA, Boston RS, Payne GA (2008) Diverse inhibitors of aflatoxin biosynthesis. Appl Microbiol Biotechnol 78:559–572

Hua SS (2001) Inhibitory effect of acetosyringone on two aflatoxin biosynthetic genes. Lett Appl Microbiol 32:278–281

Hua SS, Grosjean OK, Baker JL (1999) Inhibition of aflatoxin biosynthesis by phenolic compounds. Lett Appl Microbiol 29:289–291

Jayashree T, Subramanyam C (1999) Antiaflatoxigenic activity of eugenol is due to inhibition of lipid peroxidation. Lett Appl Microbiol 28:179–183

Joseph GS, Jayaprakasha GK, Selvi AT, Jena BS, Sakariah KK (2005) Antiaflatoxigenic and antioxidant activities of *Garcinia* extracts. Int J Food Microbiol 101:153–160

Juglal S, Govinden R, Odhav B (2002) Spice oils for the control of co-occurring mycotoxin-producing fungi. J Food Prot 65:683–687

Kabak B, Dobson ADW, Var I (2006) Strategies to prevent mycotoxin contamination of food and animal feed: a review. Crit Rev Food Sci Nut 46:593–619

Krishnamurthy YL, Shashikala J (2006) Inhibition of aflatoxin B_1 production of *Aspergillus flavus*, isolated from soybean seeds by certain natural plant products. Lett Appl Microbiol 43:469–474

Kumar R, Mishra AK, Dubey NK, Tripathi YB (2007) Evaluation of *Chenopodium ambrosioides* oil as a potential source of antifungal, antiaflatoxigenic and antioxidant activity. Int J Food Microbiol 115:159–164

Lee SE, Mahoney NE, Campbell BC (2002) Inhibition of aflatoxin B_1 biosynthesis by piperlongumine isolated from *Piper longum* L. J Microbiol Biotechnol 12:679–682

Mabrouk SS, El-Shayeb NM (1980) Inhibition of aflatoxin formation by some spices. Z Lebensm Unters Forsch 171:344–347

Mabrouk SS, El-Shayeb NMA (1992) Inhibition of aflatoxin production in *Aspergillus flavus* by natural coumarins and chromones. World J Microbiol Biotechnol 8:60–62

MacRae WD, Towers GHN (1984) Biological activities of lignans. Phytochemistry 23:1207–1220

Madhyastha MS, Bhat RV (1984) *Aspergillus parasiticus* growth and aflatoxin production on black and white pepper and the inhibitory action of their chemical constituents. Appl Environ Microbiol 48:376–379

Mahmoud AL (1994) Antifungal action and anti-aflatoxigenic properties of some essential oil constituents. Lett Appl Microbiol 19:110–113

Mahmoud AL (1999) Inhibition of growth and aflatoxin biosynthesis of *Aspergillus flavus* by extracts of some Egyptian plants. Lett Appl Microbiol 29:334–336

Mahoney N, Molyneux RJ (2004) Phytochemical inhibition of aflatoxigenicity in *Aspergillus flavus* by constituents of Walnut (*Juglans regia*). J Agric Food Chem 52:1882–1889

Mahoney N, Molyneux RJ, Campbell BC (2000) Regulation of aflatoxin production by naphthoquinones of walnut (*Juglans regia*). J Agric Food Chem 48:4418–4421

Masood A, Dogra JVV, Jha AK (1994) The influence of colouring and pungent agents of red Chilli (*Capsicum annum*) on growth and aflatoxin production by *Aspergillus flavus*. Lett Appl Microbiol 18:184–186

Masood A, Ranjan KS (1991) The effect of aqueous plant extracts on growth and aflatoxin production by *Aspergillus flavus*. Lett Appl Microbiol 13:32–34

Miller JD, Fielder DA, Dowd PF, Norton RA, Collins FW (1996) Isolation of 4-acetylbenzoxazolin-2-one (4-ABOA) and diferuloylputrescine from an extract of *Gibberella* ear rot-resistant corn that blocks mycotoxin biosynthesis, and the insect toxicity of 4-ABOA and related compounds. Biochem Syst Ecol 24:647–658

Mohanlall V, Odhav B (2006) Biocontrol of aflatoxins B_1, B_2, G_1, G_2, and fumonisin B_1 with 6,7-dimethoxycoumarin, a phytoalexin from *Citrus sinensis*. J Food Prot 69:2224–2229

Molyneux RJ, Mahoney N, Kim JH, Campbell BC (2007) Mycotoxins in edible tree nuts. Int J Food Microbiol 119:72–78

Nesci A, Gsponer N, Etcheverry M (2007) Natural maize phenolic acids for control of aflatoxigenic fungi on maize. J Food Sci 72:M180–M185

Norton RA (1997) Effect of carotenoids on aflatoxin B_1 synthesis by *Aspergillus flavus*. Phytopathology 87:814–821

Norton RA (1999) Inhibition of aflatoxin B_1 biosynthesis in *Aspergillus flavus* by anthocyanidins and related flavonoids. J Agric Food Chem 47:1230–1235

Paranagama PA, Abeysekera KH, Abeywickrama K, Nugaliyadde L (2003) Fungicidal and anti-aflatoxigenic effects of the essential oil of *Cymbopogon citratus* (DC.) Stapf. (lemongrass) against *Aspergillus flavus* Link. Isolated from stored rice. Lett Appl Microbiol 37:86–90

Prasad G, Sahay SS, Masood A (1994) Inhibition in aflatoxin biosynthesis by the extract of *Amorphophallus campanulatus* (OL) and calcium oxalate. Lett Appl Microbiol 18:203–205

Payne GA, Brown MP (1998) Genetics and physiology of aflatoxin biosynthesis. Annu Rev Phytopathol 36:329–362

Paster N, Juven BJ, Harshemesh H (1988) Antimicrobial activity and inhibition of aflatoxin B_1 formation by olive plant tissue constituents. J Appl Bacteriol 64:293–297

Pinto MM, Gonçalez E, Rossi MH, Felício JD, Medina CS, Fernández MJB, Simoni IC (2001) Activity of the aqueous extract from *Polymnia sonchifolia* leaves on growth and production of aflatoxin B_1 by *Aspergillus flavus*. Braz J Microbiol 32:127–129

Rasooli I, Fakoor MH, Yadegarinia D, Gachkar L, Allameh A, Rezaei MB (2008) Antimycotoxigenic characteristics of *Rosmarinus officinalis* and *Trachyspermum copticum* L. essential oils. Int J Food Microbiol 122:135–139

Rasooli I, Razzaghi-Abyaneh M (2004) Inhibitory effects of thyme oils on growth and aflatoxin production by *Aspergillus parasiticus*. Food Control 15:479–483

Razzaghi-Abyaneh M, Allameh A, Tiraihi T, Shams-Ghahfarokhi M, Ghorbanian M (2005) Morphological alterations in toxigenic *Aspergillus parasiticus* exposed to neem (*Azadirachta indica*) leaf and seed aqueous extracts. Mycopathologia 159: 565–570

Razzaghi-Abyaneh M, Shams-Ghahfarokhi M, Eslamifar A, Schmidt OJ, Gharebaghi R, Karimian M, Naseri A, Sheikhi M (2006a) Inhibitory effects of Akacid (plus) on growth and aflatoxin production by *Aspergillus parasiticus*. Mycopathologia 161:245–249

Razzaghi-Abyaneh M, Shams-Ghahfarokhi M, Allameh A, Kazeroon-Shiri AM, Ranjbar-Bahadori S, Mirzahoseini H, Rezaee MB (2006b) A survey on distribution of *Aspergillus* section *Flavi* in corn field soils in Iran: population patterns based on aflatoxins, cyclopiazonic acid and sclerotia production. Mycopathologia 161:183–192

Razzaghi-Abyaneh M, Shams-Ghahfarokhi M, Yoshinari T, Rezaee MB, Nagasawa H, Sakuda S (2008) Inhibitory effects of *Satureja hortensis* L. essential oil on growth and aflatoxin production by *Aspergillus parasiticus*. Int J Food Microbiol 123:228–233

Razzaghi-Abyaneh M, Yoshinari T, Shams-Ghahfarokhi M, Rezaee MB, Nagasawa H, Sakuda S (2007) Dillapiol and apiol as specific inhibitors of the biosynthesis of aflatoxin G1 in *Aspergillus parasiticus*. Biosci Biotech Biochem 71:2329–2332

Roze LV, Calvo AM, Gunterus A, Beaudry R, Kall M, Linz JE (2004) Ethylene modulates development and toxin biosynthesis in *Aspergillus* possibly via an ethylene sensor-mediated signaling pathway. J Food Prot 67:438–447

Sakuda S, Nakamura K, Akiyama T, Takahashi Y, Muraoka Y, Kurata I (2006) Aflatoxin production inhibitor, and method for control of aflatoxin poisoning using the same. PCT Patent WO 2006137297 A1 20061228

Samuelsson G (2004) Drugs of natural origin: a textbook of pharmacognosy, 5th Edition. Swedish Pharmaceutical Press, Stockholm

Sánchez E, Heredia N, García S (2005) Inhibition of growth and mycotoxin production of *Aspergillus flavus* and *Aspergillus parasiticus* by extract of agave species. Int J Food Microbiol 98:271–279

Selvi A, Joseph GS, Jayaprakasha GK (2003) Inhibition of growth and aflatoxin production in *Aspergillus flavus* by *Garcinia indica* extract and its antioxidant activity. Food Microbiol 20:455–460

Tantaoui-Elaraki A, Beraoud L (1994) Inhibition of growth and aflatoxin production in *Aspergillus parasiticus* by essential oils of selected plant materials. J Environ Pathol Toxicol Oncol 13:67–72

Thanaboripat D, Mongkontanawur N, Suvathi Y, Ruangrattanamatee V (2004) Inhibition of aflatoxin production and growth of *Aspergillus flavus* by citronella oil. KMITL Sci J 4:1–8

Udwary DW, Casillas LK, Townsend CA (2002) Synthesis of 11-hydroxyl *O*-methylsterigmatocystin and the role of a cytochrome P-450 in the final step of aflatoxin biosynthesis. J Am Chem Soc 124:5294–5303

Wright MS, Greene-McDowelle DM, Zeringue HJ, Bhatnagar D, Cleveland TE (2000) Effects of volatile aldehydes from *Aspergillus* resistant varieties of corn on *Aspergillus parasiticus* growth and aflatoxin biosynthesis. Toxicon 38:1215–1223

Yoshinari T, Yaguchi A, Takahashi-Ando N, Kimura M, Takahashi H, Nakajima T, Sugita-Konichi Y, Nagasawa H, Sakuda S (2008) Spiroethers of German chamomile inhibit production of aflatoxin G and trichothecene mycotoxin by inhibiting cytochrome P450 monooxygenases involved in their biosynthesis. FEMS Microbiol Lett 284:184–190

Yu J, Chang P-K, Bhatnagar D, Cleveland TE (2000) Genes encoding cytochrome P-450 and monooxygenase enzymes define one end of the aflatoxin pathway gene cluster in *Aspergillus parasiticus*. Appl Microbiol Biotechnol 53:583–590

Yu J, Chang P-K, Ehrlich KC, Cary JW, Montalbano B, Dyer JM, Bhatnagar D, Cleveland TE (1998) Characterization of the critical amino acids of an *Aspergillus parasiticus* cytochrome P-450 monooxygenase encoded by ordA that is involved in the biosynthesis of aflatoxins B_1, G_1, B_2, and G_2. Appl Environ Microbiol 64:4834–4841

Yu J, Chang P-K, Ehrlich KC, Cary JW, Bhatnagar D, Cleveland TE, Payne GA, Linz JE, Woloshuk CP, Bennett JW (2004) The clustered pathway genes in aflatoxin biosynthesis. Appl Environ Microbiol 70:1253–1262

Yu J, Cleveland TE, Nierman WC, Bennett JW (2005) *Aspergillus flavus* genomics: gateway to human and animal health, food safety, and crop resistance to diseases. Rev Iberoam Micol 22:194–202

Zaika LL, Buchanan RL (1987) Review of compounds affecting the biosynthesis or bioregulation of aflatoxins. J Food Prot 50:691–708

Zeringue HJ (2000) Identification and effects of maize silk volatiles on cultures of *Aspergillus flavus*. J Agric Food Chem 48:921–925

Zeringue HJ (2002) Effects of methyl jasmonate on phytoalexin production and aflatoxin control in the developing cotton boll. Biochem Sys Ecol 30:497–503

Zeringue HJ, Brown RL, Neucere JN, Cleveland TE (1996) Relationships between C-6-C-12 alkanal and alkenal volatile contents and resistance of maize genotypes to *Aspergillus flavus* and aflatoxin production. J Agric Food Chem 44:403–407

Zeringue HJ, McCormick SP (1990) Aflatoxin production in cultures of *Aspergillus flavus* incubated in atmospheres containing selected cotton leaf-derived volatiles. Toxicon 28:445–448

Zeringue HJ, Shih BY, Bhatnagar D (2001) Effects of clarified neem oil on growth and aflatoxin B_1 formation in submerged and plate cultures of aflatoxigenic *Aspergillus* spp. Phytoparasitica 29:361–366

Chapter 20
The Myco-Agents of Bioterrorism

Nathan P. Charlton and Christopher P. Holstege

20.1 Introduction

Fungi are a kingdom of eukaryotic organisms that are ubiquitous in nature. They are present throughout the world's ecosystems, from the arid desert to the ocean floor and the Arctic Circle (Golubic et al. 2005; Qaher 2006; Pietikainen et al. 2007). They frequently cause disease in humans. When growth occurs directly on humans this is referred to as a mycoses, however, when human pathogenesis is the result of a fungal metabolite (toxin) this is called a *mycotoxicoses* (Bennett and Klich 2003). Generally, fungi come in three forms: yeasts, mushrooms and molds. There are many examples of human pathology caused by toxins from each of these three groups, ranging from ethanol intoxication to hepatic failure from amatoxin. Scientific interest in toxicity from fungal metabolites (mycotoxins) has grown recently as a result of their potential use as bioweapons. The fungi investigated for use as potential bioweapons are primarily molds, as these species have toxins that are easy to produce, have the ability to cause human disease, and are easy to disperse. A few of these toxins display superior military weapons potential as they cause acute toxicity and debilitation; others may be more useful as terrorist agents, as their ability to cause chronic toxicity may result in more acute fear than actual pathology.

20.2 History

More than 400 different mycotoxins exist (Etzel 2002). The toxic effects of some species have been documented since antiquity. Ergot alkaloids produced by *Claviceps purpurea* have resulted in significant morbidity and mortality in

C.P. Holstege (✉)
Department of Emergency Medicine, Division of Medical Toxicology, University of Virginia School of Medicine, P.O. Box 800774, Charlottesville, VA 22908-0774, USA
e-mail: ch2xf@virginia.edu

M. Rai and A. Varma (eds.), *Mycotoxins in Food, Feed and Bioweapons*,

DOI: 10.1007/978-3-642-00725-5_20,

numerous outbreaks documented since the Middle Ages. The most infamous suspected outbreak was associated with the Salem Witch Trials in the United States in 1692 (Etzel 2002). Historians have theorized that the hallucinogenic aspect of ergot alkaloids contributed to the bizarre behavior that led to the accusations of witchcraft. Mycotoxin-contaminated grains have resulted in hepatic and renal diseases in both animals and humans throughout the world (Peraica et al. 1999; Etzel 2002). More recently, mycotoxicosis have occurred under more nefarious circumstances such as the alleged "yellow rain" attacks in Southeast Asia in the late 1970s and early 1980s (Bennett and Klich 2003; Holstege et al. 2007).

20.3 Mycotoxins

Mycotoxins are the byproducts of fungal metabolism, and generally are of little or no use to the organism itself. Many molds produce recognized toxins, including *Aspergillus*, *Claviceps*, *Fusarium*, *Stachybotrys*, *Myrothecium*, *Phomopsis*, *Trichoderma*, and *Trichothecium* species (Wannemacher and Wiener 1997; Etzel 2002; Holstege et al. 2007). Of these toxins, the aflatoxins, ergot alkaloids, ochratoxins, trichothecenes, and vomitoxins are most reported to cause disease in humans; the trichothecene mycotoxins being the most discussed as potential bioweapon agents. These are the agents that will be discussed in further detail.

20.4 Trichothecenes

Trichothecene mycotoxins have been divided into four types depending on structure. However, all possess an epoxide group at the 12,13 position on the central ring that appears to mediate the toxicity of the chemical (Fig. 20.1) (Watson et al. 1984; Thompson and Wannemacher 1986). These toxins are lipid-soluble, nonvolatile and have a molecular weight in the 250–550 Da range (Wannemacher and Wiener 1997). They are heat stable, being resistant to degradation at 100 °C and requiring temperatures of 900°F for 10 min or 500°F for 30 min in order to inactivate the chemical (Wannemacher and Wiener 1997). Trichothecenes inhibit protein

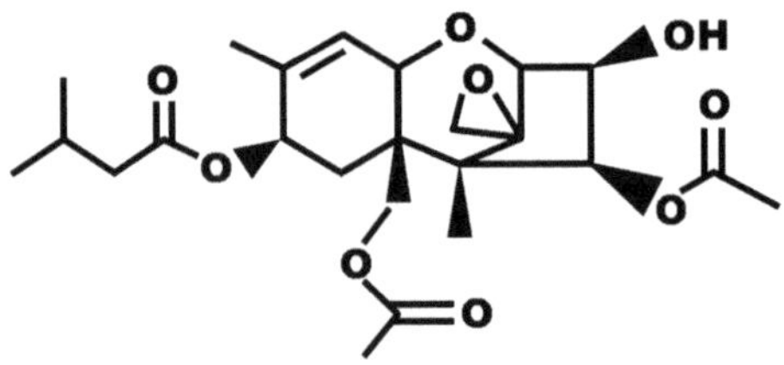

Fig. 20.1 T-2

synthesis in eukaryotic cells by binding to the 60 S subunit on the ribosome, as well as by obstructing peptidyl transferase activity, interrupting the translation process (Thompson and Wannemacher 1990; Wannemacher and Wiener 1997). This in turn affects DNA and RNA synthesis. They also cross cell membranes easily, and do not require metabolic activation in order to exert their effects (Wannemacher and Wiener 1997). The major trichothecenes include deoxynivalenol (DON or vomitoxin) and nivalenol — both of which are in the type B group — and T-2 mycotoxin, which is classified as a type A trichothecene (Bennett and Klich 2003). Following ingestion, deoxynivalenol primarily produces a gastrointestinal illness including vomiting and diarrhea. It is reported to be the most common trichothecene grain contaminant worldwide (Coulombe 1993). Multiple outbreaks from contaminated wheat and corn have occurred in China, India and the United States (Richard 2007).

T-2 mycotoxin is regarded as one of the most potent mycotoxins (Thompson and Wannemacher 1990; Wannemacher and Wiener 1997). Its mechanism of action is primarily as a vesicant agent (Table 20.1). Effects are reported to be about 400 times greater than sulfur mustard (Wannemacher and Wiener 1997). Skin irritation and vesiccation have been reported in nanogram amounts (Ueno 1984; Wannemacher and Wiener 1997). Other reported effects include nausea, vomiting, gastrointestinal hemorrhage, hemoptysis, leucopenia, and lethargy (Ueno 1984; Wannemacher and Wiener 1997; Etzel 2002; Bennett and Klich 2003). As these effects are similar to a radiation syndrome, toxic effects have been described as radiomimetic. Secondary to their lipophilic nature, the toxins are absorbed through the pulmonary tract, intestines, and skin. T-2 is most readily absorbed through inhalation, with inhalational LD50 in rats being 0.05 mg kg^{-1}, as opposed to the intravenous LD50 of 0.7–1.2 mg kg^{-1} (Wannemacher and Wiener 1997).

Trichothecenes are well-established poisons in both humans and animals. Equine disease was reported in Europe in 1931 after exposure to contaminated hay. Case reports document that horses developed stomatitis, lip edema, oral necrosis, rhinitis and conjunctivitis after eating this hay (Parent-Massin 2004; Holstege et al. 2007). This constellation of symptoms was first described as Stachybotrytoxicosis secondary to the finding of *Stachybotrys* species isolated from contaminated

Table 20.1 Characteristics of selected mycotoxins

Toxin	Molecular weight (daltons)	Solubility	Potential routes of exposure	Primary organ systems affected
T2	250–500	Lipid	Inhalation ingestion dermal	Skin gastrointestinal hematopoietic
Aflatoxin	~300	Lipid	Inhalation Ingestion	Hepatic
Ochratoxin	~400	Lipid	Inhalation Ingestion	Renal
Ergot alkaloids	300–600	Both lipid- and water-soluble forms	Ingestion	Neurologic cardiovascular

horse feed. Cases of human Stachybotrytoxicosis are also well described, occurring predominantly in farm workers who handle moldy hay (Bennett and Klich 2003). Another illness, "Akakabi Byo," or red mold disease, has been characterized by nausea, vomiting, diarrhea, anorexia, bone-marrow suppression and hemorrhage in humans (Parent-Massin 2004). Various outbreaks occurred in Japan and Korea between 1946 and 1963 following ingestion of contaminated grain (Parent-Massin 2004). Alimentary toxic aleukia is a syndrome described in Orenburg, Russia between 1942 and 1947, which affected over 10% of the population (Parent-Massin 2004). As the name suggests, intestinal hemorrhage and necrosis, along with leucopenia, was reported in those with this disease (Parent-Massin 2004; Holstege et al. 2007). It was associated with eating over-wintered corn and wheat that was contaminated with *Fusarium* and *Stachybotrys* species (Etzel 2002; Bennett and Klich 2003).

In recent history there is also evidence of intentional use of trichothecene toxins on humans. The most widely discussed are the "yellow rain" attacks which occurred in Southeast Asia in the late 1970s and in Afghanistan in the 1980s. These incidents were originally directed against the Hmong tribes in Laos and Cambodia. Reportedly, a sticky yellow substance was aerosolized by aircraft and bombs, which when falling on trees and dwellings had the sound of rain (Wannemacher and Wiener 1997; Holstege et al. 2007). Frequently encountered symptoms included skin itching and irritation, vesicles, nausea, vomiting, diarrhea, upper airway irritation and bleeding, gastrointestinal hemorrhage and death (Watson et al. 1984; Wannemacher and Wiener 1997). While others discount the idea of the use of trichothecene mycotoxins in these areas (Marshall 1983), a significant amount of evidence recorded in the literature shows that trichothecenes were present in much higher concentrations in the areas of reported "yellow rain" attacks than occurs normally in nature in those areas (Watson et al. 1984; Seiders 1986; Wannemacher and Wiener 1997).

Regardless of the likelihood of past uses of mycotoxins as military or terrorist agents, as a result of their action and chemical properties the potential does exist. Trichothecene mycotoxins, including T-2 mycotoxin, can be easily manufactured in a crystalline powder or liquid form which are highly heat stable (Creasia et al. 1987; Wannemacher and Wiener 1997). It is easily aerosolized by a variety of mechanisms including aircraft, bombs and rockets (Wannemacher and Wiener 1997). The ability to cause morbidity and mortality via parenteral, inhalational and dermal routes makes it a versatile weapon.

Detection of these mycotoxins can be accomplished though a number of different processes. The majority of methods were originally designed for mycotoxin detection in foodstuffs, but use can be transitioned to human practice. Traditional methods for detection include thin layer chromatography and gas chromatography (Maragos 2004; Zheng et al. 2006). High performance liquid chromatography combined with mass spectrometry (HPLC-MS) has a sensitivity for detection at 1 ppb (Wannemacher and Wiener 1997). Thin-layer chromatography-mass spectrometry, however, remains the most sensitive means of detection, reliably identifying picogram quantities of trichothecene toxins (Wannemacher and Wiener 1997).

Newer technologies including lateral flow tests, the enzyme-linked immunosorbent assay (ELISA) and fluorometric assays generally have been found to lack the sensitivity of the more established methods, but are more portable means of detection and are easier to use (Zheng et al. 2006). Unfortunately, these technologies have only been studied on food stores.

Treatment mainly consists of removal from the area or means of contamination. Removal of contaminated clothing is essential after airborne exposure. Washing contaminated skin with soap and water within 4–6 h removed 80–98% of T-2 toxin in study animals and prevented dermal effects (Wannemacher and Wiener 1997). Ocular exposure should be treated with large volume irrigation. Activated charcoal has been shown to bind T-2 toxin *in vitro*, and may be beneficial for decontamination of oral exposure (Wannemacher and Wiener 1997; Paterson 2006; Holstege et al. 2007). Both bentonite and zeolite have been used to decrease the oral toxicity of T-2 in rats (Harvey et al. 1994). High-dose dexamethasone (10 mg kg^{-1}) has been shown to increase survival time in rat models of T-2 toxicity delivered by the intravenous route (Shohami et al. 1987). The mechanism, however, is unknown, and the treatment has not been validated in human patients. Medical management otherwise consists of supportive care.

20.5 Aflatoxins

Aflatoxins are another class of mycotoxins considered for potential use as agents of bioterrorism. The ability for aflatoxins to serve as a true bioweapon is considered low; however, the risk of chronic toxicity makes them a potential terrorist agents. Evidence from US intelligence reports suggests that the Iraqis had manufactured and weaponized aflatoxins prior to the first Gulf War (Stone 2001, 2002; Bennett and Klich 2003). Aflatoxins may cause both acute and chronic disease. These toxins are found in nature as contaminants of nuts, cereals, and oilseeds, with peanuts, cottonseed and corn being the three crops most commonly affected (Pitt 2000; Bennett and Klich 2003). Disease produced by aflatoxins is predominantly characterized by hepatic pathology. While there are cases of acute toxicity throughout the developing world, the risk that aflatoxins pose as potential carcinogens makes them more practically a psychological terrorist agent.

Aflatoxins consist of four main toxins B1, B2, G1 and G2 (Pitt 2000). They are classified as difuranocoumarin derivatives. These agents are produced by *Aspergillus* species, the two most common being *Aspergillus flavus* and *Aspergillus Parasiticus* (Pitt 2000). The toxins are named according to their fluorescence under UV light: green (G) and blue (B) (Pitt 2000). Aflatoxins are water-insoluble, with a molecular weight of about 300 Da (Fig. 20.2). They are heat-stable chemicals, being resistant to degradation at 100°C. These agents are easily produced in the laboratory, and are usually produced in the form of a white crystalline powder (Gupta and Venkitasubramanian 1975). Aflatoxins are well-absorbed through the oral route, and primarily metabolized in the liver by cytochrome

Fig. 20.2 Aflatoxin B1

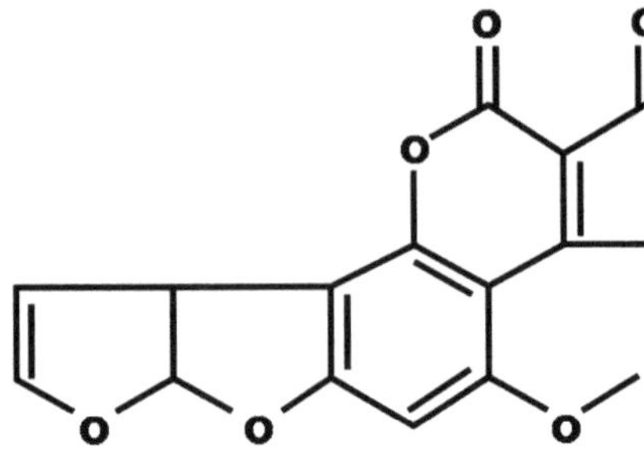

p450 (Pitt et al. 2000). After exposure through the oral, intraperitoneal or intravenous route, the toxin is concentrated in the liver. The major pathway of excretion occurs through the biliary tract; a minor amount is renally excreted (Pitt et al. 2000). Both aflatoxin B1 and its metabolites appear to bind DNA and proteins, exerting their toxic effects (Pitt 2000; Bennett and Klich 2003).

Aflatoxins have been shown to cause both acute and chronic disease in humans and animals (Table 20.1). Aflatoxins are responsible for the acute toxicity and death of 100,000 turkeys in 1960 after consumption of aflatoxin contaminated meal (Pitt et al. 2000). They are also believed to be responsible for an outbreak of acute hepatitis among 400 humans in India in 1974 (Peraica et al. 1999; Pitt 2000). This outbreak was linked to aflatoxins produced by *Aspergillus fluvus* on contaminated corn. Investigations revealed that some samples contained up to 15 mg kg^{-1} of aflatoxin (Pitt 2000). These toxins have also been implicated in acute hepatitis in Uganda, India, and Kenya (Peraica et al. 1999). Patients involved in India have estimated to have eaten 2–6 mg of aflatoxin per day (Pitt 2000; Bennett and Klich 2003). Further extrapolation estimates that the single acute lethal dose for an adult ranges from 10 to 20 mg (Pitt 2000). One case report refutes this estimated lethal dose, as a female patient acutely ingested 5.5 mg of purified aflatoxin B1 and an additional 35 mg over a 2-week period in an attempted suicide, and suffered no apparent acute or chronic effects (Peraica et al. 1999; Bennett and Klich 2003). Evidence suggests that toxicity may also occur through inhalation, and increase the risk of pulmonary carcinoma (Van Vleet et al. 2002). While evidence for reliable aerosolization in weapons form is sparse in the lay literature, aflatoxin B1 has been demonstrated to be readily aerosolized on grain dusts (Van Vleet et al. 2002).

Although acute toxicity has been documented, it is the induction of carcinoma that presents a greater risk. Aflatoxin B1 is the most biologically important aflatoxin, as it is considered the more potent natural carcinogen known (Pitt 2000; Pitt et al. 2000; Bennett and Klich 2003). In non-industrialized countries, the risk of liver cancer corresponds heavily with aflatoxin dietary intake (Pitt 2000). Hepatic carcinoma is one of the most common cancers in China, Africa and Southeast Asian countries where exposure to dietary aflatoxin is prevalent (Etzel 2002; Bennett and Klich 2003). Hepatocellular carcinoma is one of the leading causes of death in China (Glintborg et al. 2006). The risk increases during co-infection with the hepatitis B virus (Pitt 2000; Bennett and Klich 2003). The mechanism of carcinogenesis appears to be induction of mutation in p53 tumor suppressor genes (Pitt 2000). Based on prior studies, the annual number of deaths in Indonesia secondary to aflatoxin-induced liver cancer may be as high as 20,000 (Pitt 2000).

As the oral route is the most effective means of exposure, and the toxin is not water-soluble, the most likely means of exposure would be in food supplies. However, the potential for aerosolized aflatoxin does exist, and is supported by the finding of weaponized aflatoxin in Iraq (Stone 2002). Initial treatment consists of decontamination, supportive care and limiting exposure to the toxin. Multiple absorbent agents have been studied following oral exposure to aflatoxins in animals. Among the most studied are aluminosilicate compounds. In animal studies, hydrated sodium calcium aluminosilicates have shown to be protective against the deleterious effects of aflatoxin B1, including reduction of transaminitis, diminished effects of macrophages, and preservation of normal liver histology (Harvey et al. 1994). While no comparative studies are available, activated charcoal also shows efficacy in decreasing the toxicity of oral aflatoxin B1 exposure (Ademoyero and Dalvi 1983; Dalvi and McGowan 1984). Oltipraz (4-methyl-5-(2-pyraziny)-1,2-dithiole-3-thione) is an investigational drug for the treatment of aflatoxin-induced carcinoma. Originally designed as an antischistosomal agent, oltipraz modulates aflatoxin metabolism, prevents the formation of reactive epoxide species, and increases levels of glutathione (Helzlsouer and Kensler 1993; Glintborg et al. 2006). It has been shown to inhibit aflatoxin B1-induced hepatocarcinogenesis in rats (Helzlsouer and Kensler 1993). While safe in human use, oltipraz did not reduce urinary concentration of aflatoxin-induced reactive oxygen species in one human trial (Glintborg et al. 2006). However, no study on the long-term effects on human carcinogenesis has been performed. The assumption could be made that oltipraz would have a similar effect on pulmonary carcinoma after inhalational exposures. Oltipraz appears to be a reasonable treatment after ingestion of and inhalational exposures to aflatoxin. In countries where aflatoxin B1 is prevalent in food supplies, vaccination against hepatitis B has been recommended as a protective against the risk of carcinoma (Bennett and Klich 2003).

20.6 Ochratoxin A

Ochratoxin A was first isolated in 1965 from *Aspergillus ochraceus* (Bayman and Baker 2006). It has since been isolated from other *Aspergillus* species and *Penicillium verucosum* (Peraica et al. 1999; Bennett and Klich 2003; Bayman and Baker 2006). It is found in many natural products, including cereals, nuts, wines, beers, cow's milk and various other animal products (Peraica et al. 1999; Bennett and Klich 2003; Bayman and Baker 2006). In animal studies, ochratoxin has been shown to cause immunosuppressive, teratogenic, carcinogenic and nephrotoxic effects (Peraica et al. 1999; Bennett and Klich 2003; Bayman and Baker 2006). Its potential to cause acute toxicity is low, but as a result of its ability to cause chronic disease may be used to instill fear and panic in a population.

Ochratoxin A is primarily a nephrotoxic agent, producing renal disease in all animals studied (Pitt 2000; Bennett and Klich 2003). Renal tubular pathology has been found in multiple studies after exposure to the toxin. Its primary mechanisms of action appear to be inhibition of protein synthesis by competitive inhibition of phenylalanyl-tRNA synthetase and by induction of oxidation (Creppy et al. 1998; Baudrimont et al. 2001). Multiple epidemics have been linked with ochratoxins. Patients in Tunisia with renal impairment of unknown etiology were found to have high concentrations of ochratoxin A. Ochratoxin A has also been associated with porcine nephropathy, and is one of the potential causes in Balkan endemic nephropathy (Tatu et al. 1998; Peraica et al. 1999).

Ochratoxin A is a chlorinated chemical consisting of isocoumarin moiety linked by an amide bond to a phenylalanine moiety (Fig. 20.3) (Lau et al. 2000; Bayman and Baker 2006). It has a molecular weight of 403 da, and is a fat-soluble chemical (Lau et al. 2000; Pitt 2000). Ochratoxin B is the dechlorinated version; it and ochratoxin C are also found in nature, but are less common and less toxic (Lau et al. 2000; Bayman and Baker 2006). The oral LD50 of ochratoxin A in rats has been found to be 20 mg kg^{-1} (Bayman and Baker 2006).

Exposure to orchratoxin would probably occur through food stores. It is by this method that toxicity has historically been produced. However, ochratoxin is effectively absorbed via an inhalational route, and has been demonstrated to be aerosolized in grain dusts (Breitholtz-Emanuelsson et al. 1995; Skaug et al. 2001; Halstensen etal. 2004). While theorized to cause chronic toxicity, there is no conclusive evidence that inhalational exposure causes chronic disease in humans, and it is unlikely to cause acute toxicity. Furthermore, no evidence exists for effective weaponization by this means. Detection can be accomplished in similar ways to other mycotoxins, including liquid chromatography and mass spectrometry. These methods can detect toxin in both food samples and blood samples (Lau et al. 2000). Both paroxicam and aspartame have been studied as potential treatments for ochratoxin-induced nephropathy. Aspartame is a structural analog of both ochratoxin and phenylalanine. It is a widely used sweetener and generally well-tolerated

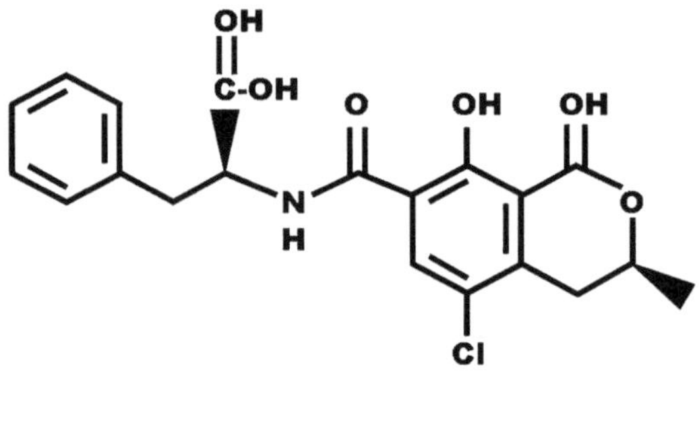

Fig. 20.3 Ochratoxin A

in humans. In animal studies aspartame (given 25 mg kg^{-1} every 48 h) appears to prevent the nephrotoxin effects of ochratoxin (Creppy et al. 1998; Baudrimont et al. 2001). Proposed mechanisms include preventing the binding of ochratoxin to plasma proteins, enhancing ochratoxin elimination and enhancing ochratoxin metabolism (Creppy et al. 1998). Paroxicam also appears to reduce the nephrotoxicity of the toxin while increasing elimination of free ochratoxin in the urine (Baudrimont et al. 1995). Catechins, which are phenolic compounds with antioxidant properties, have had limited study in treating ochratoxin-induced disease. Ochratoxins have been shown to promote free radical formation and this is hypothesized to contribute to their toxicity. *In vitro* studies demonstrate the ability of catechins to reduce reactive oxygen species; however, this has not been shown in animals or human experiments (Costa et al. 2007). As most antioxidants have a low side-effect profile, the risk-to-benefit ratio would be low, making cathechins a practical therapeutic option. Other treatment would be primarily supportive, with decontamination after acute aerosol or oral exposure and removal from the toxic environment.

20.7 Ergotism

Throughout history, two forms of ergotism have been described: gangrenous and convulsive (Merhoff and Porter 1974; Etzel 2002; Bennett and Klich 2003). Ergot alkaloids are produced by *Claviceps* species, primarily *Claviceps purpura* and *Claviceps fusiformis*. These species infect a variety of grains; however, rye is the predominant host (Woolf 2000; Favretto et al. 2007). *Claviceps* species also produce other biogenic amines including histamine, acetylcholine, tyramine, and isoamylamine (Woolf 2000). The ability of ergot alkaloids to cause human disease has been demonstrated consistently throughout history, with outbreaks of ergotism occurring in Europe, Scandinavia and Russia (Merhoff and Porter 1974). Ergotism is believed to date back to antiquity, with earliest recorded references dating back to the Assyrians (De Costa 2002; Etzel 2002). Ergot alkaloids posses both serotonin and alpha adrenergic activity (Woolf 2000; Murphy et al. 2007). By these mechanisms, they are potent smooth muscle constrictors. Their vasoconstrictive properties are often used for medical treatment, and use in childbirth dates back to the 1500s (De Costa 2002). Gangrenous ergotism is caused by peripheral vasoconstriction. Feelings of cold limbs with paresthesias, pain and burning in the extremities are common symptoms; autoamputation may eventually result (Woolf 2000). Convulsive ergotism is characterized by hallucinations, mania, delirium, vertigo headaches and painful muscle contractions (De Costa 2002). This effect is related to the serotonin agonist properties of some ergot alkaloids, as they are derived from lysergic acid similar to LSD (Woolf 2000; De Costa 2002).

The indole ring is common to all ergot alkaloids (Fig. 20.4) (Bennett and Klich 2003). They generally have a molecular weight in the 300–600 Da range,

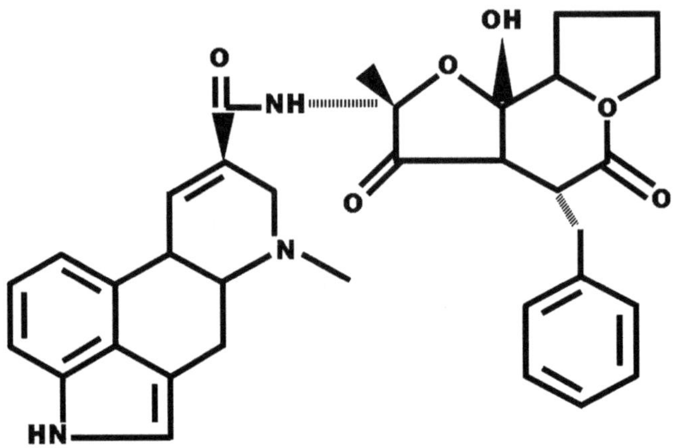

Fig. 20.4 Ergotamine

and consist of both water-soluble and -insoluble forms (Lehner et al. 2005). Ergot alkaloids are relatively heat labile; the majority of the toxin is destroyed during baking or cooking (Peraica et al. 1999; Favretto et al. 2007). Historically, ergotism has resulted after ingestion of the alkaloids. One species forms ergotamine, proved to be lethal after oral administration of 26 mg over the course of a few days (Favretto et al. 2007). Injection of much smaller doses may be fatal, especially in those with prior history of cardiovascular disease. Inhalation exposure seems to be an inefficient method of inducing toxicity; however, evidence for toxicity by this method does exist (Strange et al. 1998). Ergot alkaloid can be detected in the urine through a method known as the Van Urk test (Woolf 2000). When sulfuric acid containing p-dimethyl aminobenzaldehyde is mixed with the urine and the ferric chloride is added, an intense blue color results, indicating the presence of ergot alkaloids (Woolf 2000). High-performance liquid chromatography has been used to detect ergot alkaloids in blood samples (Favretto et al. 2007). Mass spectrometry and liquid chromatography-mass spectrometry are used for quantitative analysis in both blood and food samples (Favretto et al. 2007).

There is no consensus on recommendation for treatment. Vasoconstrictive effects can be treated with vasodilators such as nitroglycerin, nitroprusside, peripheral alpha adrenergic blockers and dihydropyridine class calcium channel blockers (Garcia et al. 2000; Woolf 2000; Kim et al. 2005). Thrombosis may be treated with platelet inhibitors, anticoagulants such as heparin or enoxaparin and, in limb-threatening cases, thrombolytics (Merhoff and Porter 1974; Woolf 2000). Prostaglandins also appear to be a good choice, as they cause both vasodilation and have antiplatelet effects (Garcia et al. 2000). Symptoms of convulsive ergotism would be treated by removal from the source, decontamination and benzodiazepines for seizures and psychiatric symptoms.

20.8 Conclusion

Mycotoxins are a legitimate threat as bioweapons or terrorist agents. They possess the proper qualities to make them effective for this purpose: they are generally heat stable, are easily manufactured, easy to disperse, and may produce disease in humans. The trichothecenes have the greatest ability to cause acute disease. Deoxynivalenol has resulted in outbreaks of gastrointestinal illness. T-2 mycotoxin, with its ability to act as a vesicant, along with the possibility of systemic toxicity and death, would be the most useful agent as a true bioweapon. Ergot alkaloids may produce acute disease; however, toxicity would be anticipated only after repeat or chronic dosing, making the potential to cause significant toxicity limited. The aflatoxins and ochratoxins demonstrate substantial risk of chronic health effects, but their ability to cause acute disease is much lower. Introduction into the food or water supply, or exposure through aerosolization, would cause enough concern over acute and chronic health risk to induce panic in a population. Treatment strategies for all mycotoxin exposures are primitive and in need of further research. The primary modality of treatment is removal from the source and decontamination. As the future risk of mass exposure remains high, further research must be initiated to develop better methods of decontamination and treatment.

References

Ademoyero AA, Dalvi RR (1983) Efficacy of activated charcoal and other agents in the reduction of hepatotoxic effects of a single dose of aflatoxin b1 in chickens. Toxicol Lett 16(1–2): 153–157

Baudrimont I, Murn M, Betbeder AM, Guilcher J, Creppy EE (1995) Effect of piroxicam on the nephrotoxicity induced by ochratoxin A in rats. Toxicology 95(1–3):147–154

Baudrimont I, Sostaric B, Yenot C, Betbeder AM, Dano-Djedje S, Sanni A, Steyn PS, Creppy EE (2001) Aspartame prevents the karyomegaly induced by ochratoxin A in rat kidney. Arch Toxicol 75(3):176–183

Bayman P, Baker J (2006) Ochratoxins: A global perspective. Mycopathologia 162(3):215–223

Bennett JW, Klich M (2003) Mycotoxins. Clin Microbiol Rev 16(3):497–516

Breitholtz-Emanuelsson A, Fuchs R, Hult K (1995) Toxicokinetics of ochratoxin A in rat following intratracheal administration. Nat Toxins 3(2):101–103

Costa S, Utan A, Cervellati R, Speroni E, Guerra MC (2007) Catechins: natural free-radical scavengers against ochratoxin A-induced cell damage in a pig kidney cell line (LLC-PK1). Food Chem Toxicol 45(10):1910–1917

Coulombe RA Jr (1993) Biological action of mycotoxins. J Dairy Sci 76(3):880–891

Creasia DA, Thurman JD, Jones LJ, Nealley ML, York CG, Wannemacher RW, Bunner DL (1987) Acute inhalation toxicity of T-2 mycotoxin in mice. Fund Appl Toxicol 8(2):230–235

Creppy EE, Baudrimont I, Betbeder AM (1998) How aspartame prevents the toxicity of ochratoxin A. J Toxicol Sci 23(suppl 2):165–172

Dalvi RR, McGowan C (1984) Experimental induction of chronic aflatoxicosis in chickens by purified aflatoxin B1 and its reversal by activated charcoal, phenobarbital, and reduced glutathione. Poult Sci 63(3):485–491

De Costa C (2002) St Anthony's fire and living ligatures: a short history of ergometrine. The Lancet 359(9319):1768–1770

Etzel RA (2002) Mycotoxins. J Am Med Assoc 287(4):425–427

Favretto DP, Frison GP, Vogliardi S, Ferrara SD (2007) Highly specific quantification of ergotamine in urine, blood, and hair samples by liquid chromatography-tandem mass spectrometry. Ther Drug Monit 29(3):325–332

Garcia GD, Goff JM, Hadro NC, O'Donnell SD, Greatorex PS (2000) Chronic ergot toxicity: A rare cause of lower extremity ischemia. J Vasc Surg 31(6):1245–1247

Glintborg B, Weimann A, Kensler TW, Poulsen HE (2006) Oltipraz chemoprevention trial in Qidong, People's Republic of China: Unaltered oxidative biomarkers. Free Radical Biol Med 41(6):1010–1014

Golubic S, Radtk G, Le Campion-Alsumard T (2005) Endolithic fungi in marine ecosystems. Trends Microbiol 13(5):229–235

Gupta S, Venkitasubramanian TA (1975) Production of aflatoxin on soybeans. Appl Microbiol 29 (6):834–836

Halstensen AS, Nordby KC, Elen O, Eduard W (2004) Ochratoxin A in grain dust — estimated exposure and relations to agricultural practices in grain production. Ann Agric Environ Med 11 (2):245–254

Harvey RB, Kubena LF, Elissalde MH, Corrier DE, Phillips TD (1994) Comparison of two hydrated sodium calcium aluminosilicate compounds to experimentally protect growing barrows from aflatoxicosis. J Vet Diagn Invest 6(1):88–92

Helzlsouer KJ, Kensler TW (1993) Cancer chemoprotection by oltipraz: experimental and clinical considerations. Prev Med 22(5):783–795

Holstege CP, Bechtel LK, Reilly TH, Wispelway BR, Dobmeier SG (2007) Unusual but potential agents of terrorists. Emergen Med Clin N Am 25(2):549–566

Kim MD, Lee G, Shin SW (2005) Ergotamine-induced upper extremity ischemia: a case report. Korean J Radiol 6(2):130–132

Lau BP, Scott PM, Lewis DA, Kanhere SR (2000) Quantitative determination of ochratoxin A by liquid chromatography/electrospray tandem mass spectrometry. J Mass Spectrom 35(1):23–32

Lehner AF, Craig M, Fannin N, Bush L, Tobin T (2005) Electrospray[+] tandem quadrupole mass spectrometry in the elucidation of ergot alkaloids chromatographed by HPLC: screening of grass or forage samples for novel toxic compounds. J Mass Spectrom 40(11):1484–1502

Maragos CM (2004) Emerging technologies for mycotoxin detection. Toxin Rev 23(2):317–344

Marshall E (1983) Bugs in the yellow rain theory. Science 220(4604):1356–1358

Merhoff GC, Porter JM (1974) Ergot intoxication: historical review and description of unusual clinical manifestations. Ann Surg 180(5):773–779

Murphy NG, Benowitz NL, Goldschlager N (2007) Cardiovascular toxicology. In: Shannon MW, Borron SW, Burns MJ (eds) Haddad and Winchester's clinical management of poisoning and drug overdose, 4th edn. Saunders Elsevier, Philadelphia, pp 133–166

Parent-Massin D (2004) Haematotoxicity of trichothecenes. Toxicol Lett 153(1):75–81

Paterson RRM (2006) Fungi and fungal toxins as weapons. Mycol Res 110(9):1003–1010

Peraica M, Radic B, Lucic A, Pavlovic M (1999) Toxic effects of mycotoxins in humans. B World Health Organ 77(9):754

Pietikainen A, Kytoviita MM, Husband R, Young JP (2007) Diversity and persistence of arbuscular mycorrhizas in a low-Arctic meadow habitat. New Phytologist 176(3):691–698

Pitt JI (2000) Toxigenic fungi: which are important? Med Mycol 38(1 Supp 1):17–22

Pitt JI, Basílico JC, Abarca ML, Lopez C (2000) Mycotoxins and toxigenic fungi. Med Mycol 38(1 suppl 1):41–46

Qaher AM (2006) Biodiversity of the genus *Fusarium* in saline soil habitats. J Basic Microbiol 46(6):480–494

Richard JL (2007) Some major mycotoxins and their mycotoxicoses — an overview. Int J Food Microbiol 119(1–2):3–10

Seiders BAB (1986) Canadian "yellow rain" research: does it weaken the case? Science 234 (4776):528

Shohami E, Wisotsky B, Kempski O, Feuerstein G (1987) Therapeutic effect of dexamethasone in T-2 toxicosis. Pharm Res 4(6):527–530

Skaug MA, Eduard W, Stormer FC (2001) Ochratoxin A in airborne dust and fungal conidia. Mycopathologia 151(2):93–98

Stone R (2001) Treaty compliance: down to the wire on bioweapons talks. Science 293 (5529):414–416

Stone R (2002) Peering into the shadows: Iraq's bioweapons program. Science 297(5584): 1110–1112

Strange K, Pohlmeier H, Lubbesmeyer A, Gumbinger G, Schmitz W, Baumgart P (1998) Vascular ergotism through inhalation of grain dust. Dtsch Med Wochenschr 123(51–52):1547–1550

Tatu CA, Orem WH, Finkelman RB, Feder GL (1998) The etiology of Balkan endemic nephropathy: still more questions than answers. Environ Health Perspect 106(11):689–700

Thompson WL, Wannemacher RW (1986) Structure-function relationships of 12,13-epoxytrichothecene mycotoxins in cell culture: Comparison to whole animal lethality. Toxicon 24 (10):985–994

Thompson WL, Wannemacher RW (1990) *In vivo* effects of T-2 mycotoxin on synthesis of proteins and DNA in rat tissues. Toxicol Appl Pharmacol 105(3):483–491

Ueno Y (1984) Toxicological features of T-2 toxin and related trichothecenes. Fund Appl Toxicol 4(2, Part 2):S124–S132

Van Vleet TR, Klein PJ, Coulombe, RA (2002) Metabolism and cytotoxicity of aflatoxin b1 in cytochrome p-450-expressing human lung cells. J Toxicol Environ Health, Part A 65(12):853–867

Wannemacher RW, Wiener SL (1997) Trichothecene Mycotoxins. In: Sidell FR, Takafuji ET, Frantz DR (eds) Textbook of Military Medicine, Part 1. Office of the Surgeon General at TMM Publications, Washington, pp 655–676

Watson SA, Mirocha CJ, Hayes AW (1984) Analysis for trichothecenes in samples from Southeast Asia associated with "Yellow Rain". Fund Appl Toxicol 4(5):700–717

Woolf A (2000) Witchcraft or Mycotoxin? The Salem Witch Trials. Clin Toxicol 38(4):457–460

Zheng MZ, Richard JL, Binder J (2006) A review of rapid methods for the analysis of mycotoxins. Mycopathologia 161(5):261–273

Chapter 21
The Weaponisation of Mycotoxins

Robert Russell Monteith Paterson and Nelson Lima

21.1 Introduction

Mycotoxins as weapons is a serious issue. The word "weaponisation" in the title of this chapter is poor grammatically, although it is useful as it is generally understood. Crucially, access to accurate information is required to enable appropriate responses for potential threats. The high interest created by a recent review of fungi and toxins as weapons indicated that further publications in the field are desirable: Paterson (2006a) remained at number 1 in Science Direct's "Top 25 Hottest Articles" (*Mycological Research*) for a year. Citation numbers put it fourth since 2006: Holstege et al. (2007) is instructive as it indicates just how seriously the threat is taken in the United States of America. The authors focus on trichothecene mycotoxins and particularly T-2 toxin. Of course, the reasons for the topicality were the mass attacks on citizens which have occurred this century, and the claim that aflatoxins had the potential to be used by Iraq. The recent attacks in the USA using anthrax spores via internal post also caused a great deal of concern: massive casualties have been predicted from anthrax released into very large cites. Dohnal et al. (2007) are also concerned with T-2 toxin. Latxague et al. (2007) focuses on anticrop bioterrorism and bioweapons against the agricultural sector. They appear to be more concerned with whole organisms, rather than purified mycotoxins per se. However, it is difficult to obtain a list of the fungi with which Latxague et al. are concerned, no doubt for security reasons. Mycotoxin-producing fungi need to be on the list, as do plant pathogens. Also, countries where crops are developed almost as a monoculture are at particular risk from natural pathogens [e.g. the fungus *Ganoderma* and the oil palm crop (Paterson et al. 2009)]. Pohanka et al. (2007) considered the issue of developing bioassays to detect mycotoxins.

R.R.M. Paterson (✉)
Micoteca da Universidade do Minho, IBB/Centre for Biological Bioengineering, Campus de Gualtar, University of Minho 4710-057, Braga, Portugal
e-mail: russell.paterson@deb.uminho.pt

M. Rai and A. Varma (eds.), *Mycotoxins in Food, Feed and Bioweapons*,
DOI: 10.1007/978-3-642-00725-5_21, © Springer-Verlag Berlin Heidelberg 2010

Finally, Casadevall and Pirofski (2006) provided a well-argued assessment concerning why human pathogenic fungi could usefully be given greater consideration as biowarfare and bioterrorism agents.

A recent example of a biosecurity lapse in Egham, UK demonstrates why complacency cannot be permitted: poor infrastructure and inadequate funding led to accidental release of the foot and mouth disease virus. Possible causes of the breach may have involved misaligned effluent pipes and tree-root damage to the pipes, compounded by an increase in water caused by floods (HSE 2007). Obviously, those working with toxigenic fungi and fungal toxins need to exercise appropriate care. It is now absolutely necessary to consider the issue as it relates to the field of weapons.

The topic of fungal toxins as weapons is no longer taboo as it has changed to one of public interest and media attention. The remainder of this chapter discusses the toxins and not the fungi per se (i.e. the biochemical rather than the biological) although this tends to overlap when the problems with the taxonomy of the toxin producing fungi are considered (Paterson et al. 2004, 2006). One might expect that the fungi which produce the toxins are well characterised. However, this is not the case. Conversely, this lack of knowledge may work as an advantage in terms of security as it is not always straightforward to select toxin-producing isolates on the basis of current taxonomy. It makes more difficult (a) the application of legal restrictions on the use or export of the fungi, and (b) tracing particular strains used in toxin production.

Public pressure exists on authorities to assess if occurrence of food contamination are nefarious acts (Elad 2005). The economic consequences of simply reacting to a potential attack can be huge, if what was experienced after the recent attacks in the USA is an example (Lenain et al. 2002). Furthermore, a sombre assessment of the dangers of bioweapons to the United States of America is provided by Bailey (2001). It is much more appropriate to focus on prevention, followed by readiness and response. The balance is to declassify essential information with a view to preventing aggressive acts. A compromise has to be drawn: it is self-defeating to ignore the subject given the large amount of public information already available.

Biosecurity guidelines are essential reading for those working with toxigenic fungi (Tucker 2003). They have highly significant implications which could limit research in the field, including determining which individuals do the work. However, just how representative the views are of this publication is unclear when the disclaimer states, "The views expressed in this report are those of the author alone. They do not necessarily reflect the views of the United States Institute of Peace". A public debate is required. Some well-known reports have claimed recently that aflatoxins were placed in warheads for use by Iraqis, although the effect of such a limited amount of aflatoxins dispersed in this manner would be minimal (Paterson 2006a). On the other hand, *Aspergillus flavus* may be more dangerous as a human pathogen than is generally realised (Hedayati et al. 2007).

There is nothing to be gained from being coy about discussing this issue. There is a great deal of information in scientific papers, published journals, newspapers, and the World Wide Web. The most comprehensive, and well considered source of information/best practices is the Australia Group (http://www.australiagroup.net). This is a body which meets every year and represents numerous nations whose aim is, inter alia, to control the spread of chemical and biological weapons. For example, it is used by the UK Department of Trade and Industry for export control of a wide range of potentially and obviously dangerous material with possible utility as weapons. The web site is essential reading for anyone involved in the field. The mycotoxins considered are diacetoxyscirpenol, T-2 toxin, HT-2 toxin, and aflatoxin B_1.

Information is becoming increasingly available on these bioweapons (Bennet and Klich 2003; Miller et al. 2005; Paterson and Lima 2005; Stark 2005; Paterson 2006a); in addition to the dubious mixture of the informative (Locasto et al. 2004) and illegitimate material on the World Wide Web. A great deal is known about botulinin from *Clostridium botulinum* which is the most toxic compound in the world (human lethal dose 0.2–2.0 $\mu g\ kg^{-1}$), and so there is little point in not discussing fungal toxins as weapons which are, after all, less toxic. To obtain some level of calibration at the extremes, mycotoxins are (a) not as dangerous as nuclear weapons and (b) more dangerous than teargas. Ease of conversion to a weapon is a crucial factor (i.e. "weaponization"). It is crucial that rational discussion appears in reputable journals, books and media.

This chapter does not concern primarily fungi that cause disease per se. Obvious growth of fungi on animals is called mycosis and they are primary pathogens e.g. *Histoplasma capsulatum* (Bennet and Klich 2003). An excellent overview of fungi as weapons is provided in Casadevall and Pirofski (2006). Dietary, respiratory, dermal and other exposures to mycotoxins are called mycotoxicosis, and this area is more relevant to the creation of weapons. However, Hedayati et al. (2007) reported the aflatoxin-producing fungus *A. flavus* as the second most serious *Aspergillus* for causing human and animal infections (*Aspergillus fumigatus* is the first). The importance of this fungus increases in regions with a dry and hot climate. Consequently, the potential as a bioweapon may have been underestimated.

Furthermore, the use of fungi in technologies (e.g. biocontrol) requires revision because of the current heightened security awareness. There is an apparent similarity between fungal biocontrol agents (FBCA) and weapons, in that toxin-producing fungi are mass-produced and, for example, sprayed onto crops. This also has ramifications for the health and safety of those who use these organisms for mass production and/or in non-sterile conditions, which raises the question, what are the natural levels of fungi (Gonçalves et al. 2006) and toxins (Paterson 2007a) in the environment? Interestingly, Bucheli et al. (2008) described the occurrence of deoxynivalenol and zearalenone in river water which is of relevance to this topic. Pharmaceuticals from fungi are also relevant to the discussion. The difference between a compound being a toxin or a drug may be a shift in a decimal point of concentration and/or a change in a simple moiety. How these compounds are classified depends to some extent on the prevailing "climate". For example, mycophenolic acid, ergot alkaloids, penicillin and perhaps patulin, can be either.

21.2 Fungal Toxins and Metabolites as Weapons

There has always been great concern about toxins from the macro fungi (e.g. mushrooms), often from accidental consumption of the fruiting body. However, scientific endeavour began in mycotoxins per se with the discovery of aflatoxins in the 1960s. It is an extremely complex field due to its multidisciplinary nature. General discussions on mycotoxins are dealt with elsewhere in the current book, and Venâncio and Paterson (2007) can be consulted.

HACCP protocols have been developed to prevent unintentional contamination of food with mycotoxins and fungi. However, further control and analytical steps may be required for intentional contamination. The universality of applying HACCP has been questioned in any case (Sperber 2005). Mycotoxins are (a) below microbiological, some phytotoxins and phycotoxins and (b) above anthropogenic contaminants, pesticide residues, and food additives in terms of acute health risks. Significantly, they are the highest chronic risk factor in the diet (Kuiper-Goodman 2004). However, chronic effects are of little interest to weapon manufacturers. Furthermore, there may be fungal metabolites which are more toxic than mycotoxins (e.g. aflatoxins) although they are not normally detected in the environment (see Cole and Schweikert 2003a, b; Cole et al. 2003). These may be revealed through natural product screenings for drugs where toxic compounds are removed from screens at early stages in the process. Organisations such as culture collections (i.e. biological resource centres) that work with numerous fungi need to be more aware of this fact and improve biosecurity measures accordingly.

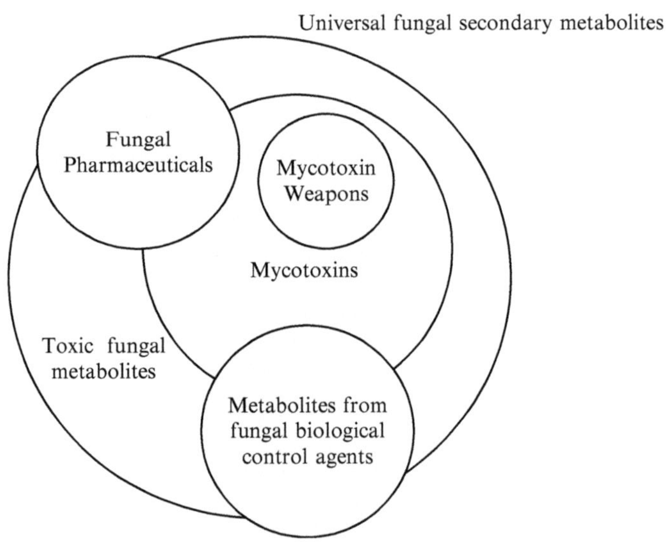

Fig. 21.1 Venn diagram of the relationship between fungal metabolites in terms of toxicity. The sizes of the circles are in proportion to the actual number of compounds only in a general manner

Table 21.1 Approximate classification of some toxins of fungi

	Weapon	Mycotoxin	Pharmaceutical	FBCA toxin	Other entomopathogen toxin
Aflatoxins	+	+[a]			+
Ochratoxin A		+[a]			+
Cytochalasins		+		+	
Beauvericin		+		+	
Eniantins		+		+	
Destruxins				+	
Oosporein				+	
Moniliformin		+		+	
Efrapeptins				+	
Beauveriolides				+	
Amanita phalloides toxins	+	+[b]			
Patulin		+[a]	+[c]		
Mycophenolic acid		+	+		
Penicillin		+	+		
T-2 toxin	+	+[a]			
Ergot alkaloids		+[a]	+		

FBCA, fungal biological control agent
[a]Regulated in food
[b]Arguably mycotoxins as the whole fungus in ingested
[c]As originally investigated

The overlapping relationships between mycotoxins, pharmaceuticals, FBCA toxins, and fungal biochemical weapons are represented diagrammatically in Fig. 21.1. Even the same compound can be represented in different fields. This is a reason why a sound knowledge of which fungi produce which toxins (i.e. fungal chemotaxonomy) is crucial. Presently there are only a few metabolites considered to be mycotoxins from potentially thousands. There are probably more toxic metabolites from FBCA than there are mycotoxins using this narrow definition as only a few (i.e. three) of them (Table 21.1) can be considered as mycotoxins. However, a minute percentage of toxin fungal metabolites have been considered seriously as weapons: two (aflatoxins and T-2 toxin) are obviously mycotoxins whereas one is a toxin from a macroscopic fungus, i.e. α-amanitin. Holstege et al. (2007) mentions a rather wide range of compounds from fungi including trichothecenes in general, although in practice only T-2 toxin is considered in depth. Paterson and Lima (2005) listed the ergot alkaloids (see CAST 2003) as other possibilities.

21.2.1 Fungal Toxins

Many fungi are not pathogens although they do produce toxins (Cole and Schweikert 2003a, b; Cole et al. 2003) and fungal toxins as weapons or "expressions of discontent" are taken seriously (Paterson 2006a). Here we will consider the fungal toxins intrinsically as weapons which could be employed by governments and/or small groups of individuals. An individual could use them in a revenge attack on an employer for example. The supposed governmental deployments have ranged from the thin evidence of the use of T-2 toxin by the Soviet Union, to the development of aflatoxins by Iraq, the efficacy of which is minimal (Bennet and Klich 2003). It is worth pointing out that T-2 toxin does appear to be a valid weapon. It is axiomatic to state that any government would be interested in developing such weapons if they had desirable characteristics as defined by that government.

The factors which are fundamental for producing a serviceable biochemical weapon are (a) efficient manufacture, (b) ease of conversion to a weapon ("weaponization"), (c) longevity of the organism or toxin in storage, (d) efficient dispersal, and (e) stability when exposed to the environment. Other factors are concealment and ability to obtain the toxin or organism.

In the early twentieth century, biotoxins were investigated militarily, and were rejected because of the difficulty of conversion to weapons. For example, it has been stated that the US has no current offensive biological weapons capability. Interest has been stimulated by developments in biotechnology (e.g. "black biology" (Casadevall and Pirofski 2006). They are less expensive than nuclear and conventional chemical weapons and may appeal to countries or terrorist organizations where cost is an important issue (Locasto et al. 2004).

Biotoxins need to be produced from precursors. For example, a growth medium containing carbon and nitrogen as the predominant precursors must be inoculated with a live fungus in a suitable vessel (e.g. a bioreactor). The toxin is produced as a consequence of the metabolism of the fungus and is then purified to some degree. This can be undertaken in large bioreactors, hence producing a high yields of toxin. So the toxin is a chemical and has no living component. This implies that these compounds need to be classified as chemical weapons. For example, they will not spread from person to person beyond the locality of attack as would infectious diseases. So there is surely a need to reclassify these compounds. A dangerous scenario can be imagined where authorities respond to a biotoxin as a biological attack (perhaps involving isolating large numbers of people to prevent disease spread) whereas a much more localised concentration of effort may be more appropriate. Furthermore, in a British Broadcasting Cooperation web site it is mentioned that 30,000 kg of fungus was produced as a bioweapon to spray on crops (http://news.bbc.co.uk/1/hi/world/americas/1618420.stm). Presumably a similar quantity could be achieved if required to produce a large amount of fungal toxin. Finally, one simply has to consider how much penicillin has been produced from initially minute yields to realise that a massive amount of toxin as a weapon could be produced.

The threat of fungal toxins and mycotoxins were made apparent in Garber et al. (2005), Miller et al. (2005), and Stark (2005). The current terrorist tactics have shifted attention to protection of food supplies and consequently it is a world-wide concern for the twenty-first century. The impact on society could be anything from low to "catastrophic". A huge range of actions and programmes are being developed and implemented to prevent, deter, and respond to potential attacks.

For example, (a) enhanced laboratory capability, (b) advanced tracking, (c) increased examinations, (d) better surveillance, (e) more training, (f) recovery plans, and (g) new medical treatments.

Mycological/chemical sampling and detection methodologies need to be improved to meet the new concerns. Inactivation of mycotoxins and decontamination of food plants requires urgent consideration. Also, foods have to be ranked in terms of vulnerability to attack, as do the risks to people when foods are intentionally contaminated. The development of algorithms to differentiate natural from unnatural food contamination (Paterson and Lima 2005) is required: international and national collaborations are essential. Finally, biomarkers in humans and animals need to be further developed with particular relevance to metabolomics.

21.2.1.1 α-Aminitin

α-Aminitin from *Amanita phalloides*, amongst others, is a major concern as it is extremely toxic, water-soluble, and heat-stabile (Garber et al. 2005). However, mass production would presumably be limited to solid substrate bioreactors if basidiocarps were used, or conventional bioreactors if mycelium can produce the compound. The background counts of α-amanitin and T-2 toxin were useful as a demonstration of what is required to be done to distinguish abnormal from normal concentrations of biotoxins. In some cases the background was higher than the spiked samples, which is interesting. In some ways Stark (2005) is simultaneously alarming and reassuring in what is possibly an overly dramatic paper, at least in the introduction. The author mentions that large-scale tactical weapons are impractical but that the sabotage arena is suitable for mycotoxins; this is slightly reassuring, in that antidotes do already exist for some mycotoxins.

21.2.1.2 Aflatoxins

The concept of liver cancer from aflatoxin as a battlefield weapon beggars belief, and the reports of acute toxicity are uncertain. However, the threat of the use of the compound may have a psychological effect on a potential enemy. It is not widely known that *A. flavus* is the second most serious *Aspergillus* disease (see above) and cannot be discounted as a potential weapon, which was not emphasised in Paterson (2006a).

21.2.1.3 T-2 Toxin

Exposure to T-2 toxin of a few milligrams is potentially lethal. The concept of mass-production of the compound may be impractical simply from the point of view of obtaining sufficient growth medium to grow the fungus (Stark 2005). Whether another cheaper and plentiful medium could be used is a possibility, although what this might be is uncertain (although large amounts of biomass of fungi have been produced (see above).

21.3 An Example

A revenge attack by a disgruntled employee is a possibility, for example by contaminating a water distribution system (Mays 2004) and so it remains worthwhile to illustrate the problem further:

Paterson et al. (1997) first reported the natural occurrence of any fungal secondary metabolite in water after attempting to produce aflatoxin in water in a bioreactor. The control water, which contained aflatoxins B_2, G_2 was from a water tank used to serve the laboratory where the work took place. An *A. flavus* strain was isolated from the control water, indicating how the aflatoxins may have arisen. However, it is not known if this was from the natural growth of a fungus or from intentional addition from a disgruntled employee. The water was used for a variety of purposes such as washing benches, cleaning floors, and supplying showers. So the levels of aflatoxins would have tended to accumulate over the years and be inhaled. Only a survey of similar water tanks would indicate if this contamination was normal or deliberate. It may be worth mentioning here that zearalenone and other metabolites from *Fusarium graminearum* were demonstrated to be capable of being produced in water (Paterson 2007b). Also, similar concentrations of zearalenone and deoxynivalenol have been detected in river water (Bucheli et al. 2008). In addition, T-2 toxin can be used as a food- or water-borne (Paterson and Lima 2005) poison. Currently, T-2 toxin is the only biologically active toxin effective through (a) dermal exposure, (b) respiratory and (c) gastrointestinal (GI) portals. Tissues involved in high cellular turnover (e.g. GI and respiratory epithelium, bone marrow cellular elements) are the most susceptible (Locasto et al. 2004).

21.4 Water as a Vector

Drinking or non-drinking water may be effective media for mycotoxin dispersal as a weapon and worthy of a separate section. The threat from contaminated drinking water is obvious. In the case of non-drinking water, the toxin could be spread by water from a shower and then inhaled. Work places where high volumes of water are employed, such as farms or car washes, could be susceptible. As an example, stored water which was demonstrated to contain aflatoxins was used to clean laboratories, and contaminated dust could be spread amongst the workers (see above). Furthermore, drinking water for animals may be at a considerably higher level of risk than that for human consumption. Toxins in water are possible naturally, unnaturally and from inoculation of fungi (Paterson and Lima 2005; Paterson 2006a, 2007a, b; Bucheli et al. 2008). However, it is interesting that high concentrations of the mycotoxin ochratoxin A (OTA) were degraded in a model waste water treatment system and that the activated sludge/microbial consortium remained effective (Nogueira et al. 2007), indicating that water treatment plants could cope with a threat from biotoxins to some extent.

21.5 Fungal Biological Control Agents

There are some interesting parallels between fungal weapons and FBCA. FBCA technology needs to be reassessed in terms of safe use (Skrobek et al. 2005), which is relevant especially in the current high-security climate. For example, the procedures involve applying natural fungal pathogens to crops in the field or in storage, to control insect pests or disease. They have been advocated by some as an effective and, more particularly, an environmentally sound means for controlling pests, disease and undesirable organisms. An immediate concern about these preparations is how safe they are in terms of toxin production (Strasser et al. 2000; Skrobek et al. 2005). After all, these could affect the general population and workers producing FBCA. They have been introduced or tested disproportionately in developing countries where, ironically, the mycotoxin problem is more acute. In a large study on risk assessment Strasser et al. (2000) did not consider compounds produced by FBCA which are already known mycotoxins. For example, cytochalasins are not discussed, although they are produced by the FBCA, *Metarhizium anisopliae*, which is considered by the authors. Cytochalasin D in particular is very toxic. If it is satisfactory to use the fungi mentioned because the levels in the environment would be low, is it equally safe to use *A. flavus* which produces aflatoxins, or *Aspergillus ochraceus* which produces OTA (Paterson 2006a)? Futhermore, there are reports where the FBCA forms a symbiosis with the plant to be protected (Wagner and Lewis 2000). In biocontrol terms this is considered to be desirable as it may provide long-term protection. In terms of persistence in the environment it is worrying.

An example of an FBCA is the use of *Fusarium oxysporum* (i.e. *F. oxysporum* f. sp. *exythoroxilum*) to kill coca plants in some Latin American countries (Connick et al. 1998; de Vries 2000). Interestingly, the idea is to introduce, rather than cure, a disease. The ultimate objective is to stop the manufacture of cocaine. Table 21.2 indicates the toxins associated with the fungus and it appears that the effect of these had not been considered sufficiently. The concepts of what constitutes a species in this taxon are complex, and toxin production from individual special forms has simply not been clarified satisfactorily. The technology has a great deal of similarity to what would be required to apply a bioweapon. After all, mycotoxins have been ranked as the most important chronic risk factor in the diet above pesticide residues, synthetic contaminants, plant toxins and food additives (Bennet and Klich 2003). They are also considered to be more acutely toxic than pesticides. Those interested in applying this technology need to collaborate more fully with those who know

Table 21.2 Some toxins or secondary metabolites of potential biocontrol fungi

Entomopathogens	Secondary metabolites/toxins
Aspergillus flavus[a]	Aflatoxins
Aspergillus ochraceus[b]	Ochratoxin A
Metarhizium anisopliae	Cytochalasin C, D, helvolic acid, destruxins
Beauvaria bassiana	Beauvericin, dipicolinic acid, oosporein, isoleucylisoleucyl anhydride, cyclo-(L-isoleucyl-L-valine, cyclo-(L-alanyl-L-proline), bassianolide
Beauvaria brongniartii	Beauverolide L, La
Verticillium lecanii	Helvolic acid, bassianolide
Paecilomyces fumosoroseus	Beauverolide L, La
Colletotrichum gloeosporioides	Gloeosporone
Trichoderma harzianum	Koninginin A,C, peptaibols, harzianum, cyclonerodiol, trichorzianines A,B
Penicillium oxalicum	Oxalic acid, oxaline, secalonic acid
Fusarium oxysporum	Moniliformin, hydroxylated fumonisin C1, fumonisin C4, fusaric acid, benzoic acid, enniatins, fusaric acid, fusarin C, ipomeamarones, sambutoxin.A

[a]Not used
[b]Not used. Pathogen of *Ceratitis capitata* (Castillo et al. 2000) amongst other insects. Well-known producer of ochratoxins

how to control the secondary metabolism that produces the toxins, although control of production in the field will be difficult in a predictable manner.

21.6 Taxonomy

The first thing to mention is that fungal taxonomy is in an immature state (Burnett 2003; Santos et al. 2009). Fungi have not received as high a level of attention as other kingdoms (plants, bacteria, animals, etc.). Species concepts are confused largely from the use of inconsistent characters. Many of the asexual isolates have been given species status when this may be inappropriate (see Paterson et al. 2006). There is the issue of stating that mycotoxins are not produced from species when what is meant is that they were not detected (Paterson et al. 2004). The fungus may produce small, difficult-to-detect amounts, or large amounts under other specific circumstances. Paterson et al. (2004) suggested identifying fungi to an easily recognised morphological character (e.g. a conidiophore) and undertaking a biochemical analysis for the toxin of interest. In the case of OTA-producing penicillia, an identification of *Penicillium* OTA " + " may be obtained. Interestingly, a somewhat similar scheme has been reported previously for *A. flavus* (Cotty 1989). Strains that produced only detectable aflatoxin Bs were designated "B" whereas those that produced detectable aflatoxins B and G were designated "BG". An approach as described in Paterson et al. (2004) may be useful as there are many very similar taxa within the *A. flavus* group (Hedayati et al. 2007). The situation is similar for trichothecene-producing fusaria where various taxa can produce biotoxins for possible use as weapons (Glenn 2007).

The history and current state of the taxonomy of mycotoxin-producing fungi is of relevance to the weapons issue. For example, (a) authorities may need to trace particular cultures used for research or in the mass production of mycotoxigenic fungi, and (b) if fungi were used to contaminate commodities it would be necessary to determine if the fungi produced particular toxins. There is often a great deal of confusion concerning species concepts and what is meant when a fungal species is described as "producing a particular mycotoxin": it is not as simple as "only this species produces this toxin".

Furthermore, PCR methods are used increasingly for identification and other taxonomically related purposes for fungi. The limitations of this technique may not have been considered adequately (Paterson 2006b) and certainly the lack of internal amplification controls in most previous studies requires attention (Paterson 2007c, 2008). Also, the method by which the fungi are grown has not been considered where mutagenic compounds may be produced in culture and affect detrimentally the DNA of the fungi of interest (Paterson et al. 2008; Paterson and Lima 2009). Too often the technique has been used by those whose primary skill is not in biochemistry or molecular biology. They appear to have simply followed published methods although adapted superficially to their particular situation, without fully considering the ramifications of so doing, and alternative methods would be useful (see Santos et al. 2009).

21.7 Genetically Modified Fungi (GMO)

It is possible to speculate about altered strains which could be more virulent and/or produce higher yield of toxins than the wild type. The use of "black biology" was discussed above. Countries with large resources could perhaps develop such strains. It may be something of a worst-case scenario. However, considering the developments in genetics it is probably only a matter of time before such an organism exists, if it does not already. In developed countries, transfer of cultures from culture collections to second parties involves the completion of a "material transfer agreement" which forbids genetic modification of the received cultures unless containment is increased to accommodate the GMO. However, it is not clear what GMOs mean. Does it specifically require that genetic information is transferred from one taxon to another? A useful example is how the titre of penicillin has been vastly increased from the original meagre amounts through strain improvement techniques, and this could conceivably be undertaken for toxin production.

21.8 Security of Laboratories and Obtaining Pure Mycotoxins

These issues are discussed in depth in Tucker (2003) although some controversial points are raised, such as which individuals and laboratories could be allowed to work in the field, and whether the degree of control possible in the USA can be applied to other countries. It needs to be determined what degree of risk pertains to each situation. Put crudely, does working with certain pathogenic bacteria equate in safety terms with working with fungi (Casadevall and Pirofski 2006)? Is the risk from the fungus (i.e. biological) or the toxin produced (i.e. chemical)? If it is the toxin then would not chemical security procedures be more appropriate and which have been established longer? Some relevant concerns are raised in WHO (2003). Recent global events and more local events such as those in Egham, UK (see earlier in this chapter), have underlined the need to design laboratories and the materials they contain in a way that will protect people, livestock, the environment and agriculture. However, there are distinctions between laboratory biosecurity and biosafety. Biosecurity measures prevent intentional release, loss, misuse, theft and diversion of pathogens and toxins. In contrast, biosafety measures are containment procedures implemented to prevent unintentional exposure of pathogens/ toxins or accidental release. Indeed, security precautions need to become routine laboratory practice, according to WHO (2003).

In the OECD (2007) report on biological resource centres (BCRs) (i.e. culture collections), it is stated that many are entrusted with the maintenance and exchange of hazardous biological resources. The report makes some self-evident statements such as, "staff should have relevant qualifications, training and competence to carry out their duties". Also, it contains such tautological sentiments as "To achieve quality assurance in BRCs, best practices for quality are clearly needed". Biosecurity could be enhanced at BRCs if they limited activities to the maintenance of organisms and did not become involved in, for example, research and development where the organisms have to be grown and extracted in large scale. However, this may not find favour with most BRCs. The menace of bioterrorism has changed the geo-political landscape and BRCs need to make special efforts to prevent loss or theft. Facilities have to be protected and to promote a sense of security. Often a risk assessment of fungi and toxins is involved, and the assigning of work activities occurs at an appropriate level of security suitable to the risk involved. This implies that the workers are aware of the toxins produced by the various fungi they hold, which at least implies awareness of current literature on the subject. It needs to be recognised that novel toxins can be discovered, and that compounds referred to as mycotoxins are not necessarily the most dangerous. An apparently safe, or unknown, fungus in term of toxin production may produce high yield of a novel toxin. Therefore, the undertaking of so-called bioprospecting projects (i.e. screening unknown fungi for bioactivity) (see the Iwokrama project in Paterson 2008) needs to be carefully considered from safety perspectives. A form of pre-toxicity screening is surely required, perhaps by a biological assay method. Grading security risks from high to low is dependent on the knowledge or motivation of the assessor or manager involved. The potential of fungi to produce dangerous chemical compounds needs to be assessed. Some further procedures may include methods for regular decontamination of work places. Before working with fungi, researchers should undertake a literature search on secondary metabolites known to be produced from the fungi.

However, the high costs estimated for a UK culture collection in the area of technical compliance of 100,000 to 340,000 Euros (OECD 2007) did not take account of even some of these points.

In addition, workers attempting to purchase pure mycotoxins from the chemical companies will realise that it has become more difficult. Often legitimate proof of use is required. Presumably this is because of security and not from an increased level of concern for workers' health per se. There is more paper work and security surrounding sending toxigenic cultures *inter*-laboratories. Concern with respect to health and safety is mostly related to the mass production of fungi and especially dried conidia which can be so easily inhaled. Similarly, a great deal of care is required when handling the purified and dried toxin preparations.

21.9 Mycotoxicosis Treatment

Supportive therapies for mycotoxicosis consist of improved diet and hydration of patients (Locasto et al. 2004), which are fairly obvious. Taking super activated charcoal orally may be effective if toxins are swallowed (e.g. T-2 toxin), and indeed the route of entry and dose indicate the clinical course for T-2. From a detailed study of OTA toxicity and activation metabolism of aflatoxin B1, it was discovered that the sweetener aspartame is very protective against OTA intoxication, and that Oltipraz effectively protects against AFB1 acute toxicity and carcinogenicity (Stark 2005). Oltipraz has been tested in China on populations exposed to aflatoxins (Bennet and Klich 2003). Some strains of *Lactobacillus* effectively bind dietary mycotoxins and may be an effective treatment. However, it is noted that management of fungal-related weapons is not discussed in Shannon (2004).

21.10 Mycotoxin Decontamination

Biotoxins from fungi would be difficult to remove from food and water (Paterson and Lima 2005). The methods devised by Castegnaro et al. (1991) at least would be effective for the mycotoxin-contamination of environments such as rooms. The most common procedure for decontamination is washing with bleach which effectively oxidises most aflatoxin and some other mycotoxins (Stark 2005). Potassium permanganate under alkaline conditions appears to be effective for a wider range of mycotoxins and for more situations than bleach, points which have been overlooked. The use of an enzyme to degrade the toxin might be feasible technically but is probably not yet applicable as a routine or emergency procedure. Finally, Sharpira (2004) provides extensive details on decontamination of foods.

21.11 Some Priorities

It needs to be recognised that it is the low molecular weight toxins from fungi that present the biggest threat. It is not the fungus that is the direct threat, apart from the remote possibility of a genetically engineered one causing unconstrained damage (although it is true that fungi have been underestimated as weapons). The use of fungal plant pathogens to devastate crops is a serious threat. An understanding is needed of what are normal levels of fungi and toxins in the environment (Strasser et al. 2000; Gonçlaves et al. 2006, Paterson 2006a, 2007a; Bucheli et al. 2008). The acute (e.g. T-2 toxin) or chronic (e.g. aflatoxins) nature of each mycotoxin needs to be established. Better methods for analysing the toxins are required. Fortunately, methods for multimycotoxin analysis based on chromatography exist (Paterson and Lima 2005). The single method procedures for hundreds of compounds are of particular value and standardised protocols could be based on these. PCR methods for the fungi can be employed (Paterson 2006b, c) although these methods have been compromised by the lack of suitable controls (Paterson 2007c) and optimal protocols for growth of the fungi for testing (Paterson and Lima 2009). There are vast amounts of data on the levels of the more well-known mycotoxins in a variety of foods. CAST (2003) and Venâncio and Paterson (2007) are excellent source materials. However, there are more data from other surveys. It is worthwhile listing those compounds which are water-soluble, as this will be a crucial factor in water and food contamination. Determining which foods would be expected to be contaminated with particular mycotoxins and which would normally not be expected is essential information.

Furthermore, it is crucial to appreciate the uncertainties in mycotoxin analysis (CAST 2003; Whitaker and Johansson 2005). For example, samples of corn contaminated with aflatoxin at (a) 10 ng g^{-1} and (b) 10,000 ng g^{-1} are estimated to vary in a repeated subsequent analysis by (a) 0 to 33.9 ng g^{-1} and (b) 8,992–11,008 ng g^{-1} respectively. One can therefore immediately understand the problem of deciding if a sample was intentionally contaminated. It should go without saying that practical classifications of the fungi that produce the toxins are required. There is a strict requirement to be able to unequivocally identify those isolates from commodities that produce toxins of relevance. A novel scheme is discussed in Paterson et al. (2004, 2006) where it is required to identify consistent morphological characters and then analyse for the toxin of relevance.

21.12 Future Trends

The authors of this chapter predict that there will be more compounds considered as mycotoxins within 10 years. Mycotoxins will become acceptable only at ever-decreasing concentrations tending towards background levels, and increasing numbers will be shown to be toxic and present in different foods. So the mycotoxin circle will become bigger in terms of Fig. 21.1. We predict that the number of FBCA will decrease. Also, compounds from these may begin to be considered more seriously as mycotoxins (e.g. destruxins) and consequently result in fewer FBCA. The trend for weapons is difficult to predict. It may be that they will begin to be considered as "not effective". Alternatively, they could expand into the "non-mycotoxin toxins" (e.g. Cole and Schweikert 2003a, b; Cole et al. 2003). This type of activity is reported in the literature but the compounds are not usually found, or investigated, in food.

21.13 Conclusions

A previous review which considered fungi and toxins from fungi as potential weapons created high interest. The demand for such work was the recent aggressive attacks on innocent citizens and concomitant increased security by governments. We need to be able to prevent rather than react to such events. The Australia Group deliberations are essential reading for anyone working in the field (http://www.australiagroup.net). Bioweapons, mycotoxins, FBCA and even pharmaceuticals need to be considered in the context of the new paradigm. None of the toxins are as (a) toxic as botulinin toxin from *C. botulinum*, and (b) dangerous as nuclear weapons. However, they are more dangerous than, for example, teargas. A toxin may be considered as a pharmaceutical and vice versa simply by a small change in concentration or a moiety. Fungal toxins of use as weapons may be defined as any toxic compound from fungi which could be "weaponised". The current list of fungal toxins as biochemical weapons is small although awareness is growing of the threats they may pose. T-2 toxin is perhaps the biggest concern. A clear distinction is required between the biological (fungus) and chemical (toxin) aspects of the issue. Various factors need to be considered and not simply overall toxicity or notoriety. Ease of "weaponisation" is important. The classification of toxins as potential biological weapons appears anomalous as they are chemicals and it is suggested here that they be considered as chemical weapons. There is an obvious requirement to be able to trace the fungi and compounds which are produced in the environment and to know when concentrations are abnormal. Many FBCA produce toxins and so the use of these preparations requires additional consideration.

The chemotaxonomy and identification of the mycotoxin producing fungi needs to be reconsidered. There is a great need to be able to link consistently toxin production with particular fungi isolated from commodities. On the positive side, it is possible to treat mycotoxicosis and to decontaminate mycotoxins. There is considerable confusion and inconsistency surrounding the topic of bioweapons which requires to be assessed in an impartial and scientific manner. It is fundamental to be able to differentiate between abnormal and normal concentrations of toxins or fungi in the food/water supply.

Acknowledgements R.R.M. Paterson is funded by grant SFRH/BPD/34879/2007 from Fundação para a Ciência e a Tecnologia, Portugal.

References

Bailey KC (2001) The biological and toxin weapons threat to the United States. National Institute for Public Policy, Fairfax, VA, pp 1–16

Bennet JW, Klich M (2003) Mycotoxins. Clin Microbiol Rev 16:497–516

Bucheli TD, Wettstein FE, Hartmann N, Erbs M, Vogelgsang S, Forrer H-R, Schwarzenbach RP (2008) *Fusarium* mycotoxins: Overlooked aquatic micropollutants? J Agric Food Chem 56:1029–1034

Burnett J (2003) Fungal populations and species. Oxford University Press Inc, New York, pp 1–348

Casadevall A, Pirofski L-A (2006) The weapon potential of human pathogenic fungi. Med Mycol 44:689–696

CAST (2003) Mycotoxins: Risks in plant, animal, and human systems. Council for Agricultural Science and Technology, Ames, IA, USA, pp 1–199

Castegnaro M, Barek J, Fremy JM, Lafontaine M, Miraglia M, Sansone EB, Telling GM (1991) Laboratory decontamination and destruction of carcinogens in laboratory wastes. Some mycotoxins. IARC Scientific Publications no. 113. International Agency for Research on Cancer, Lyon

Castillo M-A, Moya P, Hernández E, Primo-Yúfera E (2000) Susceptibility of *Ceratitis capitata* Wiedemann (Diptera: Tephritidae) to entomopathogenic fungi and their extracts. Biol Contr 19:274–282

Cole RJ, Schweikert MA (2003a) Handbook of secondary fungal metabolites, vol 1. Academic Press, San Diego, pp 1–1006

Cole RJ, Schweikert MA (2003b) Handbook of secondary fungal metabolites, vol 2. Academic Press, San Diego, pp 1–891

Cole RJ, Jarvis BB, Schweikert MA (2003) Handbook of secondary fungal metabolites, vol 3. Academic Press, San Diego, pp 1–672

Connick Jr WJ, Daigle DJ, Pepperman AB, Hebber KP, Lumsden RD, Anderson TW, Sands DC (1998) Preparation of stable, granular formulation containing *Fusarium oxysporum* pathogenic to narcotic plants. Biol Contr 13:79–84

Cotty P (1989) Virulence and cultural characteristics of two *Aspergillus flavus* strains pathogenic on cotton. Phytopath 79:808–814

de Vries GE (2000) *Fusarium* considered to kill coca plants. Trends Plant Sci 5:417

Dohnal V, JeŽková A, Kuča K, Jun D (2007) T-2 toxin: Occurrence and detection | [T-2 toxin: Výskyt a detekce. Cesk Slovenska Farm 56:174–177

Elad D (2005) Risk assessment of malicious biocontamination of food. J Food Prot 68:1302–1305

Garber EAE, Eppley RM, Stack ME, McLaughlin MA, Park DL (2005) Feasibility of immunodiagnostic devices for the detection of ricin, amanitin, and T-2 toxin in food. J Food Prot 68:1294–1301

Glenn AE (2007) Mycotoxigenic *Fusarium* species in animal feed. Anim Feed Sci Technol 137:213–240

Gonçalves AB, Paterson RRM, Lima N (2006) Survey and significance of filamentous fungi from tap water. Int J Hyg Environ Health 209:257–264

Hedayati MT, Pasqualotto AC, Warn PA, Bowyer P, Denning DW (2007) *Aspergillus flavus*: human pathogen, allergen and mycotoxin producer. Microbiol 153:1677–1692

Holstege CP, Bechtel LK, Reilly TH, Wispelweyc BP, Dobmeier SG (2007) Unusual but potential agents of terrorists. Emerg Med Clin N Am 25:549–566

HSE (2007) Initial report on potential breaches to biosecurity at the Pirbright site. Health and Security Executive, UK, pp 1–5
Kuiper-Goodman T (2004) Risk assessment and risk management of mycotoxins in food. In: Magan N, Olsen M (eds) Mycotoxins in food detection and control. Woodhead Publishing Ltd, Cambridge, UK, pp 3–27
Latxague E, Sache I, Pinon J, Andrivon D, Barbier M, Suffert F (2007) A methodology for assessing the risk posed by the deliberate and harmful use of plant pathogens in Europe. EPPO Bull 37:427–435
Lenain L, Bonturi M, Koen V (2002) The fallout from terrorism: security and the economy. OECD Observer 231/232:9–11
Locasto DA, Allswede M, Stein TM (2004) CBRNE – T-2 Mycotoxin. *e-medicine* June. http://www.emedicine.com/emerg/topic890.htm cited 20 Mar 2008
Mays LW (2004) Water supplies systems security. AWWA Products, Denver, pp 1–492
Miller AJ, Hileman CL, Droby S, Paster N (2005) Science and technology based countermeasures to foodborne terrorism: Introduction. J Food Prot 68:1253–1255
Nogueira R, Estevinho I, Abrunhosa L, Mendonc C, Machado P, Carballa M, Brito AG, Venâncio A (2007) Assessing the degradation of ochratoxin A using a bioassay: the case of contaminated winery wastewater. Water Sci Technol 56:55–61
OECD (2007) Best practice guidelines for biological resource centres. OECD Publishing, pp 1–115
Paterson RRM (2006a) Fungi and fungal toxins as weapons. Mycol Res 110:1003–1010
Paterson RRM (2006b) Identification and quantification of mycotoxigenic fungi by PCR. Proc Biochem 41:1467–1474
Paterson RRM (2006c) Primers from the isoepoxydon dehydrogenase gene of the patulin biosynthetic pathway to indicate critical control points for patulin contamination of apples. Food Contr 17:741–744
Paterson RRM (2007a) Aflatoxins contamination in chilli samples from Pakistan. Food Contr 18:817–820
Paterson RRM (2007b) Zearalenone production and growth in drinking water inoculated with *Fusarium graminearum*. Mycol Prog 6:109–113
Paterson RRM (2007c) Internal amplification controls have not been employed in diagnostic fungal PCR, hence potential false negative results. J Appl Microbiol 102:1–10
Paterson RRM (2008) Fungal enzyme inhibitors as pharmaceuticals, toxins, and scourge of PCR. Curr Enzym Inhib 4:46–59
Paterson RRM, Lima N (2005) Fungal contamination of drinking water. In: Lehr J, Keeley J, Lehr J, Kingery III TB (eds) Water Encyclopedia. Wiley, New York, pp 1–7
Paterson RRM, Lima N (2009) Mutagens manufactured in fungal culture may affect DNA/RNA of producing fungi. J Appl Microbiol 106:1070–1080
Paterson RRM, Kelley J, Gallagher M (1997) Natural occurrence of aflatoxins and *Aspergillus flavus* (LINK) in water. Lett Appl Microbiol 25:435–436
Paterson RRM, Venâncio A, Lima N (2004) Solutions to *Penicillium* taxonomy crucial to mycotoxin research and health. Res Microbiol 155:507–513
Paterson RRM, Venâncio A, Lima N (2006) A novel identification system based on 318 Penicillia strains using the isoepoxydon dehydrogenase gene and patulin production. Rev Iberoam Micol 23:155–159
Paterson RRM, Sariah M, Lima N, Zainal Abidin MA, Santos C (2008) Mutagenic and inhibitory compounds produced by fungi affect detrimentally their PCR-based diagnosis and phylogenetic analyses. Curr Bioactive Comp 4: 245–257
Paterson RRM, Moen S, Lima N (2009) Feasability of producing oil palm with altered lignin to control *Ganoderma* disease. J Phytopathol (in press)
Pohanka M, Jun D, Kuca K (2007) Mycotoxin assays using biosensor technology: A review. Drug Chem Toxicol 30:253–261

Santos C, Paterson RRM, Venâncio A, Lima N (2009) Filamentous fungal characterisations by matrix-assisted laser desorption/ionisation time of flight mass spectrometry. J Appl Microbiol (in press)

Shannon M (2004) Management of infectious agents of bioterrorism. Clin Pediatric Emerg Med 5:63–71

Sharpira R (2004) Control of mycotoxins in storage and techniques for their decontamination. In: Magan N, Olsen M (eds) Mycotoxins in food detection and control. Woodhead Publishing Ltd, Cambridge, UK, pp 190–213

Skrobek A, Boss D, Défago G, Butt TM, Maurhofer M (2005) Evaluation of different biological test systems to assess the toxicity of metabolites from fungal biocontrol agents. Toxicol Lett 161:43–52

Sperber WH (2005) HACCP does not work from farm to table. Food Contr 16:511–514

Stark A-A (2005) Threat assessment of mycotoxins as weapons: molecular mechanisms of acute toxicity. J Food Prot 68:1285–1293

Strasser H, Vey A, Butt TM (2000) Are there any risks in using entomopathogenic fungi for pest control, with particular reference to the bioactive metabolites of *Metarhizium, Tolypocladium* and *Beauveria* species? Biocontrol Sci Technol 10:717–735

Tucker JA (2003) Biosecurity: Limiting terrorist access to deadly pathogens. Peaceworks no. 52. United States Institute of Peace, Washington, pp 1–49

Venâncio A, Paterson R (2007) The challenge of mycotoxins. In: McElhatton A, Marshall RJ (eds) Food safety – a practical and case study approach. Springer, New York, pp 24–47

Wagner BL, Lewis LC (2000) Colonization of corn, *Zea mays*, by the entomopathogenic fungus *Beauveria bassiana*. Appl Environ Microbiol 66:3468–3473

Whitaker TB, Johansson AS (2005) Sampling uncertainties for the detection of chemical agents in complex food matrices. J Food Prot 68:1306–1313

WHO (2003) Laboratory biosafety manual, 2nd edn (revised). Interim guidelines. WHO, Geneva, pp 1–99

Chapter 22
Masked Mycotoxins and Mycotoxin Derivatives in Food: The Hidden Menace

Chiara Dall'Asta, Gianni Galaverna, Arnaldo Dossena, Stefano Sforza, and Rosangela Marchelli

22.1 Introduction

Conjugated or masked mycotoxins first came to attention in the mid-1980s, mainly from clinical observations on animals affected by mycotoxicoses fed with apparently low-level mycotoxin-contaminated feed. The unexpected toxicity was attributed to the presence of conjugated forms of mycotoxins, possibly generated by the plant metabolism, which were undetected by the analytical methods commonly used for unmodified mycotoxins. These derivatives could be hydrolysed to the precursor toxins in the digestive tracts of animals, or could exert toxic effects comparable to those imputable to free mycotoxins. It soon became evident that mycotoxins can be structurally modified not only by interactions with plants, but also by food processing technologies, and also that the derivatives formed in the latter case may preserve the original toxicity of their precursors.

The occurrence of masked mycotoxins formed by plant metabolism or mycotoxin derivatives due to food processing will be reported in this chapter, whereas data on derivatives formed by animal metabolism will not be included, since they have been extensively reported in the literature (Galtier 1998). Indeed, although metabolic plant-derived transformations and process-induced modifications of mycotoxins have been studied for many years, they are surprisingly scarcely mentioned in the literature, at least not for all important mycotoxins. In this chapter, the term "masked mycotoxins" will be specifically intended for mycotoxins conjugated to other molecules (i.e. proteins, carbohydrates, etc.) and which can be released after ingestion.

Mycotoxin derivatives may have a very different chemical behaviour from the precursor, including polarity and solubility. Thus, they can easily escape routine

G. Galaverna (✉)
Dipartimento di Chimica Organica e Industriale, Viale G.P. Usberti 17A, Campus Universitario, Università di Parma, I-43100, Parma, Italy
e-mail: gianni.galaverna@unipr.it

M. Rai and A. Varma (eds.), *Mycotoxins in Food, Feed and Bioweapons*,
DOI: 10.1007/978-3-642-00725-5_22, © Springer-Verlag Berlin Heidelberg 2010

analyses, since their extraction may be difficult using usual solvents; they could be lost in the clean-up process, or they may be characterized by a different chromatographic behavior or by different spectroscopic properties, which hamper their detection. Moreover, standards for these substances are usually not commercially available, making their identification and quantification a hard task. Last, but not least, they are generally not regulated by legislation, so no legal limits have been established, and are not an imperative issue for either food producers or control agencies.

In any case, conjugated mycotoxins must not be neglected, since they can be hydrolysed in the gastrointestinal tract, releasing the toxins, or can have toxic properties themselves. Humans and animals consuming mycotoxin-contaminated crops are therefore exposed not only to native mycotoxins, but also to their derivatives formed either by plants or by food processing technologies: today, the toxicity of these compounds is still largely unknown.

22.2 Masked Mycotoxins and Mycotoxin Derivatives Arising from the Interactions of Moulds with Plants

Plants protect themselves from xenobiotics and residues according to two major mechanisms. The former is by physical compartmentalization which segregates these compounds in specific organs, tissues, organelles or extracytoplasmatic space, where they can only minimally affect the normal metabolism of the plant. The latter is by chemical compartmentalization, according to which mycotoxin reactive functional groups are reduced or conjugated to other cell components, giving rise to products with a lower toxicity (Conn 1985; Sandermann 1992). Both mechanisms are elicited by plant-microorganism interaction, and may play a fundamental role in plant resistance against several virulence factors.

Although plant compartmentalization was originally studied for pesticide degradation, evidence for the metabolic transformation of a mycotoxin was first obtained from studies on the production of deoxynivalenol (DON) in field corn infected with *Fusarium graminearum*. Mycotoxin concentrations in corn reached a maximum and then declined during the growing season, probably due to the breakdown of DON by plant enzymes (Miller and Young 1985).

Similar results have been reported also for ochratoxin A (OTA): a decrease of the mycotoxin was observed after wheat and rye had been stored for more than 6 months, whereas OTA content remained constant for even 2 years when highly contaminated crops were maintained in sterile conditions (Wolff 1995). Thus, enzymes play a fundamental role in mycotoxin degradation or masking processes during storage.

As it will be shown below, plant metabolites have been identified so far for zearalenone, deoxynivalenol and ochratoxin A. Moreover, there is some evidence for the compartmentalization of fumonisins in plants. Generally, cell cultures have been used for the isolation and structural identification of mycotoxin metabolites. However, although cell cultures generally represent the metabolic reactions of the whole plant, they may eventually not perform all of the possible reactions occurring in plants, and may allow a different quantitative distribution. Nevertheless, several studies have demonstrated that mycotoxins, once formed by plant-inhabiting fungi, may be metabolised in the plant, producing derivatives, conjugates and non-extractable residues.

Unfortunately, very few reports have been published so far providing evidence for the occurrence of masked mycotoxins in naturally contaminated plants, and for their toxicity and metabolic role in mammals.

22.2.1 Zearalenone

Zearalenone (ZEN) is a well-known mycotoxin produced by different *Fusarium* species in grains. ZEN exerts its toxic action by binding to estrogen receptors, thus inducing an estrogenic response in livestocks. As a contaminant of feeding stuffs, zearalenone is mainly harmful for swine and horse, although several studies have reported its estrogenic activity also in humans (Zinedine et al. 2007). Clinical observations of suspected cases of ZEN mycotoxicoses do not always correlate with the low mycotoxin contents found in the ingested feeds. As a possible explanation, several authors have suggested the co-occurrence in feed of other estrogenic compounds such as ZEN conjugates (Gareis et al. 1990). Moreover, comparative analyses of cereals pretreated with or without β-glucosidase showed a significant increase of ZEN concentration in most samples in the former case, suggesting the occurrence of ZEN-glycoside conjugates (Zinedine et al. 2007).

Since mycotoxins are produced in food during the mutual competition of several fungi which colonize the medium, several studies have been performed in order to investigate the ability of microorganisms to metabolize zearalenone to give derivatives. Once mycotoxins are released in the medium, they may actually be taken up by other fungal cells as carbon sources, and may be metabolically converted into new substances. Indeed, the in vitro investigation of such model systems may be helpful for metabolite isolation and characterization.

The ability of plants to convert zearalenone to more polar derivatives has mainly been studied using cell-suspension cultures of plants usually affected in the field by *Fusarium* sp. infection. The ZEN transformation by cell-suspension cultures of *Zea mays* to give rise to α- and β-zearalenol (ZOLs), as well as to their β-glucosides, was reported for the first time by Zill et al. (1990). Moreover, during prolonged incubation of maize cell cultures, ZEN was shown to associate with a number of different cell-wall components, primarily with lignin but also with starch and hemicellulose. Since ZEN is strongly bound to lignin and hemicellulose, the free toxin may not be released during digestion from these structures. However, starch conjugates may be broken down by the action of α-amylase in the saliva and small intestine of mammals.

The occurrence of zearalenone-4-glucoside (Z4G) in wheat was definitely proved by Schneweis et al. (2002). This report is a milestone in the masked mycotoxin field, since it described for the first time the occurrence of a conjugated mycotoxin in naturally contaminated cereals, thus confirming all the speculations about the ability of cereals to metabolize mycotoxins, which were previously based on indirect or *in vitro* evidence. First, Z4G was synthesised to be used as the analytical standard for identity confirmation; then, a suitable mass spectrometric method was developed in order to detect its occurrence in wheat. Thereafter, the authors were able to detect Z4G in 10 out of 24 naturally contaminated samples at a level comparable to the zearalenone contamination in the same samples.

22.2.2 Deoxynivalenol

Deoxynivalenol (DON) and Nivalenol (NIV) are the most important trichothecene mycotoxins in cereal production, and are produced by several *Fusarium* species: DON is actually the most diffuse mycotoxin in grains. The main deoxynivalenol producers *F. graminearum* and *Fusarium culmorum* are responsible for two of the main diseases affecting crops worldwide: fusarium head blight (FHB) in wheat and fusarium ear rot in maize. The topic of masked deoxynivalenol was studied for over 3 decades, and particular attention was paid to its role as virulence factor in the FHB pathogenesis.

One of the first observations suggesting the occurrence of a metabolic transformation of deoxynivalenol in wheat was that FHB resistant wheat cultivars had a lower *F. graminearum* biomass and a lower concentration of DON than the susceptible ones (Miller et al. 1983; Scott et al. 1984).

Suspension cultures of the resistant cultivars incubated with sublethal concentration of ^{14}C-DON were able to metabolize this compound (Miller and Arnison 1986). Moreover, in a yeast-raised bread made from contaminated wheat flour, the concentration of DON increased over 100% (Young et al. 1984). The most plausible explanation was that the toxin had been metabolized by wheat to a DON derivative, which could be transformed back to DON during the process of bread-making. In addition, partial resistance of wheat to FHB has been suspected to be mediated by the formation of a DON conjugate (Miller and Young 1985).

As already described for zearalenone derivatives, deoxynivalenol derivatization was studied by using maize or wheat cell-suspension cultures. Also in this case, the conversion of DON to a more polar derivative, which was isolated and identified as DON-3-glucoside by means of NMR and mass spectrometry measurements, was observed (Sewald et al. 1992). Deoxynivalenol-3-O-β-glucopyranoside (D3G) had already been synthesized and characterized by NMR and mass spectrometry by Savard (1991).

Recently, Berthiller et al. (2005a, 2005b) developed a powerful LC-MS/MS method for the determination and identification of DON glucosides, which was applied to several wheat samples which had been treated at anthesis with deoxynivalenol (Berthiller et al. 2005a, 2005b). The presence of a DON glucoside was detected in all the analysed samples, and was identified as deoxynivalenol-3-glucoside by comparison of its mass spectrum with the one obtained for the synthesized authentic specimen (Dall'Asta et al. 2005).

The first report on the natural occurrence of a glucoside of deoxynivalenol in *Fusarium*-infected wheat and maize was presented by Berthiller et al. (2005b), who studied wheat varieties characterised by different FHB resistance treated with *Fusarium* spp. at anthesis. After harvest, all 56 treated wheat samples were shown to contain DON and D3G. Moreover, the authors reported the occurrence of DON-3-glucoside in several naturally DON-contaminated samples of maize and wheat, in a range from 4% to 12% of DON concentration. On the base of these data, Lemmens et al. (2005) investigated the hypothesis that resistance to deoxynivalenol was a major resistance factor in the FHB resistance complex of wheat. In particular, 96 lines with a variable resistance to FHB were treated with DON or inoculated with *Fusarium* spp. The collected data allowed to postulate a correlation among resistance to the toxin and resistance to spread of FHB ($r = 0.74$, $P < 0.001$). Moreover, the DON-resistant phenotype was closely associated with an important FHB resistance quantitative trait locus (QTL), *Qfhs.ndsu-3BS*, which had been previously identified as governing resistance to spread of symptoms in the ear, and which might either encode a DON-glucosyl-transferase or regulate the expression of this enzyme. The authors concluded that resistance to DON is important in the FHB resistance complex, and clarified the role played by deoxynivalenol in the fusarium head blight pathogenesis.

Although very good results have been obtained in the last few years to understand the biological role of deoxynivalenol-3-glucoside in grains, further investigations should be carried out to clarify the biological role of this compound in animals and humans, as well as on other trichothecenes such as nivalenol which might be involved in similar metabolic pathways. Moreover, it would be interesting to investigate the fate of this compound after digestion by mammals, in order to verify if D3G should be regarded as potentially hazardous for human and animal health.

22.2.3 Ochratoxin A

As already reported for other mycotoxins, the toxic effects of ochratoxin A also cannot be completely explained on the basis of its known biochemical sites of action: thus, ochratoxin metabolites or derivatives may be considered to act synergistically, giving a significant contribution to the overall toxicity of the contaminated stuff.

Plant metabolism of ochratoxin A was first studied using cell-suspension cultures of wheat and maize incubated with a sublethal dose of ^{14}C-OTA (Ruhland et al. 1996a). In addition to ochratoxin α (OTα), the main metabolites isolated were (4R)- and (4S)-4-hydroxyochratoxin A. In addition, β-glucosides of both isomers were also found in large amounts. Ochratoxin α is commonly regarded as non-toxic, whereas hydroxy-ochratoxin A is an immunosuppressant as effective as OTA itself. The toxicity of the other derivatives is still unknown.

Ruhland et al. (1996b) investigated the ability of crops to metabolize ochratoxin A, using cell-suspension cultures of several crop plants: the isolated derivatives were the same for all the tested vegetables, and the conversion of OTA was nearly complete, although the quantitative distribution differed strongly depending on the plant. OTA transformation occurred also in intact vegetables and germinating cereals after spiking with ochratoxin A. Also in this case, the same metabolites occurring in cell-suspension cultures were detected (Ruhland et al. 1997).

No further studies have been performed to date to check whether these derivatives really occur in food commodities. Thus, also in this case further investigations are strongly required in order to evaluate the occurrence of the masked derivatives in naturally OTA-contaminating commodities.

22.3 Transformations of Mycotoxins During Food Processing

Generally, mycotoxins are thermally stable compounds, and therefore they are largely unaffected during most food processing operations. The effect of food processing on mycotoxins has been the subject of a great number of papers in the last 2 decades, and relevant reviews have already been published about this topic (Saunders et al. 2001; Castells et al. 2005; Bullerman and Bianchini 2007).

Apart from sorting to remove contaminated parts or units, levels of mycotoxins in contaminated commodities may be reduced or redistributed by food processing procedures. Food processes have variable effects on mycotoxins, with those utilizing higher temperatures showing greater effects; however, reduction is only partial and mycotoxins are not completely eliminated. In some cases, only masking effects or even release of hidden mycotoxins are the results of food processing.

22.3.1 Fumonisins

The effects of thermal processing on the chemical structure and toxicity of fumonisins have been recently reviewed by Humpf and Voss (2004). Baking, frying and extrusion cooking of corn at high temperature (>190°C) have been found to reduce fumonisin concentration in food, depending on the process conditions. However, it is not known if the reduced concentration is due to thermal decomposition of fumonisins or to their binding to proteins, sugars or other compounds in the food matrix.

The main reaction occurring to fumonisins during thermal treatments, and in particular when alkaline conditions are applied, is the hydrolysis of the tricarballylic esters at C-14 and C-15, giving rise to the partially hydrolysed or totally hydrolysed fumonisins (Dombrink-Kurtzman et al. 2000). These derivatives seem to be as toxic as the native mycotoxin; thus, they should be regarded as a possible hazard for consumer health (Voss et al. 2001).

Another important reaction occurring in heat-treated food involves fumonisin B1 and reducing sugars to form N-(Carboxymethyl) fumonisin B_1, according to a Maillard-type reaction followed by an oxidation step. This reaction seems to be favoured by alkaline pH and temperature. Although the impact of this derivative on consumer health is not known, the substitution with a carboxymethyl group should inhibit the interaction of fumonisin B1 with ceramide synthase, which is the main cause of toxicity.

The low recoveries of fumonisins often reported for corn flakes might have been due to the extraction procedure used (De Girolamo et al. 2001) or to the substantial loss of free fumonisins observed during processing (Pineiro et al. 1999; Castelo et al. 2001; Meister 2001). No hydrolysis products were observed in extruded samples, so it was concluded that hydrolysis was not the degradative reaction. The brown colour observed in extruded samples could have rather indicated that FB1 had reacted with sugars giving a Maillard-type reaction (Pineiro et al. 1999; Castelo et al. 2001). On the other side, eventual Maillard-type products could not be detected by the traditional HPLC method, which requires derivatization of the amino group, since the latter is no longer available for derivatization.

Moreover, several studies have reported low recovery values of fumonisin in different matrices such as rice flour, cornstarch, cornmeal and even glucose (Kim et al. 2002). This fact was ascribed to the existence of interactions of FB1 with food macroconstituents such as protein or starch, which lead to non-detectable "hidden forms". Thus, according to these hypotheses, the loss of fumonisins during heat processing may be ascribed to binding to proteins (Shier et al. 1997; Shier and Abbas 1999; Resch and Shier 2000; Shier 2000). The postulated linkages should be cleaved by alkaline hydrolysis, releasing HFB1. The binding of fumonisin to proteins has been hypothesised on the basis of several pieces of experimental evidence. In an experiment with ^{3}H-labelled FB1 added to corn meal dough, after roasting, only 37% of the radioactivity was extractable; 51% of the extracted radioactivity consisted of FB1 and HFB1. Another 46% of the radioactivity was extractable with 1% sodium dodecylsulphate (SDS), a detergent used to dissolve proteins. This fraction treated under alkaline conditions released fumonisin as HFB1.

In order to study the binding of fumonisins to matrix components in thermal-treated food, model experiments have been performed (Seefelder et al. 2003a). Fumonisin B1 and HFB1 were incubated with β-D-glucose and sucrose (as models for mono- and disaccharides), with methyl β-D-glucopyranoside (as a model for starch), and with the amino acid derivatives N-α-acetyl-L-lysine methyl ester and BOC-L-cysteine methyl ester (as models for proteins). The incubation of D-glucose with FB1 or HFB1 resulted in the formation of Amadori rearrangement products. Conjugates were found following the reaction of FB1 with sucrose, methyl-β-D-glucopyranoside and the amino acid derivatives. The stable reaction product formed by heating methyl-β-D-glucopyranoside with fumonisin B1 was purified, and identified by NMR spectroscopy as the diester of the fumonisin tricarballylic acid side chains with methyl β-D-glucopiranoside. On the basis of these model experiments, it was concluded that fumonisins are able to bind to polysaccharides and proteins via their two tricarballylic acid side chains, and binding to starch occurs in much higher yield as compared to proteins, as shown with amino acid derivatives.

Nevertheless, recently the studies about hidden fumonisins focused on the protein-bound fraction. The occurrence of protein-bound fumonisins in commercial cornflakes was demonstrated by Kim et al. (2003), who used an indirect determination method: samples of retail corn flakes were analysed for both free fumonisins and protein-bound fumonisins, which were extracted with SDS and measured as HFB1 after alkaline hydrolysis. An average of 2.6 times higher content of FB1 as bound form was found. The method proposed by Kim was improved by Park et al. (2004) and applied to 30 retail samples of heat-processed corn foods: after extraction of the protein-bound FB1 with a 1% SDS, the surfactant was removed by complexation with methylene blue, and the sample underwent alkaline hydrolysis. In order to measure the total bound FB1, the sample itself was hydrolysed under alkaline conditions. In both cases, clean-up was accomplished by solid-phase extraction. Bound FB1 was found in all samples at significative levels. However, although successful, this procedure is cumbersome and time-consuming, and it presents several drawbacks, such as a complicated chromatographic separation because of methylene blue residues in the samples. Recently, further evidence for supporting the hypothesis of a covalent binding between fumonisins and the protein fraction in maize flour has been reported (Dall'Asta et al. 2008): the four main protein fractions were separated using the Osborne protocol, and analysed after alkaline hydrolysis for the presence of bound fumonisin. The occurrence of HFB1 was detected in both prolamins and globulins, whereas no bound fumonisins were found in albumins and glutelin fractions.

In conclusion, the issue of hidden fumonisins requires further investigations about the reaction sites involved in the binding phenomena, and the biological role of the conjugated fumonisins in terms of toxicity and bioavailability. Moreover, nothing is known at the moment about masking effects exerted by plants on fumonisins during "in field" contamination: further studies are required also in this case in order to get a full picture of the hidden fumonisin issue in corn-based food.

22.3.2 Trichothecenes

Many studies have been devoted to the fate of trichothecenes during grain processing (Scott et al. 1983; Abbas et al. 1985; Hazel and Patel 2004).

Surveillance of retail foods and drinks of cereal origin demonstrated that trichothecenes do survive the production processes employed, since trichothecenes are relatively heat-stable chemicals. Several authors reported the isomerisation of deoxynivalenol to iso-DON during baking. Isomerisation occurred to the extent of 3–13% of the DON present in the sample; levels were higher in the crust than in the crumbs (Greenhalgh et al. 1984). The stability of DON and NIV to thermal treatments performed under mild alkaline conditions have been extensively investigated by Bretz et al. (2005, 2006) using *in vitro* model systems.

After optimization of the heating condition for obtaining maximum yield, four degradation products were isolated and characterised for NIV, and six for DON.

The presence of these degradation products in commercially available foods was investigated: as far as nivalenol derivatives are concerned, degradation forms were detected only in one of the samples, most probably due to the low contamination of foodstuff with native NIV. On the other hand, DON derivatives were detected in 29–66% of the samples. The toxicity of these degradation products was studied by cell culture experiments, leading to the conclusion that they are less cytotoxic than the parent NIV and DON.

A more problematic issue is the behaviour of masked mycotoxins during processing of naturally contaminated cereals. Only few studies have been published in this field, although some evidence has been reported about the partial releasing of the native mycotoxins during technological treatments.

An unexpected increase of DON levels during yeast doughnut production was reported by Young et al. (1984). In this study, the DON content was compared to that found in the wheat flour used for processing. The increase of the DON level in the final product was recorded only for doughnuts, which were the only products involved in yeast fermentation. It was postulated that the increase might have been due to the enzymatic conversion of some precursors into DON; nevertheless, the identity of such precursors was not further investigated.

Very recently, Scudamore et al. (2008) examined the behaviour of deoxynivalenol, nivalenol and zearalenone during extrusion of naturally contaminated wholemeal wheat flour, using pilot scale equipment. Factors considered during the experiments were temperature and moisture content. Concentrations of the three mycotoxins experienced little change by extrusion, although the amount of deoxynivalenol decreased at the lowest moisture content. However, this effect did not appear to be temperature-dependent, and it was suggested that the apparent loss was either due to binding of the mycotoxin to food components or to inability to extract the residue. Under some conditions, concentration of the mycotoxins, particularly nivalenol and zearalenone, were higher after extrusion. Moreover, the trend for residual NIV and ZEN values, which increased, going from 140°C to 180°C, is contrary to what would be expected from chemical degradation. The answer to these changes may lie with the efficiency of the extraction solvent used for removing mycotoxins in the extruded materials, with another hypothesis that bound mycotoxins were released under the extrusion conditions — or may be a combination of both.

It must be noted that current EC legislation relies on processing to reduce DON concentrations: in the light of these results, it seems unlikely to achieve the reduction for some products such as snacks made by simple extrusion of wheat and/or maize flour.

22.4 Toxicity of Mycotoxin Derivatives

Toxicity of glycosilated or other conjugated mycotoxins can obviously emerge via the release of the free toxins during the gastrointestinal process (Gareis et al. 1990); however, mycotoxin derivatives can also be toxic per se. The literature regarding the toxicity of plant metabolism-derived or process-derived mycotoxin derivatives is extremely scarce. The only partial exception is represented by reports on fumonisin, in particular fumonisin B1, as the formation of HFB1 by alkaline treatments is well-known since quite a long time. Reduced derivatives of zearalenone, common metabolic by-products, have also been studied up to a certain degree. Finally, a few studies on the toxicity of ochratoxin A derivatives are also present in the literature. Thus, it is evident that studies on the metabolic fate and the potential toxicity of masked mycotoxins and mycotoxin derivatives are extremely urgently required, in order to provide safety regulations for the consumption of such derivatives.

As extensively described above, fumonisins are probably the most likely mycotoxins to be modified through reactions linked to food processing in basic environments. As native forms, fumonisins are well-known for causing a wide spectrum of diseases (liver and renal toxicity and carcinogenicity, neurotoxicity, induction of pulmonary oedema, etc.) (Desai et al. 2002). The molecular basis for their toxicity mainly relies on their ability to inhibit ceramide synthase and, as a consequence, to alter the sphingolipid metabolism (Merrill et al. 1993). Several contradictory results have recently been reported concerning the toxicity of modified fumonisins. HFB1 is a weaker inhibitor of ceramide synthase than FB1 (Van der Westhuizen et al. 1998), but it is also a substrate for acylation by ceramide synthase itself. Moreover, HFB1 is more readily absorbed by intestinal cells, and seems to promote cell proliferation (Caloni et al. 2002). A recent study reported that both HFB1 and palmitoyl-HFB1 may act as ceramide synthase inhibitors, but only the precursor FB1 was able to induce apoptosis in human proximal tubule-derived cells (Seefelder et al. 2003b). To the best of our knowledge, no data exist in the literature concerning toxicity of protein- or carbohydrate-conjugated fumonisins.

With regard to the toxicity of zearalenone and its derivatives, their bioactivity is mainly related to the ability of zearalenone to interact with oestrogen receptors, as demonstrated by a number of in vitro or *in vivo* experiments (Kuiper et al. 1998). Using oestrogen-dependent human breast cancer cells, the oestrogenic activity of zeralenone and its derivatives were compared using 17β-oestradiol as a positive control, confirming an oestrogenic potency in the order: α-zearalenol (which was found to be the main hepatic metabolite in pigs) > α-zearalanol > zearalenone > β-zearalenol (Malenkinejad et al. 2005). To the best of our knowledge, no data are available on the toxicity of glycosilated zearalenone.

22.5 Conclusions

In the last few years it has clearly emerged that, in mycotoxin-contaminated commodities, many structurally related compounds generated by plant metabolism or by food processing can co-exist together with the native toxins. Since many mycotoxin derivatives are probably still unknown, the *in vivo* search for metabolism-derived or process-derived modified mycotoxins has to be seen as a primary issue. Model reactions should be developed in order to mimic in the closest possible way the potential structural transformations occurring to mycotoxins, and new specific analytical procedures should also be devised for the determination of conjugated mycotoxins. The toxicity of these derivatives is mostly unknown, although several papers have indicated a wide range of potential risks, spanning from non-toxic compounds to compounds which are even more toxic than the native molecules. As soon as new toxicological data will emerge, together with new specific analytical procedures, the new mycotoxin derivatives should be included in the legal limits, while at the same time plant biologists, food technologists and food chemists will have the possibility to improve their procedures in order to transform, biologically or technologically, the dangerous toxins in non-toxic compounds.

References

Abbas HK, Mirocha CJ, Pawlosky RJ, Pusch DJ (1985) Effect of cleaning, milling, and baking on deoxynivalenol in wheat. Appl Environ Microbiol 50:482–486

Berthiller F, Krska R, Dall'Asta C, Lemmens M, Adam G, Schuhmacher R (2005a) Determination of DON-3-glucoside in artificially and naturally contaminated wheat with LC-MS/MS. Mycotoxin Res 21(3):205–208

Berthiller F, Dall'Asta C, Schuhmacher R, Lemmens M, Adams G, Krska R (2005b) Masked mycotoxins: determination of a deoxynivalenol glucoside in artificially and naturally contaminated wheat by HPLC-MS/MS. J Agr Food Chem 53:3421–3425

Bretz M, Knecht A, Göckler A, Humpf H-U (2005) Structural elucidation and analysis of thermal degradation products of the fusarium mycotoxin nivalenol. Mol Nut Food Res 49:309–316

Bretz M, Beyer M, Cramer B, Knecht A, Humpf H-U (2006) Thermal degradation of the *Fusarium* mycotoxin deoxynivalenol. J Agr Food Chem 54:6445–6451

Bullerman LB, Bianchini A (2007) Stability of mycotoxins during food processing. Int J Food Microbiol 119:140–146

Caloni F, Spotti M, Pompa G, Zucco F, Stammati A, De Angelis I (2002) Evaluation of Fumonisin B1 and its metabolites absorption and toxicity on intestinal cell line Caco-2. Toxicon 40:1181–1188

Castells M, Marin S, Sanchis V, Ramos AJ (2005) Fate of mycotoxins in cereals during extrusion cooking: a review. Food Addit Contam 22:150–157

Castelo MM, Jackson LM, Hanna MA, Reynolds BH, Bullerman LB (2001) Loss of fumonisin B1 in extruded and baked corn-based foods with sugars. J Food Sci 66:416–421

Conn EE (1985) Chemical conjugation and compartmentalization: plant adaptions to toxic natural products. UCLA Symposium on Molecular and Cellular Biology, New Series. Cell Mol Biol Plant Stress 22:351–365

Dall'Asta C, Berthiller F, Schumacher R, Adam G, Lemmens M, Krska R (2005) DON-glycosides: characterization of synthesis products and screening for their occurrence in DON-treated wheat sample. Mycotoxin Res 21:123–127

Dall'Asta C, Galaverna G, Aureli G, Dossena A, Marchelli R (2008) A LC/MS/MS method for the simultaneous quantification of free and masked fumonisins in corn and corn-based products. World Mycotoxin J 1(3):1–10

De Girolamo A, Solfrizzo M, von Holst C, Visconti A (2001) Comparison of different extraction and clean-up procedures for the determination of fumonisins in maize and maize-based food products. Food Addit Contam 18:59–67

Desai K, Sullards MC, Allegood J, Wang E, Schmelz EM, Hartl M, Humpf HU, Liotta DC, Peng Q, Merril AH (2002) Fumonisins and fumonisin analogs as inhibitors of ceramide synthase and inducers of apoptosis. Biochim Biophys Acta 1585:188–192

Dombrink-Kurtzman MA, Dvorak TJ, Barron ME, Rooney LW (2000) Effect of nixtamalization (alkaline cooking) on fumonisin-contaminated corn for production of masa and tortillas. J Agr Food Chem 48:5781–5786

Galtier P (1998) Biological fate of mycotoxins in animals. Rev Med Vet 149(6):549–554

Gareis M, Bauer J, Thiem J, Plank G, Grabley S, Gedek B (1990) Cleavage of zearalenone-glycoside, a masked mycotoxin, during digestion in swine. J Vet Med B 37(3):236–240

Greenhalgh R, Gilbert J, King RR, Blackwell BA, Startin JR, Shepherd MJ (1984) Synthesis, characterization and occurrence in bread and cereal products of an isomer of 4-deoxynivalenol (vomitoxin). J Agr Food Chem 32:1416–1420

Hazel CM, Patel S (2004) Influence of processing on trichothecene levels. Toxicol Lett 153:51–59

Howard PC, Churchwell MI, Couch LH, Margues MM, Doerge DR (1998) Formation of N-(carboxymethyl)-fumonisin B1. Following the reaction of fumonisin B1 with reducing sugars. J Agr Food Chem 46:3546–3557

Humpf HU, Voss KA (2004) Effects of thermal food processing on the chemical structure and toxicity of fumonisin mycotoxins. Mol Nutr Food Res 48(4):255–269

Kim E-K, Scott PM, Lau BP-Y, Lewis D (2002) Extraction of fumonisins B1 and B2 from white rice flour and their stability in white rice flour, cornstarch, cornmeal, and glucose. J Agr Food Chem 50:3614–3620

Kim EK, Scott PM, Lau BP-Y (2003) Hidden fumonisin in corn flakes. Food Addit Contamin 20:161–169

Kuiper GGJM, Lemmen JG, Carlsson B, Corton JC, Safe SH, van der Saag PT, van der Burg P, Gustafsson JA (1998) Interaction of estrogenic chemicals and phytoestrogens with estrogen receptor beta. Endocrinology 139(10):4252–4263

Lemmens M, Scholz U, Berthiller F, Dall'Asta C, Koutnik A, Schumacher R, Adam G, Buerstmayr H, Mesterhazy A, Krska R, Ruckenbauer P (2005) The ability to detoxify the mycotoxin deoxynivalenol colocalizes with a major quantitative trait locus for Fusarium Head Blight resistance in wheat. Mol Plant Microbe Interact 18:1318–1324

Malenkinejad H, Maas-Bakker RF, Fink-Gremmels J (2005) Bioactivation of zearalenone by porcine hepatic biotransformation. Vet Res 36:799–810

Meister U (2001) Investigations on the change of fumonisin content of maize during hydrothermal treatment of maize. Analysis by means of HPLC methods and ELISA. Eur Food Res Technol 213:187–193

Merrill AH, Vanechten G, Wang E, Sandhoff K (1993) Fumonisin-B(1) inhibits sphingosine (sphinganine) N -acyltransferase and de-novo sphingolipid biosynthesis in cultured neurons in-situ. J Biol Chem 268(36):27299–27306

Miller JD, Arnison PG (1986) Degradation of deoxynivalenol by suspension cultures of Fusarium Head Blight resistant wheat cultivar Frontana. Can J Plant Pathol 8:147–150

Miller JD, Young JC (1985) Deoxynivalenol in an experimental *Fusarium graminearum* infection of wheat. Can J Plant Pathol 7:132–134

Miller JD, Young JC, Trenholm HL (1983) Fusarium toxins in field corn. I. Time course of fungal growth and production of deoxynivalenol and other mycotoxins. Can J Bot 6:3080–3087

Park JW, Scott PM, Lau BP-Y, Lewis DA (2004) Analysis of heat-processed corn foods for fumonisins and bound fumonisins. Food Addit Contam 21:1168–1178

Pineiro MS, Miller J, Silva G, Musser S (1999) Effect of commercial processing on fumonisin concentrations of maize-based foods. Mycotoxin Res 15:2–12

Resch P, Shier WT (2000) The fate of fumonisin during thermal food processing. Lebensmittel-chemie 54:31–44

Ruhland M, Engelhardt G, Schaefer W, Wallnoefer PR (1996a) Transformation of the mycotoxin ochratoxin A in plants: 1. Isolation and identification of metabolites formed in cell suspension cultures of wheat and maize. Nat Toxins 4:254–260

Ruhland M, Engelhardt G, Wallnoefer PR (1996b) Transformation of the mycotoxin ochratoxin A in plants: 2. Time course and rates of degradation and metabolite production in cell suspension cultures of different crop plants. Mycopathologia 134:97–102

Ruhland M, Engelhardt G, Wallnoefer PR (1997) Transformation of the mycotoxin ochratoxin A in artificially contaminated vegetables and cereals. Mycotoxin Res 13:54–60
Sandermann H (1992) Plant metabolism of xenobiotics. Trends Biochem Sci 17:82–84
Saunders DS, Meredith FI, Voss KV (2001) Control of fumonisin: effect of processing. Environ Health Perspect 109:333–336
Savard ME (1991) Deoxynivalenol fatty acid and glucoside conjugates. J Agr Food Chem 39:570–574
Schneweis I, Meyer K, Engelhardt G, Bauer J (2002) Occurrence of zearalenone-4-β-D-glucopyranoside in wheat. J Agr Food Chem 50:1736–1738
Scott PM, Lau PY, Kanhere SR (1983) Effects of experimental flour milling and bread-baking on retention of deoxynivalenol (vomitoxin) in Hard Red spring wheat. Cereal Chem 60:421–424
Scott PM, Nelson K, Kahnere SR, Karpinski KF, Hayward S, Neish GA, Teich AH (1984) Decline in deoxynivalenol (vomitoxin) concentrations in 1983 Ontario winter wheat before harvest. Appl Environ Microbiol 48:884–886
Scudamore KA, Guy RCE, Kelleher B, MacDonald SJ (2008) Fate of the fusarium mycotoxins deoxynivalenol, nivalenol and zearalenone during extrusion of wholemeal wheat grain. Food Addit Contamin 25:331–337
Seefelder W, Knecht A, Humpf H-U (2003a) Bound fumonisin B1: analysis of fumonisin-B1 glyco and amino acid conjugates by liquid chromatography-electrospray ionization-tandem mass spectrometry. J Agr Food Chem 51:5567–5573
Seefelder W, Humpf HU, Schwerdt G, Freudinger R, Gekle M (2003b) Induction of apoptosis in cultured human proximal tubule cells by fumonisins and fumonisin metabolites. Toxicol Appl Pharmacol 192:146–153
Sewald N, von Gleissenthal JL, Schuster M, Mueller G, Aplin RT (1992) Structure elucidation of a plant metabolite of 4-desoxynivalenol. Tetrahedron: Asymmet 3:953–960
Shier WT (2000) The fumonisin paradox: a review of research on oral bioavailability of fumonisin B1, a mycotoxin produced by *Fusarium moniliforme*. J Toxicol 19:161–187
Shier WT, Abbas HK (1999) Current issues in research on fumonisins, mycotoxins which may cause nephropathy. J Toxicol – Toxin Rev 18:323–335
Shier WT, Abbas HK, Badria FA (1997) Structure-activity relationships of the corn fungal toxin fumonisin B1: implications of food safety. J Nat Toxins 6:225–242
Van der Westhuizen L, Shephard GL, Snyman SD, Abel S, Swanevelder S, Gelderblom WCA (1998) Inhibition of sphingolipid biosynthesis in rat primary hepatocytes cultures by fumonisin B1 and other structurally related compounds. Food Chem Toxicol 36:497–503
Voss KA, Poling SM, Meredith FI, Bacon CW, Saunders DS (2001) Fate of fumonisins during the production of fried tortilla chips. J Agr Food Chem 49:3120–3126
Wolff J (1995) Occurrence of mycotoxins in cereals. Getreide Mehl und Brot 49:139–147
Young JC, Fulcher RG, Hayhoe JH, Scott PM, Dexter JE (1984) Effect of milling and baking on deoxynivalenol (vomitoxin) content of eastern Canadian wheats. J Agr Food Chem 32:659–664
Zill G, Engelhardt G, Wohner B, Wallnoefer PR (1990) The fate of the *Fusarium* mycotoxin zearalenone in maize cell suspension cultures. Mycotoxin Res 6:31–40
Zinedine A, Soriano JM, Molto JC, Manes J (2007) Review on the toxicity, occurrence, metabolism, detoxification, regulations and intake of zearalenone: An estrogenic mycotoxin. Food Chem Toxicol 45(1):1–18

Index

A
Absorption, 135
Acetosyringone, 341
3-Acetyldeoxynivalenol, 13
15-Acetyl deoxynivalenol, 13
Achromobacter xylosoxidans, 331
Acid-treatment, 212
Activated carbon (AC), 208, 215
Acute toxic effects, 132
AFB_1 adducts, 97
AFB_1–8,9-exo-epoxide, 94
AF-DNA adducts, 25
Aflatoxicol, 46
Aflatoxin, 14–16, 25–26, 39, 65, 77, 114, 115, 123, 124, 131, 148, 164, 236, 241, 242, 357, 367
Aflatoxin B1 (AFB1), 16, 93, 190, 242
Aflatoxin B2, 16
Aflatoxin G1, 16
Aflatoxin G2, 16
Akakabi Byo, 356
Alcohol, 99, 102–105
Alimentary toxic aleukia, 356
Alkaloids, 331
Allergenic proteins, 132
Alternaria, 21
Amanita phalloides, 373
Amanitin, 371
Ammi visnaga L., 340
Ammoniation, 212
Analysis of mycotoxins, 149–150
Anethum graveolens L., 330, 340
Anthraquinone, 65, 72
Antibiobodies, 29, 282–284
Antifungal agents, 280
Antimicrobial agents, 281–282, 285
Antioxidants, 87
  butylated hydroxyanisole (BHA), 183, 184
  butylated hydroxytoluene (BHT), 183, 184
  effect on growth fumonisin production, 183
  effects on hydrolytic enzyme, production by *Fusarium,* 184
  propylparaben, 183
  trihydroxybutyrophenone, 183, 184
Antiviral activity, 285
Apiaceae, 342
Apiol, 340
Apoptosis, 301
Arachis hypogaea L., 341
Aromatase, 137
Aromatic compounds, 331
Arrays of biosensors, 32
Aspergillus, 21, 65–69, 131, 237–239, 357
  A. candidus, 244
  A. flavus, 78, 164, 170–172, 239, 240, 245, 329, 368
  A. fumigatus, 232, 238, 244
  A. ochraceus, 148, 359
  A. parasiticus, 78, 170, 329
Aspergillus differential medium, 22
Aspergillus section *Flavi,* 329
Atmospheric pressure chemical ionization (APCI), 267

Atmospheric pressure photo ionization (APPI), 267
Australia Group, 369

B
Bacillus subtilis, 331
Bacteria, 279
Baled silage, 51
Balkan endemic nephropathy (BEN), 307, 314, 360
Barley, 148
Beauvericin, 181
Bentonite, 208, 218
Bergapten, 340
Bile, 136
Binding of fumonisins to matrix components, 391
Bioactivity, 394
Biocompetitive microorganisms, 204
Biocontrol, 369
Biological resource centres, 370
Biomedicine, 281
Biopreservative agents, 280
Bioprospecting, 378
Biosafety, 378
Biosecurity, 368, 378
Biosensors, 31
Biotechnological applications, 279
Biotransformation, 136
B lymphocyte, 263
Body weight, 308

C
Calcium, 41–43, 82
cAMP-phosphodiesterase, 340
Carcinogenic, 133
Carcinogenicity, 265
Carvacrol, 341
Catechin, 342
Chemotypes, 256
Chlorine, 213
Chlorogenic acid, 340
Cholestyramine, 219
Chromatin, 70, 71
Cirrhosis, 100, 102
Citrinin, 313
Claviceps, 361
Cloning vectors, 279
Coccidioides immitis, 68
Co-contamination, 253, 259
Colony-forming units (CFUs), 279
Combinatorial peptides, 14
Conidiation, 81
Corn, 40–42, 44–48
Corn. *Véase* maize, 189
Culmorum, 166
Coumarins, 340
Cytochalasins, 375

D
Deoxynivalenol (DON), 13, 26, 143, 150–152, 164, 165, 168, 242, 253, 294, 355, 388
Destruction, 44
Detection of trichothecenes, 296–297
Detoxification, 40
Development, 71, 72
Dietary iron overload, 93, 99
3,4-Digalloyl quinic acid, 340
5,7-Dihydroxychromone, 341
Dillapiol, 330
Dioctatin A, 331
Disadvantages, 46–47
DON-3-glucoside, 388
DON producers, 256
Dothistromin, 68, 71
Dough, 40, 41, 43–46, 48

E
Early split, 202
Ear rot prevention
  biocontrol, 192
  fungicides, 191, 192
  trangenic corn, 191
Ecological, 45
Electronic noses, 32
Electrospray ionization (ESI), 267
ELISA kits, 28–29
ELISA microplate reader, 29
Endocrinal system, 139
Endocrine disrupters (EDs), 138–139
Endocrine organs, 262
Environmental, 139

Enzyme-linked immunosorbent assay (ELISA), 296
Epidemiological marker, 279
ERα, 140
ERβ, 140
Ergot alkaloids, 361
Ergotism, 361
Eriodictyol, 341
Essential oils (EOs), 331
Esterified glucomannan (EG), 219
Estrogenic properties, 138
Estrogens receptors (ERs), 134
Evolution, 69, 70, 72
Extraction, 208
Extrusion, 43, 45

F
FDK, 169
Ferulic acid, 341
Flavobacterium aurantiacum, 214
Flavonoids, 331
Fluorescence, 23
Fluorescent polarization, 31
Food processing, 390
Fumonisin B1 and reducing sugars, 391
Fumonisins, 26, 114, 124, 131, 164, 390
  biological effect, 189–190
  biomarkers, 180
  biosynthesis, 182, 190
  dietary exposure, 180
  effects on public health, 180
  losses in animal production, 180
  methods of detection, 189
  occurrence of, 185–187
  prevention and control, 191–192
  problems in the trading of agricultural commodities, 180
  producers Fusarium species, 180–181
  regulations, 193
  stability during processing, 187–189
Fungi, 113, 114, 116, 119, 122, 125
  spoilage, 53
  visible fungal growth, 53
Fungicides, tebuconazole, 204
Fusarenon-X, 13
Fusariotoxins, 133
Fusarium, 21, 114–117, 124, 132, 237–239
  AFLPs, 181
  dlamini, 181
  F. culmorum, 147, 168
  F. graminearum, 147, 166, 168, 374
  F. oxysporum, 375
  F. poae, 147, 155
  F. proliferatum, 241
  F. sporotrichioides, 155
  F. verticillioides, 171, 241
  mating populations, 180, 181
  molecular markers, 181
  nygamai, 181
  proliferatum, 179, 181–185
  subglutinans, 181
  teleomorph, 181
  vegetative compatibility groups (VCG), 181
  verticillioides, 179–185, 188, 190–192
Fusarium head blight (FHB), 167, 202, 253

G
Ganoderma, 367
Gastrointestinal tract, 133
Gene cluster, 65, 68–72, 78
Gene duplication, 69, 70
Genetically modified fungi, 377
Glucans, 286
Grain samples of winter, 148
Granular cells, 136
Gulf War, 357

H
Halo of inhibition, 276
Harvesting, 201, 204
Hazard analysis critical control point (HACCP), 370
Head blight syndrome, 166–170
Hematopoietic, 262
Hepadnaviridae, 94
Hepatitis B virus, 97–98
Hepatitis C virus, 98–99
Hepatocarcinogenesis, 98–99, 102–105
Hepatocarcinogens, 94
Hepatocellular carcinoma, 93
Herbal medicine, 331
High performance liquid chromatography, 27

Histone H2A, 71
HM-1, 281, 284
HPLC–mass spectrometry, 27–28
HPLC–tandem mass spectrometry, 27–28
3β-HSD, 137
17β-HSD, 137
HT-2, 133
HT toxin, 165
HT-2 toxin, 293
Hydrated sodium calcium aluminosilicate (HSCAS), 215–218
Hydrodistillation, 343
Hydrolysed fumonisins, 390
4-Hydroxybenzoic acid, 342
Hyperestrogenism, 135

I
Idiotypic vaccination, 282–283, 286
IgA nephropathy, 262
Immunoaffinity columns, 29
Immunobiological effects, fumonisin B1
  interleukins, 190
  peritoneal macrophages, 190
Immunoglobulins, 312
Immunomodulatory activity, 285
Inactivation, 40
Industrial, 279
Insecticides, 204, 205
Interactions, 140
Interleukin-2, 140
Internal amplification controls, 377
International Agency for Research on Cancer (IARC), 309
Intraspecific differentiation, 279
Iraq, 359
Irradiation, 206, 209

J
JECFA risk, 268
Joint FAO/WHO Expert Committee on Food Additives (JECFA), 308

K
Khellin, 340
Killer Abs (KAbs), 282
Killer mimotope, 285
Killer peptides (KPs), 284–286
Killer strains, 275, 280
Killer system, 276
Killer toxins (KTs), 275, 276, 280
KT-biotyping, 277
KTRs, 284–286

L
Lateral flow immunochromatography, 29–31
LC/TOF-MS, 266
LD_{50}, 260
Lethal doses, 308
Leukoencephalomalacia, 180, 189
Level of detection, LOD, 27
Light dependence, 83
Lime, 41
Limestone, 41
Linoleic acid, 86
Lipid peroxidation, 300
Liquid chromatography–mass spectrometry (LC–MS), 266
Livestock, mycotoxicoses, 52
Low-observed-adverse-effect level (LOAEL), 261, 310
Low-observed-effect level (LOEL), 261
Luteinizing hormone, 134

M
Maize, 40–44
  fumonisin accumulation, 184, 191
  fungal contamination, 179, 186
  fusarium ear rot, 191
  indigenous foods, 179
  South America production, 179–193
Masked mycotoxins, 385
Matricaria recutita L., 342
Medicinal plants, 331
Metarhizium anisopliae, 375
Microbial biotyping, 276
Microbicidal Abs, 284
Microbioassay, 345
Mimic, 6
Mimic template, 9, 10, 12
Mode of action, 299–301
Molecular imprinted solid phase extraction, 5–7

Molecular imprinting, 32
Molecular imprinting of mycotoxins, 7
Molecularly imprinted polymers, 4–13
Moniliformin, 12
Monoclonal, 283
Morphotypes, 85
Moulds, 139, 279
MRL, 320, 321
Multiplex RT-PCR, 26
Multivalent vaccines, 286
Mutagenic properties, 131
Mycoflora, 237
Mycoflora incidence, 244
Mycology, 254
Mycoses, 353
Mycosphaerella pini, 68
Mycotoxicoses, 52, 353
Mycotoxicosis treatment, 379
Mycotoxigenic fungi
  immunosuppressive, 58
  mycophenolic acid, 58
  mycotoxins produced by *Penicillium roqueforti* and *P. paneum,* 56–59
  neurotoxic, 57
  patulin, 56–57
  Penicillium roqueforti and *P. paneum* taxonomy, 54
  physiology of *Penicillium roqueforti* and *P. paneum,* 54–56
  PR toxin, 57
  roquefortine, 57–58
Mycotoxin decontamination, 379
Mycotoxin-enzyme conjugate, 29
Mycotoxin incidence, 242–245
Mycotoxins, 39, 40, 46, 47, 139, 229, 232, 353, 354, 367
Mycotoxins in grain, 158
Myristicin, 340

N
Natural killer cells (NK) activity, 265
Neural tube defects, 180, 190
Nitrate, 86
Nivalenol, 253–270, 294, 355, 388
NIV producers, 256
Nixtamalizacíon, 39–48
Nixtamalization, 187
No-observed-adverse-effect level (NOAEL), 261, 310

O
Oats, 148
Occurrence, fumonisins
  Argentina, 181, 186, 188, 191
  Brazil, 180, 185, 186
  Chile, 185
  Venezuela, 184
Ochratoxigenic *Aspergillus,* 26
Ochratoxin, 113, 114, 155–157
Ochratoxin A (OTA), 6–17, 147, 237, 242, 243, 307, 359, 389
OECD, 378, 379
Oesophageal cancer, 180, 189
249 of the p53, 96
Oil palm, 367
O-methylsterigmatocystin, 340
Organic anion transport (OAT), 313
Oxidative stress, 88, 340
Ozone (O_3), 212

P
*Pa*KT, 282
*Pa*KTR, 284
Patulin, 26
p-Coumaric acid, 341
Penicillium, 21, 237–239
  P. paneum, 53
  P. roqeforti, 53
Penicillium roqueforti
  propagules, 55
  spore viability, 55
Pet food, 244
Petroselinum crispum (Mill.), 340
pH, 47
Phenolic compounds, 205
Phenylpropanoids, 340
Phosphorescence, 23
Phylogenetic relationship, 68
Physical separation
  cleaning, 206
  density segregation, 207
  milling, 208
Phytochemicals, 205

Phytoestrogens, 133–134
Phytopathogenic, 286
Pichia, 281
Pigment, 65, 70, 72
Piperine, 341
Piperlongumine, 341
Piper longum L., 341–342
Pipernonaline, 341
Piperoctadecalidine, 342
Pistacia vera L., 340
Plant-microorganism interaction, 386
Polyketide, 65, 70, 72
Polymerase chain reaction, 25–27
Polythene stretch-film, visibly damaged, 53
Post-harvest management
  antagonistic microorganims, 205
  modified atmospheres, 205
  storage conditions, 204
Pre-harvest management
  crop rotation, 203
  fertilization, 203
  field management, 203
  irrigation, 203
  resistance, 203
  tillage, 203
Prevention and control, 301
Probiotic bacteria, 219
Process, 42
Production of T-2 toxin, 294–296
Progesterone, 135
Protein-bound fumonisins, 392
Protein synthesis, 300
Protocatechuic acid, 342
Provisional tolerable daily intake (PTDI), 315
Provisional tolerable weekly intake (PTWI), 316
Prunus dulcis (Mill.), 342
P450scc, 137
Psoralene, 340
Pulmonary edema syndrome in pigs, 180

Q
Quinic acid, 340

R
Radiolabel, 44–46
Rapid one-step assay (ROSA), 29
Reactive oxygen, 98
Recombinant, 283
Red mold disease, 356
Reducing sugars, 213
Reduction,
Reverse transcription-polymerase chain reaction, 25
Rye, 148

S
Saccharomyces cerevisiae, 214
Salem Witch Trials, 354
Satureja hortensis L., 341
Scientific Committee for Food of the European Commission (SCF), 316
Sclerotia, 85
Secondary metabolites, 131
Secretion-signal sequence, 280
Signaling, 71
Signaling pathway, 81
Significance, 46
Silage
  aerobically unstable, 52
  anaerobic conditions, 51
  butyric acid, 55
Silica gel, 343
S/N, 267
Sodium bisulfite ($NaHSO_3$), 213
Solid state ssDNA odor sensors, 33
Spiroethers, 342
Stachybotrys, 355
Stachybotrytoxicosis, 355
Streptomyces spp., 331
Surface plasmon resonance, 31

T
T-2, 131
Taxonomy, 368
Teratogenic properties, 131
Terpenes, 331
Terpenoids, 331
The daily tolerable intake, 268
The provisional standard, 259
Therapeutic, 282, 283, 285
Thymol, 341
Thymus vulgaris, 341

T lymphocyte, 263
T-2 mycotoxin, 355
Tortilla, 40, 41, 45, 46
Toxicity, 297–298, 393
Toxic secondary metabolites, 52
Toxin regulations, 164
Toxin resistance, 165
Transcriptional activation, 80
Transgenic plants, 281
Trichothecene mycotoxins, 354
Trichothecenes, 28, 114, 124, 253, 292, 392
Triticale, 148
T-2 toxin, 26, 152–155, 292, 367
Type A, 257
Type B, 257
Type of trichothecenes, 292–293
Types, 42–43

U
Undesirable substances, 136
Uridine diphosphate glucuronic acid, 137
Uridine diphosphate glucuronyltransferase, 137
UV light, 207

V
Vanillic acid, 341, 342
Versicolorin, 68
Visnagin, 340
Vomitoxin, 355

W
Warfare, 298
Water, 374
Weapons, 367
Wheat, 148
WHO, 378
Williopsis, 281
Wine-making, 280
Woodchuck hepatitis virus, 94
Woodchucks, 97

X
Xanthotoxin, 340
Xenobiotic, 138

Y
Yeast killer phenomenon, 275
Yeasts, 208, 219, 278
Yellow rain, 354
Yellow rain attacks, 356

Z
Zearalenol, 141
α-Zearalenol, 11, 137
β-Zearalenol, 11, 137
Zearalenone, 10–12, 26, 139, 152, 164, 185, 232, 374, 387
  chemical structure, 115–116
  derivatives, 115–116
  detoxication, 122–124
  guidelines, 117–118
  occurrence, 116–117
  properties, 115–116
  toxic effects, 118–122
ZEN-glycoside conjugates, 387
Zeolites, 218

MIX
Papier aus verantwortungsvollen Quellen
Paper from responsible sources
FSC® C105338

If you have any concerns about our products,
you can contact us on
ProductSafety@springernature.com

In case Publisher is established outside the EU,
the EU authorized representative is:
Springer Nature Customer Service Center GmbH
Europaplatz 3, 69115 Heidelberg, Germany

Printed by Libri Plureos GmbH
in Hamburg, Germany